Innovative Manufacturing Engineering and Energy
IManEE2024

International Conference ImanEE2024 – Innovative Manufacturing Engineering & Energy, 23-25 October 2024, Athens, Greece

Editors

Angelos Markopoulos[1]

[1]National Technical University of Athens School of Mechanical Engineering, Athens, Greece

Peer review statement

All papers published in this volume of "Materials Research Proceedings" have been peer reviewed. The process of peer review was initiated and overseen by the above proceedings editors. All reviews were conducted by expert referees in accordance to Materials Research Forum LLC high standards.

Published under License by **Materials Research Forum LLC**
Millersville, PA 17551, USA

Published as part of the proceedings series
Materials Research Proceedings
Volume 46 (2024)

ISSN 2474-3941 (Print)
ISSN 2474-395X (Online)

ISBN 978-1-64490-336-0 (Print)
ISBN 978-1-64490-337-7 (eBook)

Distributed worldwide by

Materials Research Forum LLC
105 Springdale Lane
Millersville, PA 17551
USA
https://mrforum.com

Manufactured in the United State of America
10 9 8 7 6 5 4 3 2 1

Table of Contents

Preface

The 28th edition of the Innovative Manufacturing Engineering & Energy International Conference (IManEE 2024) was held in Athens, Greece, organized by the School of Mechanical Engineering of the National Technical University of Athens (Greece). This marked the second time the conference was hosted by the National Technical University of Athens, following IManEE 2020. Although the 2020 edition was originally planned to take place in Athens, it transitioned to an online format due to the COVID-19 pandemic, which severely impacted international travel and imposed total lockdowns in many countries. This shift led to the first virtual edition of the conference. Building on this experience, IManEE 2024 adopted a hybrid format, allowing for both on-site and online participation. This approach aligns with the conference's core mission: to bring researchers and scientists together using all available technological means.

Looking back, the inaugural conference of the series was organized in May 1996 in Iaşi, Romania. Since 1999, the conferences have been alternately hosted by Iaşi and Chişinău, facilitated through collaboration between the Machine Manufacturing Departments of the "Gheorghe Asachi" Technical University of Iaşi (Romania) and the Technical University of Moldova in Chişinău (Republic of Moldova). Recognizing global trends and demands in manufacturing engineering, the conference was renamed "Innovative Manufacturing Engineering" in 2013 and later evolved into "Innovative Manufacturing Engineering and Energy." Over the years, other departments from universities in Romania (Braşov, Bucureşti, Cluj, Galaţi, and Piteşti) and abroad (Serbia, Greece) have contributed to organizing the event. The conference has since become a significant platform for regular participants to exchange ideas and foster collaborations.

This year's conference attracted researchers from 17 countries, including Bulgaria, Brazil, the Czech Republic, Georgia, Germany, Greece, Hungary, Moldova, Poland, Portugal, Romania, Slovakia, Tunisia, Türkiye, Ukraine, the United Kingdom, and the United States. Six plenary speeches on compelling topics in manufacturing engineering and energy enriched the conference and provided a foundation for meaningful discussions. The plenary sessions featured distinguished members of the academic community. These included Professor N. Michailidis from Aristotle University of Thessaloniki, Greece, who delivered a presentation on "3D Printing Multifunctional Materials and Advanced Composites." Professor Spyros Papaefthymiou from the National Technical University of Athens, Greece, explored "A Journey from Ancient Greece to Industry 4.0." Dr. Ilias Georgakopoulos-Soares from Penn State University, USA, discussed "From 3D Printing to Biomedical Applications." On the following day, Professor Kai Cheng from Brunel University London, UK, presented on "Ultraprecision Machining of Freeform Surfaced Components and Devices in the Context of Innovative Sustainable Manufacturing." Professor Alina Minea from the Gheorghe Asachi Technical University of Iaşi, Romania, elaborated on "Nanoparticle-Enhanced Fluids: Synthesis, Properties, and Applications in the Manufacturing Industry." Finally, on the last day, Professor Panagiotis Kyratsis from the University of Western Macedonia, Greece, introduced "Manufacturing Computationally Designed Wearables via 3D Printing."

The success of the conference was made possible through the generous support of the Section of Manufacturing Technology at the School of Mechanical Engineering, National Technical University of Athens and the sponsors of the event, namely Anima, Raycap, Lino, CNC Solutions and Measurements. Special thanks are extended to the conference organizers, members of the Scientific Committee, reviewers, and, most importantly, all the participants, whose collective efforts contributed to the success of IManEE 2024.

The organizing committee wishes all participants continued success and looks forward to reuniting in the near future at the next IManEE conference.

Professor Angelos P. Markopoulos,
National Technical University of Athens
School of Mechanical Engineering
Section of Manufacturing Technology
Athens, Greece

Committees

Scientific Committee

Ádám JACSÓ, Budapest University of Technology and Economics, Hungary
Adriana MUNTEANU, Gheorghe Asachi Technical University of Iaşi, România
Agathoklis KRIMPENIS, Hellenic Mediterranean University, Greece
Agnieszka SKOCZYLAS, Lublin University of Technology, Poland
Ágota BÁNYAI, University of Miskolc, Hungary
Aleksandar VENCL, University of Belgrade, Serbia
Alexandru Cătălin FILIP, Transilvania University of Braşov, România
Alexandru KECHAGIAS, University of Thessaly, Greece
Alexandru SOVER, Ansbach University of Applied Sciences, Germany
Alexei TOCA, Technical University of Moldova, Moldova Republic
Anamaria FEIER, Politehnica University Timisoara, România
Anastasios TZOTZIS, University of Western Macedonia, Greece
Andrei Marius MIHALACHE, Gheorghe Asachi Technical University of Iaşi, România
Andrzej DZIERWA, Rzeszów University of Technology, Poland
Andrzej WRÓBEL, Silesian University of Technology, Poland
Angelos P. MARKOPOULOS, National Technical University of Athens, Greece
Anna NOCIVIN, Ovidius University of Constanta, România
António GONÇALVES-COELHO, NOVA School of Science and Technology, Portugal
António MOURÃO , Universidade NOVA de Lisboa, Portugal
Apostolos KORLOS, International Hellenic University, Greece
Arik SADEH, Holon Institute of Technology, Israel
Athanasios MANAVIS, University of Western Macedonia, Greece
Aurel Mihail TITU, Lucian Blaga University of Sibiu, România
Aurel TULCAN, Politehnica University Timisoara, România
Bahman AZARHOUSHANG, Furtwangen University, Germany
Calin NEAMTU, Technical University of Cluj-Napoca, România
Carmen BUJOREANU, Gheorghe Asachi Technical University of Iaşi, România
Carol SCHNAKOVSZKY, Vasile Alecsandri University of Bacău, România
Catalin DUMITRAS, Gheorghe Asachi Technical University of Iaşi, România
Catalin FETECAU, Dunărea de Jos University of Galaţi, România
Catalina MAIER, Dunărea de Jos University of Galaţi, România
Christophe SAUVEY, Université de Lorraine, France
Christos SPITAS, University of Nottingham, Ningbo, China
Constantin BUNGAU, University of Oradea, România
Constantin CARAUSU, Gheorghe Asachi Technical University of Iaşi, România
Cristian TURC, Politehnica University Timisoara, România
Csaba FELHŐ, University of Miskolc, Hungary
Daniel GHICULESCU, University Politehnica of Bucharest, România
Daniel MINDRU, Gheorghe Asachi Technical University of Iaşi, România
Daniela POPESCU, Gheorghe Asachi Technical University of Iaşi, România
Daniela-Monica IORDACHE, University of Pitesti, România
Dimitris MANOLAKOS, National Technical University of Athens, Greece
Dimka VASILEVA, Technical University of Varna, Bulgaria
Dorin LUCA, Gheorghe Asachi Technical University of Iaşi, România
Dragos CHITARIU, Gheorghe Asachi Technical University of Iaşi, România
Dumitru NEDELCU, Gheorghe Asachi Technical University of Iaşi, România
Dumitru OLARU, Gheorghe Asachi Technical University of Iaşi, România
Eduard Laurentiu NITU, University of Pitesti, România

Committees

Scientific Committee

Experimental study on the determination of reduced compressive properties of FFF-printed enhanced PLA specimens

Emmanouil L. PAPAZOGLOU[1,a], Panagiotis KARMIRIS-OBRATAŃSKI[2,b], Evangelos NIKOLIDAKIS[3,c], Nikolaos E. KARKALOS[1,d] and Angelos P. MARKOPOULOS[1,e*]

[1]Section of Manufacturing Technology, School of Mechanical Engineering, National Technical University of Athens, Heroon Polytechniou 9, 15780 Athens, Greece

[2]Department of Manufacturing Systems, Faculty of Mechanical Engineering and Robotics, AGH University of Cracow, 30-059 Cracow, Poland

[3]Technical University of Crete, School of Production Engineering & Management, Micromachining & Manufacturing Modeling Laboratory, University campus, Building E4.026-Kounoupidiana, 73100 Chania, Greece

[a]mlpapazoglou@gmail.com, [b]karmiris@agh.edu.pl, [c]kahthkournas@gmail.com, [d]nkark@mail.ntua.gr, [e]amark@mail.ntua.gr

Keywords: Compression Test, Fused Filament Fabrication, PLA Pro, Reduced Mechanical Properties

Abstract. Mechanical parts used in industrial applications should be able to comply not only with rigorous requirements regarding dimensional accuracy and surface quality, but also possess adequate mechanical strength in order to achieve a long service life. Additive manufacturing (AM) processes are considered nowadays an alternative way to produce high-performing parts with complex geometries and different types of materials with high efficiency. Especially in the case of polymers, Fused Filament Fabrication (FFF) is a well-established choice for creating high quality parts at an affordable cost when optimal parameters are used for the printing process. Although a lot of studies have been carried out in order to optimize the parameters of FFF process for the increase of individual mechanical properties, often neglecting the effect of infill density on these properties, it is proposed that the use of reduced mechanical properties, by normalizing them by part weight, can provide more generalized findings and deeper insight into the correlations between process parameters and mechanical behavior. Thus, in the present work, compression test specimens were fabricated by FFF technique under various infill density and layer height values for three different infill patterns, and the fundamental compressive properties such as Young's Modulus, strength and absorbed energy were determined. After statistical analysis was carried out, the role of each parameter on the improvement of reduced mechanical properties was identified.

Introduction

AM processes have already gained an important place in Manufacturing Technology due to their special characteristics and important advantages in comparison with conventional manufacturing methods. These processes, adopting the bottom-up concept aim to facilitate the fabrication of parts with rather complex geometries, with specific material properties and structure, minimize material waste, and are capable to be used for customized production of parts with a relatively low cost. Moreover, AM can be used to process almost every type of material such as metals, ceramics, polymers and composites. Especially, for polymer and polymer composites processing, the FFF process has been established as an affordable and efficient method for producing parts with a considerable range of mechanical properties including engineering and advanced materials [1].

The capabilities of FFF process to produce parts with appropriate mechanical properties are significantly dependent on the choice of suitable process parameters [1]. For that reason, several authors have carried out experimental works in order to determine the effect of process parameters on the mechanical properties of printed specimens. Al Tamimi et al. [2] carried out a comprehensive experimental work, considering six different process parameters and studied their effect on various mechanical properties. Especially, for the compressive UTS, they found out that infill density is the most significant parameter, followed by wall thickness, whereas for the compressive Young's modulus, the most important parameters were printing speed and temperature. Yadav et al. [3] determined the effect of different infill patterns on mechanical properties of PLA specimens. They created specimens with seven different patterns and four different densities and observed that specimens with Hilbert curve pattern exhibited the highest compressive strength, which exceeded the strength obtained by the second best pattern by 40 MPa. Maszybrocka et al. [4] used gradient infill for PLA samples and investigated the effect of infill density, as well as number of different gradient zones regarding compressive behavior of the printed samples. Kholil et al. [5] focused on the effect of layer thickness on compressive strength of PLA samples, indicating that it is inversely correlated with strength. Zhu et al. [6] carried out experiments designed by orthogonal array with three parameters, namely layer height, infill density and printing speed and found out the most important parameter for compressive strength was infill density, followed by layer height and printing speed. Elmrabet and Siegkas [7] concluded that the determination of mechanical properties of 3D printed materials by using traditional standards may be considerably affected by dimensional variations of the specimens, causing inhomogeneity in the mechanical properties. Dixit and Jain [8] created lattice structures by FFF technique and determined by ANOVA method that layer thickness was the most important parameter, followed by infill density, whereas printing speed was found to be statistically insignificant for the compressive strength.

Brischetto and Torre [9] analyzed the repeatability of mechanical properties for PLA specimens using statistical method and concluded that the printing process was repeatable. Mercado-Colmenero et al. [10] carried out experiments both on specimens and mechanical parts for their compression properties and also implemented numerical models for the calculation of various responses which are important during the design phase of parts. Bakhtiari et al. [11] investigated the effect of various printing parameters such as layer height, extrusion width, nozzle temperature and printing speed on the compressive strength, compressive modulus, specific strength, toughness, failure strain, hysteresis loss and residual strain. Layer height and printing speed were found to be the first and second best parameters overall. Netto et al. [12] fabricated gyroid lattice structures and investigated the effect of lattice cell size and wall thickness regarding the compressive behavior of these structures. The cell size was identified as more important parameter than the wall thickness.

Although many studies have been carried out regarding most of the compressive mechanical properties of FFF printed specimens, the effect of the weight of the specimens, representing the amount of material of the specimen on the mechanical properties has only rarely been considered in relevant studies [13, 14], by calculating the specific or reduced mechanical properties of specimens instead of the nominal ones. However, given that in most cases the printed specimens are not printed in 100% infill density, taking into consideration the amount of material used for the parts is rather important for a comprehensive evaluation of mechanical properties of printed specimens.

Thus, in this work, the compressive properties, both nominal and reduced, of specimens produced from enhanced PLA material (PLA Pro) are investigated by means of experimental and statistical methods in order to determine the correlation between process parameters and material properties. The significance of each process parameter is identified and suggestions on the appropriate range of process variables for improved material properties of specimens in respect to their weight, as well, are presented.

Materials and Methods

In the present work, it is intended to determine the correlation between process parameters and compressive mechanical properties of the PLA PRO specimens both nominal and reduced. Especially, the study of reduced properties, i.e. mechanical properties normalized by weight, can reveal important details about the contribution of different parameters to the mechanical strength of printed parts, taking into consideration the material consumption as well.

Three different process parameters, namely infill density, infill pattern and layer height were considered and the experiment was designed by L9 Taguchi orthogonal array for three factors at three levels each. The selection of these parameters was based on suggestions from the relevant literature [15] regarding compression tests and the range of the factors' values was comprehensive enough to determine their influence on mechanical strength.

The experimental specimens were printed in a commercial FFF 3D printer, Flashforge Creator 3, after slicing was performed from the available CAD files in dedicated software. The filament used in the experiments was an improved variety of PLA, termed as PLA pro (Flashforge, Zhejiang, China), which exhibits higher toughness and impact strength, minimal warping and improved layer adhesion than standard PLA, as per the manufacturer's specifications. The filament diameter was 1.75 mm whereas the other constant settings for printing were: nozzle temperature of 220°C, platform temperature of 45°C, nozzle diameter of 0.4 mm, 2 mm wall thickness and travel speed of 60 mm/s. According to the ASTM D695-15 standard, five specimens were printed for each different combination of parameters and average values of the measured mechanical properties are reported. The responses of the experimental tests were the compressive modulus of elasticity (CME), compressive yield strength (CYS) and toughness (CT).

The compression tests were performed on an Instron 4482 universal testing machine with a crosshead speed of 2 mm/min and the force-displacement curves were subsequently obtained through Bluehill software. Finally, the determination of mechanical properties by the experimental curves was achieved by analysis in MATLAB software.

Results and Discussion

After the experiments were carried out and the results were analyzed by appropriate statistical methods, the main effects and relative importance of each process parameter regarding CME were determined and depicted in Fig. 1. For the nominal CME, it can be clearly observed that the increasing infill density has a positive effect as the increase of infill density from 20 to 80%, gradually leads to an increase of CME. Regarding the infill pattern, it was found that the triangular pattern is slightly superior to the hexagonal one, with the gyroid pattern being clearly inferior to both of them, despite its optimized structure, as it seems to be rather inefficient to withstand vertical loads, which occur during compression. Finally, the increase of layer height leads to lower CME values as it is related to the improvement of bonding between layers and reduction of voids. Based on these results and statistical analysis, it was determined that the most important parameter is infill pattern, followed by infill density and layer height. Thus, it is essential to choose a suitable infill pattern which can ensure the increased strength of the printed part, apart from choosing the appropriate infill density and layer height, at their highest and lowest levels, respectively.

The analysis of the main effects of process parameters on reduced CME (abbreviated as R-CME) reveals additional aspects regarding their influence on the mechanical properties of printed specimens. Although the triangular pattern and low layer height are proven efficient also from the perspective of used material for the specimens and the relative importance of the three parameters is the same as for the nominal CME, the infill density is inversely correlated with R-CME. This finding implies that although a higher amount of material was used at larger infill densities, this additional amount did not contribute to the actual improvement of the printed parts but also led to lower R-CME probably due to the existence of additional defects and voids in a larger amount of material. Thus from a more general point of view, it is essential to reach a suitable compromise between the required strength of the parts and the amount of material used.

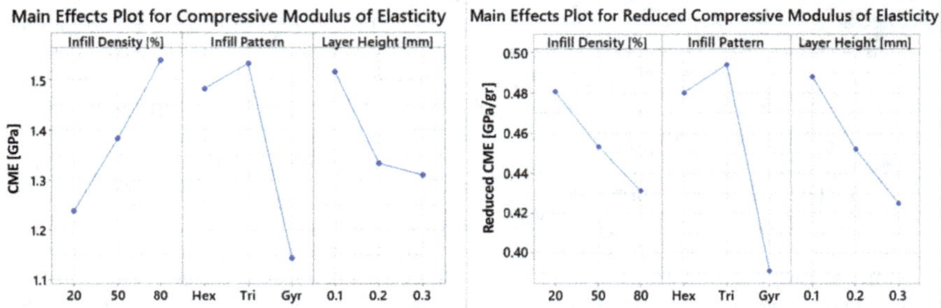

Level	Infill Density [%]	Infill Pattern	Layer Height [mm]	Level	Infill Density [%]	Infill Pattern	Layer Height [mm]
1	1.237	1.483	1.517	1	0.4810	0.4806	0.4886
2	1.382	1.533	1.333	2	0.4533	0.4944	0.4521
3	1.540	1.143	1.309	3	0.4314	0.3908	0.4250
Delta	0.303	0.390	0.207	Delta	0.0497	0.1036	0.0636
Rank	2	1	3	Rank	3	1	2

Fig.1. Main effects plot for nominal and reduced CME.

The effect of the three process parameters on CYS is similar to the effect on CME. Specimens with higher density, triangular pattern and small layer height have considerably higher strength than other specimens. A slight difference from CME results is that the negative correlation of layer height and CYS is reversed after a layer height value of 0.2 mm, probably due to the combined effect of other parameters as well. Based on the results, depicted in Fig.2, the infill pattern is again identified as the most significant parameter, followed by infill density and layer height. Thus, the parts which have near full density, adequate bonding between the layers and a pattern which provides better load distribution and stability of the printed structure, exhibit the highest nominal strength.

On the hand, the effect of these variables on the reduced CYS (abbreviated as R-CYS) is slightly different, especially for infill density. More specifically, the increase of infill density does not contribute to increased strengthening of the part but rather to a small reduction of R-CYS and no further variation after an infill density value of 50%. The statistical analysis shows that the infill pattern is again the most important parameter, followed by layer height, while the effect of infill density is rather negligible compared to the other two variables. Thus, it is crucial to adopt the appropriate infill pattern and layer height, as well as a moderate value of infill density in order to obtain the required strength without excessive material consumption.

appropriate range of process variables for improved material properties of specimens in respect to their weight, as well, are presented.

Materials and Methods

In the present work, it is intended to determine the correlation between process parameters and compressive mechanical properties of the PLA PRO specimens both nominal and reduced. Especially, the study of reduced properties, i.e. mechanical properties normalized by weight, can reveal important details about the contribution of different parameters to the mechanical strength of printed parts, taking into consideration the material consumption as well.

Three different process parameters, namely infill density, infill pattern and layer height were considered and the experiment was designed by L9 Taguchi orthogonal array for three factors at three levels each. The selection of these parameters was based on suggestions from the relevant literature [15] regarding compression tests and the range of the factors' values was comprehensive enough to determine their influence on mechanical strength.

The experimental specimens were printed in a commercial FFF 3D printer, Flashforge Creator 3, after slicing was performed from the available CAD files in dedicated software. The filament used in the experiments was an improved variety of PLA, termed as PLA pro (Flashforge, Zhejiang, China), which exhibits higher toughness and impact strength, minimal warping and improved layer adhesion than standard PLA, as per the manufacturer's specifications. The filament diameter was 1.75 mm whereas the other constant settings for printing were: nozzle temperature of 220°C, platform temperature of 45°C, nozzle diameter of 0.4 mm, 2 mm wall thickness and travel speed of 60 mm/s. According to the ASTM D695-15 standard, five specimens were printed for each different combination of parameters and average values of the measured mechanical properties are reported. The responses of the experimental tests were the compressive modulus of elasticity (CME), compressive yield strength (CYS) and toughness (CT).

The compression tests were performed on an Instron 4482 universal testing machine with a crosshead speed of 2 mm/min and the force-displacement curves were subsequently obtained through Bluehill software. Finally, the determination of mechanical properties by the experimental curves was achieved by analysis in MATLAB software.

Results and Discussion

After the experiments were carried out and the results were analyzed by appropriate statistical methods, the main effects and relative importance of each process parameter regarding CME were determined and depicted in Fig. 1. For the nominal CME, it can be clearly observed that the increasing infill density has a positive effect as the increase of infill density from 20 to 80%, gradually leads to an increase of CME. Regarding the infill pattern, it was found that the triangular pattern is slightly superior to the hexagonal one, with the gyroid pattern being clearly inferior to both of them, despite its optimized structure, as it seems to be rather inefficient to withstand vertical loads, which occur during compression. Finally, the increase of layer height leads to lower CME values as it is related to the improvement of bonding between layers and reduction of voids. Based on these results and statistical analysis, it was determined that the most important parameter is infill pattern, followed by infill density and layer height. Thus, it is essential to choose a suitable infill pattern which can ensure the increased strength of the printed part, apart from choosing the appropriate infill density and layer height, at their highest and lowest levels, respectively.

The analysis of the main effects of process parameters on reduced CME (abbreviated as R-CME) reveals additional aspects regarding their influence on the mechanical properties of printed specimens. Although the triangular pattern and low layer height are proven efficient also from the perspective of used material for the specimens and the relative importance of the three parameters is the same as for the nominal CME, the infill density is inversely correlated with R-CME. This finding implies that although a higher amount of material was used at larger infill densities, this additional amount did not contribute to the actual improvement of the printed parts but also led to lower R-CME probably due to the existence of additional defects and voids in a larger amount of material. Thus from a more general point of view, it is essential to reach a suitable compromise between the required strength of the parts and the amount of material used.

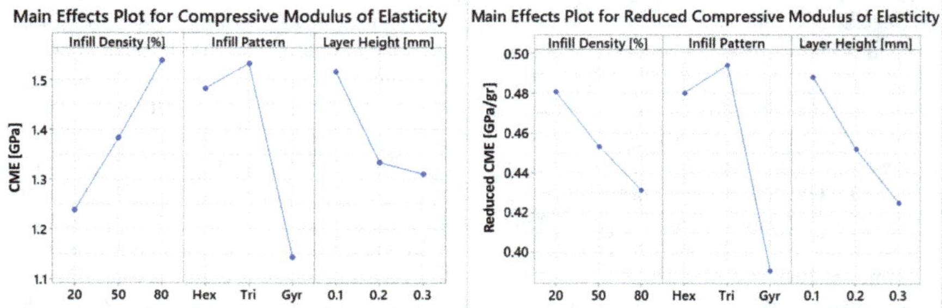

Level	Infill Density [%]	Infill Pattern	Layer Height [mm]
1	1.237	1.483	1.517
2	1.382	1.533	1.333
3	1.540	1.143	1.309
Delta	0.303	0.390	0.207
Rank	2	1	3

Level	Infill Density [%]	Infill Pattern	Layer Height [mm]
1	0.4810	0.4806	0.4886
2	0.4533	0.4944	0.4521
3	0.4314	0.3908	0.4250
Delta	0.0497	0.1036	0.0636
Rank	3	1	2

Fig.1. Main effects plot for nominal and reduced CME.

The effect of the three process parameters on CYS is similar to the effect on CME. Specimens with higher density, triangular pattern and small layer height have considerably higher strength than other specimens. A slight difference from CME results is that the negative correlation of layer height and CYS is reversed after a layer height value of 0.2 mm, probably due to the combined effect of other parameters as well. Based on the results, depicted in Fig.2, the infill pattern is again identified as the most significant parameter, followed by infill density and layer height. Thus, the parts which have near full density, adequate bonding between the layers and a pattern which provides better load distribution and stability of the printed structure, exhibit the highest nominal strength.

On the hand, the effect of these variables on the reduced CYS (abbreviated as R-CYS) is slightly different, especially for infill density. More specifically, the increase of infill density does not contribute to increased strengthening of the part but rather to a small reduction of R-CYS and no further variation after an infill density value of 50%. The statistical analysis shows that the infill pattern is again the most important parameter, followed by layer height, while the effect of infill density is rather negligible compared to the other two variables. Thus, it is crucial to adopt the appropriate infill pattern and layer height, as well as a moderate value of infill density in order to obtain the required strength without excessive material consumption.

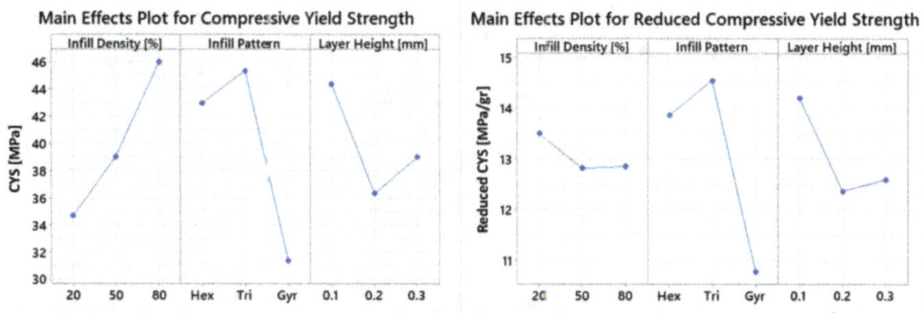

Level	Infill Density [%]	Infill Pattern	Layer Height [mm]
1	34.70	42.98	44.41
2	39.04	45.40	36.33
3	46.02	31.37	39.01
Delta	11.32	14.02	8.08
Rank	2	1	3

Level	Infill Density [%]	Infill Pattern	Layer Height [mm]
1	13.49	13.85	14.20
2	12.81	14.54	12.36
3	12.85	10.76	12.59
Delta	0.68	3.78	1.84
Rank	3	1	2

Fig.2. Main effects plot for nominal and reduced CYS.

Toughness is another important material property apart from CME and CYS, which is directly related to the capability of energy absorption by the printed parts before their fracture and is both affected by material strength and ductility. Toughness was appropriately estimated by integrating the stress-strain curve of the experimental results. The analysis of the results in the main effects plot, depicted in Fig. 3, shows that the correlation of infill density and infill pattern with CT are almost the same as with CME and CYS. Higher infill density increases the toughness of the part due to more available material to absorb the exerted amount of energy and the efficient distribution of stress achieved by some infill patterns can also lead to higher toughness, as well as the increased adhesion provided by lower layer heights. However, some notable differences can also be observed as the infill pattern plays a less important role than infill density in that case, and the trend between layer height and CT is not monotonous. The latter observation, along with the fact that the layer height is the least significant parameter, indicates that the importance of layer height is less important regarding toughness of the printed parts and no specific recommendation can be determined for its appropriate values in terms of toughness. Generally, from these findings it is recommended to use a higher infill density and triangular infill pattern for improvement of the toughness of the part.

Regarding the reduced values of toughness (termed as R-CT), only the infill pattern can affect R-CT considerably and both infill density and layer height have a negligible role in altering the toughness of the specimens. However, a small difference can be observed in the case of R-CT in terms of infill density, as contrary to the CME and CYS, the increase of infill density can contribute to a small extent to the increase of R-CT.

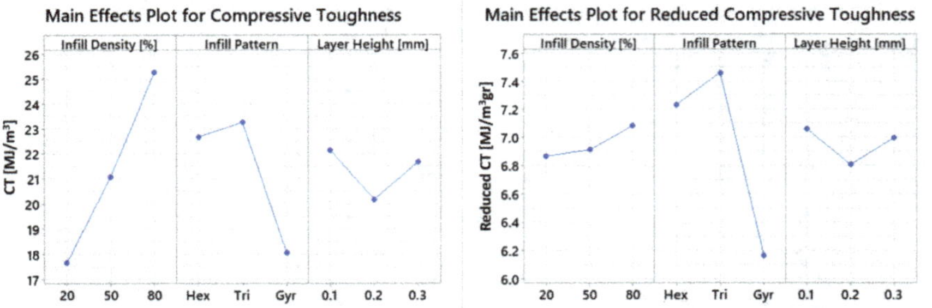

Level	Infill Density [%]	Infill Pattern	Layer Height [mm]
1	17.64	22.67	22.14
2	21.07	23.27	20.18
3	25.27	18.04	21.66
Delta	7.63	5.23	1.96
Rank	1	2	3

Level	Infill Density [%]	Infill Pattern	Layer Height [mm]
1	6.865	7.232	7.060
2	6.910	7.461	6.804
3	7.084	6.167	6.995
Delta	0.219	1.294	0.255
Rank	3	1	2

Fig.3. Main effects plot for nominal and reduced CT.

Conclusions

In this work, an experimental investigation was carried out regarding the determination of the impact of three important process factors on the compressive mechanical properties of printed parts by FFF method. Both nominal and reduced properties were analyzed and several useful conclusions were able to be drawn.

The nominal CME is primarily affected by the infill density and infill pattern. Thus apart from the anticipated positive influence of higher infill density, it is crucial to choose an infill pattern that can ensure the required CME of the parts, especially for demanding applications. Layer height of the parts plays a less important role, particularly by improving the interlayer adhesion when lower values are chosen. Similarly, the nominal CYS is affected in the same way by the three parameters. In the case of CT, infill density is more important than other factors and layer height is barely affecting CT.

The analysis of reduced values provides additional insight to the role of each process factor to the improvement of the mechanical behavior of parts. It was found that, from the perspective of weight-normalized properties, the role of infill density is inferior to that of the infill pattern and layer height regarding R-CME and R-CYS, as it does not contribute to further improvement of properties compared to the additional material used. Finally, regarding R-CT, only infill pattern can offer a noticeable improvement, with the other two factors being practically non-significant.

References

[1] M. Doshi, A. Mahale, S.K. Singh, S. Deshmukh, Printing parameters and materials affecting mechanical properties of FDM-3D printed parts: perspective and prospects, Mater. Today Proc. 50 (2021) 2269–2275. https://doi.org/10.1016/j.matpr.2021.10.003

[2] A.A. Al-Tamimi, A. Pandžić, E. Kadrić, Investigation and prediction of tensile, flexural, and compressive properties of Tough PLA material using definitive screening design, Polymers 15(20) (2023) 4169. https://doi.org/10.3390/polym15204169

[3] P. Yadav, A. Sahai, R.S. Sharma, Strength and surface characteristics of FDM-based 3D printed PLA parts for multiple infill design patterns, J. Inst. Eng. (India): C 102 (2021), 197-207. https://doi.org/10.1007/s40032-020-00625-z

[4] J. Maszybrocka, M. Dvorak, G. Nowakowska, P. Osak, B. Losiewicz, The influence of the gradient infill of PLA samples produced with FDM technique on their mechanical properties, Materials 15(4) (2022) 1304. https://doi.org/10.3390/ma15041304

[5] A. Kholil, E. Asyaefudin, N. Pinto, S. Syaripuddin, Compression strength characteristics of ABS and PLA materials affected by layer thickness on FDM, J. Phys.: Conf. Ser. 2377 (2022) 012008. https://doi.org/10.1088/1742-6596/2377/1/012008

[6] Y. Zhu, Y. Gao, J. Jiang, H. Gu, S, Lv, H. Ni, X. Wang, C. Jia, Study on effects of FDM 3D printing parameters on mechanical properties of polylactic acid, IOP Conf. Ser.: Mater. Sci Eng. 688 (2019) 033026. doi:10.1088/1757-899X/688/3/033026

[7] N. Elmrabet, P. Siegkas, Dimensional considerations on the mechanical properties of 3D printed polymer parts, Polym. Test. 90 (2020) 106656.

[8] N. Dixit, P.K. Jain, Effect of Fused Filament Fabrication process parameters on compressive strength of thermoplastic polyurethane and polylactic acid lattice structures, J. Mater. Eng. Perform. 31 (2022) 5973-5982. https://doi.org/10.1007/s11665-022-06664-0

[9] S. Brischetto, R. Torre, Tensile and compressive behavior in the experimental tests for PLA specimens produced via Fused Deposition Modeling technique, J. Compos. Sci. 4(3) (2020) 140. https://doi.org/10.3390/jcs4030140

[10] J.M. Mercado-Colmenero, M.A. Rubio-Paramio, M. D. la Rubia-Garcia, D. Lozano-Arjona, C. Martin-Doñate, A numerical and experimental study of the compression uniaxial properties of PLA manufactured with FDM technology based on product specifications, Int J Adv Manuf Technol 103 (2019) 1893-1909. https://doi.org/10.1007/s00170-019-03626-0

[11] H. Bakhtiari, M. Nikzad, M. Tolouei-Rad, Influence of three-dimensional printing parameters on compressive properties and surface smoothness of polylactic acid specimens, Polymers 15(18) (2023) 3827. https://doi.org/10.3390/polym15183827

[12] J.J. Netto, M. Sardinha, M. Leite, Influence of the cell size and wall thickness on the compressive behaviour of fused filament fabricated PLA gyroid structures, Mech Mater 195 (2024), 105051. https://doi.org/10.1016/j.mechmat.2024.105051

[13] A. Forés-Garriga, M. A. Pérez, G. Gómes-Gras, G. Reyes-Pozo, Role of infill parameters on the mechanical performance and weight reduction of PEI Ultem processed by FFF, Mater Des 193 (2020) 108810. https://doi.org/10.1016/j.matdes.2020.108810

[14] X. Chen, Q. Ji, J. Wei, H. Tan, J. Yu, P. Zhang, V. Laude, M. Kadic, Light-weight shell-lattice metamaterials for mechanical shock absorption, Int J Mech Sci 169 (2020), 105288.

[15] L. Auffray, P.A. Gouge, L. Hattali, Design of experiment analysis on tensile properties of PLA samples produced by fused filament fabrication, Int J Adv Manuf Technol 118 (2022) 4123–4137. https://doi.org/10.1007/s00170-021-08216-7

Effect of orientation on the bending properties of a carbon fibre reinforced thermoplastic

Katarina MONKOVA[1,2,a *], Peter Pavol MONKA[1,2,b], Gyula SZABO[3,c], Kamel MEHDI[4,5,d]

[1]Technical University of Kosice, Faculty of Manufacturing Technologies with a seat in Presov, 080 01 Presov, Slovakia

[2]Tomas Bata University in Zlin, Faculty of Technology, Nam. T. G. Masaryka 5555, 760 01 Zlin, Czech Republic

[3]Obuda University, Donat Banki Faculty of Mechanical and Safety Engineering, Nepszfnhaz u 8, 1081 Budapest, Hungary

[4] Preparatory Institute for Engineering Studies El Manar (IPEIEM), University of Tunis EL Manar (UTM), PB. 244, Tunis 2092, Tunisia

[5]Production and Energetics Laboratory (LMPE), Engineering National High School of Tunis (ENSIT), Mechanics, University of Tunis (UT), 5 Avenue Taha Hussein, P.B. 56, Bab Mnara, Tunis 1008, Tunisia

[a]katarina.monkova@tuke.sk, [b]peter.pavol.monka@tuke.sk, [c]szabo.gyula@bgk.uni-obuda.hu; [d]kamel.mehdi@ipeiem.utm.tn

Keywords: Bending Properties, Plastic, Carbon Fibre, Additive Manufacturing, Sample Orientation

Abstract. Additive manufacturing is becoming a very rapidly developing production technology. It provides a wide range of combinations of material processing and used techniques, including the possibility of using different types of 3D printing machines. In order to know the properties of the part and take them into account in the design of the product, it is necessary to know the behaviour of the sample after loading, which was produced under the given specific conditions. The manuscript aims to study the bending characteristics of samples made of plastic reinforced with carbon fibres using additive technology. The samples of the same type were fabricated by FDM (Fused Deposition Modelling) using a MakerBot Method X printer. The experimental investigation was carried out using a Zwick 1456 testing machine. The experimentally tested samples were oriented in two different ways regarding the layers building. The measured data were evaluated and compared with the numerically obtained values. The results showed a strong effect of the sample orientation on the bending properties.

Introduction

Currently, thanks to the development of technologies, a large number of new materials are created, the properties of which need to be known before their application, especially when it comes to stressed components, the safety and reliability of which is critical for the operation of equipment. Knowing the properties of the component is all the more important if it is made of plastic using an additive approach since the heterogeneity of the properties affects its behaviour in operation. [1,2]

A specific category of materials creates the reinforced materials to which also Nylon 12 CF belong. Nylon 12 Carbon Fiber (Nylon 12CF) material consists of Nylon 12 reinforced with small pieces of cut carbon fibre for better bending strength, making this material one of the best of all materials in terms of stiffness-to-weight ratio. used in the manufacture of FDM (Fused Deposition Modelling) components. [3] Relatively good strength and stiffness allow this material to replace metals in some possible applications, for example replacing heavy metal tools with lighter and

Materials Research Proceedings 46 (2024) 8-14 https://doi.org/10.21741/9781644903377-2

more ergonomic tools made of carbon fibre-reinforced plastics. [4] Its unique properties make it ideal for use in many applications across a wide range of industries, including automotive, consumer goods, recreational manufacturing and aerospace. [5]

Only a very limited number of studies have dealt with the characteristics of this material. A comprehensive study regarding the effect of printing conditions on the quality of produced samples was provided by Korol et al. [6], however, an investigation of the effect of orientation on the bending properties was not presented in the paper, so the authors decided to deal with the topic.

Production and testing conditions

Totally 6 samples with sizes 20 x 20 x 250 mm were made by means of 3D printing machine type MakerBot Method X using the FDM technique. The next parameters were chosen based on the manufacturer's recommendation as well as a preliminary test and they were: nozzle diameter 0.4 mm; both the temperature of the basement and the temperature of the workspace area were 65 °C; the print speed was 35 mm/s and the layer thickness was set to 0.15 mm. The heat treatment of the samples was performed at 80 °C according to the 3D printer producer's recommendation for the given material.

Three-point bending tests were carried out by the Zwick 1456 testing machine at an ambient temperature of 20 °C with a humidity of 50 % according to the ISO 178:2019 standard. [7]

The samples were placed on supporting pins with a supporting span of 200 mm (Fig. 1). The orientation of the three samples during the experimental testing corresponded to the orientation of the sample formation (Fig. 1a), while three samples were oriented perpendicular to the layering plane (i.e., parallel to the direction of movement of the pusher as shown in Fig. 1b). The radius of the pusher was 5 mm and the speed of the cross-head was set at 20 mm/min.

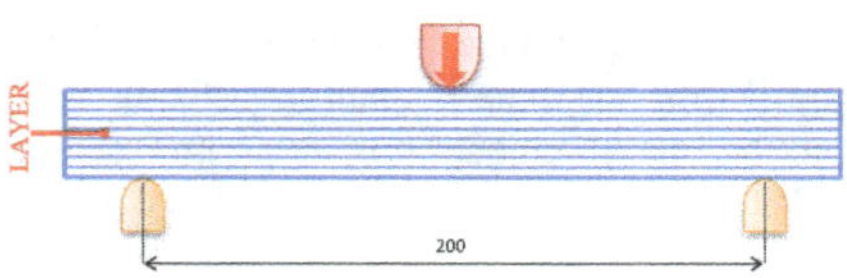

a) parallel orientation

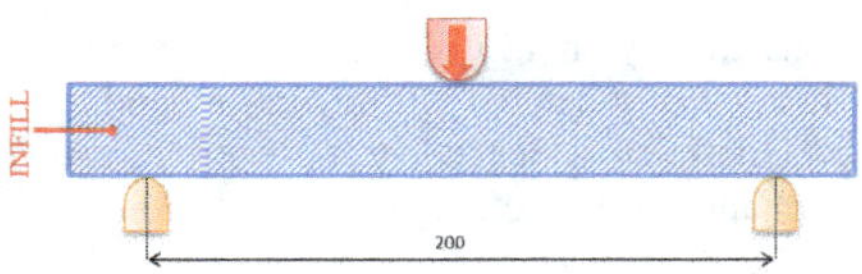

b) perpendicular orientation

c) real view

Figure 1. Position of a sample during testing.

Results and discussion

Within the preliminary study, the filament was tested by quasistatic tensile test according to ASTM D638 [8] standard to confirm its properties given by the datasheet. The experiments were realized employing a Testometric X350-5 testing machine (Fig. 2a) at an ambient temperature of 19 °C and a cross-head speed of 50 mm/min. Six pieces of fibre with a length of 100 mm were used, and the measured data are shown in Fig. 2b. The results (elongation at break 1.9 % and tensile strength 85.35 MPa) showed that the properties of the thermoplastic polymer correspond to the properties listed in the material sheet. [9,10]

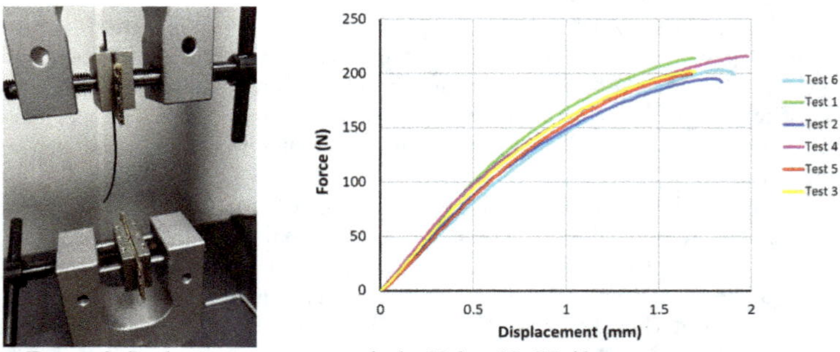

Figure 2. Preliminary test to verify the Nylon 12 CF filament properties

Using the Nylon 12 CF filament, the samples described in the previous section were produced, which were subjected to three-point bending testing. During the experimental investigation, the force-deflection dependencies were recorded and the measurements for the same sample orientation were plotted together, as shown in Fig. 3.

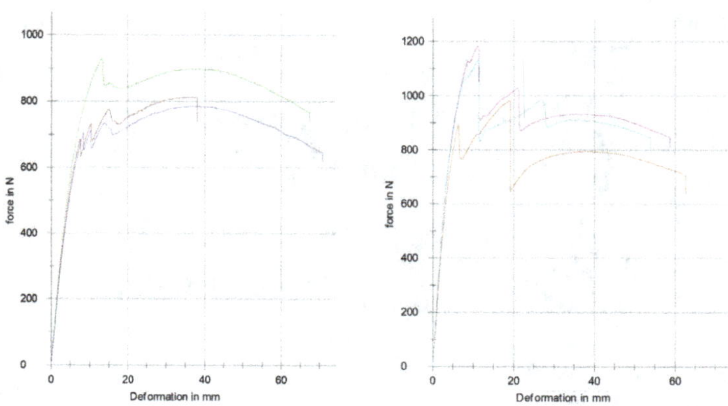

a) parallel position b) perpendicular position

Fig. 3 Measured force-deflection dependencies

The obtained data were statistically processed and based on them, the flexural yield strength and flexural elastic modulus were calculated according to the equations (1) and (2) respectively [11,12]

$$\sigma = \frac{M}{I} z \quad \text{(MPa)} \tag{1}$$

$$E = \frac{F_e l^3}{48 u_e I} \quad \text{(MPa)} \tag{2}$$

where σ is the resulting stress at a distance "z" from the neutral axis, I is the area moment of inertia, l is a supporting span and u_e is elastic deflection.

Moreover, the absorbed energies at break were also computed based on equation (3) which is represented by the area under the curve representing the force-deflection dependence: [13]

$$U = \int_0^l \frac{M^2}{2EI} dx. \tag{3}$$

Average achieved values are listed in Table 1 and it can be seen that the difference between Young´s modulus is very high.

Table 1 Measured and calculated values of flexural characteristics

Orientation of the sample with respect to the layering plane	Elastic force (N)	Deflection at elastic force (mm)	Flexural modulus (MPa)	Yield strength (MPa)	Total absorbed energy (J)
Parallel	852	14.1	755	31.95	45.32
Perpendicular	1100	9.5	1408	40.125	47.85

Based on the obtained results it is clear that there are relatively big differences not only in flexural modulus (more than 85 %) but also in yield strength values (25 %), however, the total absorbed energy in both cases of orientation is almost the same.

The experimental data were compared with the results solved by means of numerical analyses. Given that the beam is simple, a linear idealization was applied to the software PTC Creo 10. The beam with the constraints and a force load is shown in Fig. 4a, while Figures 4b and 4c show the results for the parallel orientation of the beam, and Figures 4d, and 4e show the results at the perpendicular beam orientation. They are summarized in Table 2, where also the comparison with experimental data can be found.

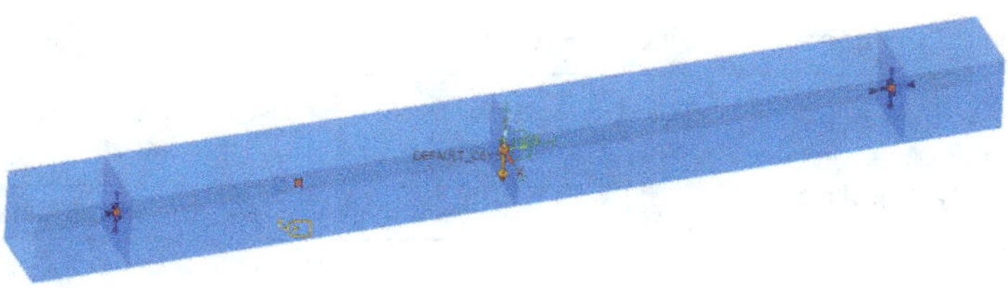

a) load set and constraints of FEA

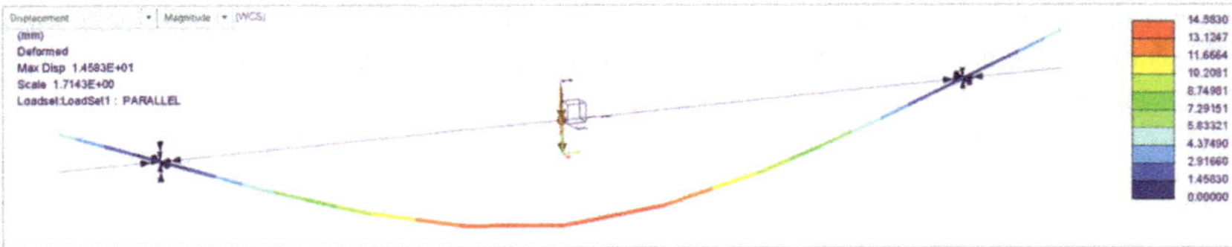

b) displacement at parallel orientation of a sample during testing

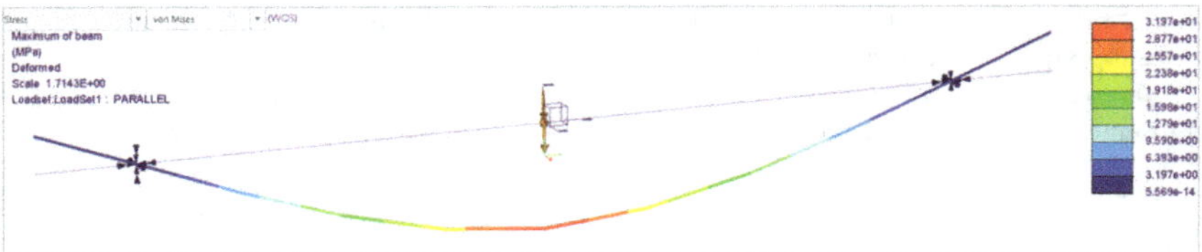

c) Stress von Mises at parallel orientation of a sample during testing

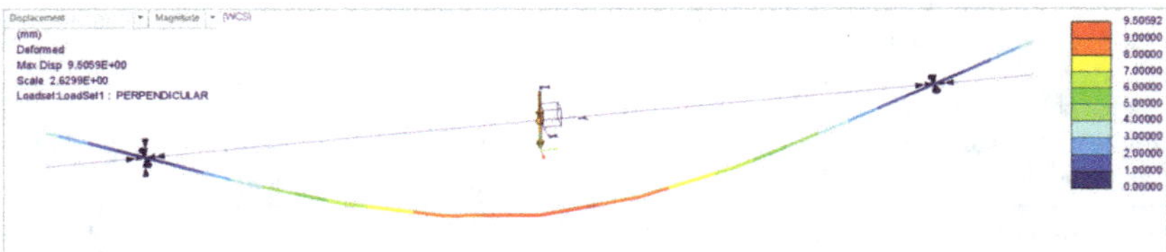

d) displacement at perpendicular orientation of a sample during testing

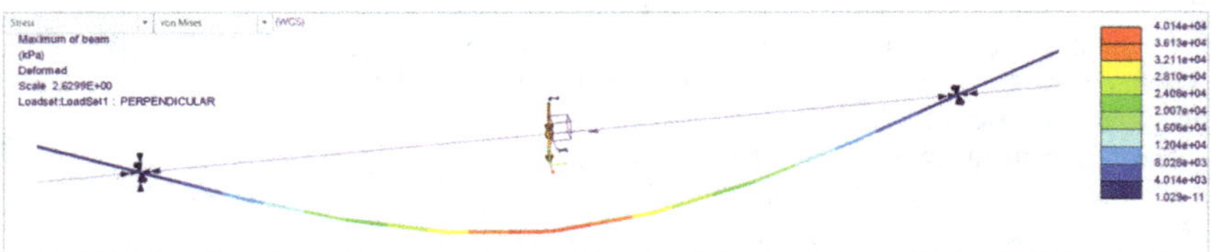

e) Stress von Mises at the perpendicular orientation of a sample during testing.

Fig. 4 Results of numerical analyses.

Table 2 Comparison of the measured/calculated data with numerical analysis results

Orientation of the sample with respect to the layering plane	Deflection at elastic force (mm)		Flexural yield strength (MPa)	
	FEM	Measured	FEM	Calculated
Parallel	14.58	14.1	31.97	31.95
Perpendicular	9.82	9.5	40.14	40.125

Based on Table 2, it is clear that the numerical analysis results are in high accordance with measured (calculated) values, however, the values of the samples´ flexural properties are very far from the values related to the filament flexural properties mentioned in datasheets. Stratasys FDM-Nylon 12 CF data sheet provides the flexural modulus values of the filament in two axes of 10.3 and 2.07 GPa [9,14] (obtained according to the ASTM D790 test method), and Ricoh Europe PLC [15] provides the value of a flexural modulus 10.62 GPa. It indicates that values of the Nylon 12 CF filament properties can not be used for the sample body behaviour simulation in a software application and should be first verified experimentally, also due to a big variety of material, production, testing and processing aspects that it is necessary to take into account during numerical analysis.

Summary

Additive technology is a rapidly developing way of the various parts production for a wide spectrum of usage and applications. This approach to manufacturing can provide the components with a lot of benefits, including sustainability, material savings and improving the environmental conditions. However, there are still a lot of challenges that need to be addressed and solved related to this technology.

Within the manuscript, the effect of sample orientation during testing regarding to the orientation during its production was studied. Six specimens produced at the same conditions were subjected to the three-point bending testing at two different orientations (rotated 90° to each other).

The results showed a big difference between the flexural modulus for the specimens oriented in parallel (755 MPa) and perpendicular (1408 MPa) ways regarding the layering during 3D printing, as well as relatively big differences between yield strengths (31.95 MPa at parallel orientation and 40.125 MPa at perpendicular orientation) but the amount of absorbed energies was almost the same (45.32 J at parallel orientation and 47.85 J at perpendicular orientation).

The obtained experimental values were in good agreement with the results obtained by the numerical analysis, however, the values indicate that the fibre characteristics are different from those characterizing the behaviour of the fabricated body/specimen.

In conclusion, it can be concluded that not only the orientation during 3D printing with respect to the subsequent load has a great influence on the mechanical properties of the components, but also other conditions such as the printing strategy, pre-, and post-processing of samples and the like, which subsequently influence the resulting properties of the part, also play a large role. and can affect its reliability and service life. In the near future, the authors would like to continue the research and study the effect of other aspects on the Nylon 12 CF behaviour under loading.

Acknowledgement

This research was funded by the Ministry of Education, Science, Research and Sport of the Slovak Republic by grants APVV-19-0550, KEGA 032TUKE-4/2022, KEGA 042TUKE-4/2025 and by the Erasmus+ EU program for education, training, youth and sport through grant ERASMUS+ 2021-1-PL01-KA220-HED-000031182.

References

[1] Mitrovic, N.; Milosevic, M.; Momcilovic, N.; Petrovic, A.; Miskovic, Z.; Sedmak, A.; Popovic, P. Local Strain and Stress Analysis of Globe Valve Housing Subjected to External Axial Loading. In Key Engineering Materials; Trans Tech Publications Ltd.: Zurich, Switzerland, 2014; Volume 586, pp. 214–217.

[2] Monkova, K. et. al.; The Effect of a Face Wall on a Cellular Structure During Bending. In: Smart Cities: Importance of Management and Innovations for Sustainable Development. Mobility IoT 2023. EAI/Springer Innovations in Communication and Computing. Springer, Cham. 2024; https://doi.org/10.1007/978-3-031-56533-5_12

[3] Christodoulou, T.I. et al.; An experimental investigation of the mechanical properties of fused filament fabricated nylon-carbon fiber composites, Cutting & Tools in Technological System, 2024, Edition 100; doi:10.20998/2078-7405.2024.100.11

[4] De Toro, E. V.; Sobrino, J. C.; Martínez, A. M. and Eguía, V. M.; Analysis of the influence of the variables of the fused deposition modeling (FDM) process on the mechanical properties of a carbon fiber-reinforced polyamide, Procedia Manuf., vol. 41, pp. 731–738, 2019, doi: 10.1016/j.promfg.2019.09.064

[5] Vanca, J. et al.; Investigation of the Influence of Orientation on the Tensile Properties of 3D Printed Samples with Gyroid Structure. In Proceedings of the 2022 13th International Conference on Mechanical and Aerospace Engineering (ICMAE), Bratislava, Slovakia, 2022; 526–531.

[6] Korol, M. et al.; Researching on the Effect of Input Parameters on the Quality and Manufacturability of 3D-Printed Cellular Samples from Nylon 12 CF in Synergy with Testing Their Behavior in Bending. Polymers 2024, 16, 1429. https://doi.org/10.3390/polym16101429.

[7] ISO 178:2019 Standard; Plastics Determination of Flexural Properties. ISO: Geneva, Switzerland, 2019.

[8] ASTM D638; Standard Test Method for Tensile Properties of Plastics. ASTM International: West Conshohocken, PA, USA, 2022.

[9] FDM Nylon 12 Carbon Fiber Data Sheet; 2018 Stratasys Ltd.; available on https://rapidaxis.com/wp-content/uploads/2022/10/FDM-Nylon12CF.pdf; 2024/02/15

[10] Korol, M.; Research on the adaptation of the characteristics of cellular structures for the distribution of stress in the components, PhD thesis; 2024, FMT TUKE, 94.

[11] Basan, R., Marohnić, T., Marković, E., Evaluation of fatigue parameters estimation methods with regard to specific ranges of fatigue lives and relevant monotonic properties, Procedia Structural Integrity, 42, 2022, 655-662, https://doi.org/10.1016/j.prostr.2022.12.083.

[12] Grbović, A.; Kastratović, G.; Božić, Ž.; Božić, I.; Obradović, A.; Sedmak, A.; Sedmak, S. Experimental and numerical evaluation of fracture characteristics of composite material used in the aircraft engine cover manufacturing. Eng. Fail. Anal. 2022, 137, 106286. https://doi.org/10.1016/j.engfailanal.2022.106286.

[13] Bellini, C.; Borrelli, R.; Di Caprio, F.; Di Cocco, V.; Franchitti, S.; Iacoviello, F.; Mocanu, L.P.; Sorrentino, L. Manufacturing process effect on the bending characteristics of titanium-lattice/FRP hybrid structures. Procedia Struct. Integr. 2022, 42, 196–201. https://doi.org/10.1016/j.prostr.2022.12.024.

[14] Data-Sheet-EN-A4-FDM-Nylon-12CF-1; 2019 Ricoh Europe PLC; available on https://rapidfab.ricoh-europe.com/wp-content/uploads/2019/10/Ricoh-TDS-FDM-Nylon-12CF-Web-Final.pdf; 2024/02/15

[15] Nylon 12 CF Fused Deposition Modelling (FDM), available on https://rapidfab.ricoh-europe.com/wp-content/uploads/2019/10/Ricoh-TDS-FDM-Nylon-12CF-Web-Final.pdf; TDS-FDM-Nylon-12CF | 09/19-v1; 2019 Ricoh Europe PLC; 2024/02/15

Materials Research Proceedings 46 (2024) 15-22 https://doi.org/10.21741/9781644903377-3

A deep learning approach for automated inspection of 3D printed orodispersible films

E. TSINTAVI[1,a], E. STATHATOS[2,b], P. BENARDOS[2,c *] and D.M. REKKAS[1,d]

[1]Section of Pharmaceutical Technology, Department of Pharmacy, National & Kapodistrian University of Athens, Panepistimiopolis, Athens 15784, Greece

[2]Manufacturing Technology Laboratory, School of Mechanical Engineering, National Technical University of Athens, Heroon Polytechniou 9, Athens 15772, Greece

[a]eltsintavi@pharm.uoa.gr, [b]mstatha@mail.ntua.gr, [c]pbenard@mail.ntua.gr, [d]rekkas@pharm.uoa.gr

Keywords: Pharma 4.0, Quality Control, Deep Neural Network, Additive Manufacturing, Personalized Dosage Form, Pharmaceutical Compounding

Abstract. This work aims at demonstrating the applicability of additive manufacturing and machine vision technologies for the on-demand manufacture of customized medicinal products at the point of need, while conforming to fundamental quality guidelines such as cGood Manufacturing Practices (cGMPs), Quality by Design (QbD) as well as Quality Risk Management (QRM) and assuring product quality. The adopted approach combines a 3D printing process for orodispersible films and a deep neural network model in order to develop an automated quality control system for defect detection. The orodispersible films were manufactured using an off-the-shelf 3D printer equipped with a modified semisolids printhead, with Warfarin Sodium ($C_{19}H_{15}NaO_4$) selected as the active pharmaceutical ingredient. The dose could be accurately regulated by varying the rectangular film's dimensions. Three different defect types were considered, corresponding to layer shifting, incomplete prints and inclusion of bubbles within the film, respectively. A pretrained convolutional neural network (GoogleNet) was fine-tuned to perform the defect classification of the 3D printed films, which after data augmentation and hyperparameter tuning, achieved an accuracy of up to 100% in detecting conforming and not conforming products.

Introduction

The marketed pharmaceutical products follow the needs of the average population not considering parameters that can lead to variability in the therapeutic effect. Variations of weight, race, age or sex of a patient can lead to the necessity of dose tuning and thus dose customization, especially when they are addressed to either pediatric or geriatric populations or even more frequently to the use of medicine with narrow therapeutic range. The "*one-size fits all*" application of the marketed products is not always desirable and/or suitable for every patient, leading to the option of precision medicine, which are developed to alter the release profiles and customize the dosages according to the personal needs of the patient [1].

The process followed by pharmacists in order to prepare medicinal products in domestic pharmacies is called pharmaceutical compounding. Even though this methodology is generally applied, galenic preparations show high variability in their quality as it depends on the skill and experience of the pharmacist as well as on the capabilities of the available equipment and even regulatory constraints, e.g. licensing of the facilities. Moreover, globally or within European countries, for the compounded drug formulations there are no uniform manufacturing methods and are not required to be checked for their safety, effectiveness, preparation methods nor for their proper labeling, disposal or cleaning of the equipment as is the case with the marketed products.

[2]. Even though European and American regulators have issued relevant instructions and directives for pharmaceutical products prepared in pharmacies [3-5] that state the existence of a quality assurance system, risk assessment studies, in-process controls and traceability-transparency during manufacture in order to reduce the associated risks, there are inadequacies regarding the quality control and assurance of the galenic preparations.

Considering the above, it is apparent that the current practice regarding the production of personalized medicinal products is characterized by several challenges in terms of its application, especially in cases where the active pharmaceutical ingredients (APIs) have a limited therapeutic index and/or adjustable doses. These challenges can potentially be addressed with a technology such as additive manufacturing allowing easy, adjustable, inexpensive and fast modifications in both the amount and release of the API according to the patient's personalized needs and thus offer the preferred therapeutic outcome, at the point and time of need [6]. Combining the advantages of product flexibility and customization with an automated quality control system which includes analytical methods and pharma 4.0 tools can assure the proper production process as well as product quality and therefore, patient safety. In this paper, the emphasis is placed on defect detection of orodispersible films that are produced through 3D printing, since these defects directly influence the dose that a patient eventually receives. A deep neural network model is trained with images that were obtained through an experimental procedure in order to detect these defects.

Materials and methods

Gel preparation. The "hot/cold" technique was used for the preparation of Hypromellose gels as proposed by Colorcon®. In order to avoid lumping, this method proposes that 1/3 of the water quantity is heated to ~90°C, Methocel^{TM} powder is dispersed and its particles are wetted with constant agitation for around 30 min. Subsequently, the 2/3 of the water quantity are cooled down to ~10°C and added slowly to the dispersion while mixing. The agitation continues for 30 min at ambient temperature to achieve complete solubilization. Warfarin Sodium ($C_{19}H_{15}NaO_4$) was selected as the active pharmaceutical ingredient (API) along with three target weight concentrations (2.5%, 5.0% and 7.5% w/w). The desired gel quantity was measured, and the API was dissolved. The content of Hypromellose was kept constant at 3% w/w in all formulations as its viscosity allowed both the extrusion from the syringe and the formation of a defined square shape.

Film preparation. Two different sizes of square films (25x25mm and 15x15mm) were printed with the different gels that had the target weight concentrations of the API resulting in different doses in the printed films. Subsequently, the 3D printed wet films were left to ambient temperature for one day to dry until stable weight and thus, water content, and form the orodispersible films. The characterization of the dry films included quality characteristics such as appearance and shape, dimensions, weight, thickness and %assay. Examples of wet and dry films are shown in Fig. 1. For demonstration purposes, red color was added in the formulation.

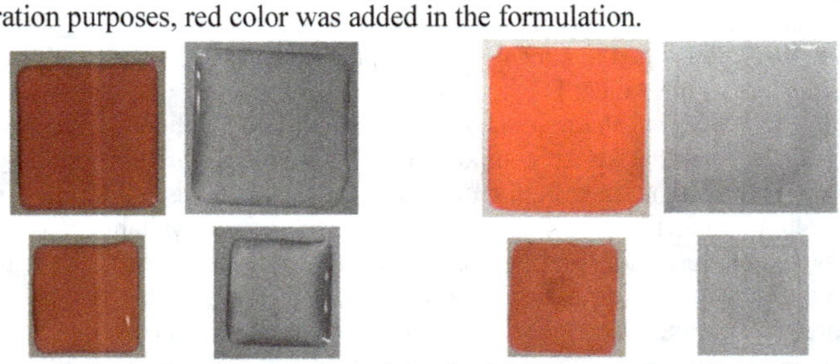

Figure 1: Examples of wet (left) and dry (right) 3D printed orodispersible films.

3D printing of orodispersible films. The printing of the orodispersible films was carried out using a commercial 3D printer, Zmorph SX 2.0 [7], which can be equipped with interchangeable printheads. For printing semisolids, the commercial configuration consists of a 100ml syringe with an adjusted plastic tip at its end, a piston and a suitable support base. However, as the films require low material quantities and controlled material extrusion to achieve their desirable detailed shape, it was necessary to modify the commercial configuration. Thus, the 100ml syringe was replaced by a 10 ml syringe and the plastic tip by a G26 needle, which were attached and supported through custom designed adapters (Fig. 2). The syringe was loaded with approximately 8 ml of gel and the material was then extruded on the printing surface by automatic operation of the syringe piston to produce the film according to the corresponding CAD file and relevant G-code. Due to the modified printhead configuration, all printing parameters had to be adjusted accordingly (Table 1) so as to achieve successful extrusion in terms of film shape and size.

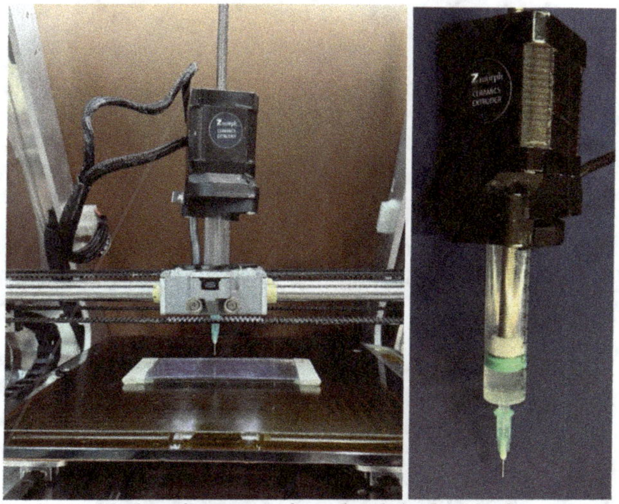

Figure 2: (Left) Modified configuration of printhead, (right) 10 ml syringe equipped with a G26 needle used for material extrusion.

Table 1: Printing parameters of orodispersible films

General settings	2D extrusion settings
Printhead: Chocolate\Ceramics	Layer count: 2
Syringe: 10.0 ml	Layer height: 0.1 mm
Nozzle temperature: Ambient	Path width: 3 mm
Bed temperature: Ambient	Travel speed: 120 mm/s
Needle: G26	Print speed: 12 mm/s
	Extrusion order: Proximity
	Retraction: OFF

Orodispersible film defects. The possible defects that can occur in the films during the printing process consist of bubble inclusions, filling gaps, distorted film shapes and/or dimensions due to movement of the printing base or over-extrusion of material, and incomplete films due to inadequate raw material. Some representative examples of films with such defects are shown in Fig. 3.

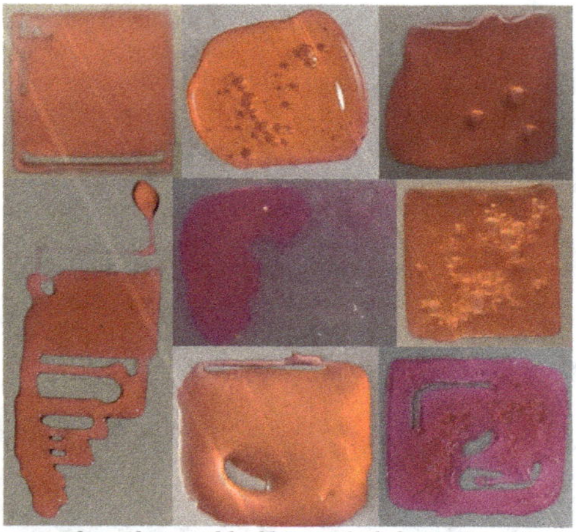

Figure 3: Wet 3D printed orodispersible films with defects. For demonstration purposes red color was added in the film formulation.

From the set of possible defects, the three most probable to occur were selected to be included in the development of the automated quality control system; a) films with distorted shape due to misplacement/movement of the printing base and/or layer shifting (Type I), b) incomplete printed films due to not enough material (Type II) and c) formation of bubbles due to trapped air in the gel (Type III). Each defect class is presented in Fig. 4 against an orodispersible film with no defects.

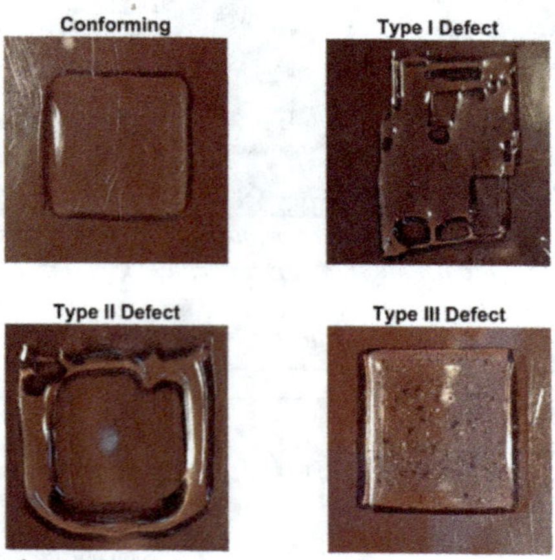

Figure 4: Examples of considered defect types compared to conforming orodispersible film.

Machine Vision. A deep neural network-based machine vision system was developed to characterize the 3D-printed films and classify them into one of the four available classes, as defined earlier. The model was designed not only to categorize the films as "conforming" or "not conforming" but also to identify the specific type of defect, if present.

This was achieved through transfer learning using a pretrained convolutional neural network, GoogleNet, which is based on the inception architecture [8]. MATLAB R2022a was utilized to fine-tune the neural network according to the requirements of film characterization. This approach significantly simplified the training process, saving time (minutes instead of hours), reducing the necessary training data (from thousands to dozens), and allowing training on a standard PC. As is typical in a transfer learning process, the earlier layers of the pretrained network, which capture general features like edges and textures, were frozen to retain their learned weights. This allows the model to leverage previously learned representations from large datasets. Only the final layers, see Fig. 5, which are more specific to the classification task, were retrained to adjust to the new classes.

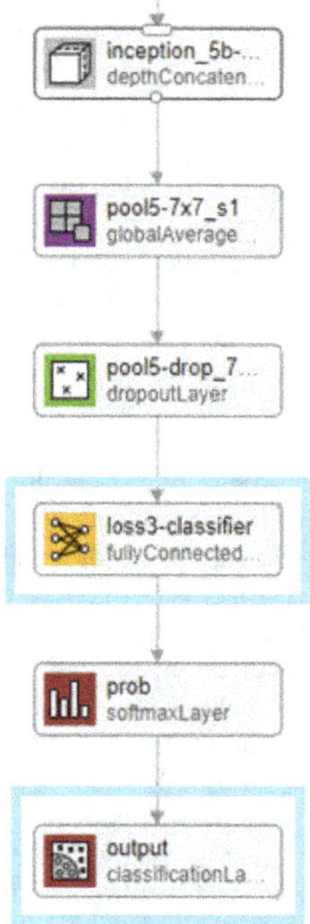

Figure 5: Final GoogleNet neural network layers. The modified layers according to the needs of the film categorization are marked in blue boxes.

Data augmentation techniques [9], including random rotations by 90 degrees and translations by 30 pixels, were applied to increase the dataset size. The dataset was then split into training, validation, and testing subsets using a 70/15/15 ratio. It is important to note that no augmented images were included in the testing sets – only the original, raw images were used to ensure unbiased evaluation. Table 2 lists the number of images in each class before and after augmentation.

Materials Research Proceedings 46 (2024) 15-22 https://doi.org/10.21741/9781644903377-3

Table 2: Number of images in each class before and after data augmentation.

Number of images	Conforming	Type I	Type II	Type III	Total
Original	14	15	14	8	51
After Augmentation	35	35	35	25	130

The neural network was fine-tuned using the stochastic gradient descent with momentum (SGDM) method [10]. Early stopping callbacks with a patience parameter were employed to prevent overfitting. Various values for the patience parameter and the learning rate, as well as different batch sizes and epoch limits, were explored to balance the trade-off between training speed and stability.

Results and Discussion

The results were obtained by training the system with a limited number of raw images, as shown in Table 2. Despite the small dataset, the neural network was able to accurately identify the type of defect in most cases. However, there were occasional misclassifications between Type I and Type II defects, due to the wide variability, even in in-class appearance. This is an inherently difficult problem related to the specific samples included in the training and testing sets.

To address these challenges, the dataset was randomly divided into training, validation, and testing sets ten times to better assess the model's performance. The percentile split was 70/15/15 (rounded to 90/20/20 images), and 10 different neural networks were trained. The mini-batch size was set to 30 to ensure the dataset was fully utilized without exclusions, and the maximum number of epochs was capped at 20. The training process employed an early stopping mechanism, which uses a patience parameter. This parameter controls how many consecutive epochs the model is allowed to continue training after the validation loss stops improving. If the validation loss does not improve for a set number of epochs (as defined by the patience value), the training process stops early to prevent overfitting. The patience parameter was tested at two levels: 3 (low patience) and 5 (high patience), to examine its effect on performance.

Learning rates of 1e-3 (fast) and 1e-4 (slow) were also tested. It was anticipated that a slower learning rate combined with higher patience would produce the best results. The key focus was achieving 100% defect detection accuracy. While it was acceptable for the network to confuse similar defect types, false negatives and false positives were unacceptable.

The results confirmed that, with a fast learning rate and low patience, there was one false negative and two false positives. With fast learning and higher patience, the model produced one false negative but no false positives. Finally, with slow learning and higher patience, the network achieved 100% accuracy in defect recognition, with no false positives or negatives. Table 3 summarizes results from the testing set for the best combination of hyperparameters. As can be seen, from the 10 neural networks, there are two, no 5 and no 8, which exhibit up to 100% prediction accuracy in both the validation and testing subsets.

Materials Research Proceedings 46 (2024) 15-22 https://doi.org/10.21741/9781644903377-3

Table 3: Results for neural networks with slow learning rate (1e-4) and high patience (5).

Predicted Label	Correct Label	Run Number	Validation Accuracy	Testing Accuracy
Type II Defect (Finished Gel)	Type I Defect (Misplaced)	1	90%	90%
Type II Defect (Finished Gel)	Type I Defect (Misplaced)	1	90%	90%
Type I Defect (Misplaced)	Type II Defect (Finished Gel)	2	85%	95%
Type I Defect (Misplaced)	Type II Defect (Finished Gel)	3	80%	90%
Type I Defect (Misplaced)	Type II Defect (Finished Gel)	3	80%	90%
Type I Defect (Misplaced)	Type II Defect (Finished Gel)	4	100%	95%
NO ERRORS	NO ERRORS	5	100%	100%
Type II Defect (Finished Gel)	Type I Defect (Misplaced)	6	90%	90%
Type II Defect (Finished Gel)	Type I Defect (Misplaced)	6	90%	90%
Type II Defect (Finished Gel)	Type I Defect (Misplaced)	7	95%	85%
Type I Defect (Misplaced)	Type II Defect (Finished Gel)	7	95%	85%
Type I Defect (Misplaced)	Type II Defect (Finished Gel)	7	95%	85%
NO ERRORS	NO ERRORS	8	100%	100%
Type II Defect (Finished Gel)	Type I Defect (Misplaced)	9	95%	90%
Type II Defect (Finished Gel)	Type I Defect (Misplaced)	9	95%	90%
Type II Defect (Finished Gel)	Type I Defect (Misplaced)	10	100%	95%
		Means:	**93.5%**	**93.0%**

Based on the achieved accuracy, it is evident that the augmented dataset of the 3D printed film images is more than sufficient for the investigated problem and the considered defect types. Of course, if a new defect type was to be introduced, then new images containing the new defect type would have to be collected and processed with the same augmentation techniques in order to re-train the neural network.

Conclusions

To address the quality challenges faced during compounding activities in Community and Hospital Pharmacies an approach for automated quality control of compounded drug products prepared at the point of need was developed. This approach focused on defect detection in images of orodispersible films by means of a convolutional neural network, data augmentation and hyperparameter tuning. The main conclusions can be summarized as follows:

- Additive manufacturing can be employed for the preparation of small batches of customized dosage forms according to the principle of "on-demand manufacturing".

- Using a limited number of images, and thus a small data set, a neural network was developed that was able to not only identify "conforming" and "not-conforming" produced films but also classify the type of defect with up to 100% accuracy.
- Thus, during a production batch, any acquired film image can provide information on whether the manufacturing process was successful or not and, if not, identify the reason for failure.

The presented approach combined with other analytical techniques could serve as the basis for the preparation of fully customized pharmaceutical forms at the point and when in need, while complying with the quality guidelines and ensuring traceability throughout the entire production process.

References

[1] Tan, Y.J.N., Yong, W.P., Low, H.R., Kochhar, J.S., Khanolkar, J., Lim, T.S.E., Sun, Y., Wong, J.Z.E., Soh, S., 2021. Customizable drug tablets with constant release profiles via 3D printing technology. Int. J. Pharm. 598, 120370. https://doi.org/10.1016/j.ijpharm.2021.120370

[2] J. Gudeman, M. Jozwiakowski, J. Chollet, M. Randell, Potential risks of pharmacy compounding, Drugs R D 13 (2013) 1–8. https://doi.org/10.1007/s40268-013-0005-9

[3] Council of Europe, Resolution CM/Res(2016) on quality and safety assurance requirements for medicinal products prepared in pharmacies for the special needs of patients, (2016). Information on: https://search.coe.int/cm?i=090000168065c132

[4] U.S. Department of Health and Human Services Food and Drug Administration Center for Drug Evaluation and Research, Registration of Human Drug Compounding Outsourcing Facilities Under Section 503B of the FD & C Act Guidance for Industry (2014).

[5] U.S. Food and Drug Administration, FDA's Human Drug Compounding Progress Report - Three Years After Enactment of the Drug Quality and Security Act. (2017). Information on: https://www.fda.gov/drugs/human-drug-compounding/fdas-human-drug-compounding-progress-report-three-years-after-enactment-drug-quality-and-security

[6] M. Attaran, The rise of 3-D printing: The advantages of additive manufacturing over traditional manufacturing. Bus. Horiz. 60 (2017) 677–688. https://doi.org/10.1016/j.bushor.2017.05.011

[7] Zmorph, Zmorph - 3D Printers that deliver, (2022). Information on: https://zmorph3d.com/about-us/

[8] C. Szegedy, W. Liu, Y. Jia, P. Sermanet, S. Reed, D. Anguelov, D. Erhan, V. Vanhoucke, A. Rabinovich, Going deeper with convolutions, arXiv. (2014) https://doi.org/10.48550/arXiv.1409.4842

[9] C. Shorten, T.M. Khoshgoftaar, A survey on Image Data Augmentation for Deep Learning. J Big Data 6 (2019) 60. https://doi.org/10.1186/s40537-019-0197-0

[10] Y. Liu, Y. Gao, W. Yin, An improved analysis of stochastic gradient descent with momentum. arXiv. (2020) https://doi.org/10.48550/arXiv.2007.07989

XYZ calibration cube - A misleading tool for achieving print accuracy

Vasile ERMOLAI[1, a *], Marius-Ionuţ RÎPANU[1, b], Vasile MERTICARU Jr.[1, c], Alexandru-Ionuţ IRIMIA[1, d], Marius-Andrei MIHALACHE[1, e], Alexandru SOVER[2, f], Nicolae-Răzvan MITITELU[1, g], Ionuţ-Mădălin PIŞTA[1, h]

[1]"Gheorghe Asachi" Technical University of Iaşi, Department of Machine Manufacturing Technology, Blvd. Dimitrie Mangeron 59A, Iaşi, 700050, Romania

[2]Ansbach University of Applied Science, Faculty of Technology, Residenzstraße 8, Ansbach, 91522, Germany

[a]vasile.ermolai@academic.tuiasi.ro, [b]marius-ionut.ripanu@academic.tuiasi.ro, [c]vasile.merticaru@academic.tuiasi.ro, [d]alexandru-ionut.irimia@student.tuiasi.ro, [e]marius-andrei.mihalache@academic.tuiasi.ro, [f]a.sover@hs-ansbach.de, [g]nicolae-razvan.mititelu@student.tuiasi.ro, [h]ionutpista@yahoo.com

Keywords: Fused Filament Fabrication, Material Extrusion Additive Manufacturing, XYZ Cube, Calibration Cube, Print Accuracy

Abstract. Various factors influence the dimensional accuracy in Fused Filament Fabrication (FFF). The kinematic system is a major influential factor, as the motion precision depends on the implemented solution. Kinematics have a direct influence on equipment costs. The FFF 3D printing market has abundant equipment options, from low-cost to high-end solutions. The filament extrusion precision and the shrinkage induced by the thermal contractions are other influential factors for the resulting parts' quality. Nowadays, the XYZ cube is the calibration benchmark. This cube is used to adjust the dimensional accuracy of the resulting parts. The deficiency with the calibration cubes is that they allow only the measurement of external dimensions, which makes the calibration useless when tolerancing the internal features or scaling. This paper discusses the limitations of the accuracy calibration method using the XYZ cube and possible solutions to overcome them.

Introduction

Fused Filament Fabrication (FFF) is a Material Extrusion (MEX) Additive Manufacturing (AM) technology that uses molten thermoplastic polymer-based materials to directly build parts based on a 3D model [1-3].

The FFF technology basics were set by Adrian Boyer and his research team in 2004 through the RepRap (i.e., self-replicating machine) project. After the Stratasys patent termination for the core technology, Fused Deposition Modeling (FDM), the consumer 3D printing market rose significantly [4, 5]. As the RepRap project was open source, the FFF 3D printer idea was taken by various companies (e.g., MakerBot) and sold with a price tag under 1,000 USD [6, 7]. Wohlers reports stated that a budget printer is under 5,000 USD [8].

Nowadays, the consumer 3D Printing market has various options, with equipment starting at a few hundred (i.e., Bambu Lab A1 Mini at 199 US [9]) and going up to thousands of USD (i.e., Ultimaker S7 at 8,400 USD [10]). To those, we can sum up the 3D printers developed by the DIY community (e.g., Voron 2.4). Those are further customized by the users [11]. Knowing that the overall print quality of the resulting parts depends on the printer's kinematics precision, the 3D printer cost reflects the build quality. This way, to address the printing quality of a given 3D printer, the users' community developed benchmark parts (e.g., 3D Benchy [12, 13]) and methodologies to improve the resulting parts' dimensional quality [14-16].

The XYZ cube is a widely used quality benchmark [17-20]. This tool helps adjust the steps/mm or how many rotations the step motor requires to extrude one mm of filament [18-20]. However, the XYZ cube is misleading as it allows only the evaluation of the external dimensions. This makes the cube futile for skew correction, internal features tolerancing, or scaling for thermal contraction compensations [21]. Thus, this paper discusses the limitations of current accuracy adjustment methods and examines potential alternatives.

Methods

Calibration cubes are probably one of the most 3D printed parts as they are used as a tool to see if equipment prints dimensional accurately. Usually, the calibration cube has an edge size of 20 to 30 mm [22-36] and is easy to recognize based on the X, Y, and Z embedded letters, which indicate the printer's axes. However, cubes can have a smaller edge size (i.e., 15 mm [18]) or significantly bigger (i.e., 100 mm [36]).

Over the years, there have been many variations of the calibration cube, each iteration aiming to provide more information about the printing process parametrization (e.g., overhangs, filament cooling, seam alignments). On the other hand, other calibration cube designs introduced aesthetics (e.g., Calicat) or interlocking (e.g., Lego calibration cube) as a feature, to increase the resulting parts' utility. Fig. 1 shows the print result of a fraction of the existing calibration cubes available on sharing platforms [37, 38] such as Thingiverse [39], Printables [40], My Mini Factory [41], Cults 3D [42], Grab CAD [43] and others.

Fig. 1. 3D printed calibration cubes made in the past years within their research.

Fifteen of the available calibration cubes are shown in Table 1. The resulting captions are screenshots of the print preview made using the Cura slicing tool. As can be observed, besides the flat, parallel surfaces (cubes 1, 4, 5, 11, and 15), these calibration cubes integrate features that give information about clearance between assembled parts (cube 6), filament retraction length and speed (cube 7), circular interpolation (cube 4, 5 8 and 10), small features size quality (cube 3, 5, 7 and 9), horizontal holes (cube 3 and 13), overhangs due to X and Y letter marks, appearance of ringing (also known as ghosting or rippling [44]) defects next to the X, Y letters or other details, or axels skewness (cube 15).

The calibration cubes help adjust the filament-feeding stepper motor's number of steps/mm. In other words, the number of rotational steps needed to feed the filament during the 1 mm travel of the printhead [45, 46]. The common practice is to print a cube, with a given edge of 20 mm, take a measurement using a caliper, and adjust the steps/mm accordingly. Then, after the motor's steps/mm adjustment, the cube is printed again, aiming for its edge nominal dimension. This process is repeated until the printer has produced a cube with the aimed level of dimensional accuracy.

In FFF parts are made layer-wise, by multiple passes of the extrusion head. This way, each layer integrates walls (also known as permitters or shells), one or more solid layers, and an internal structure (i.e., infill). A preview of FFF printed parts is illustrated in Fig. 2. Each construction element is characterized by a rectangle cross-section (*i.e., layer height by extrusion width*) with rounded corners and deposited along some extrusion paths. One misconception is that calibration cubes are needed to adjust the printhead positioning [47], the $P_{x_1y_2}, P_{x_2y_2}, \cdots, P_{x_iy_i}$ points highlighted in Fig. 2. In reality, they are used to adjust the number of steps/mm of extruded filament, to prevent under or over-extrusion [48]. The updated values are introduced in the printer firmware and depend on the used g-code flavor.

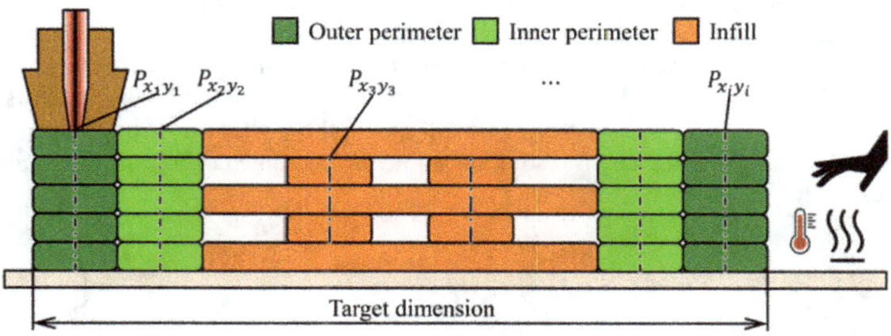

Fig. 2. Printhead positioning during the filament deposition.

Table 1. Different types of calibration cubes available on the internet. Print preview made using Ultimaker Cura 5.7.2.

1. XYZ 20mm Calibration Cube			**6.** Hinged Calibration Cube			**11.** Calibration Cube 20x20		
	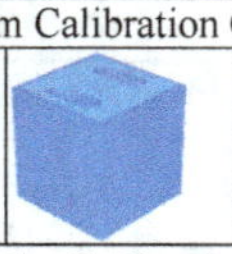	[22]		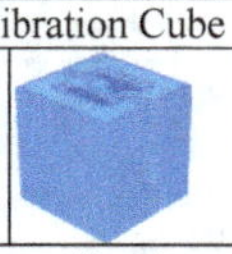	[27]	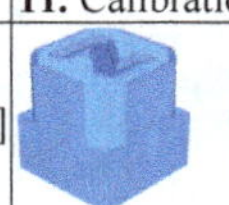	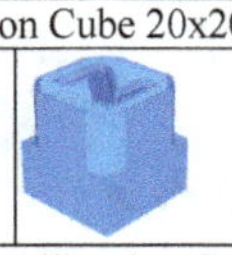	[32]
2. Cali Cat - The Calibration Cat			**7.** 3 in 1 20mm calibration cube			**12.** Detailed calibration Cube		
		[23]	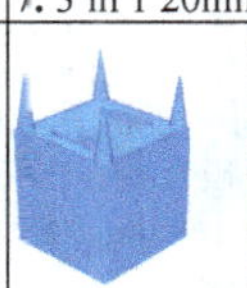		[28]			[33]
3. Filament Test Cube			**8.** XYZ Calibration Cube			**13.** Calibration Cube		
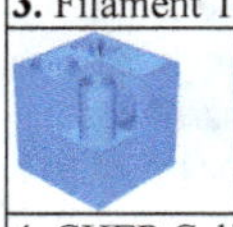	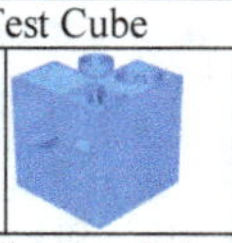	[24]		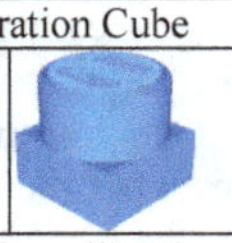	[29]	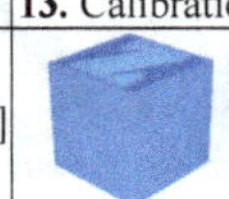	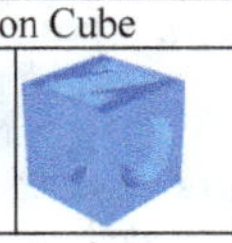	[34]
4. CHEP Calibration Cube			**9.** Test Cube for calibration			**14.** Calibration cube		

		[25]	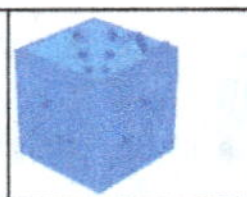	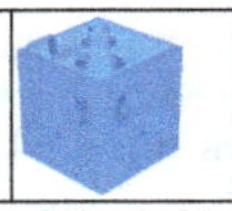	[30]		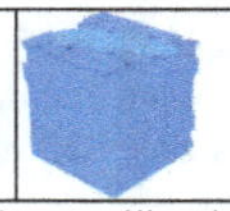	[35]
5. Lego Calibration Cube			**10.** Calibration Cube & Circle			**15.** XYZ 100mm calibration cube		
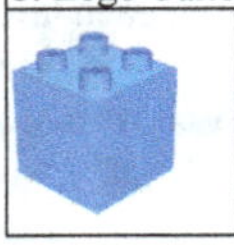	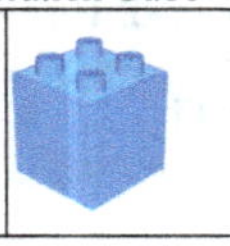	[26]	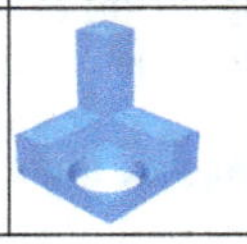	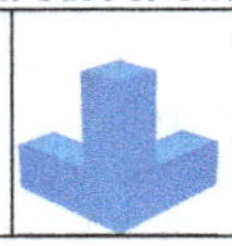	[31]	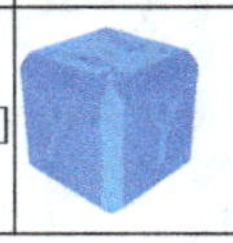	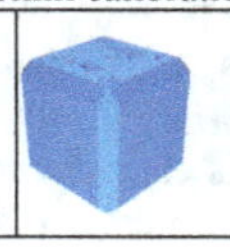	[36]

Faulty measurements

On the one hand, the XYZ cubes within the above-described methodology are misleading as the resulting measured surfaces can have one or multiple defects, such as elephant foot or inconsistent extrusion (as shown in Fig. 3). This makes outer dimension measurement flawed.

On the other hand, the XYZ cubes are misleading because of their small scale. A 20 mm edge cube adjusts the printer's "dimensional accuracy". This 20 mm cube gauge is used as a reference to calibrate all printable dimensions in the build volume. A 0.2 mm measurement flaw on the 20 mm edge block is a 1% error. The same measurement error on a 100 mm edge block is 0.2%. In other words, the smaller the size of the measured gauge, the higher the error in case of a faulty measurement. For this reason, a high-resolution caliper or micrometer should be used for a small-size gauge.

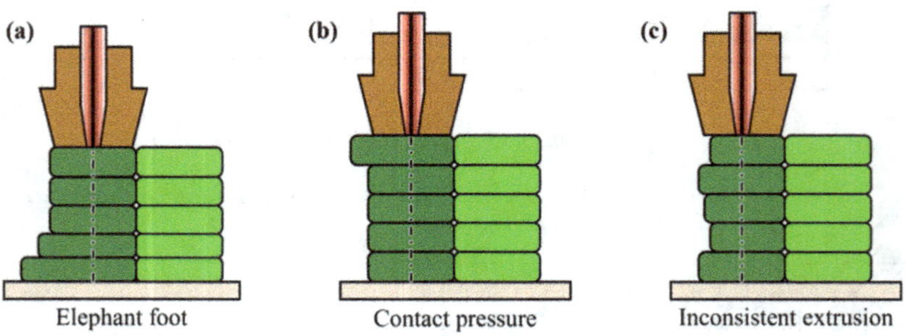

Fig. 3. Possible causes of variation for resulting dimensions: (a) Elephant foot; (b) Nozzle contact overpressure; (c) Inconsistent material extrusion volume.

When printing the XYZ cube, one or multiple defects can appear on the block's external surface, which impacts the resulting measurements of the parts. These printing defects include:

- ***Elephant foot***: the defect occurs when the gap between the nozzle and the build plate is too small [49] (i.e., smaller than the layer height). As a result, the molten material squeezes out and increases the part cross-section for the first few layers (i.e., 2-3 layers). Another cause for the elephant foot appearance is build plate over-heating, above the glass transition temperature of the printing polymer. Elephant footstep appearance can be reduced via slicing tool through the initial layer horizontal expansion setting or by creating a minimum 1x45° chamfer at the part's base;

- ***Nozzle contact overpressure:*** appears when the distance between the nozzle and the previously added layers is too small, and the material squishes out [50]. This effect is problematic when printing the outer shell because it affects the part's dimensional accuracy. The gap size between the nozzle and printed layers can influenced by material thermal expansion or over-extrusion of the previous layers or construction elements (i.e., inner perimeters, infill);

- ***Inconsistent extrusion:*** referees to the variation of the molten material flow. Inconsistent material filament diameter, and issues with the filament feeding mechanism, as the extruder melting rate is too small relative to the maximum volumetric speed are possible reasons for melt flow variation. Those issues must be solved by elimination;

- ***Ringing (Ghosting):*** this issue appears when the nozzle is unable to lay down material in a controlled manner leaving after images of a printed feature in the neighboring area [44, 52] (X and Y cube faces of Fig. 4). This effect appears only on a part's sides and caused by vibration due to over increasing the print speed (by not considering the extruder inertia), uneven belts tension or a bad frame design (i.e., not rigid enough). This issue is solvable by reducing print speed and acceleration;

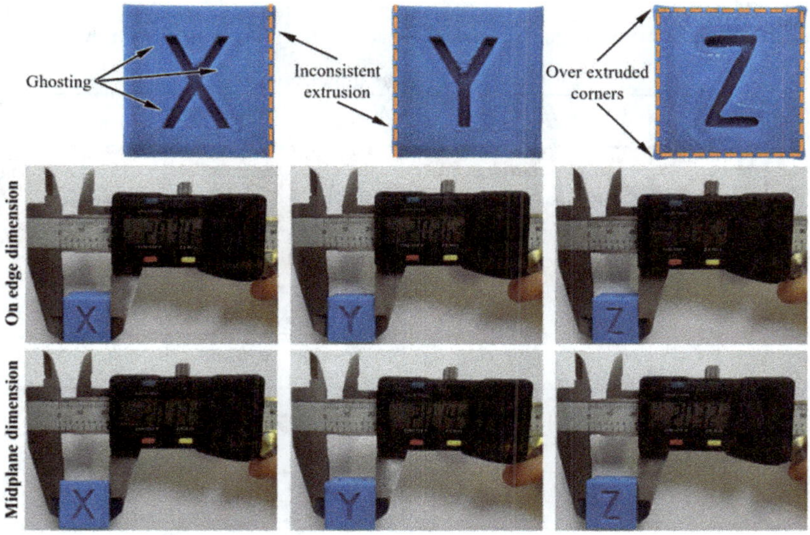

Fig. 4. Measured dimensions of an XYZ cube relative to analyzed sides and locations.

- ***Exposed seam:*** in FFF a seam refers to the start and endpoint of a perimeter [53, 54]. Depending on the material, nozzle size, and parametrization, the closing point of the outer shell can be more or less visible. In the case of a calibration cube, an over-exposed seam can increase the part size. The seam exposure is adjustable through the contour closing method, nozzle wipe distance, filament retraction, and material volume coasting;

- ***Over-extruded corners:*** Every Computer Numerically Controlled machine has a tool head whose movement is controlled through velocity and acceleration based on blank specifics and Material Removal Rate [55, 56]. Jerk Speed is specific for FFF equipment that controls the speed at which the printhead can travel through corners [54, 58]. Firstly, the Jerk Speed is set at a value greater than zero to avoid excessive material ooze when turning corners [54, 57, 58] (i.e., the over-extruded corners presented in the cube Z face of Fig. 4). Secondly, the jerk speed is always lower than the global printing speed to reduce vibrations (which lead to ghosting X and Y cube faces of Fig. 4). Over-extrusion corner formation can [54, 58] be avoided by decreasing the print speed or by creating filets (i.e., rounded corners) to replace the linear interpolation printhead movement with the circular interpolation one. For the printers introduced in the past four years, the over-extrusion corner issue was solved via input shaping, a control technique that creates a commanding signal that cancels vibrations.

Thermal contractions

During the deposition of the molten thermoplastic, the polymeric chains are rearranged along the extrusion paths. Subsequently, the melt is wetting adjacent extruded filament creating intimate contact. Simultaneously, the polymeric chains diffuse, creating interpenetrating networks (i.e., coils, entanglements, or crosslinks) promoting molecular adhesion [59, 60-62].

Thermodynamically speaking, when melted, the polymeric chains have a high degree of freedom, and a high entropy, which allows material reshaping. As the melt cools down, the polymer begins to crystallize. During this step, the polymeric chains start to lose their mobility (entropy decreases), and the chains try to occupy all available volume [62]. As a result, a volumetric decrease can occur due to thermal contractions (i.e., shrinkage), as illustrated in Fig. 5.

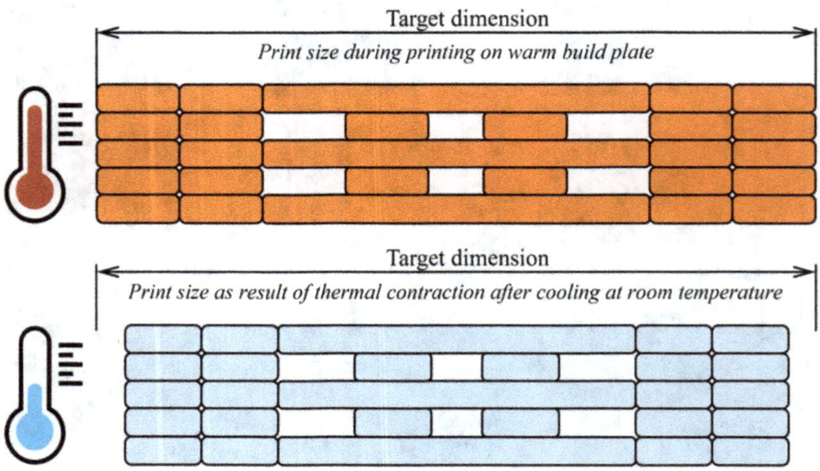

Fig. 5. Part shrinkage as a result of thermal contractions during cooling.

All polymeric materials are affected by thermal contractions. Engineering-grade polymers such as ABS, ASA, or PC require a heat chamber to reduce thermal stress and shrinkage caused by the thermal gradient resulting from the difference between printing temperature (e.g., 240-260°C for ABS) and environment (e.g., 25°C).

Skewed axes

A precisely designed and assembled cartesian 3D printer requires X, Y, and Z axes perpendicular to each other. Otherwise, the axles are skewed, and the resulting parts are deformed compared to the original design. When printing the calibration cube, instead of a part with a square cross-section (Fig. 6.a), the printer will make a part with a parallelogram-like cross-section (Fig. 6.b). In other words, the resulting parts will have diagonals of different sizes.

XYZ calibration cube - A misleading tool for achieving print accuracy

Vasile ERMOLAI[1, a *], Marius-Ionuț RÎPANU[1, b], Vasile MERTICARU Jr.[1, c], Alexandru-Ionuț IRIMIA[1, d], Marius-Andrei MIHALACHE[1, e], Alexandru SOVER[2, f], Nicolae-Răzvan MITITELU[1, g], Ionuț-Mădălin PIȘTA[1, h]

[1]"Gheorghe Asachi" Technical University of Iași, Department of Machine Manufacturing Technology, Blvd. Dimitrie Mangeron 59A, Iași, 700050, Romania

[2]Ansbach University of Applied Science, Faculty of Technology, Residenzstraße 8, Ansbach, 91522, Germany

[a]vasile.ermolai@academic.tuiasi.ro, [b]marius-ionut.ripanu@academic.tuiasi.ro, [c]vasile.merticaru@academic.tuiasi.ro, [d]alexandru-ionut.irimia@student.tuiasi.ro, [e]marius-andrei.mihalache@academic.tuiasi.ro, [f]a.sover@hs-ansbach.de, [g]nicolae-razvan.mititelu@student.tuiasi.ro, [h]ionutpista@yahoo.com

Keywords: Fused Filament Fabrication, Material Extrusion Additive Manufacturing, XYZ Cube, Calibration Cube, Print Accuracy

Abstract. Various factors influence the dimensional accuracy in Fused Filament Fabrication (FFF). The kinematic system is a major influential factor, as the motion precision depends on the implemented solution. Kinematics have a direct influence on equipment costs. The FFF 3D printing market has abundant equipment options, from low-cost to high-end solutions. The filament extrusion precision and the shrinkage induced by the thermal contractions are other influential factors for the resulting parts' quality. Nowadays, the XYZ cube is the calibration benchmark. This cube is used to adjust the dimensional accuracy of the resulting parts. The deficiency with the calibration cubes is that they allow only the measurement of external dimensions, which makes the calibration useless when tolerancing the internal features or scaling. This paper discusses the limitations of the accuracy calibration method using the XYZ cube and possible solutions to overcome them.

Introduction

Fused Filament Fabrication (FFF) is a Material Extrusion (MEX) Additive Manufacturing (AM) technology that uses molten thermoplastic polymer-based materials to directly build parts based on a 3D model [1-3].

The FFF technology basics were set by Adrian Boyer and his research team in 2004 through the RepRap (i.e., self-replicating machine) project. After the Stratasys patent termination for the core technology, Fused Deposition Modeling (FDM), the consumer 3D printing market rose significantly [4, 5]. As the RepRap project was open source, the FFF 3D printer idea was taken by various companies (e.g., MakerBot) and sold with a price tag under 1,000 USD [6, 7]. Wohlers reports stated that a budget printer is under 5,000 USD [8].

Nowadays, the consumer 3D Printing market has various options, with equipment starting at a few hundred (i.e., Bambu Lab A1 Mini at 199 US [9]) and going up to thousands of USD (i.e., Ultimaker S7 at 8,400 USD [10]). To those, we can sum up the 3D printers developed by the DIY community (e.g., Voron 2.4). Those are further customized by the users [11]. Knowing that the overall print quality of the resulting parts depends on the printer's kinematics precision, the 3D printer cost reflects the build quality. This way, to address the printing quality of a given 3D printer, the users' community developed benchmark parts (e.g., 3D Benchy [12, 13]) and methodologies to improve the resulting parts' dimensional quality [14-16].

The XYZ cube is a widely used quality benchmark [17-20]. This tool helps adjust the steps/mm or how many rotations the step motor requires to extrude one mm of filament [18-20]. However, the XYZ cube is misleading as it allows only the evaluation of the external dimensions. This makes the cube futile for skew correction, internal features tolerancing, or scaling for thermal contraction compensations [21]. Thus, this paper discusses the limitations of current accuracy adjustment methods and examines potential alternatives.

Methods

Calibration cubes are probably one of the most 3D printed parts as they are used as a tool to see if equipment prints dimensional accurately. Usually, the calibration cube has an edge size of 20 to 30 mm [22-36] and is easy to recognize based on the X, Y, and Z embedded letters, which indicate the printer's axes. However, cubes can have a smaller edge size (i.e., 15 mm [18]) or significantly bigger (i.e., 100 mm [36]).

Over the years, there have been many variations of the calibration cube, each iteration aiming to provide more information about the printing process parametrization (e.g., overhangs, filament cooling, seam alignments). On the other hand, other calibration cube designs introduced aesthetics (e.g., Calicat) or interlocking (e.g., Lego calibration cube) as a feature, to increase the resulting parts' utility. Fig. 1 shows the print result of a fraction of the existing calibration cubes available on sharing platforms [37, 38] such as Thingiverse [39], Printables [40], My Mini Factory [41], Cults 3D [42], Grab CAD [43] and others.

Fig. 1. 3D printed calibration cubes made in the past years within their research.

Fifteen of the available calibration cubes are shown in Table 1. The resulting captions are screenshots of the print preview made using the Cura slicing tool. As can be observed, besides the flat, parallel surfaces (cubes 1, 4, 5, 11, and 15), these calibration cubes integrate features that give information about clearance between assembled parts (cube 6), filament retraction length and speed (cube 7), circular interpolation (cube 4, 5 8 and 10), small features size quality (cube 3, 5, 7 and 9), horizontal holes (cube 3 and 13), overhangs due to X and Y letter marks, appearance of ringing (also known as ghosting or rippling [44]) defects next to the X, Y letters or other details, or axels skewness (cube 15).

The calibration cubes help adjust the filament-feeding stepper motor's number of steps/mm. In other words, the number of rotational steps needed to feed the filament during the 1 mm travel of the printhead [45, 46]. The common practice is to print a cube, with a given edge of 20 mm, take a measurement using a caliper, and adjust the steps/mm accordingly. Then, after the motor's steps/mm adjustment, the cube is printed again, aiming for its edge nominal dimension. This process is repeated until the printer has produced a cube with the aimed level of dimensional accuracy.

In FFF parts are made layer-wise, by multiple passes of the extrusion head. This way, each layer integrates walls (also known as permitters or shells), one or more solid layers, and an internal structure (i.e., infill). A preview of FFF printed parts is illustrated in Fig. 2. Each construction element is characterized by a rectangle cross-section (*i.e., layer height by extrusion width*) with rounded corners and deposited along some extrusion paths. One misconception is that calibration cubes are needed to adjust the printhead positioning [47], the $P_{x_1y_2}$, $P_{x_2y_2}$, $\cdots$, $P_{x_iy_i}$ points highlighted in Fig. 2. In reality, they are used to adjust the number of steps/mm of extruded filament, to prevent under or over-extrusion [48]. The updated values are introduced in the printer firmware and depend on the used g-code flavor.

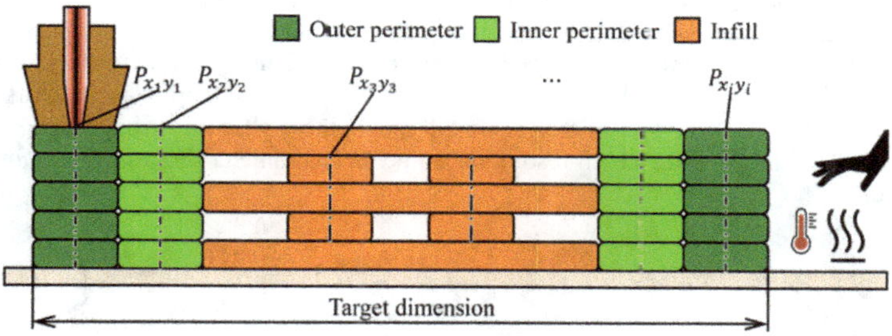

Fig. 2. Printhead positioning during the filament deposition.

Table 1. Different types of calibration cubes available on the internet. Print preview made using Ultimaker Cura 5.7.2.

1. XYZ 20mm Calibration Cube	**6.** Hinged Calibration Cube	**11.** Calibration Cube 20x20
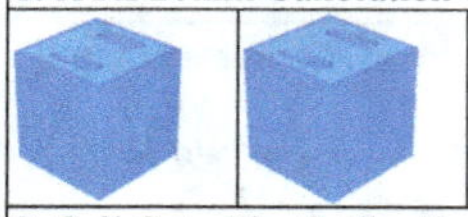 [22]	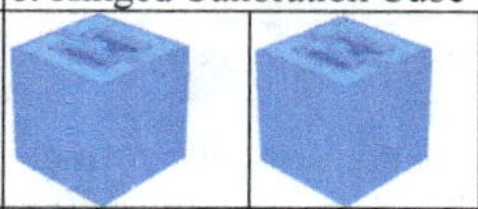 [27]	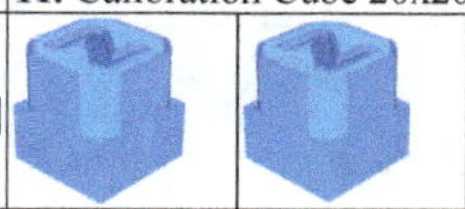 [32]
2. Cali Cat - The Calibration Cat	**7.** 3 in 1 20mm calibration cube	**12.** Detailed calibration Cube
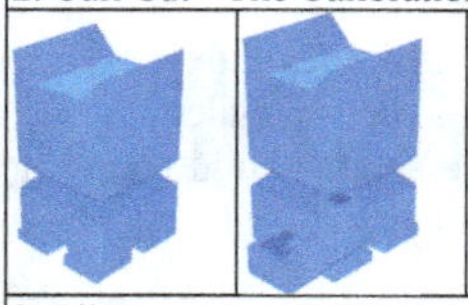 [23]	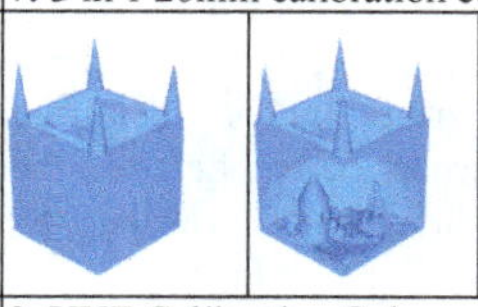 [28]	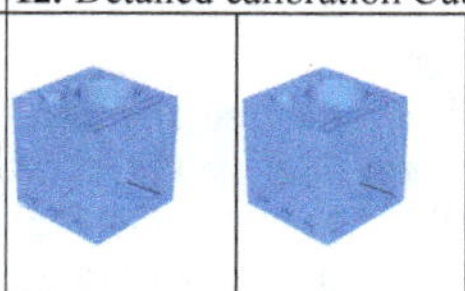 [33]
3. Filament Test Cube	**8.** XYZ Calibration Cube	**13.** Calibration Cube
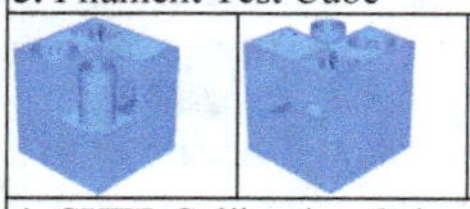 [24]	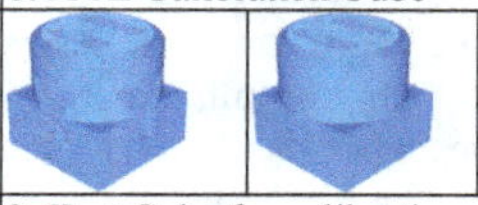 [29]	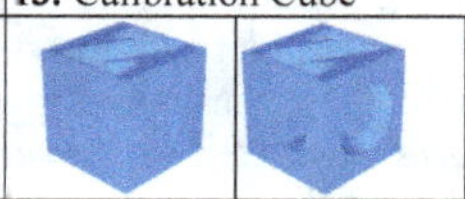 [34]
4. CHEP Calibration Cube	**9.** Test Cube for calibration	**14.** Calibration cube

		[25]	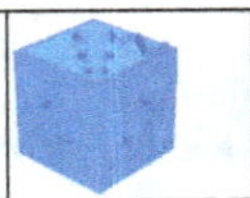		[30]			[35]
5. Lego Calibration Cube			**10.** Calibration Cube & Circle			**15.** XYZ 100mm calibration cube		
	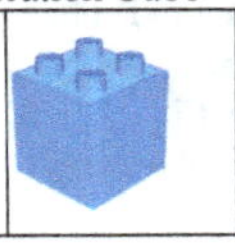	[26]	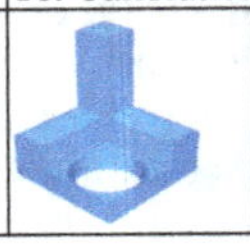	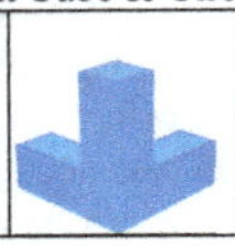	[31]	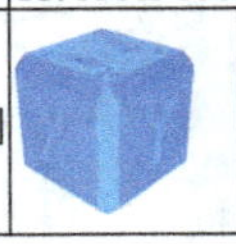	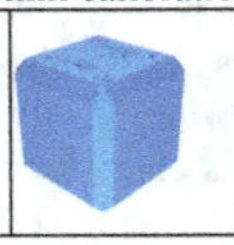	[36]

Faulty measurements
On the one hand, the XYZ cubes within the above-described methodology are misleading as the resulting measured surfaces can have one or multiple defects, such as elephant foot or inconsistent extrusion (as shown in Fig. 3). This makes outer dimension measurement flawed.

On the other hand, the XYZ cubes are misleading because of their small scale. A 20 mm edge cube adjusts the printer's "dimensional accuracy". This 20 mm cube gauge is used as a reference to calibrate all printable dimensions in the build volume. A 0.2 mm measurement flaw on the 20 mm edge block is a 1% error. The same measurement error on a 100 mm edge block is 0.2%. In other words, the smaller the size of the measured gauge, the higher the error in case of a faulty measurement. For this reason, a high-resolution caliper or micrometer should be used for a small-size gauge.

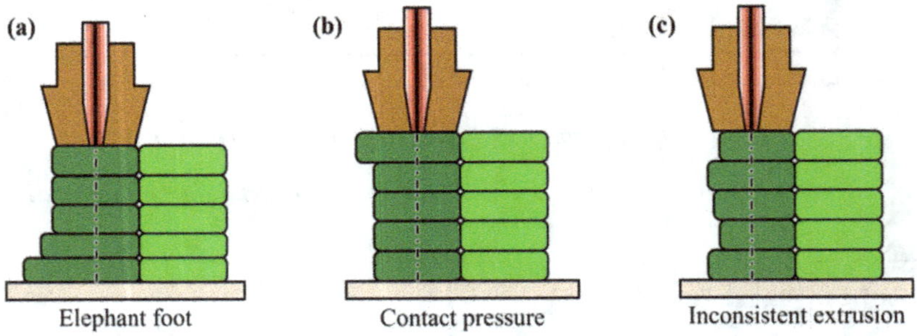

Fig. 3. Possible causes of variation for resulting dimensions: (a) Elephant foot; (b) Nozzle contact overpressure; (c) Inconsistent material extrusion volume.

When printing the XYZ cube, one or multiple defects can appear on the block's external surface, which impacts the resulting measurements of the parts. These printing defects include:

- ***Elephant foot***: the defect occurs when the gap between the nozzle and the build plate is too small [49] (i.e., smaller than the layer height). As a result, the molten material squeezes out and increases the part cross-section for the first few layers (i.e., 2-3 layers). Another cause for the elephant foot appearance is build plate over-heating, above the glass transition temperature of the printing polymer. Elephant footstep appearance can be reduced via slicing tool through the initial layer horizontal expansion setting or by creating a minimum 1x45° chamfer at the part's base;

- ***Nozzle contact overpressure:*** appears when the distance between the nozzle and the previously added layers is too small, and the material squishes out [50]. This effect is problematic when printing the outer shell because it affects the part's dimensional accuracy. The gap size between the nozzle and printed layers can influenced by material thermal expansion or over-extrusion of the previous layers or construction elements (i.e., inner perimeters, infill);

- ***Inconsistent extrusion:*** referees to the variation of the molten material flow. Inconsistent material filament diameter, and issues with the filament feeding mechanism, as the extruder melting rate is too small relative to the maximum volumetric speed are possible reasons for melt flow variation. Those issues must be solved by elimination;

- ***Ringing (Ghosting):*** this issue appears when the nozzle is unable to lay down material in a controlled manner leaving after images of a printed feature in the neighboring area [44, 52] (X and Y cube faces of Fig. 4). This effect appears only on a part's sides and caused by vibration due to over increasing the print speed (by not considering the extruder inertia), uneven belts tension or a bad frame design (i.e., not rigid enough). This issue is solvable by reducing print speed and acceleration;

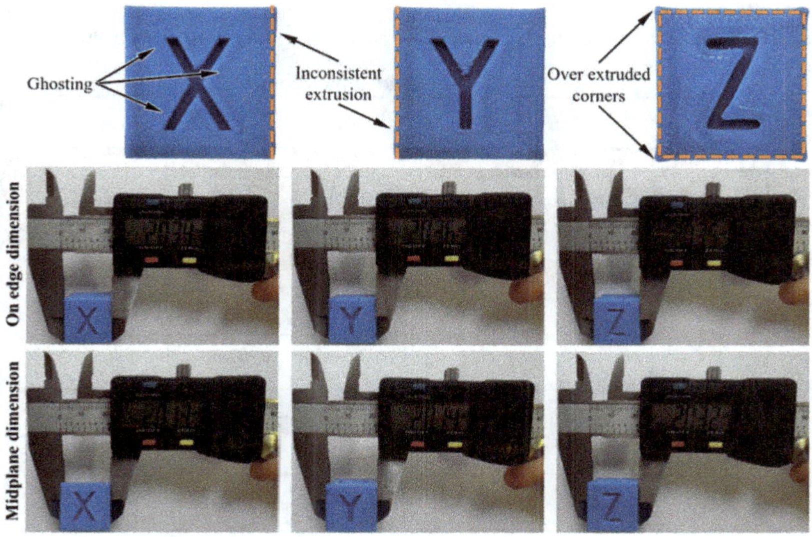

Fig. 4. Measured dimensions of an XYZ cube relative to analyzed sides and locations.

- ***Exposed seam:*** in FFF a seam refers to the start and endpoint of a perimeter [53, 54]. Depending on the material, nozzle size, and parametrization, the closing point of the outer shell can be more or less visible. In the case of a calibration cube, an over-exposed seam can increase the part size. The seam exposure is adjustable through the contour closing method, nozzle wipe distance, filament retraction, and material volume coasting;

- ***Over-extruded corners:*** Every Computer Numerically Controlled machine has a tool head whose movement is controlled through velocity and acceleration based on blank specifics and Material Removal Rate [55, 56]. Jerk Speed is specific for FFF equipment that controls the speed at which the printhead can travel through corners [54, 58]. Firstly, the Jerk Speed is set at a value greater than zero to avoid excessive material ooze when turning corners [54, 57, 58] (i.e., the over-extruded corners presented in the cube Z face of Fig. 4). Secondly, the jerk speed is always lower than the global printing speed to reduce vibrations (which lead to ghosting X and Y cube faces of Fig. 4). Over-extrusion corner formation can [54, 58] be avoided by decreasing the print speed or by creating filets (i.e., rounded corners) to replace the linear interpolation printhead movement with the circular interpolation one. For the printers introduced in the past four years, the over-extrusion corner issue was solved via input shaping, a control technique that creates a commanding signal that cancels vibrations.

Thermal contractions

During the deposition of the molten thermoplastic, the polymeric chains are rearranged along the extrusion paths. Subsequently, the melt is wetting adjacent extruded filament creating intimate contact. Simultaneously, the polymeric chains diffuse, creating interpenetrating networks (i.e., coils, entanglements, or crosslinks) promoting molecular adhesion [59, 60-62].

Thermodynamically speaking, when melted, the polymeric chains have a high degree of freedom, and a high entropy, which allows material reshaping. As the melt cools down, the polymer begins to crystallize. During this step, the polymeric chains start to lose their mobility (entropy decreases), and the chains try to occupy all available volume [62]. As a result, a volumetric decrease can occur due to thermal contractions (i.e., shrinkage), as illustrated in Fig. 5.

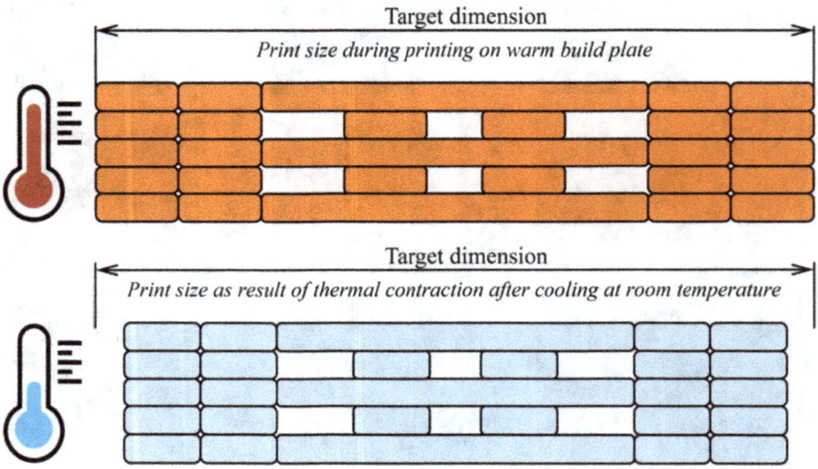

Fig. 5. Part shrinkage as a result of thermal contractions during cooling.

All polymeric materials are affected by thermal contractions. Engineering-grade polymers such as ABS, ASA, or PC require a heat chamber to reduce thermal stress and shrinkage caused by the thermal gradient resulting from the difference between printing temperature (e.g., 240-260°C for ABS) and environment (e.g., 25°C).

Skewed axes

A precisely designed and assembled cartesian 3D printer requires X, Y, and Z axes perpendicular to each other. Otherwise, the axles are skewed, and the resulting parts are deformed compared to the original design. When printing the calibration cube, instead of a part with a square cross-section (Fig. 6.a), the printer will make a part with a parallelogram-like cross-section (Fig. 6.b). In other words, the resulting parts will have diagonals of different sizes.

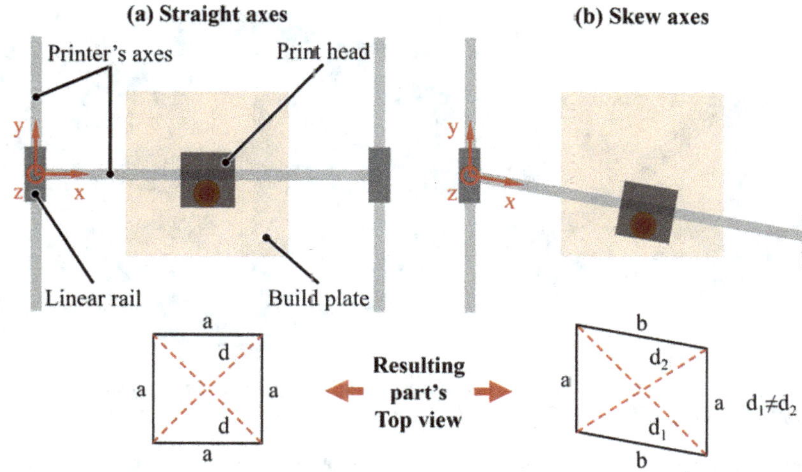

Fig. 6. Schematic of a 3D printer with (a) right-angled axles; and (b) skew axles.

On the one hand, the main limitation of the conventional calibration cube is its size. At 0.5-1° axles inline, it becomes hard to tell if the difference between the measured diagonals results from skewness or thermal contractions. On the other hand, the current cubic design with sharp edges leads to flawed measurements when using calipers. Chamfering the cube edges as shown in Design 15 from Table 1 reduces the measurement error.

When printing esthetical (i.e., non-functional) parts such as vases, figurines, or toys, the effect of skewness is hard or even impossible to notice. This statement is especially valid if the part has multiple details, textures, or multiple bodies. However, when switching to functional parts (e.g., cases, gears, brackets) mounted in a precise environment, axle skewness significantly influences the dimensional accuracy, particularly for larger parts (e.g., over 50 mm). Fig. 7 demonstrates a scenario illustrated for a 100 x 200 mm rectangle. Even at a relatively small axel incline of 0.5°, the resulting profile has a deviation of 1.745 mm. This way, a square profile will result in a parallelogram, a circular profile as an ellipse, and so on.

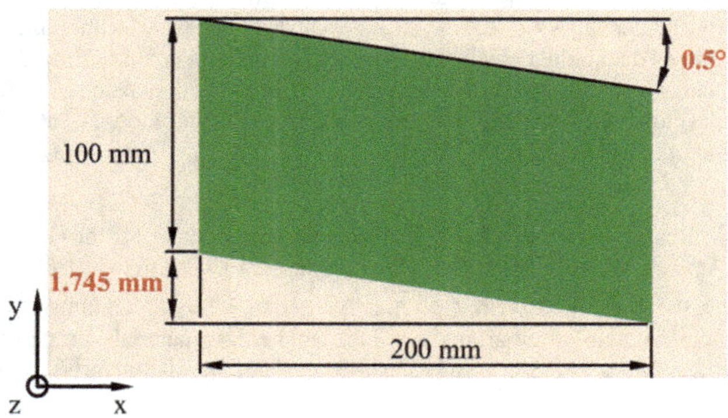

Fig. 7. Illustration of the effect of 0.5° skew axis over the resulting part cross-section profile.

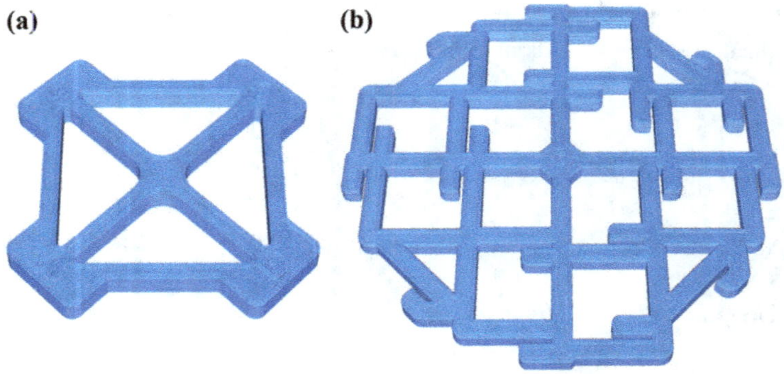

Fig. 8. Skew correction models: (a) cross-like model for diagonals measurement [63]; (b) Califlower Calibration Tool [64]. Print preview made using Ultimaker Cura 5.7.2.

Skew axle correction requires a larger model to reduce the measuring errors. Gauges with a nominal dimension of 50 mm are a good starting point in adjusting the printer's skewness. The 3D shared platforms provide reference models to correct the skew (Fig. 8.a). However, not all of them are equal regarding the provided information. An interesting approach is the Califlower (Fig.8.b) created by Adam Meadows from Vector 3D [64], which besides the skew correction, allows for dimensional error correction by measuring both external and internal features, together with scale factor calculation when printing with different materials (e.g., PLA, PETG, ABS). The results are more consistent as the Califlower has ten reference regions of 50 and 100 mm.

Conclusions

Nowadays, the FFF 3D printing market shows great equipment availability suitable for all categories of consumers, with prices ranging from a few hundred to thousands of USD. This diversity and the open-source character of the FFF printers lead to an increased, and sometimes overwhelming, number of approaches to improve the resulting 3D printed parts quality.

Dimensional accuracy is one of these quality benchmarks. Over the years, the accuracy problem was approached multiple times, but poorly understood due to the vast available information, and poor methodical documentation. Thus, the current paper proposes systematic documentation regarding misconceptions about printing accuracy calibration using the XYZ calibration cube.

The currently available XYZ cubes can be misleading when calibrating the printer's dimensional accuracy due to their small size, printing defects, and possible flawed measurements. Not least, the cubes are not suitable for tolerancing internal dimensions or evaluating axle skewness. The development of better gauge models can solve a part of those issues. The Califlower represents a good example in this direction.

References

[1] A. Dey, I.N. Roan Eagle, N. Yodo, A Review on Filament Materials for Fused Filament Fabrication. *J. Manuf. Mater.* 2021; 5(3):69. https://doi.org/10.3390/jmmp5030069

[2] A. Yadav, P. Rohru, A. Babbar, R. Kumar, N. Ranjan, J. Singh, R. Kumar, M. Gupta, Fused filament fabrication: A state-of-the-art review of the technology, materials, properties and defects. *Int. J. Interact. Des. Manuf.* **17**, 2867–2889, 2023. https://doi.org/10.1007/s12008-022-01026-5

[3] R.C. Stavarache, V. Ermolai, M.I. Ripanu, L. Andruşcă, M. Mareş, O. Dodun, Infill pattern optimization of fused filament fabrication samples for enhanced mechanical properties. *Sci. Bulletin Series C: Fascicle Mech., Tribol., Machine Manuf. Technol., 35*, 2021, 80-85.

[4] L. Romero, A. Guerrero, M.M. Espinosa, M. Jimenez, I.A. Dominguez, M. Dominguez, Additive manufacturing with RepRap methodology: current situation and future prospects. 2014 http://dx.doi.org/10.26153/tsw/15669

[5] E. Sells, S. Bailard, Z. Smith, A. Bowyer, V. Olliver, 2010, RepRap: the replicating rapid prototyper: maximizing customizability by breeding the means of production. In Handbook of Research in Mass Customization and Personalization, vol. 2 (pp. 568-580). http://dx.doi.org/10.1142/9789814280280_0028

[6] S. Bradshaw, A. Bowyer, P. Haufe, The Intellectual Property Implications of Low-Cost 3D Printing. Script Ed. 2010; 7(1): 5-31. https://script-ed.org/archive/volume-7/issue-71-1-241/

[7] M.A. Boca, A. Sover, L. Slătineanu, 2023. Short foray into the stages of conversion from 2.5D to volumetric printing. In Proc.: 5th International Conference Business Meets Technology. Valencia, 13th-15th July 2023. 13-50. https://doi.org/10.4995/BMT2023.2023.16748

[8] T.T. Wohlers, I. Campbell, O. Diegel, R. Huff, J. Kowen, Wohlers Report: 3D Printing and Additive Manufacturing Global State of the Industry. Fort Collins, Colorado, 2022.

[9] M. Mensley, (2024, September 4). The Best 3D Printers in 2024: Our Picks in 25 Categories. All3DP. Retrieved September 12, 2024, from https://all3dp.com/1/best-3d-printer-reviews-top-3d-printers-home-3-d-printer-3d/

[10] All3DP. (2024, March 15). *Professional 3D printers under $10,000 (FDM & Resin).* https://all3dp.com/1/best-professional-3d-printer-small-business/

[11] S. Frey, F. Gehrke, (2023, May 8). *The Best DIY 3D Printer Kits of 2023.* All3DP. Retrieved September 12, 2024, from https://all3dp.com/1/best-cheap-diy-3d-printer-kit/

[12] J.R. Ai, F. Peng, P. Joo, B.D. Vogt, Enhanced dimensional accuracy of material extrusion 3D-printed plastics through filament architecture. *ACS Applied Polymer Materials, 3*(5), 2021, 2518-2528. https://doi.org/10.1021/acsapm.1c00110

[13] F.R. Read, J.E. Seppala, F. Tourlomousis, J.A. Warren, N. Bakker, N. Gershenfeld, Online Measurement for Parameter Discovery in Fused Filament Fabrication. *Integr. Mater. Manuf. Innov.* 13, 2024, 541–554. https://doi.org/10.1007/s40192-024-00350-w

[14] J. Grubbs, B.C. Sousa, D.L. Cote, Establishing a Framework for Fused Filament Fabrication Process Optimization: A Case Study with PLA Filaments. *Polymers.* 15(8), 2023, 1945. https://doi.org/10.3390/polym15081945

[15] M. Alatefi, A.M. Al-Ahmari, A.Y. Al Faify, M.A. Saleh Framework for Multivariate Statistical Quality Monitoring of Additive Manufacturing: Fused Filament Fabrication Process. *Processes.* 11(4), 2023, 1216. https://doi.org/10.3390/pr11041216

[16] L. Lendvai, I. Fekete, D. Rigotti. Experimental study on the effect of filament-extrusion rate on the structural, mechanical and thermal properties of material extrusion 3D-printed polylactic acid (PLA) products. *Prog. Addit. Manuf.* 2024. https://doi.org/10.1007/s40964-024-00646-5

[17] S. Eva, A. Sover, V. Ermolai, 2023, The impact of the G-code flavour selection in FFF. En 4th International Conference Business Meets Technology 2022. Editorial Universitat Politècnica de València. 54-62. https://doi.org/10.4995/BMT2022.2022.15547

[18] L. Santana, J.L. Alves, A.D.C. Netto, A study of parametric calibration for low cost 3D printing: Seeking improvement in dimensional quality. *Materials & Design, 135*, 2017, 159-172. https://doi.org/10.1016/j.matdes.2017.09.020

[19] F. Cucinotta, G. Di Bella, M. Raffaele, F. Salmeri, A design strategy for removing the debinding and sintering gas in additive manufactured samples of a bronze/polylactic acid filament. *Adv. Eng. Mater.*, *26*(6), 2024, 2301722. https://doi.org/10.1002/adem.202301722

[20] X. Wang, S. Shujaat, E. Shaheen, R. Jacobs, Trueness of cone-beam computed tomography-derived skull models fabricated by different technology-based three-dimensional printers. *BMC Oral Health* 23, 2023, 397,. https://doi.org/10.1186/s12903-023-03104-w

[21] A.M. Mihalache, A. Hrițuc, L. Slătineanu, G. Nagîț, O. Dodun, V. Ermolai, E. Panaite, The Behavior of a 3D Printed Panel under Thermal Stress. *Macromolecular Symposia*, *404*(1), 2022. https://doi.org/10.1002/masy.202100332

[22] iDig3Dprinting. (2016, January 19). *XYZ 20mm Calibration Cube*. Thighiverse. Retrieved September 10, 2024, from https://www.thingiverse.com/thing:1278865

[23] Dezign. (2016, May 7). *Cali Cat - The Calibration Cat*. Thingiverse. Retrieved September 10, 2024, from https://www.thingiverse.com/thing:1545913

[24] Sch00f. (2017, March 10). *Filament Test Cube*. Thighiverse. Retrieved September 10, 2024, from https://www.thingiverse.com/thing:2166102

[25] Elproducts (2018, November 2018). CHEP Cube - Calibration Cube. Thighiverse. Retrieved September 10, 2024, from https://www.thingiverse.com/thing:3189377

[26] EnginEli. (2020, November 6). Lego Calibration Cube. Thighiverse. Retrieved September 10, 2024, from https://www.thingiverse.com/thing:4645349

[27] Agepbiz. (2022, December 18). Hinged Calibration Cube. Printables Retrieved September 10, 2024, from https://www.printables.com/model/74825-hinged-calibration-cube

[28] Jun Ho, L., 1102. (2020, January 28). *3in1 20mm calibration cube*. Thighiverse. Retrieved September 10, 2024, from https://www.thingiverse.com/thing:4128036

[29] FenixPrints. (2018, April 8). XYZ Calibration Cube. Thighiverse. Retrieved September 1, 2024, from https://www.thingiverse.com/thing:2882777

[30]de Baer. (2015, July 27). Test Cube for calibration. Thighiverse. Retrieved September 1, 2024, from https://www.thingiverse.com/thing:943755

[31] Grabejud. (2019, February 19). 20mm calibration cube and circle. Thighiverse. Retrieved September 10, 2024, from https://www.thingiverse.com/thing:3435944

[32] Ecjuliatti. (2024, February). Calibration Cube 20x20. Thighiverse. Retrieved September 1, 2024, from https://www.thingiverse.com/thing:6471070

[33] Ograymic. (2021, August 9). Detailed calibration Cube. Thighiverse. Retrieved September 1, 2024, from https://www.thingiverse.com/thing:4928482

[34] OneEyedOwl. (2022 December 12). Calibration Cube. Printables. Retrieved September 2, 2024, from https://www.printables.com/model/335425-calibration-cube

[35] Rubenv., 2006. (2019, June 19). XYZ Test Calibration Cube 20mm. Thighiverse. Retrieved September 2, 2024, from https://www.thingiverse.com/thing:3684033

[36] Fdumi. (2021, October 26). XYZ 100mm calibration cube / skew correction test piece. Thighiverse. Retrieved September 3, 2024, from https://www.thingiverse.com/thing:5031020

[37] J. West, G. Kuk, 2016, The complementarity of openness: How MakerBot leveraged Thingiverse in 3D printing. *Technol. Forecast. Soc. Change*, *102*, 169-181. https://doi.org/10.1016/j.techfore.2015.07.025

[38] M. Soreni-Harari, R. St. Pierre, C. McCue, K Moreno, S. Bergbreiter, Multimaterial 3D printing for microrobotic mechanisms. *Soft robotics*, *7*(1), 2020, 59-67. https://doi.org/10.1089/soro.2018.0147

[39] Thingiverse (n.d.). *Thingiverse - Digital Designs for physical Objects.* https://www.thingiverse.com/

[40] *Printables*. (n.d.). https://www.printables.com/

[41] *Discover STL files for 3D printing ideas and high-quality 3D printer models. MyMiniFactory.* (n.d.). MyMinifactory. https://www.myminifactory.com/

[42] *Cults • Download free 3D printer models • STL, OBJ, 3MF, CAD.* (n.d.). Cults 3D. https://cults3d.com/en

[43] *Free CAD designs, files & 3D models, The GrabCAD Community Library.* (n.d.). https://grabcad.com/library

[44] M.R. Caputo, M. Fernández, R. Aguirresarobe, A. Kovalcik, H. Sardon, M.V. Candal, A.J. Müller, Influence of FFF Process Conditions on the Thermal, Mechanical, and Rheological Properties of Poly(hydroxybutyrate-co-hydroxy Hexanoate). *Polymers.* 15(8), 2023, 1817. https://doi.org/10.3390/polym15081817

[45] F. Javed, S. Javed, Design Methodology and Development of an Economical 3D Printer, SAE Technical Paper 2016-01-0325, 2016, https://doi.org/10.4271/2016-01-0325.

[46] J.T. Green, I.A. Rybak, J.J. Slager, M. Lopez, Z. Chanoi, C.M. Stewart, R.V. Gonzalez, Local composition control using an active-mixing hotend in fused filament fabrication. *Additive Manufacturing Letters*, *7*, 2023 100177. https://doi.org/10.1016/j.addlet.2023.100177

[47] Nguyen, A. (2022). Hard Real-time Linux on a Raspberry Pi for 3D Printing. https://doi.org/10.31979/etd.68zc-s4dq

[48] G. Hsiang Loh, E. Pei, J. Gonzalez-Gutierrez, M. Monzón, An Overview of Material Extrusion Troubleshooting. *Appl. Sci.* 10(14), 2020, 4776. https://doi.org/10.3390/app10144776

[49] D. Vieira da Silva, L. Santana, A. Magalhães, J. Lino Alves, Parametric calibration study of fused filament fabrication printing with large nozzle diameter. *Proceedings of the Institution of Mechanical Engineers, Part E: J. of Process Mechanical Engineering*, 238(4), 2024, 1525-1536. https://doi.org/10.1177/09544089221119691

[50] L. Schmidt, K. Schricker, J.P. Bergmann, F. Hussenöder, M. Eiber, Characterization of a granulate-based strand deposition process in the FLM-method for definition of material-dependent process strategies, Rapid Prototyp. J., Vol. 25 No. 1, 2019, 104-116. https://doi.org/10.1108/RPJ-09-2017-0186

[51] C. Cardona, A.H. Curdes, A.J. Isaacs, Effects of filament diameter tolerances in fused filament fabrication. IUJUR, 2(1), 2016, 44-47. https://doi.org/10.14434/iujur.v2i1.20917

[52] C. Chadha, G. Olaivar G, M.A. Mahrous, A.E. Patterson, I. Jasiuk. Exploring the Effect of Specimen Size on Elastic Properties of Fused-Filament-Fabrication-Printed Polycarbonate and Thermoplastic Polyurethane. Materials. 17(11), 2024, 2677. https://doi.org/10.3390/ma17112677

[53] A. Heuer, J. Huether, W.V. Liebig, P. Elsner, Fused filament fabrication: Comparison of methods for determining the interfacial strength of single welded tracks. Manufacturing Review, 8, 2021, 32. https://doi.org/10.1051/mfreview/2021031

[54] Ghostkeeper. (n.d.). *GitHub - Ghostkeeper/SettingsGuide: More extensive explanations of Cura slicing settings.* GitHub. https://github.com/Ghostkeeper/SettingsGuide

[55] H.A. Hemminger, 2022. *Improvements and Characterization of Error, Speed, and Capabilities in Low-Cost Additive Manufacturing Systems.* UCLA. ProQuest ID: Hemminger_ucla_0031N_21067. Merritt ID: ark:/13030/m55v0vfq. Retrieved from https://escholarship.org/uc/item/3pw7z6d3

[56] J. Kundrák, V. Molnár, T. Makkai, T. Dági, Analysis of Material Removal Efficiency in Face Milling of Aluminum Alloy. In: Gapiński, B., Szostak, M., Ivanov, V. (eds) Advances in Manufacturing II. MANUFACTURING 2019. Lecture Notes in Mechanical Engineering. Springer, Cham, 2019. https://doi.org/10.1007/978-3-030-16943-5_34

[57] M. Moretti, A. Rossi, N.J.A.M. Senin, N. J. A. M., In-process monitoring of part geometry in fused filament fabrication using computer vision and digital twins. *Addit. Manuf. 37*, 2021, 101609. https://doi.org/10.1016/j.addma.2020.101609

[58] V. Ermolai, A. Sover, 2023, Multi-material 3D Printed Interfaces. Influencing Factors and Design Considerations. In: D.D. Cioboată, (eds) International Conference on Reliable Systems Engineering (ICoRSE) - 2023. ICoRSE 2023. Lecture Notes in Networks and Systems, vol 762. Springer, Cham. https://doi.org/10.1007/978-3-031-40628-7_11

[59] R. Freund, H. Watschke, J. Heubach, T. Vietor, Determination of Influencing Factors on Interface Strength of Additively Manufactured Multi-Material Parts by Material Extrusion. *Appl. Sci.* 9(9), 2019, 1782. https://doi.org/10.3390/app9091782

[60] A. Benarbia. V. Sobotka, N. Boyard, C. Roua, Fused filament fabrication: Numerical adhesion modeling suitable for semicrystalline polymers. Mater. Res. Proc. 28, 2023, 139-148. https://doi.org/10.21741/9781644902479-16

[61] A. Lepoivre, A. Levy, N. Boyard, V. Gaudefroy, V. Sobotka, Coalescence in fused filament fabrication process: Thermo-dependent characterization of high-performance polymer properties. *Polym. Test. 98*, 2021, 107096. https://doi.org/10.1016/j.polymertesting.2021.107096

[62] G.W. Ehrenstein, Structure of Polymeric Materials, In: Polymeric materials, Hanser, Ohio, 2001, pp. 61-141. https://doi.org/10.3139/9783446434134.004

[63] AfroPsyho. (2023, May 20). Skew Correction Made Easy! (For X & Y Axis). Printables. Retrieved September 10, 2024, from https://www.printables.com/model/485014-skew-correction-made-easy-for-x-y-axis

[64] Vector 3D. (n.d.). *Califlower Calibration Tool.* September 10, 2024, from https://vector3d.shop/products/califlower-calibration

Materials Research Proceedings 46 (2024) 35-40 https://doi.org/10.21741/9781644903377-5

Manufacturing computationally designed wearables via 3D printing

Panagiota LIGKA[1,a], Nikolaos EFKOLIDIS[1,b], Athanasios MANAVIS[1,c] and Panagiotis KYRATSIS[1,d *]

[1]Department of Product and Systems Design Engineering, University of Western Macedonia, Kila, GR50100 Kozani, Greece

[a] pligka@gmail.com, [b] nefkclidis@uowm.gr, [c] manavis.athanasios@gmail.com, [d] pkyratsis@uowm.gr

Keywords: CAD-Based Programming, Computational Design, 3D Printing, Accessory Design, Biomimetic

Abstract. Computational product design is a new trend that combines product design with CAD-based programming, thus the designer creates an application that can model a great deal of product alternatives. Every algorithm built is able to provide sophisticated geometries in 2D/3D and by changing a series of parameters to customize the design, based on the user demands and needs. The unusual geometries proposed can only be materialized by additive manufacturing technologies. The present paper follows a multidisciplinary approach by combining biomimetically inspired geometries with computational design and additive manufacturing. The transition from the digital modeling to the physical prototyping is facilitated by 3D printing, which allows for precise and customizable prototyping and production. The research asserts that the fusion of computational design and 3D printing enables wearable accessories to not only meet, but surpass conventional expectations. This innovative approach boosts human creativity but also integrates fashion and functionality, shaping a dynamic and interactive future landscape in the realm of designing products.

Introduction

The advent of 3D printing has transformed various industries, particularly in the realm of wearable technology. This research focuses on the utilization of computational design techniques and biomimetic structures, specifically employing Voronoi patterns, to fabricate wearables that are not only visually appealing but also lightweight and durable. The intersection of design, engineering, and technology fosters a new paradigm in product development, enabling the creation of customized solutions tailored to individual needs [1, 2].

When computational design is used, the geometries of the products can be far more complex and at the same time a number of alternatives can become available to the user. The traditional role that the designer plays, evolves by transferring to the end user the opportunity to interact with the designs. At the same time, the additive manufacturing technologies deal with not only prototyping but in a number of cases in the final production [3].

Methodology: Enchanted functionality

The fusion of technology and jewelry introduces the concept of "enchanted" wearables, which push beyond conventional practicality by incorporating interactive functions with captivating aesthetics, fostering emotional and imaginative engagement [4]. The present research explores the use of advanced materials, digital interfaces, and human-centered design to reshape how we perceive and interact with technology through wearables (Fig. 1). Drawing on McCarthy et al.'s framework of enchantment in design, which emphasizes aesthetics, identity, play, paradox, and transparency, this study highlights the importance of integrating ethics into aesthetic interactions

Materials Research Proceedings 46 (2024) 35-40 https://doi.org/10.21741/9781644903377-5

to create richer, more meaningful experiences [5]. Enchanted wearables blend art and engineering, delivering not only functional but truly magical experiences that enhance everyday life [6, 7, 8].

Figure 1: Initial sketches aiming in the form exploration

Methodology: Biomimetic Principles

Biomimicry serves as fundamental inspiration for the design of this wearable. Drawing from the complex yet efficient structures found in nature, the study utilizes natural models to achieve functional and aesthetic benefits. The products created could be used not only for modelling purposes but for computer aided analyses as well [9]. For example, the structure of lace lichen (Ramalina menziesii), known for its lightweight yet durable composition, was used as the basis for getting inspiration with an aim to design lightweight yet strong Voronoi lattices (Fig. 2). This approach mirrors nature's ability to optimize material use and distribute forces efficiently, ensuring both flexibility and strength. By mimicking these biological structures, the design achieves a balance between durability, adaptability, and aesthetic fluidity, making it particularly suitable for wearables that must withstand physical movement, while maintaining an organic, visually appealing form. Additionally, the porous nature of the Voronoi structure allows for airflow and comfort, making it ideal for wearable applications [10, 11].

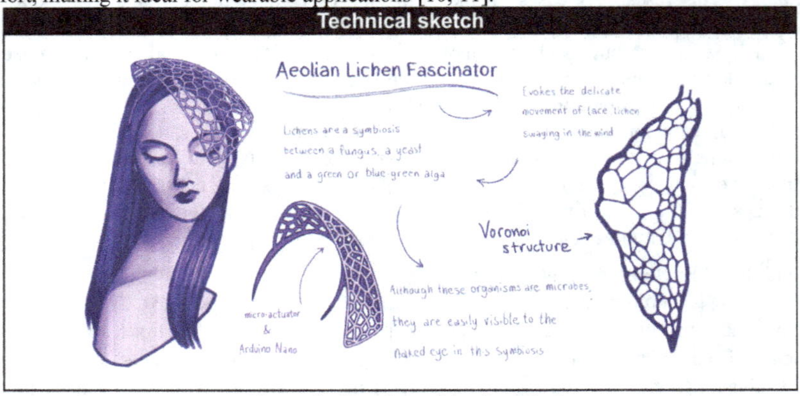

Figure 2: Sketches of the final design

Methodology: CAD Programming and Computational Design

Advanced CAD programming allows for the automation and customization of wearable designs. Designers can create parameterized forms by entering different human body dimensions, which makes it easier to produce precisely fitting wearables. This method enhances the potential for customization, as each design can be tailored to the specific measurements and preferences of the wearer [12]. Figure 3 depicts the Grasshopper™ code developed with an aim the designer to be able to produce a series of alternatives by varying the parameters set. As a result, a complete family of products can be modeled and be sent for production at a letter stage.

In this case, the code is designed to enable modifications to several key design parameters that directly influence the geometry and structure of the wearable. For example, anthropometric data, which is essential for ensuring a precise and comfortable fit of the headpiece. Additionally, control points are used to define the overall geometry, while the number of Voronoi cells can be adjusted to manage the complexity and pattern density of the design. The random seed parameter introduces variability, allowing for the generation of unique, one-of-a-kind designs, and the Voronoi mesh thickness is customizable, balancing structural integrity and visual aesthetics.

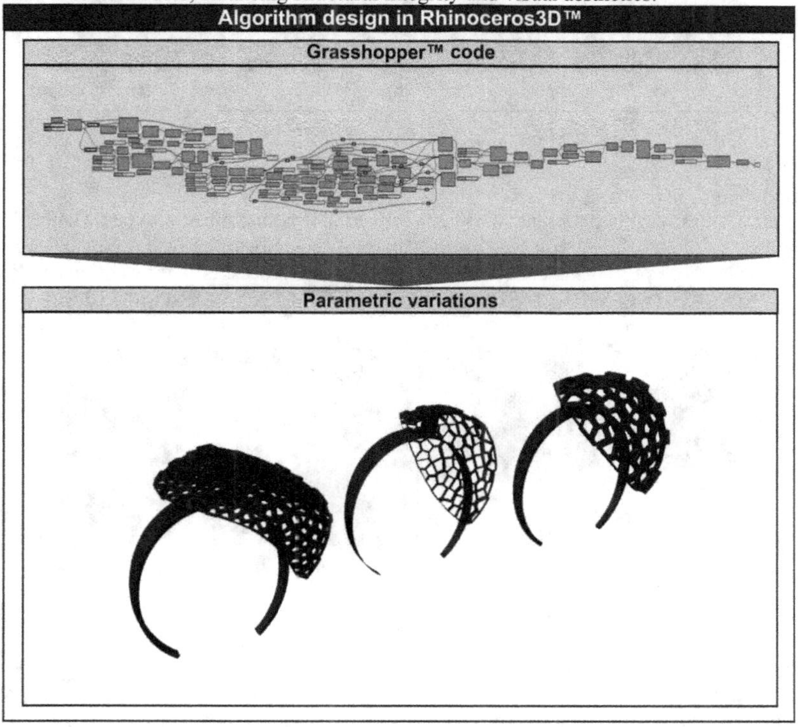

Figure 3: Algorithm development in Rhinoceros3D™

A photorealistic depiction of the wearable is deemed necessary in order to visualize the finalized design (Fig. 4). This can be achieved through computer-aided design techniques that simulate how the wearable will look and function in real life. By incorporating photorealistic renderings, designers can accurately assess the aesthetics and functionality of the product before production begins, ensuring a high-quality end result. Following this approach, computational design can support the product's brand identity and offers distinct characteristics that the end user is looking to purchase [13].

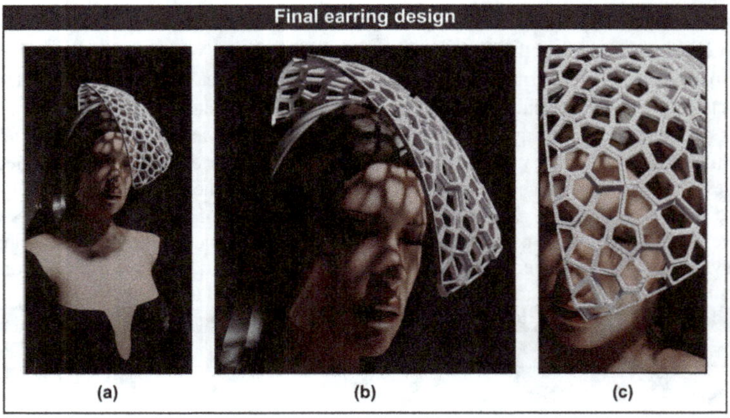

Figure 4: Photorealistic 3D depictions of the fascinator

Methodology: 3D Printing Process

3D printing allows digital representations to be converted into tangible prototypes (Figure 5). This technology facilitates the precise creation of complex geometries that would be challenging or impossible to accomplish using conventional manufacturing processes. The use of materials such

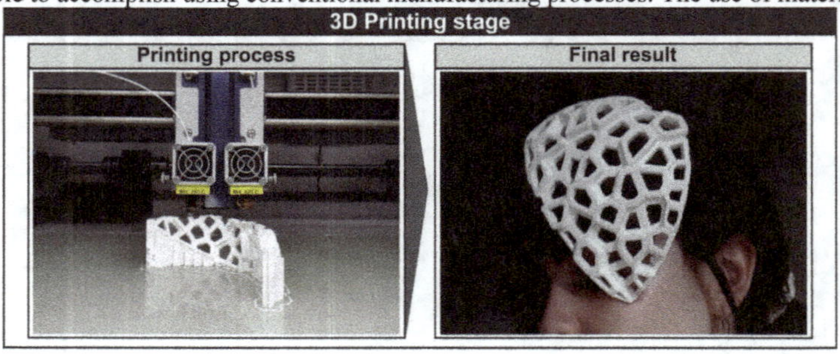

Figure 5: The manufacturing of the wearable via 3D printer

as polylactic acid (PLA) and its composites ensure that the produced outcome will be both functioning and sustainable. The 3D printing process allows for extremely fast product development, thus encouraging designers to iterate their product based on the customers' opinion and participation in the process[14].

The 3D printer used is FFF (Fused Filament Fabrication) class and is called CreatBot^{TM} D600 Pro. The dimensions of the printer are 600mmx600mmx600mm. The geometry was printed with the material NEEMA3D^{TM} PLA: EVO. Alternatively, 15-50% recycled material can be added to a PLA based mixture to promote increased environmentally friendly practices. Table 1 depicts the 3D printing parameters set.

Table 1: 3D printing parameters

Nozzle Temperature: 225 [°C]	Nozzle Diameter: 0.6 [mm]
Bed Temperature: 50 [°C]	Printing Speed: 40 [mm/sec]
Infill: 100%	Perimeter Speed: 50%
Filament Diameter: 1.75 [mm]	Cooling Fan: 100%

Results and Discussion

The integration of computational design and 3D printing has demonstrated significant potential in the development of wearables that exceed conventional expectations. The study showcases various prototypes that highlight the effectiveness of biomimetic designs in achieving optimal performance. The lightweight nature of the Voronoi structures contributes to enhanced comfort for the wearer, while the durability of the materials used ensures longevity and reliability. Furthermore, this approach emphasizes the importance of interdisciplinary collaboration. By combining expertise from design, engineering, and technology, the development process becomes more holistic, resulting in innovative solutions that push the boundaries of what is possible in wearable technology. Although using 3D technology offers advantages, i.e. easy fabrication of unusual geometries, relatively low cost of equipment purchase, use of different materials; there are several challenges to be addressed, especially when it comes to large scale production i.e. limited production rates compare to traditional molding technology, increased material availability for 3D printing, research needed for optimized manufacturing parameters when 3D printing is used. Thus, for Small and Medium Enterprises producing such customized products offers advantages, while they are introduced to niche markets and user demands.

It was demonstrated that such complex and unconventional geometries could be realized exclusively through the use of 3D printing technology, not only for prototyping but also for the actual production of fully customized products. Traditional manufacturing methods often fall short when attempting to produce intricate forms like these, making 3D printing an indispensable tool for turning ambitious design concepts into reality. In these scenarios, cost and production time are no longer the primary considerations. Instead, the focus shifts towards the customization and uniqueness of the final product, which adds significant value for the end user. Customers are often willing to invest in the higher costs and the resulted lead times associated with 3D printing in exchange for a one-of-a-kind, tailor-made item that meets their exact specifications. This shift reflects a growing market trend, where personalization and exclusivity take precedence, particularly in sectors such as fashion, accessories, and wearable technologies.

Conclusion

This research underscores the transformative power of 3D printing in the realm of wearable technology. The fusion of computational design and biomimetic principles not only enhances human creativity but also integrates fashion with functionality. As the industry continues to evolve, the potential for creating dynamic and interactive wearable products will shape the future landscape of design, offering endless possibilities for customization and innovation. The findings of this study advocate for continued exploration in this interdisciplinary field, as the collaboration between design and technology holds the key to unlocking new dimensions in wearable product development.

References

[1] P. Rosso, J.A. Gopsill, S.C. Burgess, B. Hicks, Does CAD smell like code? A mapping between violation of object-oriented programming design principles and computer aided design modelling, Proceedings of the Design Society 2 (2022) 1737-1746. https://doi.org/10.1017/pds.2022.176

[2] S.E. Yang, H. Han, J.S. Son, Recent progress in 3D printing of Bi_2Te_3-based thermoelectric materials and devices, Journal of Physics: Energy 6 (2024). https://doi.org/10.1088/2515-7655/ad3983

[3] L. Firtikiadis, A. Manavis, P. Kyratsis, N. Efkolidis, Product design trends within the footwear industry: A review, Designs 8(3) (2024) 49. https://doi.org/10.3390/designs8030049

[4] D. Rose, Enchanted Objects: Innovation, Design, and the Future of Technology. Scribner Publications, New York, 2014.

[5] P.R. Ross, C.J. Overbeeke, S.A.G. Wensveen, C.M. Hummels, A designerly critique on enchantment, Pers Ubiquit Comput 12(5) (2008) 359-371. https://doi.org/10.1007/s00779-007-0162-3

[6] K. Vones, Microjewels: Digital enchantment and new materiality, CITAR Journal 7(1) (2015) Special Issue: xCoAx 2015. https://doi.org/10.7559/citarj.v7i1.148

[7] J. McCarthy, P. Wright, The enchantments of technology, in M. Blythe, A. Monk (Eds), Funology 2: From usability to enjoyment, 2nd edition, Springer, Human–Computer Interaction Series, 2018. https://doi.org/10.1007/978-3-319-68213-6_23

[8] J. McCarthy, P. Wright, J. Wallace, A. Dearden, 2006. The experience of enchantment in human–computer interaction. Pers Ubiquit Comput. 10(5) (2006) 369-378. https://doi.org/10.1007/s00779-005-0055-2

[9] G. Alexandridis, D. Tzetzis, P. Kyratsis, Biomimicry in product design through materials selection and computer aided engineering. In: 20th Innovative Manufacturing Engineering and Energy Conference (IManEE 2016), Greece, 2016. https://doi.org/10.1088/1757-899X/161/1/012046

[10] Information on https://www.micropia.nl/en/discover/microbiology/lichens/.

[11] J. Ma, S. Zheng, Y. Fu, X. Wang, J. Qin, Z.S. Wu, The status and challenging perspectives of 3D-printed micro-batteries, Chemical Science 15(15) (2024) 5451-5481. https://doi.org/10.1039/d3sc06999k

[12] P. Minaoglou, N. Efkolidis, A. Manavis, P. Kyratsis, A review on wearable product design and applications, Machines 12(1) (2024) 62. https://doi.org/10.3390/machines12010062.

[13] A. Manavis, A. Tzotzis, A. Tsagaris, P. Kyratsis, A novel computational-based visual brand identity (CbVBI) product design methodology. Machines, 10(11) (2022) 1065. https://doi.org/10.3390/machines10111065

[14] T. Spahiu, H. Almeida, A. Manavis, P. Kyratsis, 2020. Fashion products through digital manufacturing - a case study with FDM technology, Academic Journal of Manufacturing Engineering 18(4) (2020) 63-72.

Addressing surface quality via seam alignment parametrization

Alexandru-Ionuț IRIMIA[1, a, *], Vasile ERMOLAI[1, b], Gheorghe NAGÎȚ[1, c], Marius-Andrei MIHALACHE[1, d], Marius-Ionuț RÎPANU[1, e], Răzvan-Cosmin STAVARACHE[1, f]

[1]"Gheorghe Asachi" Technical University of Iași, Department of Machine Manufacturing Technology, Blvd. Dimitrie Mangeron 59A, Iași, 700050, Romania

[a]alexandru-ionut.irimia@student.tuiasi.ro, [b]vasile.ermolai@academic.tuiasi.ro, [c]gheorghe.nagit@academic.tuiasi.ro, [d]marius-andrei.mihalache@academic.tuiasi.ro, [e]marius-ionut.ripanu@academic.tuiasi.ro, [f]razvan-cosmin.stavarache@student.tuiasi.ro

Keywords: Fused Filament Fabrication, Surface Quality, Surface Finish, Z-Seam, Seam Alignment

Abstract: Fused Filament Fabrication (FFF) is an additive manufacturing technology that uses molten thermoplastic material to build parts layer-wise. Any of the part's constituent layers is obtained through multiple passes of the extrusion heat. An external contour or perimeter and an internal geometry characterize each layer. This external contour is defined by a start-point, an extrusion path, and an end-point, which coincides with the starting point. The overlap between the extrusion path's start and the end-point is known as z-seam alignment or seam. By default, the slicing algorithms place the seams at sharp corners or hidden edges to hide possible imperfections such as blobs and zits. Unfortunately, seam hiding is more difficult for curved surfaces without linear transitions. As a result, the curved surface printed parts show visible seams of the external contour, affecting the part's aesthetics. In this regard, this study aims to reduce some of those limitations by systematically fine-tuning the seam-related parameters.

Introduction

Fused Filament Fabrication (FFF) is an additive manufacturing process that builds parts through the selective deposition of semi-molten materials in filaments with rectangular-like cross-sections [1].

Due to low-cost equipment, relatively good quality and mechanical properties, and great availability of raw materials at a fraction of cost compared to other additive technologies (e.g., SLS) FFF has become one of the most known additive manufacturing (AM) techniques [2]. Other benefits of FFF, and other AM technologies, are the complexity for free (i.e., the manufacturing cost is not affected by the part's complexity) and the reduced material waste compared to conventional manufacturing (e.g., turning, milling) [3].

As for any manufacturing process, the part's resulting quality is influenced by multiple variables such as equipment precession, process variables, material properties, or environment specifics [4-6].

In the case of FFF, the seam-related parameters are a specific category of parameters, met only for extrusion-based AM technologies. FFF builds parts from multiple elements: shells, internal structures, and solid layers. The seam is located on the part's shells, as any shell profile is characterized by a start and end-point [7, 8]. Harum et al. observed that seams become visible as the excess material is squeezed out at the closing point resulting in a poor surface finish [9]. The literature review shows that seam effects were discussed due to their effect on the surface quality. However, none of these studies documented and studied the effect of seam parameters on surface quality. This study aims to document and study the effect of seam parameters on the resulting surface quality. The chosen variables are directly influencing resulting seams visibility on parts' outer surfaces.

Seam parameters

As stated above, a seam represents the location where the start and the endpoint of a shell or wall intersect. Frequently the slicing algorithm chooses the seam location based on the layer's profile (e.g., ellipse, hexagon) and the slicing algorithm inputs (which will be further discussed below). As a rule of thumb, the seam is placed at the sharpest corner for polygonal profiles, while for curved profiles it will be placed at a random location if not specified [7-9].

The illustration presented in Fig. 1 was designed to better understand how seams are placed based on the profile. Both profiles, round (Fig. 1a) and squared (Fig. 1b) show a front seam positioning. As seen, for the round profile the seam is placed at the specific location. However, for the squared profile, the seam is located in the same plane but on the corner. In the Ultimaker Cura slicing software the seam position is controlled through the *Z Seam Alignment* parameter. The variable has multiple seam positioning options, as follows: *User Specified*, *Shortest*, *Random,* and *Sharpest Corner* [7, 8]. The illustration below simulates a *User Specified*, in this case, *Front*, *Z Seam Alignment.*

The *Seam Corner Preference* extends the above-described variable and controls how the seam will be positioned relative to the model. The variable controls how the hidden or exposed seam will be placed based on the following parametrization levels: *None, Hide Seam, Expose Seam, Hide or Expose Seam,* and *Smart Hiding* [7, 8].

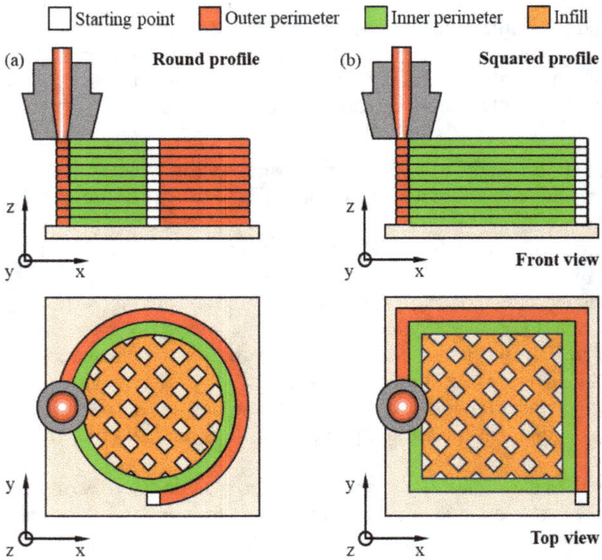

Fig. 1. Seam formation (a) circular profile vs (b) rectangular profile.

Besides the seam relative position control, the Cura slicer provides alternative ways to reduce the seam visibility on the outer surface through the *Outer Wall Wipe Distance* and *Coasting Volume* [7, 8].

Materials Research Proceedings 46 (2024) 41-48 https://doi.org/10.21741/9781644903377-6

In Cura, the *Outer Wall Wipe Distance* variable can be found in the Walls tab together with the previous two. This parameter makes the extrusion head to crossover the contour endpoint without extruding. The crossover distance is set by the user (see the dark red colored zone from Fig. 2b) and allows the printer to wipe out the nozzle against the outer wall. This way, any extra amount of molten material or over-extrusion is evenly spread across the wipe distance before going on to the next layer. This parameter becomes very useful to prevent molten material from oozing before the material filament is retracted out of the heated zone. Setting the *Outer Wall Wipe Distance* too high can lead to under-extrusion, as the molten material reservoir empties too much. Basically, in the time frame needed to switch layers and begin a new one, there is not enough time for the feeding mechanism to replenish the melt reservoir [7, 8]. This scenario can be limited by firstly printing the internal structure and by choosing an inside-to-outside wall ordering [10].

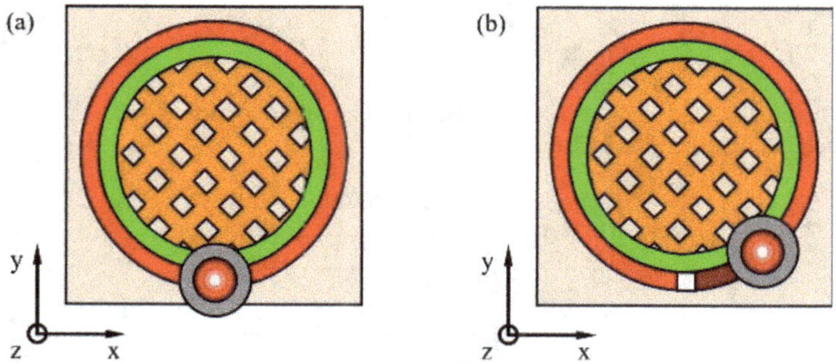

Fig. 2. Outer wall wipe distance (a) with wipe, (b) without wipe.

The *Coasting Volume* parameter can be found in the Experimental tab of the same slicing tool. This variable makes the feeding mechanism stop extruding the filament before the nozzle meets the outer contour endpoint by setting the coasting volume [7, 8, 11]. In other words, for a certain length, the nozzle will follow the deposition path without extruding as illustrated in Fig. 3.b. In a way, the use of *Coasting Volume* can be considered a pre-retraction step. Similarly to the *Outer Wall Wipe Distance*, it compensates for blobs and exposed seams formation. However, over-increasing the *Coasting Volume* can lead to under-extrusion in the seams area [7, 8]. This effect could be further amplified if combined with a null wipe distance.

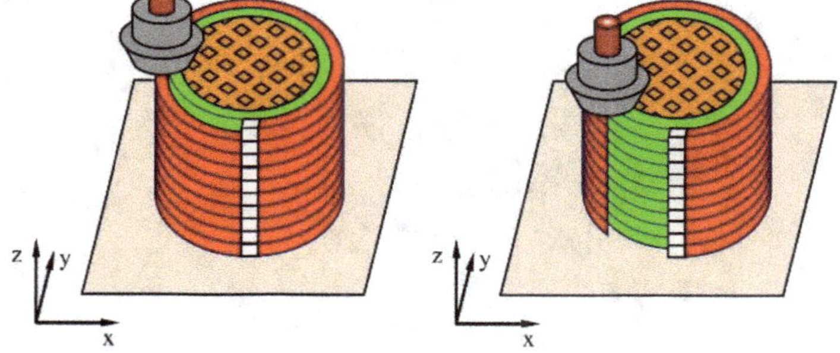

Fig. 3. Coasting Volume - (a) coasting activated, (b) coasting deactivated.

Design of experiment

The experimental research was made based on L9 (3^4) Taguchi experimental design [12-14] with four factors on three levels of variation. The chosen factors are the above-described variables, namely: *Z Seam Alignment (ZSA), Seam Corner Preference (SCP), Outer Wall Wipe Distance (WD),* and *Coasting Volume (CV).*

Based on the information shown in the previous section, it can be stated that the ZSA and SCP variables' levels of variation exceed the Taguchi L9 design matrix. However, this orthogonal matrix with the chosen levels of variation gives the highest level of variability. When simulating other Taguchi designs (i.e., L16 and L18) it was observed that some variable levels make other factor levels unavailable. This way, the L9 matrix had the highest factor variation availability.

For the chosen factor the variation levels were used as follows:

- ZSA: User Specified (US), Shortest (S) and Sharpest Corner (SC);
- SCP: None (N), Hide Seam (HS) and Expose Seam (ES);
- WD: 0, 0.2 and 0.4 mm;
- CV: 0, 0.03 and 0.06 mm^3.

According to the L9 design matrix, nine determinations were performed in total, which respect the parametrization described in Table 1.

Table 1. Experimental design Taguchi.

Run no.	ZSA	SCP	WD	CV
R1	US	N	0	0
R2	US	HS	0.2	0.03
R3	US	ES	0.4	0.06
R4	S	N	0.2	0.06
R5	S	HS	0.4	0
R6	S	ES	0	0.03
R7	SC	N	0.4	0.03
R8	SC	HS	0	0.06
R9	SC	ES	0.2	0

The influence of the above-described variables was evaluated based on the resulting seam area for custom conical specimens with a 40° incline refeed to the XY base plane. All specimens were printed on an Ultimaker 2+ only with two shells using a Fillamentum PLA Ivory material.

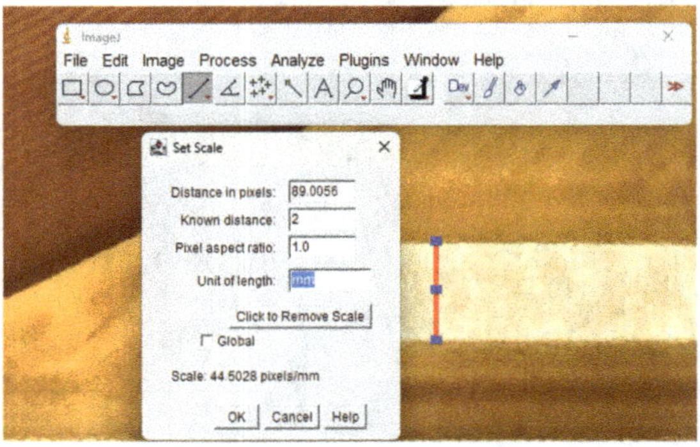

Fig. 4. Set scale to a known dimension.

The seam area was measured through visual inspection. This way, each sample was photographed and the resulting image was further analyzed using the ImageJ [15, 16]. To ensure precise measurements, ImageJ needs a reference scale. This way each of the nine samples was designed with a predefined L-shaped scale having a 2 mm gap as a reference. First, the gap was measured using a Mitutoyo caliper. Then the ImageJ measurement tool was used to set the right scale, as shown in Fig. 4.

Results

For each resulting sample, the seam area was measured using ImageJ. This was achieved by using a lasso function available in the software's analyze menu. A preview of the described methodology is presented below in Fig. 5.

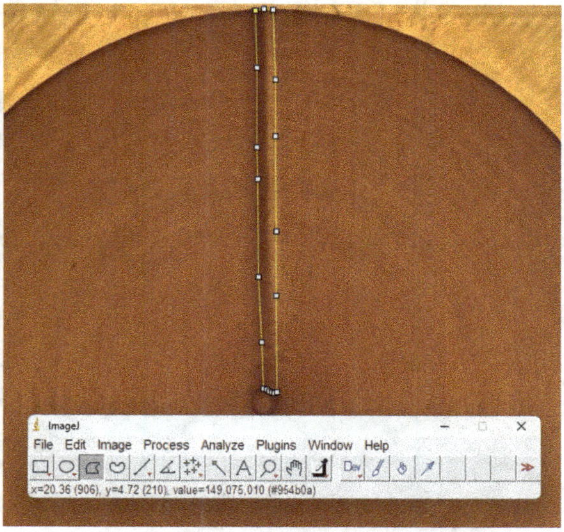

Fig. 5. Sample area measurement using ImageJ.

Since the seam area is selected by the ImageJ software's user, for the same measured seam, either by the same user or by a different one, the area value will never be identical to another determination. This happens because there is a small probability of picking the same measuring point consecutively. This way, each feature needs to be measured multiple times. The area values shown in Table 2 are the averages of five determinations. The provided results have a relatively small deviation of 0.457-1.507 mm^2.

Table 2. Seam area value resulted.

Run no.	R1	R2	R3	R4	R5	R6	R7	R8	R9
Resulting area (mm^2)	42.35	50.06	49.67	46.32	44.34	46.02	48.91	52.86	40.73
	46.43	51.09	49.15	46.70	43.87	44.02	49.20	50.77	41.64
	43.68	51.65	49.40	47.75	43.59	44.61	48.42	51.35	40.83
	44.67	51.17	49.91	46.16	42.21	45.20	47.92	51.48	40.36
	44.83	52.91	50.33	46.09	43.03	44.65	49.67	51.80	40.56

Materials Research Proceedings 46 (2024) 41-48 https://doi.org/10.21741/9781644903377-6

Average	44.39	51.38	49.69	46.60	43.41	44.90	48.82	51.65	40.82
St. dev.	1.507	1.036	0.457	0.683	0.820	0.752	0.678	0.773	0.490

The factors' variations show that, when the ZSA is used with the level *Shortes* (i.e., print head shortest travel movement relative to the absolute coordinate system), the seam orientation angle is offset by 135° in a clockwise direction as shown in Fig. 6. At first view, the *Seam Corner Preference* variable could be responsible for the R4, R5, and R6 seam offset. However, there is not enough evidence to indicate that for the Sharpesr Corner ZSA, the resulting seam will always have the location presented in Fig. 6 (i.e., R4, R5, and R6).

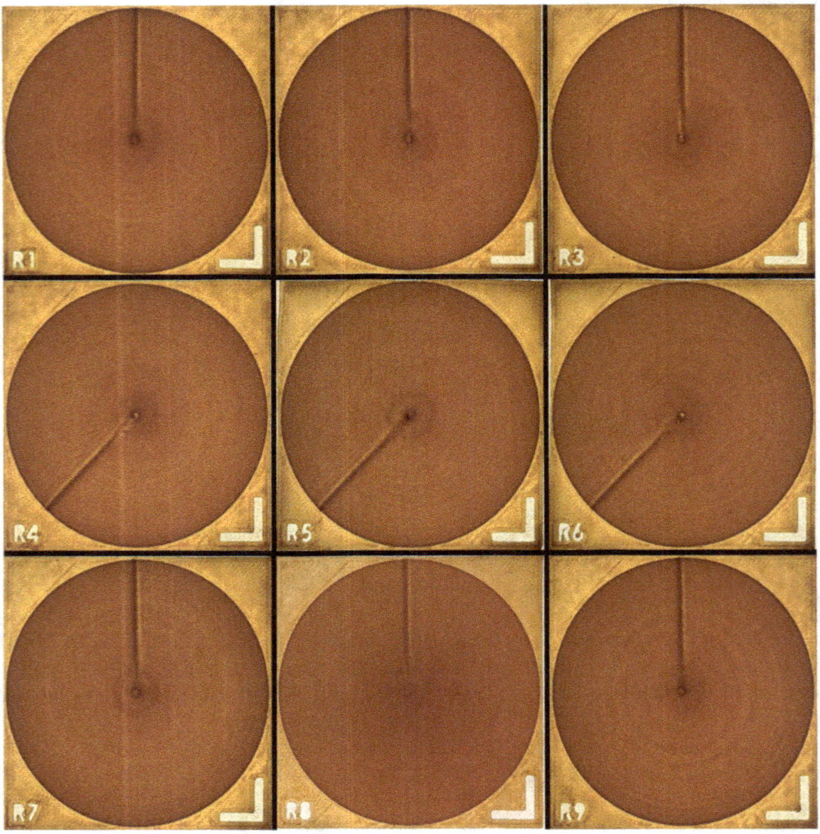

Fig. 6. Image of the Taguchi L9 resulting samples.

The resulting seam area measurements presented in Table 2 were plotted using bar charts as shown in Fig. 7. From this, results that the R9 variables' configuration provides the smallest seam area, having 40.82±0.49 mm^2. However, when judging based on aesthetics the R8 variable configuration shows better results, even if the resulting seam area is the highest.

The data provided in Table 2 and Fig.7 shows that the measured seam area varied between 43.4±0.457 mm^2 (i.e., R3) and 51.65±0.773 mm^2 (i.e., R8) with the highest standard deviation of 1.507 mm^2 (i.e., R1).

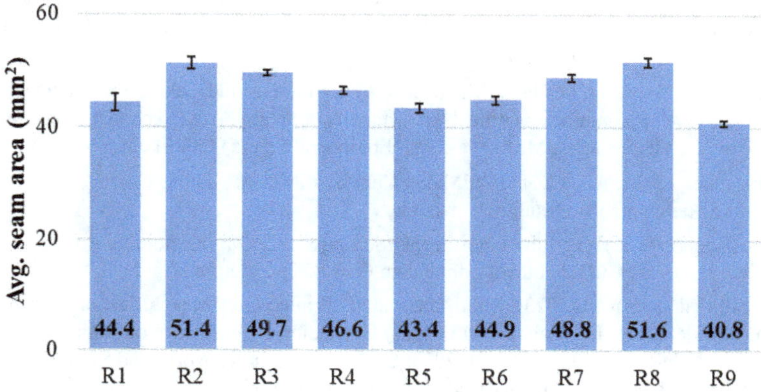

Fig. 7. Bar plot of the resulting seam area.

Conclusions

In Fused Filament Fabrication, several factors influence the seam characteristics of a part's outer surface. These factors include Z *Seam Alignment, Seam Corner Preference, Outer Wall Wipe Distance*, and *Coasting Volume*. While smaller seam areas are generally preferred for reducing exposed material on the part's surface, they don't always offer the best aesthetics. In this case, the R8 parameter configuration delivered the best outer surface quality, even though it resulted in the largest seam area.

The seam parametrization data for FFF 3D-printed parts highlight the trade-off between seam area exposure and aesthetic. Achieving a balance between the part's structural integrity and visual quality is desirable.

Future work can focus on further optimizing these variable configurations and exploring new ones, such as extrusion temperature and alternative filament materials. This approach may provide more insights into seam parametrization, helping improve seam outlook without compromising the part's strength.

References

[1] J. Kechagias, D. Chaidas, N. Vidakis, K. Salonitis, N.M. Vaxevanidis, Key parameters controlling surface quality and dimensional accuracy: a critical review of FFF process. Mater. Manuf. Proces., 37(9), 963-984 (2022). https://doi.org/10.1080/10426914.2022.2032144

[2] Krishnanand, M. Taufik, Fused filament fabrication (FFF) based 3D printer and its design: a review. Advanced Manufacturing Systems and Innovative Product Design: Select Proceedings of IPDIMS 2020, 497-505 (2021). https://doi.org/10.1007/978-981-15-9853-1_41

[3] R. B. Kristiawan, F. Imaduddin, D. Ariawan, Ubaidillah, Z. Arifin, A review on the fused deposition modeling (FDM) 3D printing: Filament processing, materials, and printing parameters. Open Engineering, 11(1), 639-649 (2021). https://doi.org/10.1515/eng-2021-0063

[4] R. Patel, C. Desai, S. Kushwah, M.H. Mangrola, A review article on FDM process parameters in 3D printing for composite materials. Materials Today: Proceedings, 60, 2162-2166 (2022). https://doi.org/10.1016/j.matpr.2022.02.385

[5] S. Restrepo, J. Jaramillo, H.A. Colorado, Fused Filament Fabrication (FFF) Additive manufacturing of Bronze-Based materials. In The Minerals, Metals & Materials Series, 105–112 (2024). https://doi.org/10.1007/978-3-031-50349-8_10

[6] V. Ermolai, A. Sover, G. Nagîţ, M. A. Boca, A.I. Irimia, Influence of Fused Filament Fabrication Raster Parameters over the Top Surface Topography. Acta Technica Napocensis-Series: Applied Mathematics, Mechanics, And Engineering, 66(5) (2023).

[7] UltiMaker. (2024, August 30). UltiMaker Cura - UltiMaker. https://ultimaker.com/software/ultimaker-cura/

[8] Ghostkeeper. (n.d.). GitHub - Ghostkeeper/SettingsGuide: More extensive explanations of Cura slicing settings. GitHub. https://github.com/Ghostkeeper/SettingsGuide

[9] V. Ermolai, A. Sover, (2023). Multi-material 3D Printed Interfaces. Influencing Factors and Design Considerations. In: Cioboată, D.D. (eds) International Conference on Reliable Systems Engineering (ICoRSE) - 2023. ICoRSE 2023. Lecture Notes in Networks and Systems, vol 762. Springer, Cham. https://doi.org/10.1007/978-3-031-40628-7_11

[10] N. Harun, N. Kasim, M. Abidin, M. Izamshah, R. Attan, H. Ganesan., A Study on Surface Roughness During Fused Deposition Modelling: A Review. JAMT, 12(1(1), 25-36 (2018).

[11] V. Ermolai, A. Sover, A. Lang, Characterisation of the shape memory effect of PET polymer by FFF 3D printing. MRP, 28 (2023). https://doi.org/10.21741/9781644902479-11

[12] S. K. Karna, R. Sahai, An overview on Taguchi method. International journal of engineering and mathematical sciences, 1(1), 1-7 (2012).

[13] A. Freddi, M. Salmon, A. Freddi, M. Salmon, Introduction to the Taguchi method. Design principles and methodologies: from conceptualization to first prototyping with examples and case studies, 159-180 (2019). https://doi.org/10.1007/978-3-319-95342-7_7

[14] A. Mitra, The Taguchi method. Wiley Interdisciplinary Reviews: Computational Statistics, 3(5), 472-480 (2011). https://doi.org/10.1002/wics.169

[15] ImageJ. (n.d.). https://imagej.net/ij/

[16] T. (n.d.) Ferreira, ImageJ User Guide - IJ 1.46R. https://imagej.net/ij/docs/guide/index.html

Shape accuracy and roughness optimization of a fashion button by utilizing 3D printing

John D. KECHAGIAS[1,a *], Nikolaos A. FOUNTAS[2,b], Stephanos ZAOUTSOS[3,c], Dimitrios CHAIDAS[1,d] and Nikolaos M. Vaxevanidis[2,e]

[1]University of Thessaly, FWSD Dept., Karditsa, Greece

[2]School of Pedagogical and Technological Education, Dept. of Mechanical Engineering Educators, Amarousion, Greece

[3]University of Thessaly, Energy Department, Larissa, Greece

[a] jkechag@uth.gr, [b] nfountas@aspete.gr, [c] szaoutsos@uth.gr, [d] dchaidas@uth.gr, [e] vaxev@aspete.gr

Keywords: Fashion, Button, Optimization, Material, Extrusion, FFF, PLA, Surface, Accuracy

Abstract. Thanks to the simplicity and low cost, the open chamber filament material extrusion process, also known as fused filament fabrication (FFF), has spread to customized applications in the fashion and textile industries. This work is an experimental optimization of a 3D printed button changing two primary parameters, the printing speed and temperature as predictors and surface and shape accuracy as performance measures. The results were processed with the analysis of means (ANOM), normalized according to the 'min-max' rule and finally optimized utilizing the weighted multi-objective approach and the 'minimum the best' case. Due to the small shape of the button and the wide range of different filament materials, this optimization study is challenging for fashion products where the design and quality are of primary interest. The higher printing temperature (225 °C) and speed (50 mm/s) optimize the process. This research mainly explores the possibility of producing quality decorative products with FFF 3D printing to widen the avenue of using cost-effective and personalized designs in the fashion industry.

Introduction

Since the garment is used in everyday life, it must be functional and protective and follow seasonal fashion [1]. Moreover, the clothing industry is a thriving area that requires frequent changes due to each season's fashion changes and trends [2]. Therefore, new technologies in the industry, such as computers and 3D printers [3, 4], are essential for developing fashion products in terms of sustainability, recyclability, and quality [5]. It is noted that clothing waste in landfills is a large part of them, contributing significantly to CO_2 emissions [6], and therefore, materials, manufacturing, usage, and circular economy are all of utmost importance [7, 8]. In the above frame, the incorporation of new technologies and, furthermore, the investigation of improving product quality by reducing parallel time and cost is critical, resulting in better sustainability indices of new fashion products [9]. Since digitization is the first step in applying new technologies in the fashion industry [10], designing the product is the first step, and then the manufacturing codes (g-codes) feed into 3D printers to realize the final garment or decorative part [11]. Process optimization, which includes tuning structural and welding parameters, is a critical and non-routine phase [12, 13]. Therefore, experimental design is an essential tool for optimizing process efficiency and product quality, and it is utilized in various domains of activity, including chemistry, biomedical, aeronautics, and automotive engineering [14–16]. The first concern in product process optimization is to define the performance measures and variable parameters [17]. This is followed by defining the experiment boundaries so that all experiments can be conducted and the results can be safely analysed with statistical tools [18, 19]. The height and volume of the parts are of considerable importance and affect both the time and quality of the 3D−printed parts [20, 21]. In the literature, an elliptical button with 15- and 10-mm diameters and a 5-mm height was optimized according to the upper surface mean-roughness (Ra) utilizing a Box-Behnken design and varying the three basic processing parameters: the layer thickness (0.1, 0.2, 0.3; mm), printing temperature (210, 220, 230; °), and printing velocity (40, 50, 60; mm/s) [22]. The 0.2-mm layer thickness optimises the upper surface mean roughness (Ra) values, while the printing temperature improves the Ra at 210 and 230 °, and the printing speed is optimized near the lower value (45 mm/s) according to the Box-Behnken fractional design. The layer thickness (LT) is a key factor for both shape accuracy and surface quality and has been extensively investigated in the literature [19]. Lower LT values benefit vertical and sloped surface quality, but this is not generalized for flat surfaces, where the material flow dominates [23]. In the case of flat surfaces, the LT optimized close to 0.2 mm, which is about half of the nozzle diameter (0.4 mm), as shown in [22]. The increase in printing speed (PS) influences the number of gaps between sequential strands, which is expected to affect the upper surface roughness [24]. Moreover, according to the literature, printing temperature (PT) is also a very influential parameter on the shape accuracy of 3D-printed parts [25]. Specifically, increasing PT increases the external and wall dimensions regarding the thin-walled cuboid parts, as investigated in [26]. On the other hand, according to the same study [26], PT is not a significant parameter for vertical surfaces, where the thin-walled parts were fabricated of PLA/wood filament, with the layer thickness and printing temperature varying.

Therefore, the printing speed and temperature were selected here to optimize the performance of a button part with a similar dimension as in [27, 28]. The layer thickness was kept constant at 0.2 mm, and the limits of the printing speed and temperature were modified in regard to [22]. The study of these two parameters, keeping the layer thickness constant at 0.2 mm, was not met in the literature for small volume parts, such as in the case of a customised fashion product (button). So, a full factorial design was utilized to identify and model the effects on the dimensional accuracy and average max-hight roughness parameter, also known as Rz. Last but not least, this work can be utilised as a good paradigm for product optimization cases in the 3D printing area for specific and intricate parts.

Materials and Methods

The test specimen is an elliptical button with 24- and 20-mm diameters, a 6-mm height and a 2-mm hole through the base (Fig. 1a). A fused filament fabrication 3D printer (Craftbot Plus 3D, Budapest) was utilised for the optimization experiments (see Fig. 1b) with 220 mm × 220 mm × 250 mm build volume. The white eSUN-PLA$^+$ 1.75±0.05 mm filament (Esun Industrial Co, Shenzhen) was selected to 3D−print all buttons (Fig. 1c). The nozzle diameter was 0.4 mm, the bed temperature was 60 ºC, and the infill density was 20%. More four (4) top, four (4) bottom layers and three (3) outline perimeters were used. The stylus-type Surftest SJ-210 (Mitutoyo, America) profilometer was utilised to measure the surfaces in a 4 mm sampling length (Lc=0.8 mm; Fig. 1d) and a micrometer with an accuracy of ±0.01 mm for the dimensional measurements (Fig. 1e).

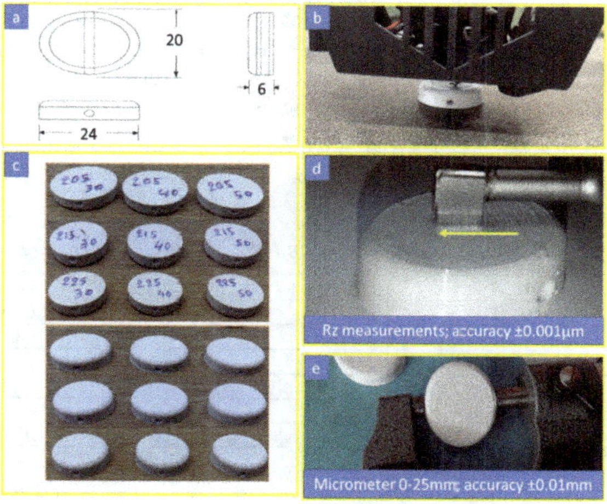

Fig. 1 (a) parts details, (b) 3D printing process, (c) all 3D printed parts, (d) direction of roughness measurements, and (e) dimensional accuracy measurements

Table 1 tabulates the measurements of all specimens, where X, Y, and Z are the measurements of each nine-part dimension in the X, Y, and Z directions, and Rz is the average of the five highest distances between the maximum peak and lowest valley in the measuring sample length Lc. The dimensional measurements are repeated in each button between three and five, but they are almost identical. In contrast, the measurements on the surface have been taken in a vertical direction to strands, as depicted in Fig. 1d, and the repeatability error was lower than ±2μm for all specimens. The Rz parameter represents the average difference between the five tallest peaks and the five deepest valleys and is often preferred instead of Ra, representing the average roughness. The parameter values' combinations follow the full factorial design as described in [29]. All experiments were performed carefully by experienced users to avoid man-material-machine-measurements (4M) errors. Note here that, more experiments will give more power to the results, but as explained in Kechagias [30], the benefits will not be considerably better.

Table 1 Dimensional and roughness measurements

No ex.	NT (°C)	PS (mm/s)	X (mm)	Y (mm)	Z (mm)	Rz (μm)
1	205	30	23.71	19.81	5.79	156.19
2	205	40	23.67	19.81	5.73	147.47
3	205	50	23.64	19.84	5.78	145.97
4	215	30	23.73	19.73	5.82	127.92
5	215	40	23.71	19.76	5.85	156.02
6	215	50	23.65	19.82	5.86	163.85
7	225	30	23.21	19.78	5.87	146.10
8	225	40	23.67	19.81	5.87	157.69
9	225	50	23.68	19.89	5.84	131.92
Average			23.63	19.81	5.82	148.13

Table 1 shows that all dimensions in X, Y, and Z are lower than the nominal dimensions. This is due to the components shrinking when the specimens' temperature drops to 20 degrees, at which point the measurements are taken. Therefore, to proceed with the statistical analysis of the results, the X, Y, and Z dimensions are transformed in the dimensionless error of the nominal dimensions

Materials Research Proceedings 46 (2024) 49-56 https://doi.org/10.21741/9781644903377-7

(DX/X, DY/Y, DZ/Z) and then extracted the average (Davm) as shown in Eq.1. Note here that this approach applies to parts with dimensions and volumes similar to those of the investigated samples (0-25 mm) and does not generalize to higher-dimensional parts due to the 3D printing path-related process, which is different from the moulding process, where the samples are formed in one step.

$$Davm = \frac{\frac{DX}{24}+\frac{DY}{20}+\frac{DZ}{6}}{3} \quad (1)$$

Table 2 Mean error (Davm) in the X, Y, and Z dimensions

No Exp.	DX/24	DY/20	DZ/6	Davm
1	0.0121	0.0095	0.0350	0.0189
2	0.0137	0.0095	0.0450	0.0228
3	0.0150	0.0080	0.0367	0.0199
4	0.0113	0.0135	0.0300	0.0183
5	0.0121	0.0120	0.0250	0.0164
6	0.0146	0.0090	0.0233	0.0156
7	0.0329	0.0110	0.0217	0.0219
8	0.0137	0.0095	0.0217	0.0150
9	0.0133	0.0055	0.0267	0.0152
Average	0.0154	0.0097	0.0294	0.0182

Results and Discussion

Figs 2a and 2b (parameter effect plots; [31]) illustrate the impact of the nozzle temperature (NT) and printing speed (PS) on the Rz and Davm regarding of the FFF 3D-printed samples, respectively. At first glance, these two figures display that the NT affects the Davm and the PS the Rz performances more.

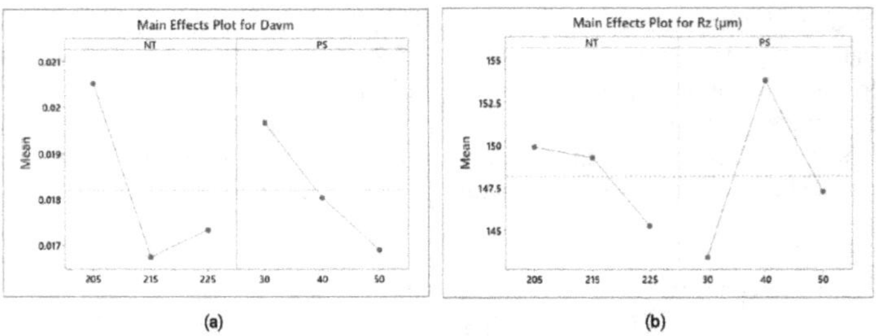

Fig. 2 MEP plots: (a) Davm and (b) Rz

In the manufacturing area [32], the objectives and performance measures do not have the same meaning. In our optimization case, the selected performances are (i) the average of the highest five distances between the maximum peak and lowest valley in the measuring line (Rz; see Table 1) and (ii) the mean deviation of the three dimensions on the X, Y, and Z axes (Davm; see Table 2). On the other hand, the objective is for both performance measures to be close to zero values. In the literature, this objective is called the lower the best [33], and the developed prediction models include terms (i.e., logarithmic) or the input and output normalized between 0 and 1 to avoid values less than zero, especially for absolute quantities such as Rz [23, 34]. Here, both two inputs (NT and PS) and two outputs (Rz and Davm) have been normalized according to the formula (see Eq. 2):

$$L_i = \frac{L_i - L_{min}}{L_{max} - L_{min}} \quad (2)$$

where L is each input or output parameter.

Table 3 tabulates all inputs and outputs normalized values (Eq. 2). Moreover, in Table 3, the last two columns present the R_{gr} (ranging grade) and the ranking (Rank) of the best combination regarding the lower the best having considered the formula described in Eq. 3, where we consider 0.3 the weight of the L_{Davm} and 0.7 the weight of the L_{Rz} metrics, following the multi-criteria decision-making approach [35].

$$R_{gr} = 0.3 \times L_{Davm} + 0.7 \times L_{Rz} \quad (3)$$

Table 3 Normalization values for inputs and outputs and ranking

L_NT	L_PS	L_Davm	L_Rz	Rgr	Rank
0	0	0.500	0.787	0,701	8
0	0.5	1.000	0.544	0,681	7
0	1	0.632	0.502	0,541	3
0.5	0	0.421	0.000	0,126	2
0.5	0.5	0.179	0.782	0,601	5
0.5	1	0.086	1.000	0,726	9
1	0	0.886	0.506	0,620	6
1	0.5	0.000	0.829	0,580	4
1	1	0.025	0.111	0,085	1

According to Table 3, the best combination of the printing speed and temperature are 225 °C and 50 mm/s. Fig. 3a gives the Rgr means, showing the best combination (225 °C and 50 mm/s) and the worst (205 °C and 40 mm/s). Additionally, Fig. 3b shows the interaction trends between printing temperature and speed, depicting antagonistic interaction; conflict lines [36].

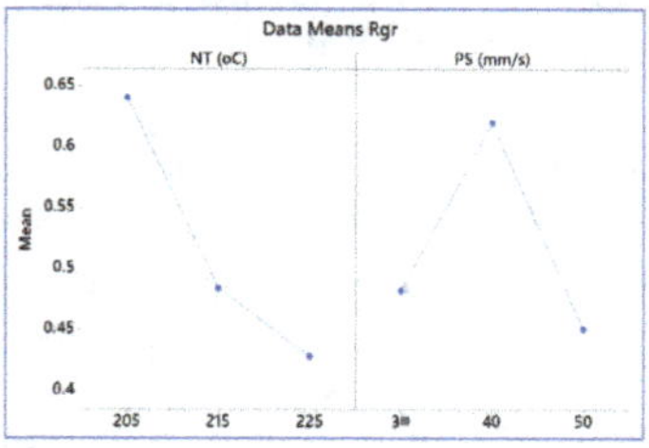

(a)

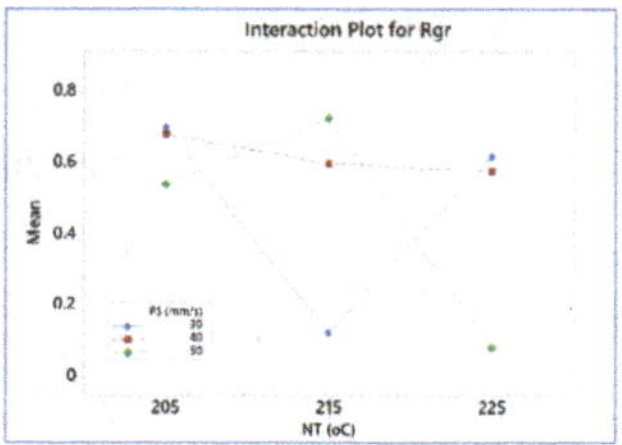

(b)

Fig. 3 (a) MEP plot for Rgr, and (b) interaction plots between printing temperature and speed

Conclusions

This work presents a multi-parameter, multi-objective case study in the fashion industry. Specifically, the nozzle temperature and printing speed were tuned for a button, giving more significance to surface roughness (0.7 weight) and lower in dimensional accuracy (0.3 weight). The higher nozzle temperature (225 °C) and speed (50 mm/s) benefit the objective scenario, showing that the higher temperature benefits the surface roughness Rz parameter the most. On the other side, the results showed that the higher printing speed benefits the dimensional deviations, giving the lowest error values.

This work is a representative optimization case study with benefits in fashion products, specifically customized small-sized buttons. Logos and other details are of the utmost importance in the fashion industry, and future work incorporating more processing parameters and performance measures is suggested.

References

[1] Y.L. Yap, W.Y. Yeong, Additive manufacture of fashion and jewellery products: a mini review. Virtual Phys. Prototyp. 9 (2014) 195-201. https://doi.org/10.1080/ 17452759.2014. 938993

[2] M. Polewka, F. Enz, M. Jennißen, E. Wirth, L. Sabantina, 3D Printing with Biomaterials-The New Sustainable Future of Textiles?, in: ECP 2023, MDPI, Basel Switzerland, pp. 59.

[3] S. Kim, H. Seong, Y. Her, J. Chun, A study of the development and improvement of fashion products using a FDM type 3D printer. Fashion and Textiles 6 (9) (2019). https://doi.org/10.1186/s40691-018-0162-0

[4] D. Fico, D. Rizzo, R. Casciaro, E.C. Corcione, A Review of Polymer-Based Materials for Fused Filament Fabrication (FFF): Focus on Sustainability and Recycled Materials. Polymers (Basel) 14 (465) (2022). https://doi.org/10.3390/polym14030465

[5] P. Saxena, P. Stavropoulos, J. Kechagias, K. Salonitis, Sustainability Assessment for Manufacturing Operations. Energies (Basel) 13 2730 (2020). https://doi.org/10.3390/ en13112730

[6] Fast Fashion Market Analysis, Size and Trends Global Forecast to 2022-2030. https://www.thebusinessresearchcompany.com/report/fast-fashion-global-market-report. Accessed 9 Jul 2022

[7] M. Koszewska, Circular Economy — Challenges for the Textile and Clothing Industry. Autex Res. J. 18 (2018) 337-347. https://doi.org/10.1515/aut-2018-0023

[8] X. Chen, H.A. Memon, Y. Wang, I. Marriam, M. Tebyetekerwa, Circular Economy and Sustainability of the Clothing and Textile Industry. Mater. Circ. Econ. 3 12 (2021). https://doi.org/10.1007/s42824-021-00026-2

[9] T. Spahiu, E. Canaj, E. Shehi, 3D printing for clothing production. J. Eng. Fiber. Fabr. 15 155892502094821 (2020). https://doi.org/10.1177/1558925020948216

[10] J.P. Manaia, F. Cerejo, J. Duarte, Revolutionising textile manufacturing: a comprehensive review on 3D and 4D printing technologies. Fashion and Textiles 10:20 (2023). https://doi.org/10.1186/s40691-023-00339-7

[11] E. Pei, J. Shen, J. Watling, Direct 3D printing of polymers onto textiles: experimental studies and applications. Rapid Prototyp. J. 21 (2015) 556-571. https://doi.org/10.1108/RPJ-09-2014-0126

[12] D. Chaidas, T. Spahiu, J. Kechagias, An Iinvestigation Of 3D Printing On Tulle Textile Changing The Platform Levelling. In: Contemporary trends and innovations in the textile industry (2023), Union Of Engineers And Textile Technicians Of Serbia, Beograd, Serbia.

[13] K. Chatterjee, T.K. Ghosh, 3D Printing of Textiles: Potential Roadmap to Printing with Fibers. Adv. Mater. 32 (2020). https://doi.org/10.1002/adma.201902086

[14] S.R. Rajpurohit, H.K. Dave, Impact strength of 3D printed PLA using open source FFF-based 3D printer. Prog. Addit. Manuf. 6 (2021) 119-131. https://doi.org/10.1007/s40964-020-00150-6

[15] I. Bianchi, A. Forcellese, T. Mancia, M. Simoncini, A. Vita, Process parameters effect on environmental sustainability of composites FFF technology. Mater. Manuf. Process. 37 (2022) 591-601. https://doi.org/10.1080/10426914.2022.2049300

[16] M.F. Vicente, M. Canyada, A. Conejero, Identifying limitations for design for manufacturing with desktop FFF 3D printers. Int. J. Rapid Manuf. 5 (2015) 116. https://doi.org/10.1504/IJRAPIDM.2015.073551

[17] A. Elkaseer, S. Schneider, S.G. Scholz, Experiment-Based Process Modeling and Optimization for High-Quality and Resource-Efficient FFF 3D Printing. Appl. Sci. 10 (2020) 2899. https://doi.org/10.3390/app10082899

[18] D.C. Montgomery, Design and Analysis of Experiments, 8th ed., Wiley, Hoboken, NJ, USA, 2012.

[19] D.C. Montgomery, C. Lynch, Optimal Experimental Designs for Hypothesis Testing with Multiple Factors: Maximizing Power for the Biological Sciences. Int. J. Exp. Des. Proc. Opt. 1 (2023). https://doi.org/10.1504/IJEDPO.2023.10061657

[20] I. Ullah, M.Wasif, M. Tufail, Analysis of shrinkage and dimensional accuracy of additively manufactured tooling for the composite manufacturing. Int. J. Interact. Des. Manuf. (2023) https://doi.org/10.1007/s12008-023-01640-x

[21] M.F.Vicente, M. Canyada, A. Conejero, Identifying limitations for design for manufacturing with desktop FFF 3D printers. Int. J. Rapid Manuf. 5 (2015) 116. https://doi.org/10.1504/IJRAPIDM.2015.073551

[22] T. Spahiu, K. Kitsakis, J.D. Kechagias, Box-Behnken design to optimise 3D printing parameters in applications for fashion products. Int. J. Exp. Des. Proc. Opt. 7 (2022) 49-61. https://doi.org/10.1504/IJEDPO.2022.131225

[23] J.D.Kechagias, S.P. Zaoutsos, An investigation of the effects of ironing parameters on the surface and compression properties of material extrusion components utilizing a hybrid-modeling experimental approach. Prog. Addit. Manuf. (2023). https://doi.org/ 10.1007/ s40964-023-00536-2

[24] I. Buj-Corral, A. Bagheri, M. Sivatte-Adroer, Effect of Printing Parameters on Dimensional Error, Surface Roughness and Porosity of FFF Printed Parts with Grid Structure. Polymers (Basel) 13 (2021) 1213. https://doi.org/10.3390/polym13081213

[25] D. Frunzaverde, V. Cojocaru, C-R. Ciubotariu, C-O Miclosina, D.D. Ardeljan, E.F. Ignat, G. Marginean, The Influence of the Printing Temperature and the Filament Color on the Dimensional Accuracy, Tensile Strength, and Friction Performance of FFF-Printed PLA Specimens. Polymers (Basel) 14 (2022) 1978. https://doi.org/10.3390/polym14101978

[26] D. Chaidas, J.D. Kechagias, An investigation of PLA/W parts quality fabricated by FFF. Mater. Manuf. Process. 37 (2022) 582-590. https://doi.org/10.1080/10426914.2021.1944193

[27] V.E. Kuznetsov, A.N. Solonin, A. Tavitov, O. Urzhumtsev, A. Vakulik, Increasing strength of FFF three-dimensional printed parts by influencing on temperature-related parameters of the process. Rapid Prototyp. J. 26 (2020) 107-121. https://doi.org/10.1108/RPJ-01-2019-0017

[28] B. Yin, Q. He, L. Ye, Effects of deposition speed and extrusion temperature on fusion between filaments in single-layer polymer films printed with FFF. Adv. Ind. Eng. Polym. Res. 4 (2021) 270-276. https://doi.org/10.1016/j.aiepr.2021.07.002

[29] J. Kechagias, Multiparameter signal-to-noise ratio optimization for end milling cutting conditions of aluminium alloy 5083. Int. J. Adv. Manuf. Technol. (2024) https://doi.org/10.1007/s00170-024-13667-9

[30] J.D. Kechagias, 3D printing parametric optimization using the power of Taguchi design: an expository paradigm. Mater. Manuf. Process. 39 (2024) 797-803. https://doi.org/10.1080/10426914.2023.2290258

[31] R. Arrigo, D. Battegazzore, G. Bernagozzi, F. Cravero, D.N. Ribero Pedraza, A. Frache, Recycled PP for 3D Printing: Material and Processing Optimization through Design of Experiment. Appl. Sci. 12 (2022) 10840. https://doi.org/10.3390/app122110840

[32] N. Costa, Design of experiments – overcome hindrances and bad practices. TQM J. 31 (2019) 772-789. https://doi.org/10.1108/TQM-02-2019-0035

[33] Z. Shakeri, K. Benfriha, N. Zirak, M. Shirinbayan, Mechanical strength and shape accuracy optimization of polyamide FFF parts using grey relational analysis. Sci. Rep. 12 (2022) 13142. https://doi.org/10.1038/s41598-022-17302-z

[34] M.S. Phadke, Quality engineering using robust design. Prentice Hall PTR, Englewood Cliffs, New Jersey, 1989.

[35] T. Samala, V.K. Manupati, J. Machado, S. Khandelwal, K. Antosz, A Systematic Simulation-Based Multi-Criteria Decision-Making Approach for the Evaluation of Semi–Fully Flexible Machine System Process Parameters. Electronics (Basel) 11 (2022) 233. https://doi.org/10.3390/electronics11020233

[36] R.T. Mushtaq, A. Iqbal, Y. Wang, M. Rehman, M.I. Petra, Investigation and Optimization of Effects of 3D Printer Process Parameters on Performance Parameters. Mater. 16 (2023) 3392. https://doi.org/10.3390/ma16093392

Peculiarities of the evolution of the microstructure and mechanical properties depending on the correction of the focal length in parts made by laser melting in a powder layer

Serhii ADJAMSKY [1,2, a], Bogdan DOVGYY [3, b], Ganna KONONENKO [2,4, c], Panagiotis KARMIRIS-OBRATAŃSKI [5, d *] and Rostyslav PODOLSKYI [2,4, e],

[1] Department of dynamics and strength of new types of transport, Institute of Transport Systems and Technologies of the National Academy of Sciences of Ukraine, 49000, Dnipro, Ukraine

[2] LLC «Additive laser technology of Ukraine», 65007, Odessa, Ukraine

[3] International Additive Manufacturing Group, 05-800, Pruszków (Polska), Poland

[4] Department of problems of deformation and thermal treatment of structural steel, Iron and Steel Institute of Z.I. Nekrasov of National Academy of Sciences of Ukraine, 49107, Dnipro, Ukraine

[5] Advanced Manufacturing Laboratory, Department of Manufacturing Systems, Faculty of Mechanical Engineering and Robotics, AGH University of Krakow,30-059 Krakow, Poland

[a] as@alt-print.com, [b] b.dovgyy@iamg.ltd, [c] perlit@ua.fm, [d] karmiris@agh.edu.pl, [e] rostislavpodolskij@gmail.com

Keywords: Focal Spot, Optical System, Dynamic Focusing, Inconel 718, Microstructure, Mechanical Properties

Abstract. Defects in finished products are a major factor contributing to the failure of parts produced through laser melting in a powder bed. Consequently, numerous studies have focused on controlling the laser melting process to achieve high-density parts without voids. In industrial applications, an optical system, specifically a three-axis dynamic focusing system, is commonly used to increase the printing area. However, this system requires focal length adjustments, which significantly influence the quality of the finished products, particularly in terms of the repeatability of mechanical properties based on the focal length offset from the center. Research has shown that variations in focal length correction result in deviations in the mechanical properties of the parts. Microstructural analysis revealed that focal length shifts lead to distortion of track boundaries, indicating changes in track crystallization due to the altered curvature of the focal spot.

Introduction

Defects in finished products are a primary cause of failure in components produced through laser melting in a powder bed [1-3]. As a result, numerous studies have been conducted to control the laser melting process in order to achieve high density and eliminate discontinuities. This has been accomplished using various analytical methods, such as optical microscopy, computed tomography [4], and in-process transmission X-ray imaging [5, 6], as well as by developing defect structure process maps [7]. Most research focuses on analyzing defect populations, their shapes, and the mechanisms behind their formation, which are now well understood. Building on this body of knowledge, many researchers have reached the desired level of process control and defect minimization. Currently, in the industrial sector of laser melting in a powder bed, new challenges are emerging, including increasing process productivity and the ability to scale up both serial and large-scale production of parts, while maintaining the high density of the finished product [8]. In terms of productivity, partial solutions have been proposed, such as adapting the focal spot shape, increasing the thickness of the powder layer, and other techniques. However, when it comes to scaling up the printing volume, certain challenges arise, particularly related to the limitations of using the F-Theta lens [9]. To address this, the industrial sector has implemented an alternative optical system, namely the three-axis focusing system (dynamic focusing), to expand the printing area. This is achieved by compensating for focal length variations through the movement of the adjustable focal lens and the integration of additional optical systems, allowing for coverage of larger distances from the vertical axes of the optical system (over 400 mm) [5, 9]. The use of this system requires careful focal length correction, which plays a crucial role in the quality of the finished products, particularly in ensuring the repeatability of mechanical properties, which can be affected by any deviation in the focal length correction from the center.

From the above, it follows that when applying dynamic focusing, it is crucial to investigate the influence of manufacturing parameters, microstructure, and mechanical properties in relation to focal length correction. Additionally, it is important to determine the specific characteristics of the structural formation of the heat-resistant alloy Inconel 718 depending on its position on the build platform.

Materials and Methods

The test samples were produced using the Alfa-280 system from LLC "ALT Ukraine," equipped with a dynamic focusing system and an argon-based protective gas blowing system with a circulation speed of 1.5 m/s. The system utilizes a 1064 nm ytterbium laser as the energy source, and the construction platform has dimensions of 280x280 mm. The material used in this study was a heat-resistant nickel-based alloy, LPBF-Inconel 718, with a particle size range of 10 to 45 μm. This multiphase alloy was selected to evaluate the effects of dynamic focusing and layer thickness on part integrity and mechanical properties, depending on the location on the build platform.

The chemical composition of the starting material, Inconel 718, in weight percent is as follows: Cr=17.82; Fe=18.03; Nb=5.50; Al=0.55; Mo=3.0; Ti=0.94; C=0.026; Si=0.07; Mn=0.08, with Ni making up the balance. The raw material was analyzed using a SEM-106 scanning electron microscope (Figure 1a) to determine the particle shape and size. The results of the granulometric analysis are shown in Figure 1b.

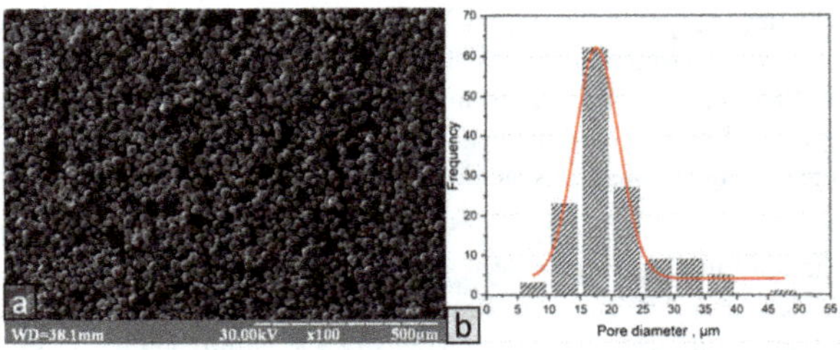

Fig. 1. Particles of Inconel 718 raw material at a magnification of 100 (a), results of particle size analysis (b)

Square Samples 5×5×10 mm, manufactured using layer thicknesses of 40 and 60 μm, and variable scanning parameters (power, scanning speed, and distance between tracks), specifically 200...260 W with a step of 30 W; 950 to 1150 mm/s with a step of 50 mm/s; and 0.09 to 0.14 mm with a step of 0.01 mm, were used for integrity studies and to determine optimal manufacturing parameters. Additionally, parameters ranging from 260 to 320 W in 30 W increments, 950 to 1150 mm/s in 50 mm/s increments, and 0.09 to 0.14 mm in 0.01 mm increments were used. Samples for tensile testing were produced with a constant layer thickness of 40 μm, using 200 W of power, a scanning speed of 1100 mm/s, and a track spacing of 0.1 mm. For samples with a 60 μm layer thickness, the parameters were 290 W of power, a scanning speed of 1100 mm/s, and a track spacing of 0.11 mm—these settings were identified as optimal based on the results of the integrity analysis.

The tensile test samples were positioned vertically on the build platform, equidistant from each other, to facilitate the analysis of property variations based on their location on the platform. Metallographic studies were conducted using Carl Zeiss light microscopes, specifically the Neophot 32 and Axiovert 200 M MAT models. Sample preparation was performed mechanically using specialized equipment from Struers, specifically the Labopol-5 model, with the application of diamond suspensions for precision grinding. Porosity in the samples was measured through metallographic analysis of polished surfaces, utilizing ImageJ software's "Particle Analysis" function. Objects smaller than 4 μm were filtered out to exclude potential defects introduced during sample preparation. Microstructure analysis was carried out using Kalling's Reagent No. 1 (1.5 g $CuCl_2$, 33 ml HCl, 33 ml ethanol, 33 ml water). Electrochemical polishing was performed using a mixture of perchloric acid and ethyl glycol (HCl 99%, $C_2H_6O_2$ 99%, 1:2 ratio). Before testing, all tensile specimens were machined on a HAAS ST10 lathe to reduce surface roughness and bring them to the standard size. Tensile tests were conducted on cylindrical specimens with a working zone diameter of 3 mm, manufactured in accordance with DSTU ISO 6892-1:2019. The tests were performed on a PHYWE testing machine, designed for rapid analysis of ultimate tensile strength and relative elongation, with a maximum load capacity of 3 tons and a traverse speed of 0.5 mm/s.

Results and Discussion

As a result of the analysis of the test samples in the polished state (Fig. 2, a, b), which were identified as samples with rational manufacturing modes, it was established that the samples made with a layer thickness of 40 microns mostly have pores of 6-8 microns in size with a small number of single discontinuities that are characteristic of non-fusion defects (lack of fusion) and have a size of 22-24 microns and 26-28 microns (Fig. 2, c). As a result of the analysis of the test sample in a polished state, made with a layer thickness of 60 μm, the majority of discontinuities is 4 μm (Fig. 2, d), which is 1.5...2 times less compared to samples made at 40 μm. But it should be noted that the relative share of these pores occupies 0.68% of the total area of the field of view compared to samples made at 40 μm with a relative share of 0.43%. This is mainly explained by the formation of discontinuities due to the insufficient bulk density of the powder when applied to the previous layer, gas capture during crystallization in the main volumes of the tracks. The decrease in the size of the predominant number of pores is due to the presence of more particles in the powder coating and the formation of voids between them, which explains the work [8].

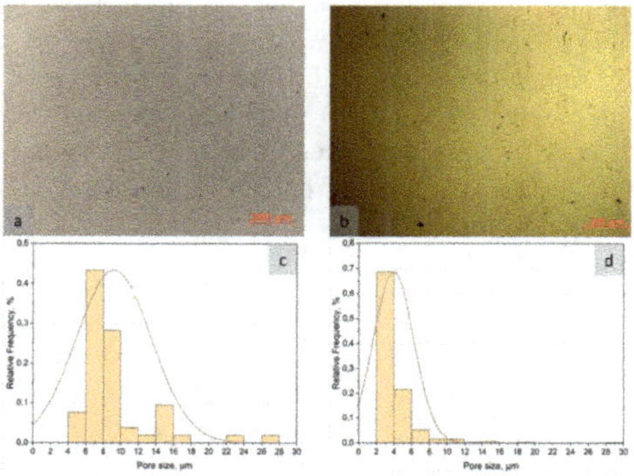

Fig. 2. Test samples in a polished state, made with a layer thickness of 40 (a) and 60 (b) μm; results of statistical analysis: relative proportion/pore size of experimental samples – 40 μm (c), 60 μm (g).

Materials Research Proceedings 46 (2024) 57-64 https://doi.org/10.21741/9781644903377-8

When analyzing the microstructure of the test samples (Fig. 3), it was established that the test samples manufactured at 40 µm (Fig. 3, a) have characteristically pronounced textural columnar grains with a length of 35-120 µm. The height of these columnar grains indicates passage through several previous layers and indicates characteristic overheating of the base metal, which can cause high internal stresses, which in turn can affect the final mechanical properties in the finished product without heat treatment. When analyzing the microstructure of test samples manufactured at a thickness of 60 µm (Fig. 3, b), it was established that the tracks have characteristic doublings in the region of the beginning of the track, which indicates the formation of a region of thermal influence (HAZ) (1-3 µm) without a characteristic the formation of columnar grains, indicating moderate heating and heat outflow without critical areas of overheating of previous layers. Nevertheless, from the results of the analysis of the works [9] it was established that the HAZ area has an effect on the mechanical properties, but it should be noted that the analysis of literature data did not reveal a positive or negative effect of this area in its presence, which is formed during the manufacture of LPBF tensile test technology in addition to microhardness.

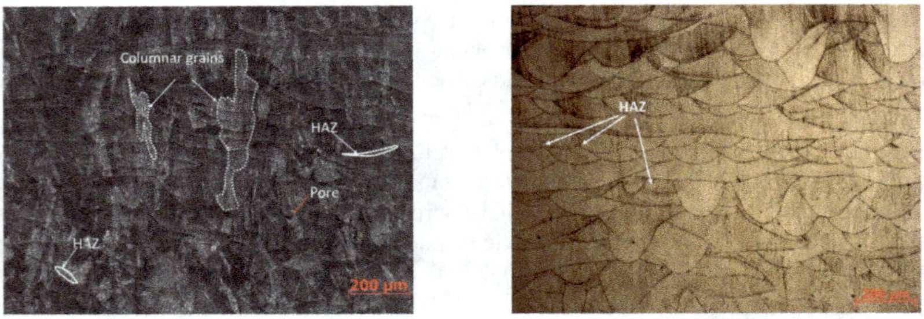

a b

Fig. 3. Microstructure of test samples, identified as rational manufacturing modes at a thickness of 40 µm (a) and 60 µm (b) of tensile test samples.

One of the main objectives of the research was to determine the influence of mechanical properties based on the location of test samples at varying distances from each other within the critical correlation range of the dynamic focusing optical system. Tensile tests were conducted, and mechanical properties were assessed, revealing the distribution of these properties according to the average values for three tests (groups 1-9) (Fig. 4). The analysis showed that the distribution of mechanical properties is generally uniform (stable). However, it should be noted that the mechanical properties of samples manufactured with a layer thickness of 60 µm were higher, with ultimate tensile strength ranging from 1098.3 to 1143.2 MPa and relative elongation between 18.6% and 32.4%. In contrast, samples produced with a 40 µm layer thickness exhibited lower values, with ultimate tensile strength ranging from 1027.6 to 1137.8 MPa and relative elongation between 12.5% and 19.1%.

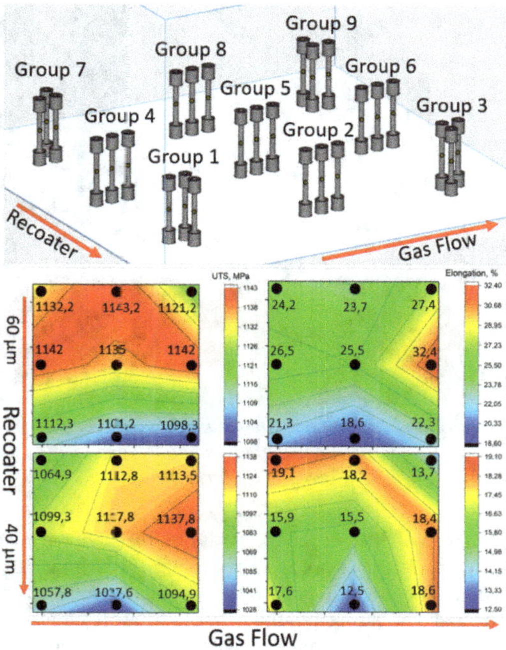

Fig. 4. Average values of the ultimate tensile strength and relative elongation of test samples manufactured according to the regimes that were identified as rational at a thickness of 40 and 60 µm, depending on the location on the platform.

Based on the analysis, it was determined that the samples in group 6 exhibited the highest values regardless of layer thickness. Notably, the ultimate tensile strength values were very close, at 1142 MPa and 1137.8 MPa for the 60 µm and 40 µm layer thicknesses, respectively. However, the relative elongation of the samples produced with a 60 µm layer thickness was approximately 1.8 times greater than that of the 40 µm samples. The samples in group 2 had the lowest values for both layer thicknesses, with ultimate tensile strength values of 1027.6 MPa and 1101.2 MPa, and relative elongation values of 12.5% and 18.6%, respectively.

Analysis of groups 2 and 6, located near the discharge hopper and the exit nozzle area, suggests that test samples in these regions of critical correlation distances within the dynamic focusing optical system tend to show lower values, likely due to the influence of factors such as flow circulation patterns and powder deficiencies. When examining the microstructure of the test samples (Fig. 5), produced with a layer thickness of 40 µm and located at varying distances from each other, notable structural differences were observed. Density analysis revealed that most pores measured between 3-5 microns, with isolated discontinuities (lack of fusion) observed at 7, 10, and 30 microns. Columnar grains were present in the center of the platform (group 5, Fig. 5c), indicating overheating. In contrast, samples located at the platform's edges showed a heat-affected zone (HAZ) with fewer or smaller columnar grains (group 4, 6, Fig. 5b, d). Additionally, samples near the bulk hopper and exit nozzle (groups 2 and 9, Fig. 5f, e) exhibited both columnar grain areas and HAZ regions, suggesting exposure to increased energy parameters.

Materials Research Proceedings 46 (2024) 57-64 https://doi.org/10.21741/9781644903377-8

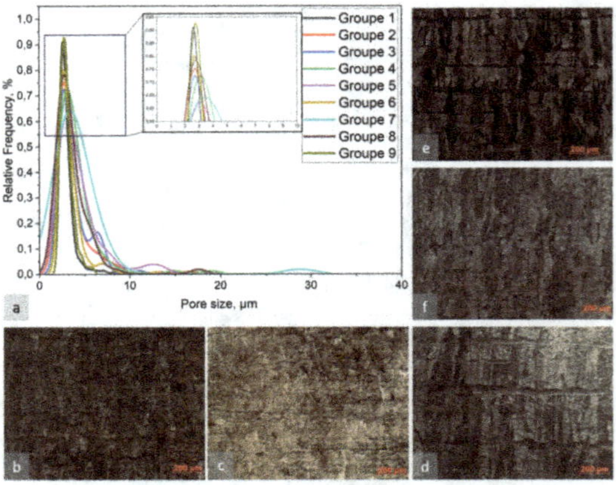

Fig. 5. The relative share and size of pores depending on the location on the platform of the test samples made at 40 μm (a); microstructure of test samples: group 4 (b), group 5 (c), group 6 (d), group 9 (e), group 2 (f)

During a comparative analysis of test samples produced with a layer thickness of 60 μm against those made with a thickness of 40 μm (Fig. 6), it was observed that the relative proportion and size of pores were greater in the 60 μm samples. The pore sizes predominantly ranged from 4 to 6 μm, with occasional discontinuities up to 20 μm. This result (Fig. 6a) indicates that when using a thicker layer and rational parameters during printing, the location of the experimental groups on the platform does not significantly affect pore size, but rather impacts the overall integrity of the base metal (i.e., total porosity). Microstructural analysis revealed that the test samples from groups 4 and 5 (Fig. 6b, c) contained a small number of columnar grains, measuring 30-80 μm, which generally corresponded to the 60 μm manufacturing layer. In contrast, samples from group 6 (Fig. 6d), located near the exit nozzle, exhibited a microstructure primarily consisting of characteristic track textures with heat-affected zone (HAZ) regions. Further analysis of group 2 samples (Fig. 6f), situated near the bulk bunker, revealed a structure made up of distinct textural arc-shaped tracks. Meanwhile, the experimental samples from group 9 (Fig. 6e) showed no significant structural changes compared to the 40 μm layer samples. These samples exhibited columnar grain structures and HAZ regions, indicating overheating in the preceding area. It is also worth noting that in the experimental samples from groups 1, 4, 7, and 3, 6, 9, slight distortions in the arcuate structure were observed, with an inclination angle of 14°, depending on the focal length correlation. However, in the 40 μm samples, no change in the inclination angle was detected. This lack of change is attributed to the thinner layer and the presence of columnar grains, which tend to obscure the detection of an arcuate structure.

Materials Research Proceedings 46 (2024) 57-64 https://doi.org/10.21741/9781644903377-8

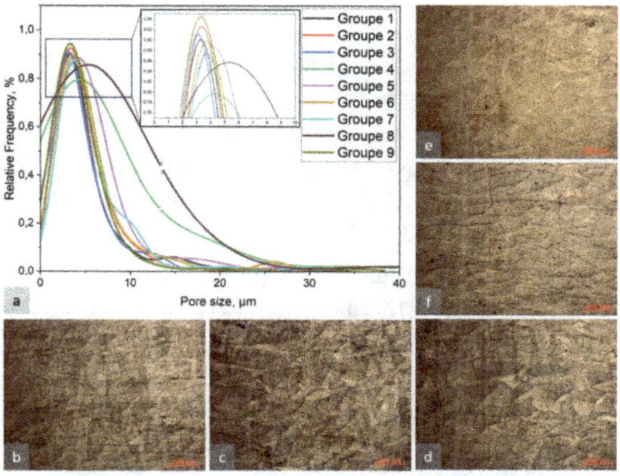

Fig. 6. The relative share and size of pores depending on the location on the platform of test samples made at 60 µm (a); microstructure of test samples: group 4 (b), group 5 (c), group 6 (d), group 9 (e), group 2 (f)

Based on the ABOVE analysis, it can be concluded that test samples produced with a layer thickness of 60 µm and optimized parameters exhibit a characteristic arc-shaped structure with a small number of columnar grains, measuring between 30-80 µm (as seen in groups 4 and 5). The analysis of samples made with a 40 µm layer thickness revealed the presence of columnar grains and heat-affected zones (HAZ), which indicate significant overheating of the previous melt layers. Thus, it was determined that both the microstructure and mechanical properties are affected by the layer thickness and the application of optimized manufacturing parameters, especially in areas with critical correlation distances related to the dynamic focusing optical system. The mechanical properties, specifically the ultimate tensile strength and relative elongation, varied depending on the sample's location on the platform with respect to these critical distances. For samples with a 60 µm layer thickness, the discrepancies were 45 MPa in tensile strength and 13.8% in relative elongation. For samples with a 40 µm layer thickness, the discrepancies were more pronounced, with a difference of 110.2 MPa in tensile strength and 6.6% in relative elongation.

Conclusions

The research aimed to explore the impact of focal length adjustments on the quality of parts produced through laser melting in a powder bed, particularly focusing on the mechanical properties and microstructure of the finished products. The research revealed that variations in the correction of the focal length, relative to the platform's critical distances for dynamic focusing, lead to deviations in the mechanical properties of the parts. Notably, samples from groups 2 and 6 were particularly influenced by external factors, such as their proximity to the bulk hopper and the outlet nozzle, respectively. Therefore, these areas warrant further investigation to fully understand the impact of these external influences. Additionally, microstructural analysis demonstrated that shifts in focal length correction result in distorted track boundaries, suggesting changes in crystallization conditions due to focal spot curvature. Additionally, when rational manufacturing parameters are applied, samples produced with a 40 µm layer thickness show the presence of columnar grains and heat-affected zones (HAZ), indicating overheating of previous layers. In contrast, samples produced at a 60 µm layer thickness exhibit a characteristic arc-shaped track structure, highlighting the influence of layer thickness on the microstructure formation.

References

[1] P. H. Chung, J. R. Zhuang, C. H. Pan, Evaluation and prediction of thermal defects in SLM-manufactured tibial components using FEM-based deep learning and statistic methods, Int J Adv Manuf Technol 134 (2024) 691–709. https://doi.org/10.1007/s00170-024-14139-w

[2] C. Du, Y. Zhao, Z. Xu, et al. Influence of In Situ Laser Polishing on Pore Defects and Mechanical Properties of IN718 Alloy Fabricated by Selective Laser Melting, J. of Materi Eng and Perform (2024). https://doi.org/10.1007/s11665-024-09678-y

[3] B. Onuike, A. Bandyopadhyay, Additive Manufacturing in Repair: Influence of Processing Parameters on Properties of Inconel 718, Mater. Lett., 252 (2019), 256–259. https://doi.org/10.1016/j.matlet.2019.05.114

[4] Darragh S. Egan, Kiera Jones, Denis P. Dowling, Selective laser melting of Ti-6Al-4V: Comparing µCT with in-situ process monitoring data, CIRP Journal of Manufacturing Science and Technology, Volume 31,= 2020, Pages 91-98, ISSN 1755-5817, https://doi.org/10.1016/j.cirpj.2020.10.004

[5] Y. Chen, S. Clar k, A.C. L. Leung, L. Sinclair, S. Marussi, R. Atwood, P.D. Lee, et al., Melt pool morphology in directed energy deposition additive manufacturing process, IOP Conf. Ser. Mater. Sci. Eng. 861 (1) (2020) p. 012012. https://doi.org/10.1088/1757-899X/861/1/012012

[6] Y. Chen, S.J. Clar k, D.M. Collins, S. Marussi, S.A. Hunt, D.M. Fenech, P.D. Lee, et al., Correlative synchrotron X-ray imaging and diffra ction of directed energy deposition additive manufacturing, Acta Mater. 209 (2021) 116777. https://doi.org/10.1016/j.actamat.2021.116777

[7] S.J. Wolff, S. Webster, N.D. Para b, B. Aronson, B. Gould, A. Greco, T. Sun, Insitu observations of directed energy deposition additive manufacturing using high-speed X-ray imaging, JOM 73 (2021) 189–200, https://doi.org/10.1007/s11837-020-04469-x

[8] Yang Qi, Hu Zhang, Xiaojia Nie, Zhiheng Hu, Haihong Zhu, Xiaoyan Zeng, A high strength Al–Li alloy produced by laser powder bed fusion: Densification, microstructure, and mechanical properties, Additive Manufacturing, Volume 35, 2020, 101346, ISSN 2214-8604, https://doi.org/10.1016/j.addma.2020.101346

[9] S. V. Adzhamskyi, G. A. Kononenko, R. V. Podolskyi, Mechanical properties of Inconel 718 alloy produced using selective laser melting technology with dynamic focusing on application surface, Mater Sci., 59 (2023), 420–425 https://doi.org/10.1007/s11003-024-00793-8

[10] M. Letenneur, A. Kreitcberg, V. Brailovski, Optimization of Laser Powder Bed Fusion Processing Using a Combination of Melt Pool Modeling and Design of Experiment Approaches: Density Control, J. Manuf. Mater. Process., 21 (2019), https://doi.org/10.3390/jmmp3010021

[11] N. Ren, J. Li, R. Zhang, et al. Solute trapping and non-equilibrium microstructure during rapid solidification of additive manufacturing, Nat Commun, 14 (2023), 7990 https://doi.org/10.1038/s41467-023-43563-x

Fabrication and characterization of TiO_2-ABS composite filaments for 3D printing: Structural analysis and evaluation of antibacterial properties

TUDOSE Ioan Valentin[1,a], ORFANOU Maria[1,b], ROMANITAN Cosmin[2,c], PACHIU Cristina[2,d], BRINCOVEANU Oana[2,e], ROSCA Irina [3,f], PASCARIU Petronela [2,3,g], BUCUR Stefan[4,h], VISKADOURAKIS Zacharias[5i] KENANAKIS George [5,j], SUCHEA Mirela Petruta [1,2,k]* and KOUDOUMAS Emmanouel[1,l]*

[1]Center of Materials Technology and Photonics, Hellenic Mediterranean University, 71410, Heraklion, Crete, Greece (A.B.)

[2]National Institute for Research and Development in Microtechnologies (IMT-Bucharest), 1 26 A, Erou Iancu Nicolae Street, P.O. Box 38-160, 023573 Bucharest Romania

[3]Petru Poni" Institute of Macromolecular Chemistry 41A Aleea Grigore Ghica-Voda, 700487 Iasi, Romania

[4]Faculty of Materials Science and Engineering, "Gheorghe Asachi" Technical University of Iasi, 67 Dimitrie Mangeron Blvd., 700050 Iasi, Romania

[5]Institute of Electronic Structure and Laser, Foundation for Research & Technology-Hellas, N. Plastira 100, 70013, Heraklion, Crete, Greece

[a]tudose_valentin@yahoo.com, [b]marorf1998@gmail.com, [c]cosmin.romanitan@imt.ro, [d]cristina.pachiu@imt.ro, [e]oana.brincoveanu@imt.ro, [f]rosca.irina@icmpp.ro, [g]pascariu.petronela@icmpp.ro, [h]bucurm.stefan@gmail.com, [i]zach@iesl.forth.gr, [j]gkenanak@iesl.forth.gr, [k] mirasuchea@hmu.gr and mira.suchea@imt.ro, [l]koudoumas@hmu.gr

Keywords: TiO_2-ABS composites, Filament Fabrication, 3D Printing, Structural Characterization, Antibacterial Activity

Abstract. This study focuses on preparing and characterizing acrylonitrile-butadiene-styrene (ABS) polymer composites reinforced with titanium dioxide (TiO_2) nanoparticles. The aim is to enhance the material's antibacterial and anti-fouling properties for potential use in 3D printing applications. The composites were synthesized by incorporating varying concentrations of TiO_2 (0.5–20wt.%) into the ABS matrix using melt compounding followed by filament extrusion. Morphological and structural properties were evaluated for filaments and printed materials using Scanning Electron Microscopy (SEM) X-ray Diffraction (XRD) analysis and Raman spectroscopy, respectively. Antibacterial efficacy was tested against *Escherichia coli*, *Staphylococcus aureus*, and *Candida albicans*, demonstrating a slight reduction in bacterial adhesion on the composite surfaces. Furthermore, incorporating TiO_2 can improve UV resistance and provide self-cleaning properties, indicating its potential for applications in medical devices, water purification components, and other high-performance 3D-printed structures.

Introduction

Metal oxide (MO) nanomaterials are commonly incorporated into polymer matrices to produce composite filaments with enhanced properties such as improved mechanical strength, thermal stability, and antimicrobial activity. ABS is a frequently used thermoplastic for 3D printing, and MO like TiO_2, ZnO, and CuO have been integrated to impart additional functionalities [1]. Fabricating these composites typically involves melt compounding techniques followed by filament extrusion. Optimization of extrusion parameters such as temperature and mixing speed is critical to ensure uniform dispersion of the MO particles, which directly affects the performance of the final printed material [2]. The properties of the composite filaments, such as thermal stability and mechanical strength, do not always translate directly to the printed parts due to changes during the 3D printing process (e.g., layer adhesion and print orientation). Recent studies show that while mechanical properties like tensile strength and modulus may be comparable, the surface roughness and porosity can vary significantly, influencing surface-dependent properties such as antibacterial performance [3,4]. The challenge lies in optimizing 3D printing parameters like raster angle, layer height, and infill density to preserve or enhance the composite properties. Characterization techniques such as SEM, XRD, and Raman spectroscopy are employed to evaluate the dispersion of MO particles in the ABS matrix and analyze the printed objects' microstructure. These techniques are critical for verifying the homogeneity and crystallinity of the composites, which influence both mechanical performance and antibacterial properties [3]. Additionally, morphological changes post-printing, like surface defects or poor interlayer bonding, can alter the expected behavior of the composites. MO such as ZnO and TiO_2 are well-known for their photocatalytic and antibacterial properties. When integrated into ABS, these composites exhibit promising results in reducing bacterial adhesion and biofilm formation. The effectiveness of these composites is primarily influenced by surface morphology, particle size, and concentration of the MO. ZnO-ABS composites, for example, have been shown to exhibit anti-fouling properties when exposed to simulated marine conditions, suggesting potential for use in aquatic environments. The presence of nanoparticles enables photocatalytic activity under UV light, leading to reactive oxygen species (ROS) generation that disrupts bacterial cell walls [3,5]. Currently, limited commercially available ABS-MO composite filaments are designed explicitly for antibacterial applications. Most solutions focus on enhanced mechanical and thermal properties rather than multifunctional use. Some academic and industrial studies have reported successfully fabricating experimental filaments (e.g., ZnO-ABS and TiO_2-ABS composites), but these materials have yet to see widespread commercial adoption [3]. The gap between laboratory-scale research and market-ready products is due to challenges in scalability, reproducibility, and regulatory compliance for materials used in healthcare and food-contact applications. Integrating MOs into ABS for 3D printing holds significant promise for producing advanced materials with tailored surface properties, including antimicrobial behavior [6]. However, standardized protocols are needed to ensure that the properties observed in the filament are maintained in the printed parts [7]. This work focuses on embedding TiO_2 powders within ABS matrices to produce “fishnet-like” structures via 3D printing, combining germicidal and photocatalytic action. ABS was chosen for its compatibility with 3D printing, mechanical properties, and durability. This novel approach aims to fabricate custom TiO_2-ABS composites using 3D printing techniques. Limited studies have explored TiO_2-ABS composites, mainly addressing their mechanical, thermal, and dielectric properties. TiO_2-ABS composite filaments are not yet commercially available, and existing studies do not extensively optimize filament fabrication and 3D printing processes. The initial challenge was to produce 3D-printable filaments and create grid structures for antifouling applications. Various TiO_2-ABS composite filaments were prepared and characterized using XRD and SEM to ensure uniform dispersion and structural integrity. These filaments were then used to 3D print grids, analyzed by XRD, SEM, and Raman spectroscopy to evaluate material transfer and printing quality. The printed grids were subjected to antibacterial tests to assess their properties.

Materials and Methods

Reagents and filament production. ABS from INEOS Styrolution. TiO_2 anatase ~30 nm particle size. The TiO_2 nanopowders and the ABS pellets were mixed in necessary proportions to achieve specific concentration and forwarded to a “Noztek Pro” high-temperature extruder to be processed at 220 °C to transform them into a cylindrical filament with a diameter of 1.75 ± 0.15 mm, suitable for 3D printing. All extrusion parameters, such as extrusion velocity and temperature, have been optimized to produce a uniform continuous cylindrical string with an overall length of ~ 5m. 3D printing of "fishnet-like" grid structures. Flat, rectangular-shaped "fishnet-like" grid structures (2×4 cm) were created using Tinkercad, a free design and 3D modeling tool provided by Autodesk Inc. A dual-extrusion FDM 3D printer was employed to produce TiO_2-ABS composite samples, utilizing previously synthesized filaments. During the FDM process, the composite filament is heated and extruded through a nozzle (0.4 mm diameter) at 230 °C onto a build platform preheated to 110 °C. The object is built layer-by-layer using a three-axis positioning system with a spatial resolution of around 100 µm in the vertical (z-axis) and 11 µm in the horizontal directions, allowing for the precise fabrication of detailed TiO_2-ABS structures as shown in Fig. 1.

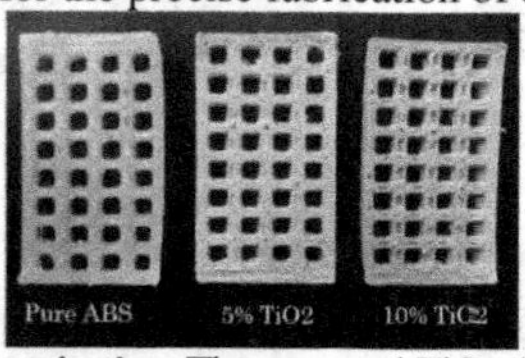

Fig. 1. Examples of the printed TiO_2-ABS structures

Characterization. The prepared TiO_2-ABS composite filaments and corresponding 3D-printed samples were structurally and morphologically characterized using SEM, XRD, and Raman spectroscopy. The antifouling performance of the materials was evaluated through controlled biological tests using specific algal species. To analyze the dispersion and surface morphology of the TiO_2 nanoparticles within the ABS matrix, SEM characterization was carried out on both the extruded filaments and the 3D printed specimens using a Nova NanoSEM 630 field emission microscope. This approach provided high-resolution imaging, which is essential for assessing the homogeneity and microstructural integrity of the composite. To avoid surface artifacts, samples were examined in a high-vacuum mode without any conductive coating. After the antifouling experiments, detailed SEM studies were also conducted using a Hitachi SU-8230 field emission microscope coupled with an Oxford energy-dispersive X-ray spectrometer to detect any surface modifications induced by the biofilms. In this case, a 7 nm layer of Pt/Pd was applied to the surfaces using an Agar Automated Sputter-Coater to improve imaging resolution. XRD analysis was performed with a Rigaku Smart Lab X-Ray Diffraction System to determine the crystalline phases present in the TiO_2-ABS composite. The measurements were conducted in grazing incidence mode, scanning the 2θ range from 10° to 60° at a speed of 5°/min. This setup provided insights into the crystalline structure and phase stability of the embedded TiO_2 particles using the ICDD (International Center for Diffraction Data) database for peak identification. Raman spectroscopy was utilized for further structural characterization. The spectra were recorded at room temperature using a WiTec Raman spectrometer (Alpha-SNOM 300 S,) with a 532 nm laser as the excitation source. The laser beam was focused to a spot size of ~1.0 µm, allowing for high-resolution mapping of the composite's surface. Raman mapping (25 µm x 25 µm) was achieved with a lateral resolution of ~0.8 µm, and spectra were collected pixel-by-pixel, using a 600 grooves/mm grating for accurate analysis and data processing and visualization were performed with the WiTec Project Five software. The microbiological evaluation involved using Tryptic Soy Broth, bacteriological agar, and Sabouraud Dextrose Broth. The antimicrobial properties of ABS-TiO_2 (0.5-20%) composites were assessed via disk diffusion against *S. aureus* ATCC 25923, *E. coli* ATCC 25922, and *C. albicans* ATCC 10231. Bacterial and yeast strains were cultured and adjusted to 0.5 McFarland standards. Samples sterilized under UV light were placed on inoculated agar plates and incubated. Inhibition zones were measured using Image J, and triplicate experiments validated the results. Anti-adherence tests followed a modified JIS Z2801:2000 protocol. Samples were incubated with bacteria on tryptic soy agar for one month. Post-incubation, the cells were collected, diluted, and analyzed for viability using the CellTiter 96® AQueous One Solution Cell Proliferation Assay. Measurements were taken using a microplate reader, and results were expressed as the percentage of cell viability relative to control.

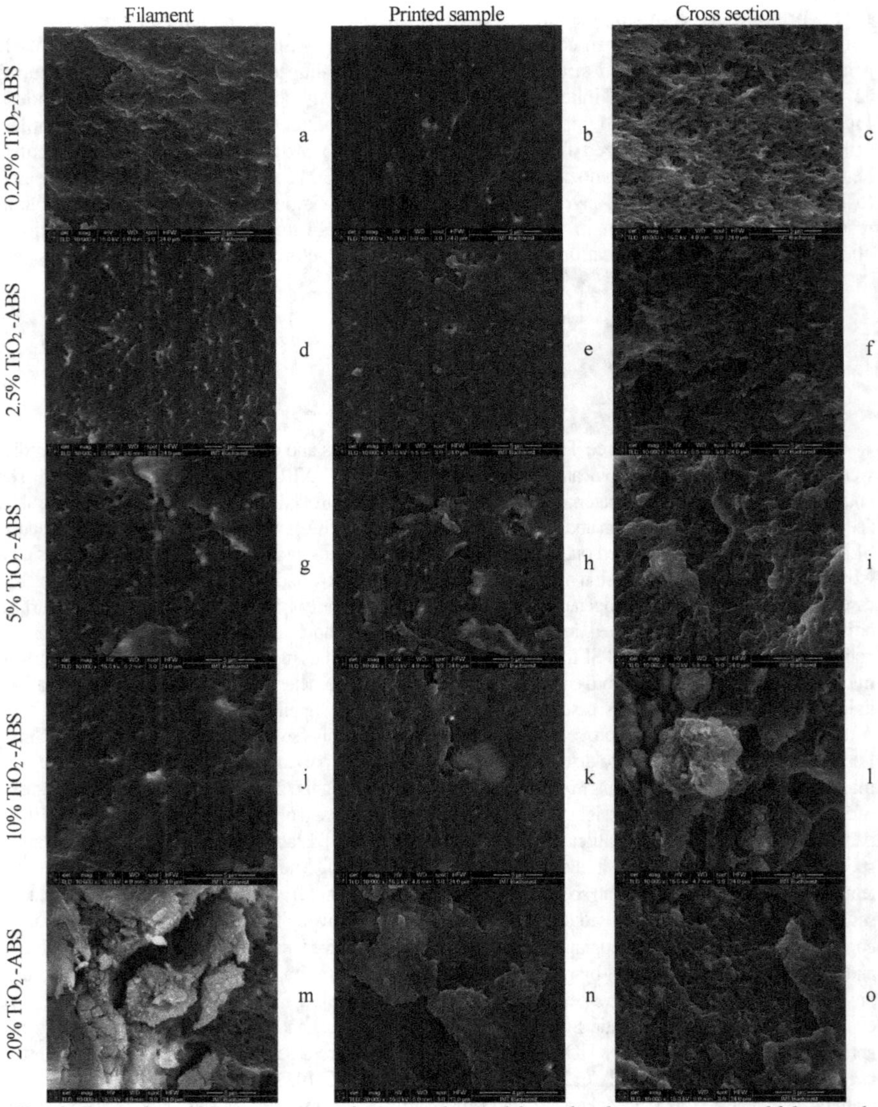

Fig. 2. Examples of SEM images, scale 5 μm, obtained from the characterization of fabricated TiO_2-ABS composite materials and printed TiO_2-ABS structures surface and cross-section.

Results and discussions

SEM Characterization. Fig. 2 presents an example of SEM images obtained from the characterization of fabricated TiO_2-ABS composite materials with different TiO_2 content filaments and structures. Based on the SEM images, one can analyze the surface and internal structures of TiO_2-ABS composites for both filaments and 3D printed samples, considering the effects of increasing TiO_2 concentration on morphological and structural properties. This analysis will compare the filament and printed material to highlight changes the 3D printing process introduced. The SEM images of the filaments show varying surface textures and particle distribution patterns depending on TiO_2 concentration For 0.5% TiO_2-ABS filament, the TiO_2 nanoparticles are sparsely distributed, with a smooth overall polymer surface and minimal surface defects, indicating good dispersion at low concentrations. At 2.5% TiO_2, a slight surface roughening appears; as more TiO_2 is incorporated, small clusters of TiO_2 are visible, but they remain relatively uniform in distribution. A more pronounced surface roughness with a visible agglomeration of TiO_2 nanoparticles and the increase in surface asperities suggests some degree of phase separation between

ABS and TiO_2 at 5% TiO_2. For filaments with 10% and 20% TiO_2, large aggregates are visible, significantly altering the surface morphology; the surface becomes irregular and rough, indicating that at higher concentrations, the polymer matrix struggles to distribute the TiO_2 particles evenly; such agglomeration can lead to stress points, potentially affecting mechanical properties. The SEM images of the printed surfaces exhibit a clear impact of the 3D printing process as layer-by-layer deposition during printing results in more distinct interlayer boundaries and surface irregularities compared to the filament samples. At low concentrations (0.5% and 2.5%), the printed surface morphology is still relatively smooth, similar to the filament. However, as the concentration increases, there are marked changes. At higher TiO_2 concentrations (10% and 20%), the printed surfaces show pronounced roughness and particle agglomeration, more so than in the filament samples, suggesting that during the thermal processing of 3D printing, TiO_2 particles may have migrated, leading to uneven distribution. The presence of large agglomerates can result in surface defects and poor interfacial bonding between printed layers. *The Cross-Sectional Morphology* of printed parts at low TiO_2 concentrations (0.5% and 2.5%) shows relatively homogeneous internal structures with good interlayer bonding. The distribution of TiO_2 appears uniform, indicating minimal impact on print quality. There is a noticeable increase in voids and poor layer adhesion, particularly at 10% and 20% concentrations. Large TiO_2 clusters are observed, disrupting the polymer matrix and suggesting potential mechanical weaknesses. These agglomerates create stress points, possibly contributing to the observed mechanical failure points in high-concentration printed samples.

XRD Characterization. The XRD patterns show distinct diffraction peaks corresponding to TiO_2 (Anatase), with strong reflections at (101), (004), (200), (105), (204), (116), and (215) planes, which are consistent with the anatase phase of TiO_2. These reflections become more pronounced as the TiO_2 concentration increases. The characteristic broad, amorphous hump at ~15°–25° in the pure ABS sample corresponds to the amorphous nature of the polymer matrix. This peak remains visible but diminishes in relative intensity as the TiO_2 concentration increases, indicating that the crystalline TiO_2 is taking up a more significant fraction of the material. For 0.5% and 2.5% TiO_2, the intensity of the TiO_2 (101) peak at ~25° is relatively low, suggesting good dispersion of the nanoparticles within the ABS matrix. The weak signal indicates a small crystallite size or a low degree of aggregation. The absence of other strong TiO_2 peaks at higher angles in these samples further supports that the TiO_2 particles are well distributed and may interact with the polymer chains, preventing extensive crystallization. For medium TiO_2 concentrations (5%), the TiO_2 peaks become more defined, with the (101) peak at ~25° and additional peaks appearing at ~37°, ~48°, and ~55°, indicating the onset of crystalline domain formation. This suggests that as the TiO_2 content increases, some degree of particle aggregation or local phase separation is occurring, leading to more prominent crystalline reflections. At 10% and 20% TiO_2 content the TiO_2 peaks are sharp and intense, indicating large crystalline domains or significant particle agglomeration. The relative intensity of these peaks compared to the amorphous hump of ABS shows that the composite is becoming increasingly dominated by the crystalline TiO_2 phase. This phase separation could lead to a heterogeneous distribution of mechanical and surface properties. For the filament, the anatase TiO_2 phase peaks (e.g., (101) at ~25.3°, (200) at ~48.0°, (204) at ~55.1°) are visible even at low conc., indicating well-defined crystalline phases. Increasing TiO_2 content increases the peak intensities proportionally, suggesting larger crystalline domains or enhanced crystallinity. The printed samples show sharper TiO_2 peaks at higher concentrations (10% and 20%), indicating potential particle migration and aggregation during printing. The ABS amorphous hump (~15°–25°) diminishes in intensity with increasing TiO_2 conc., as observed in the filament samples, but the effect is more pronounced in printed parts.

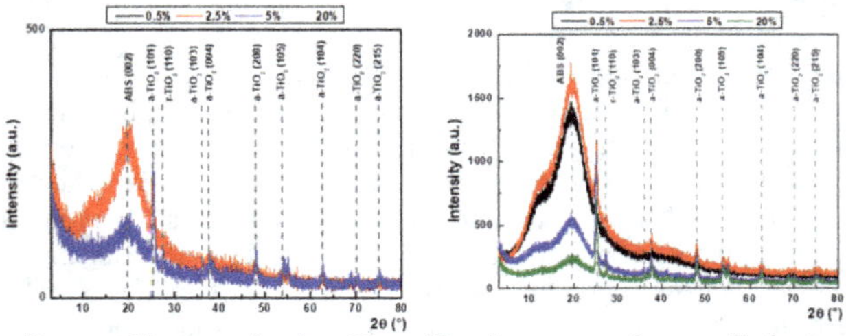

Fig. 3. XRD diffractograms obtained from the characterization of fabricated TiO_2-ABS composite materials with different TiO_2 content: a) filaments and b) printed TiO_2-ABS structures.

Raman Characterization. From Fig. 4, one can observe distinct variations in the intensity and position of key peaks corresponding to ABS and TiO_2 components as the content of TiO_2 increases from 0.5% to 20%. The pure ABS sample exhibits typical Raman peaks at ~636 cm^{-1}, ~1001 cm^{-1}, and 1600-1800 cm^{-1}, which are associated with the aromatic rings, C-H bending, and stretching vibrations of the styrene-butadiene backbone. These peaks serve as a reference for comparison across the composite samples. The addition of TiO_2 introduces new peaks around ~143 cm^{-1}, ~398 cm^{-1}, ~515 cm^{-1}, and ~640 cm^{-1}, which are indicative of the anatase phase of TiO_2. These peaks become more pronounced and sharper with increasing TiO_2 conc., confirming the successful incorporation and distribution of the filler in the ABS matrix. The intensity of these TiO_2-specific peaks, especially at ~143 cm^{-1}, increases notably in the 10% and 20% samples, indicating a higher volume fraction and potential agglomeration at higher conc. Minor shifts in the ABS backbone peaks, especially in the 1600-1800 cm^{-1} range, suggest potential interactions between the TiO_2 nanoparticles and the polymer chains. This could be due to hydrogen bonding or physical interfacial interactions between the matrix and the filler. The uniformity of the TiO_2-related peak intensities in lower conc. (0.5% and 2.5%) suggests a more homogeneous distribution. At higher loadings (10% and 20%), the broadening of these peaks could indicate particle clustering, which may negatively affect mechanical properties in the printed material.

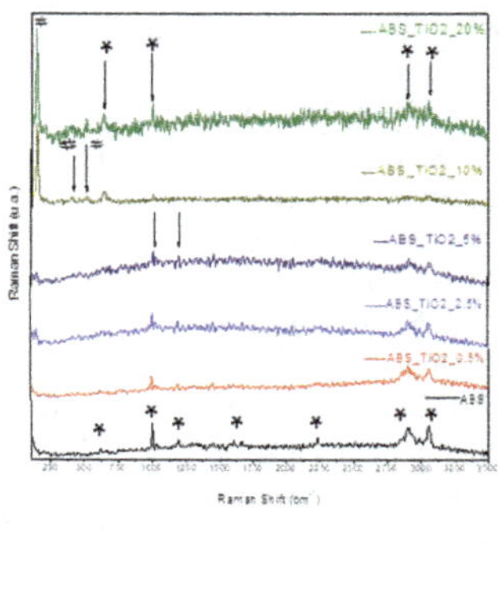

Fig. 4. Raman spectra obtained from the characterization of fabricated TiO_2-ABS composite materials with different TiO_2 content, printed TiO_2-ABS structures.

Materials Research Proceedings 46 (2024) 65-73 https://doi.org/10.21741/9781644903377-9

Antibacterial. The disk diffusion assay shows that none of the ABS-TiO_2 samples presented antimicrobial activity against the reference strains after 24 hours of incubation at 36±1 °C (see Fig. 5). The anti-adherence test quantified only the planktonic microbial cells, not attached to an eventual biofilm formed at the material's surfaces. ABS-TiO_2 materials, regardless of conc. (Fig. 6), were not efficient against the Gram-negative bacterial strain but efficient against the Gram positive represented by *S. aureus*, the relative cell viability being up to 88% after incubation. For the Gram-negative strain, represented by *E. coli*, the materials were quite efficient, destroying up to 80% of the relative cell viability. In the case of *C. albicans* yeast, only the higher conc. of TiO_2 (10 and 20%) lead to a significant decrease in cell viability.

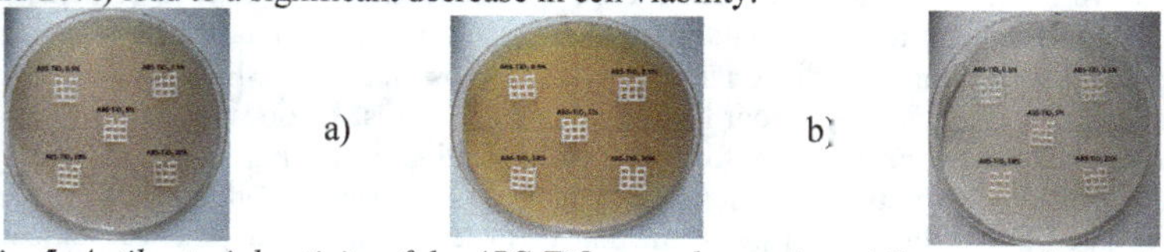

a) b) c)

Fig. 5. Antibacterial activity of the ABS-TiO_2 samples against a) S. aureus, b)E. coli, c) C. albicans.

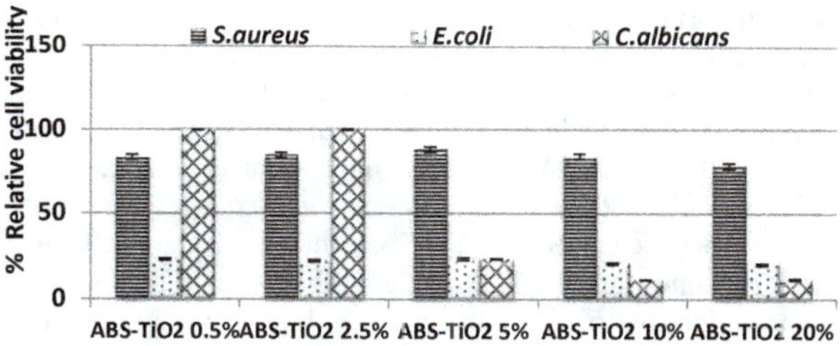

Fig. 6. Antimicrobial activity of ABS-TiO_2 samples against the reference strains of S. aureus, E. coli, and C. albicans.

Correlation with Structural and Morphological Features. The antibacterial activity shows distinct trends that can be correlated with the TiO_2content, surface morphology, and internal structure observed in the SEM and XRD characterization. The inhibition rate against *S. aureus* is consistently high across all TiO_2 concentrations, peaking at 5% TiO_2 (88.53%), which suggests that the anatase TiO_2 in the ABS matrix is effective against Gram-positive bacteria, likely due to the oxidative stress induced by ROS generated on the TiO_2 surface. At 5% TiO_2, the SEM images show a moderately rough surface with good dispersion of TiO_2 nanoparticles, which may provide optimal interaction between the bacterial cell walls and the TiO_2, enhancing antibacterial activity. Antibacterial efficacy decreases slightly at higher content (10% and 20%), which could be due to increased particle agglomeration observed in the SEM images, leading to reduced surface area and fewer active sites. Effectiveness against *E. coli* (Gram-Negative Bacteria) is much lower than for *S. aureus*, remaining below 24% for all concentrations. The outer membrane of Gram-negative bacteria, which provides additional protection against the ROS generated by TiO_2, may have a role as the relatively smooth surfaces at low content (0.5% and 2.5%) might not offer enough surface roughness for adequate bacterial adhesion and subsequent inactivation. Even at higher content, where large agglomerates of TiO_2 are formed, which may hinder uniform interaction between *E. coli* cells and the active TiO_2 surfaces is more or less the same. Effectiveness against *C. albicans* (Yeast) is highest at low concentrations (0.5% and 2.5%) with near-total inhibition (99.90% and 100%, respectively). However, the effectiveness drops drastically at 5%, 10%, and 20% concentrations, with inhibition falling to as low as ~11%. The superior performance at low MO content correlates with the uniform and well-dispersed TiO_2 observed in the SEM images. The absence of large aggregates ensures that a higher surface area is exposed, maximizing ROS generation. As the concentration increases, the formation of large agglomerates reduces the number of active sites, limiting the interaction between the yeast cells and the TiO_2 particles. The XRD patterns confirm this trend, where the sharp peaks at higher NO conent indicate larger crystalline domains that may reduce the availability of reactive surfaces. The smooth surfaces observed in low-MO content samples (0.5% and 2.5%) are more effective against *C. albicans* but less effective against *E. coli*. This suggests that surface roughness is crucial in bacterial adhesion and subsequent inactivation. SEM images of high-concentration samples (10% and 20%) show significant TiO_2 agglomeration, resulting in uneven surface morphology and lower effective surface area for ROS generation. Consequently, the antibacterial performance against all three strains is reduced, highlighting the importance of maintaining a homogeneous TiO_2 dispersion for optimal antibacterial activity. The cross-sectional SEM images of printed samples at high TiO_2 content show poor layer adhesion and voids. These structural defects could contribute to reduced antibacterial performance by providing protective niches for bacteria to evade ROS exposure. Additionally, these voids and poor interlayer bonding may reduce the mechanical integrity of the printed parts, making them less suitable for practical antibacterial applications where surface integrity is crucial.

Summary

The comparison between filament and printed materials reveals significant differences in surface morphology and internal structure. Filaments have smoother surfaces, while 3D printing introduces roughness and interlayer defects due to thermal and mechanical stresses. TiO_2 particles, which are well-dispersed at low concentrations (0.5% and 2.5%) in filaments, tend to agglomerate in printed parts at higher MO content (10% and 20%), leading to rough surfaces and poor adhesion. Cross-sectional SEM images show voids and discontinuities in high-concentration samples, indicating weaker mechanical properties. The antibacterial performance is strongly influenced by TiO_2 content and distribution. Low concentrations (0.5% and 2.5%) show high efficacy against *C. albicans* due to uniform TiO_2 distribution, but performance drops at higher concentrations (10% and 20%) due to agglomeration. The antibacterial effect against *S. aureus* is high across all concentrations, peaking at 5%, suggesting Gram-positive bacteria are more susceptible to ROS. In contrast, *E. coli* shows low susceptibility due to its protective outer membrane. 3D printing enables complex geometries with TiO_2-ABS composites, but alters material properties compared to filaments. Controlling TiO_2 dispersion and surface morphology is crucial for maintaining antibacterial properties. Adjusting printing parameters like temperature, speed, and infill can reduce agglomeration and improve adhesion, optimizing structural and antibacterial performance. Thus, 3D printing can create functional TiO_2-ABS composites, but careful optimization is required to balance mechanical integrity and antibacterial efficiency.

Acknowledgment

HMU contribution to this work was partially supported by NATO Science for Peace and Security Programme, grant G5868. IMT contribution was partially supported by PNRR/2022/C9/MCID/I8 CF23/14 11 2022 contract 760101/23.05.2023 financed by the Ministry of Research, Innovation and Digitalization in "Development of a program to attract highly specialized human resources from abroad in research, development, and innovation activities" within the – PNRR-IIIC9-2022 - I8 PNRR/2022/Component 9/investment 8 and by the Romanian Ministry of Research, Innovation and Digitalization through the µNanoEl, Cod: 23 07 core Programme.

References

[1] M.S. Chavali, M.P. Nikolova, Metal oxide nanoparticles and their applications in nanotechnology. SN Appl. Sci. 1 (2019) 607. https://doi.org/10.1007/s42452-019-0592-3

[2] A. Bouranta et al., 3D Printed Metal Oxide-Polymer Composite Materials for Antifouling Applications. Nanomaterials. 12 (2022) 917. https://doi.org/10.3390/nano12060917

[3] J. Tungtrongpairoj et al., Processing and modeling of 3D-printed mill scale strengthened acrylonitrile butadiene styrene composites, Int. J. Adv. Manuf. Technol. 131 (2024) 1567–1586. https://doi.org/10.1007/s00170-024-13037-5

[4] T. Abdullah et al., 3D Printing of Metal/Metal Oxide Incorporated Thermoplastic Nanocomposites with Antimicrobial Properties. Front. Bioeng. Biotechnol. 8 (2020) 568186. https://doi.org/10.3389/fbioe.2020.568186

[5] S. Park et al., 3D printing of polymer composites: Materials, processes, and applications. Matter. 5 (2022) 43-76. https://doi.org/10.1016/j.matt.2021.10.018

[6] N. Thakur et al., A critical review on the recent trends of photocatalytic, antibacterial, antioxidant and nanohybrid applications of anatase and rutile TiO2 nanoparticles, Sci. of The Total Envir, 914 (2024). 169815,https://doi.org/10.1016/j.scitotenv.2023.169815

[7] L. Li,; et al., Physicochemical and Photocatalytic Properties of 3D-Printed TiO2/Chitin/Cellulose Composite with Ordered Porous Structures. Polymers 14(2022), 5435. https://doi.org/10.3390/polym14245435

Innovative Manufacturing Engineering and Energy - IManEE2024 Materials Research Forum LLC
Materials Research Proceedings 46 (2024) 74-81 https://doi.org/10.21741/9781644903377-10

A novel self-assessment technique of 3D printer accuracy

Dimitrios KRYFOS[1,a*], Christos KALLIGEROS[2,b], Konstantinos KARAGIANNIS[2,c], Vasilios SPITAS[2,d], Nickolas SAPIDIS[1,e]

[1]Department of Mechanical Engineering, University of Western Macedonia, ZEP 50100 Kozani, Greece

[2]Laboratory of Machine Design, National Technical University of Athens, 9 Iroon Polytechniou, 15780, Zografou, Greece

[a]dimkrifos@gmail.com, [b]ckalligeros@mail.ntua.gr, [c]kostas.app@hotmail.com, [d]vspitas@mail.ntua.gr, [e]nsapidis@uowm.gr

Keywords: 3D Printing, Benchmark Layout, Additive Manufacturing, Geometrical Accuracy, Self-Assessment

Abstract. As Additive Manufacturing Technologies continue to evolve, the need for accuracy in 3D printing methods increases. To evaluate the accuracy of 3D printers and/or the 3D printing process, several benchmark layouts have been proposed in the literature. However, these layouts are typically very complex and do not allow the average 3D printer user to quickly reach a clear conclusion about printer accuracy. This study introduces a novel self-assessment technique for evaluating the accuracy of 3D printers. The proposed benchmark layout consists of only three main 3D-printed components. By leveraging geometric relationships and amplifying deviations through inclined surfaces, this method enables the measurement of the cylindricity of a 3D-printed cylinder without the need for specialized equipment. The layout can be easily printed and used to assess the accuracy of the printer. Measurements from the layout were compared with those obtained from a CMM and a caliper. While this new method may not capture the full three-dimensional form error of a cylinder as accurately as a CMM, it provides a practical measure of the apparent diameter, which is critical for successful assembly operations. This simple, cost-effective technique offers a valuable tool for both amateur and professional users to assess and compare the accuracy of different 3D printing processes and machines.

Introduction

Additive manufacturing (AM) technologies have significantly evolved since their inception in the late 1980s, offering various processes for creating physical parts directly from 3D CAD models [1]. Specifically, these technologies can fabricate solid free-form and/or conventional geometries using layer manufacturing methods such as 3D printing, which has advanced considerably in recent decades [2-4]. Compared to conventional manufacturing methods (i.e., casting, machining), 3D printing reduces prototyping time and process complexity.

As these technologies continue to advance, there is an increasing need for methods to evaluate and compare the performance of different 3D printing processes and machines. This need is particularly crucial as 3D printing transitions from primarily prototyping applications to the production of functional end-use parts [5,6]. In terms of dimensional accuracy, the use of a common benchmark layout for quantitatively comparing the results of 3D printing procedures is significant [7,8].

Several researchers have developed benchmarking parts for additive manufacturing processes and machines to evaluate the quality of the printing process. These studies focus on the accuracy of a 3D printing machines and processes. Key geometrical features include rectangular bosses and pockets, cylinders, conical bosses, and inclined surfaces, as shown in different proposed benchmark layouts [17].

Materials Research Proceedings 46 (2024) 74-81 https://doi.org/10.21741/9781644903377-10

Some of these benchmark layouts are shown in Table (1). Specifically, several different blocked artifacts have been designed and manufactured by 3D printing machines. These artifacts mainly consist of cylinders, planes, and holes whose dimensions and positions can be measured [12,14]. More advanced layouts primarily have a table arrangement. Upon the table, main features (i.e., holes, cylinders, pyramids, etc.) are placed. Additionally, Spitaels et al. [9,10] have developed an Adaptive Benchmark Artifact which can be used for different Additive Manufacturing processes. ISO/ASTM 52902:2019 [15] proposes a significant benchmark method for evaluating the manufacturing process. Different artifacts in specific positions and angles on the manufacturing table of the 3D printer result in a solid conclusion about the accuracy of the machine [10-17].

Benchmark layouts have also been developed for comparing the accuracy of different AM processes. To accomplish this, several important criteria needed to be satisfied [1,8]: i) the printing volume of the final layout should be smaller than the minimum printing volume of the smallest 3D printer, ii) the use of features that can produce results for every AM method, iii) short fabrication time, iv) minimal material usage, and v) printability by every AM method.

There is a significant need in the printing community for a benchmark layout that can be applied to different AM processes and machines. As the number of measurements increases, there is a need to simplify the accuracy measurement procedure of the layout. The novelty of the proposed layout lies in its simplicity and ease of fabrication for both amateur and professional users.

Table 1:Benchmark Artifacts for AM processes and machines

Spitaels et al. [10]	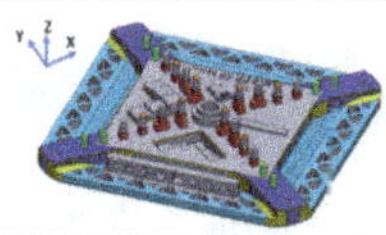
ISO/ASTM 52902:2019 [15]	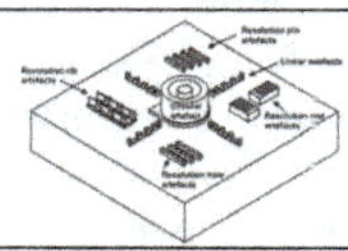
Tanveer et al. (2008) [12]	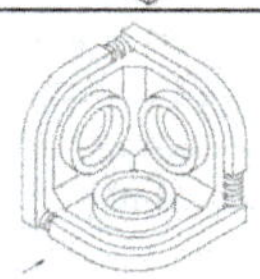
Vorkapic et al. [13]	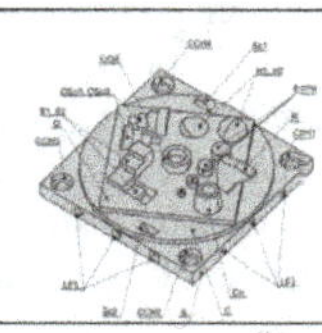
Yang et al. (2016) [19]	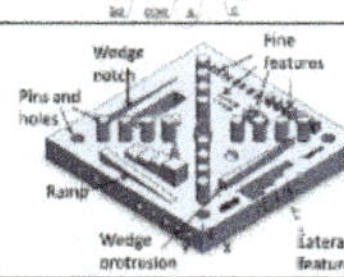

Materials Research Proceedings 46 (2024) 74-81 https://doi.org/10.21741/9781644903377-10

Proposed Concept

Operating Principle. The proposed layout uses three main components: the Base, the Slider, and a Cylinder, as shown in Figure 1. The Slider has a V-grooved slot in which the Cylinder is placed. As shown in the figure, the circumference of the Cylinder makes contact with the inclined surface of the Base (purple line) at a single point, while it also touches the V-grooved slot of the Slider at two points. Due to manufacturing errors from the 3D printing process, changing the orientation of the Cylinder inside the Slider (as depicted in Figure 1) alters its apparent diameter, causing a vertical movement of the Cylinder within the V-grooves of the Slider. This, in turn, moves the Slider laterally along the straight surface of the Base (green line). The lateral movement of the Slider is directly linked to the diameter change of the Cylinder, essentially amplifying the deviation through the use of the inclined surfaces. By repeating this procedure for different orientations of the Cylinder, a maximum and minimum diameter can be calculated, thereby determining its cylindricity. Furthermore, the arrangement ensures that measurements are not affected by warping phenomena, as the printed area of the Cylinder is prevented from touching the inclined surface of the Base.

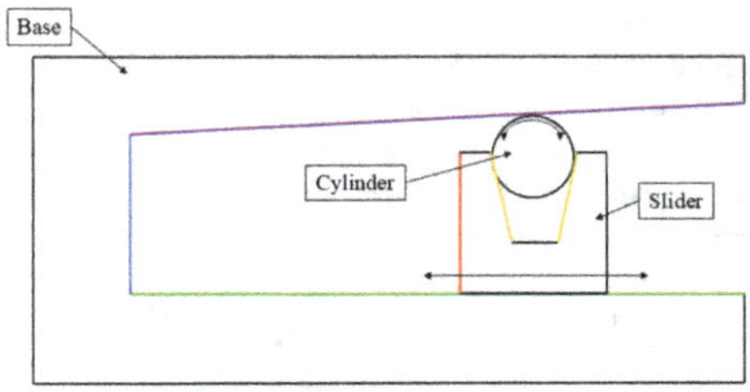

Fig. 1: Proposed Layout

Mathematical Model. To quantify the deviations of the cylinder surface, a mathematical model has been developed. This model aims to establish a correlation between the true diameter of the Cylinder and its position relative to the Slider and the Base.

The model is based on the following geometrical considerations:

The perimeter of the Cylinder makes point contact with the inclined surfaces of the Slider (highlighted in yellow in Figure 1) at two specific points, denoted as A and B in Figure 2.

As the Cylinder-Slider assembly moves within the Base, it makes a point contact with the inclined surface of the Base (highlighted in purple in Figure 1) at point C, as illustrated in Figure 2.

The apparent diameter of the cylinder is calculated from these contact points (A, B, and C).

This process is repeated eight times for eight equidistant positions of the cylinder along its circumference. As mentioned above, due to manufacturing errors in the Cylinder, the diameter values will vary, allowing us to assess its cylindricity. The diameter of the Cylinder (eq.1) can be expressed as a function of some constant dimensions of the proposed layout, as illustrated in Figure 2. These constant dimensions are denoted by red characters in the figure.

Materials Research Proceedings 46 (2024) 74-81 https://doi.org/10.21741/9781644903377-10

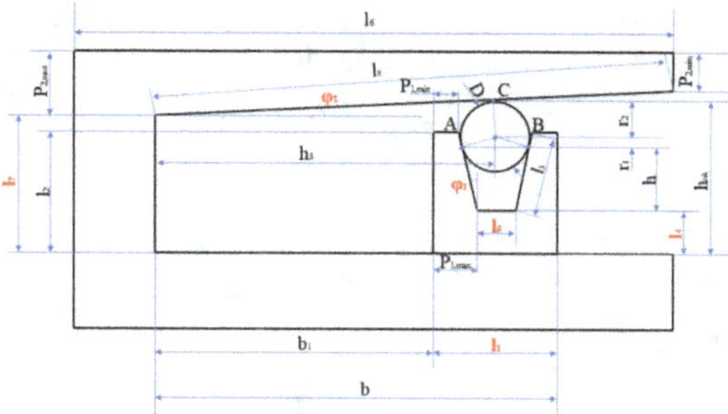

Fig. 2: Dimensions of the Layout

$$D = \frac{b_1 + l_1 - \left(\frac{l_1}{2} + \frac{l_4 - \frac{l_5}{2tan\varphi_1} - l_7}{tan\varphi_2} \right)}{\sigma} \tag{1}$$

where σ is a shape factor and is defined as:

$$\sigma = \frac{cos\varphi_2(1 + cos\varphi_2 \cdot sin\, sin\, \varphi_1)}{2 \cdot sin\varphi_1 \cdot sin\varphi_2} + \frac{sin\varphi_2}{2} \tag{2}$$

The cylindricity of the Cylinder is then calculated as:

$$c = d_{max} - d_{min} \tag{3}$$

where d_{max}, d_{min} is the maximum and the minimum measured diameter.

As a result, the measured diameter of the Cylinder can be calculated by simply measuring the length b_1 between the vertical surface of the Base (blue line in Figure 1) and the vertical surface of the Slider (red line in Figure 1). The sensitivity of the layout is primarily influenced by the angle of the interior slopes of the Slider (yellow lines in Figure 1). The angle of the interior upper surface of the Base (purple line in Figure 1) affects the final measurement of the diameter accordingly. Based on the equation for the shape factor σ, an inverse relationship can be inferred between the steepness of the slope and the sensitivity of the layout. Specifically, as the slope angle increases, the sensitivity of the layout to dimensional variation decreases.

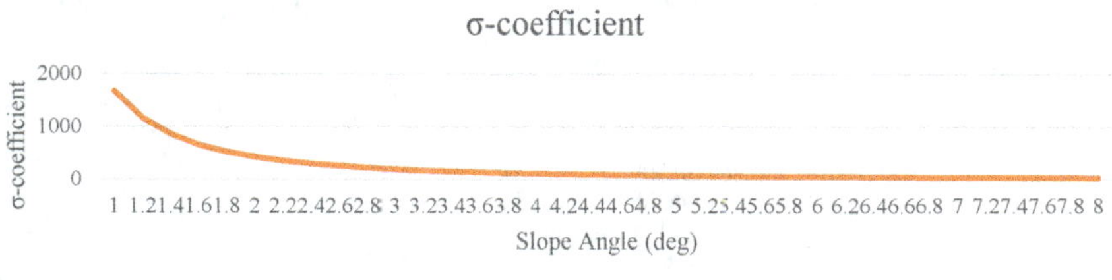

Fig. 3: σ-coefficient sensitivity

Experimental Evaluation

Printing Parameters. The entire benchmark layout and test specimens were designed using Solidworks software. The resulting designs were exported as STL files, which were subsequently imported into CURA slicing software for pre-processing. The sliced models were then converted into G-code files and transferred to a Creality Ender 3 fused deposition modeling (FDM) 3D printer for fabrication. The specific printing parameters employed in this study are detailed in Table 2.

Table 2: Printing Parameters

Filament Material	PLA (1,75 mm)
Bed Temperature [°C]	50
Nozzle Temperature [°C]	200
Printing Speed [m/s]	50
Infill [%]	20
Layer Height [mm]	0.2
Wall Layers	4
Retraction	Activated
Retraction Distance [mm]	5
Supports	Not Needed
Build Plate Adhesion	No

Evaluation of the Assembly using CMM. To evaluate the dimensions of the primary components (Base and Slider) of the assembly, a Coordinate Measuring Machine (CMM) was employed. With a measuring error ranging between ±6 μm, the CMM served as the reference measurement system for this study. All requisite error measurements were generated using the Dimensional Geometric & Tolerancing (DG&T) tools within the PC-DMIS software suite. The comprehensive measurement results for the Base and Slider components are presented in Tables 2 and 3, respectively, and they were incorporated into the corresponding variables of the mathematical model.

Fig. 4: Parts Evaluation

Table 3: Slider Measurements

Surface Tolerances	**Deviation**
Slope Angle [deg]	92.57
Horizontal Surface Flatness [mm]	0.43
Vertical Surface Flatness [mm]	0.09
Inclined Surface Flatness [mm]	0.26
Surface Tolerance	**Deviation**
Right Slope Angle [deg]	92.43
Left Slope Angle [deg]	92.47
Vertical Surface Flatness [mm]	0.11
Horizontal Surface Flatness [mm]	0.10

Experimental Validation. To assess the accuracy of the benchmark layout, twenty (20) cylindrical specimens were fabricated using additive manufacturing. These specimens were subsequently measured using a Coordinate Measuring Machine (CMM), a caliper and the proposed benchmark layout. The measurement protocols for each method are detailed below.

The cylindrical specimens were measured using a CMM at the Machine Design Laboratory (MD-Lab) at the National Technical University of Athens. Each specimen was secured to the

Materials Research Proceedings 46 (2024) 74-81 https://doi.org/10.21741/9781644903377-10

CMM table using a binder clip mounted on a vise, as illustrated in Figure 5. This fixture method ensured stability and accessibility for the CMM's touch probe throughout the measurement process. Thirty points were sampled along the height of the surface of each cylinder (Figure 5). The PC-DMIS software was used to calculate the surface error of the cylinder in accordance with ISO 1101 standards.

Additionally, the tested parts were measured using a caliper with a maximum error of ±20μm at 100mm of measurement and a resolution of 0.01mm, according to the manual of the manufacturer. A specific height, approximately at the middle of the height of the Cylinder, was selected to measure the diameter of the Cylinder. Four different angular positions were chosen for diameter measurement. Using equation (3), the cylindricity of the Cylinders was calculated.

The same cylindrical specimens were then evaluated using the proposed benchmark layout. Each Cylinder was carefully positioned on the Slider, ensuring that the surface of the Cylinder made contact with the Slider without introducing deformations that could affect diameter measurements. The Slider, with the attached Cylinder, was then moved along the horizontal surface of the Base (green line in Figure 1). The sliding motion was halted when the Cylinder made initial contact with the inclined surface of the Base. At this point, the lateral distance between the Slider and the Base was recorded. The exact diameter of the Cylinder, and consequently its cylindricity, were calculated using the equations previously described in this study.

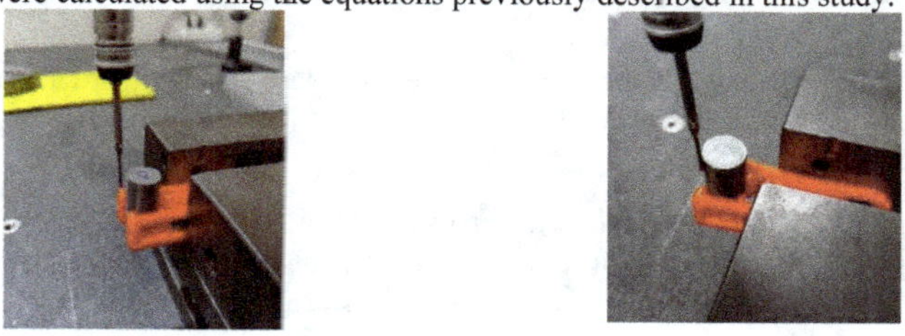

Fig. 5: Cylinder Measurement

A comparative analysis was conducted to evaluate the measurements obtained from the three distinct methodologies. The results from all three methods were consolidated into a single graphical representation, as illustrated in Figure 6. In this graph, the CMM data are represented by blue-colored bars and the proposed layout results by yellow-colored bars. To facilitate direct comparison there is a line, showing the difference between the proposed layout and CMM results. Notable in this visualization is the consistent discrepancy between the proposed layout and CMM measurements, which maintains an approximate value of 0.025 mm across all specimens.

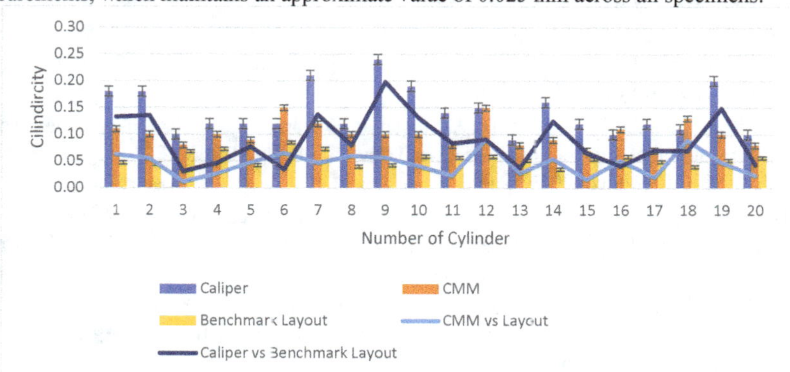

Conclusion
Based on the lack of standardized methods for defining the accuracy of 3D printers, there is a need for designing and manufacturing benchmark layouts to calculate it. Previous artifacts are difficult to measure using common measuring tools, making the whole process complicated. The proposed layout is a simple, self-assessed design in which the accuracy of the 3D printer can be calculated. The measurements taken by a CMM capture points along the circumference of the cylinder in three-dimensional space. These points represent a geometry that can have any shape in space, including variations in curvature and dents at any height, allowing the CMM to accurately calculate the actual cylindricity. The proposed layout, however, does not measure all vertical cross-sections of the cylinder in every orientation. Instead, it measures three contact points that may belong to different heights and provide an equivalent apparent diameter. Although this setup does not determine the theoretical form error as accurately as a CMM, it is particularly useful in practice. In assembly applications, this apparent diameter measured by this setup is critical for successful assembly. The proposed layout accounts for the error in inductivity of the 3D printed surfaces. To overcome this error, the suggested solution is to place the cylinder on the slider and measure the correct distance when the cylinder's surface touches the inclined surface of the base. Therefore, while the form error calculated by the setup may slightly differ from CMM measurements, it may be a more representative measurement in terms of assembly requirements.

Acknowledgement
The authors thank very much the University of Western Macedonia for partial financial support for pursuing and publishing this research.

References
[1] S. Moylan, J. Slotwinski, A. Cooke, K. Jurrens and M. A. Donmez, An additive manufacturing test artifact, Journal of Research of the National Institute of Standards and Technology,119 (2014) 429-459, https://doi.org/10.6028%2Fjres.119.017

[2] M. Kumar, R. Ramakrishnan and A. Omarbekova, 3D printed polycarbonate reinforced acrylonitrile-butadiene-styrene composites: Composition effects on mechanical properties, microstructure and void formation study, Journal of Mechanical Science and Technology, 33 (11) (2019) 5219-5226. https://doi.org/10.1007/s12206-019-1011-9

[3] Chadha, U., Abrol, A., Vora, N. P., Tiwari, A., Shanker, S. K., & Selvaraj, S. K., Performance evaluation of 3D printing technologies: a review, recent advances, current challenges, and future directions. *Progress in Additive Manufacturing*, 7(5), (2022) 853-886, https://doi.org/10.1007/s40964-021-00257-4

[4] Leach, R. K., Bourell, D., Carmignato, S., Donmez, A., Senin, N., & Dewulf, W., Geometrical metrology for metal additive manufacturing. *CIRP annals*, *68*(2), (2019) 677-700, https://doi.org/10.1016/j.cirp.2019.05.004

[5] Haleem, A., Javaid, M., Khan, R. H., & Suman, R., 3D printing applications in bone tissue engineering. Journal of clinical orthopaedics and trauma, 11, (2020) 118-124, https://doi.org/10.1016/j.jcot.2019.12.002

[6] Al-Meslemi, Y., Ferreira, K., Mehdi-Souzani, C., Obaton, A. F., Nouira, H., & Anwer, N. Quality Control for Additive Manufacturing. In Springer Handbook of Additive Manufacturing, (2023) 797-819, https://doi.org/10.1007/978-3-031-20752-5_47

[7] C. Cajal, J. Santolaria, J. Velazquez, S. Aguado and J. Albajez, Volumetric error compensation technique for 3D printers, Procedia Engineering, 63 (2013) 642-649, https://doi.org/10.1016/j.proeng.2013.08.276

[8] P. Minetola, L. Iuliano and G. Marchiandi, Benchmarking of FDM machines through part quality using IT grades, Procedia CIRP, 41 (2016) 1027-1032, https://doi.org/10.1016/j.procir.2015.12.075

[9] Spitaels, L., Rivière-Lorphèvre, E., Demarbaix, A., & Ducobu, F. Adaptive benchmarking design for additive manufacturing processes. Measurement Science and Technology, 33(6), (2022) 064003, https://doi.org/10.1088/1361-6501/ac5877

[10] Spitaels, L., Nieto Fuentes, E., Rivière-Lorphèvre, E., Arrazola, P. J., & Ducobu, F., A systematic method for assessing the machine performance of material extrusion printers. Journal of Manufacturing and Materials Processing, 8(1), (2024) 36, https://doi.org/10.3390/jmmp8010036

[11] de Pastre, M. A., Tagne, S. C. T., & Anwer, N., Test artefacts for additive manufacturing: A design methodology review. *CIRP Journal of Manufacturing Science and Technology, 31*, (2020) 14-24, https://doi.org/10.1016/j.cirpj.2020.09.008

[12] Tanveer, M. Q., Suhaib, M., & Haleem, A., A New 3D Benchmarking Artifact to Evaluate Dimensional Accuracy and Geometric Tolerancing of Additive Manufacturing Technique. In *Recent Advances in Mechanical Engineering: Select Proceedings of NCAME 2019,* (2020) 261-273, https://doi.org/10.1007/978-981-15-1071-7_22

[13] Vorkapic, N., Pjevic, M., Popovic, M., Slavkovic, N., & Zivanovic, S., An additive manufacturing benchmark artifact and deviation measurement method. Journal of Mechanical Science and Technology, 34, (2020) 3015-3026, https://doi.org/10.1007/s12206-020-0633-2

[14] Santos, V. M. R., Thompson, A., Sims-Waterhouse, D., Maskery, I., Woolliams, P., & Leach, R., Design and characterisation of an additive manufacturing benchmarking artefact following a design-for-metrology approach. Additive Manufacturing, 32, (2020) 100964, https://doi.org/10.1016/j.addma.2019.100964

[15] ISO/ASTM 52902:2019(E) – Additive Manufacturing – Test Artifacts – Geometric Capability Assessment of Additive Manufacturing Systems, 2019.

[16] H. Yang, J. C. Lim, Y. Liu, X. Qi, Y. L. Yap, V. Dikshit, W. Y. Yeong and J. Wei, Performance evaluation of ProJet multimaterial jetting 3D printer, Journal Virtual and Physical Prototyping, 12 (1) (2017) 95-103, https://doi.org/10.1080/17452759.2016.1242915

[17] Rebaioli, L., & Fassi, I., A review on benchmark artifacts for evaluating the geometrical performance of additive manufacturing processes. *The International Journal of Advanced Manufacturing Technology*, *93*, (2017) 2571-2598, https://doi.org/10.1007/s00170-017-0570-0

Investigation of the vibrational characteristics of tzouras 3D printed top plates

Spyros BREZAS[1,a,*], Helen PAPADAKI[1,b], Markos KATSIPIS[1,c], Despoina GRIGORIOU[1,d], Yannis ORPHANOS[1,e], Makis BAKAREZOS[1,f], Nektarios A. PAPADOGIANNIS[1,g], Evaggelos KASELOURIS[1,h], Vasilis DIMITRIOU[1,i]

[1]Laboratory of Physical Acoustics and Optoacoustics, Department of Music Technology and Acoustics, Hellenic Mediterranean University, Greece

[a]sbrezas@hmu.gr, [b]ddk134@edu.hmu.gr, [c]ddk233@edu.hmu.gr, [d]ta1970@edu.hmu.gr, [e]yorphanos@hmu.gr, [f]bakarezos@hmu.gr, [g]npapadogiannis@hmu.gr, [h]vagfem@hmu.gr, [i]dimvasi@hmu.gr

Keywords: Vibration, Musical Instrument, 3D Printing, Modal Measurements, Finite Element Method

Abstract. This work describes an integrated method for the investigation of the vibrational characteristics of 3D printed tzouras top plates. The top plates vary in thickness, in tandem to the dimensions of a wooden instrument. The frequency response of the manufactured top plates is determined by modal measurements and the vibration modes are visualized using holographic imaging and further compared to finite element simulation results. The measurements reveal insights applicable to the manufacturing of the instrument by alternative and modern materials. Further analysis includes measuring the vibrational characteristics of a functional wooden tzouras. To gradually implement the transition from traditional to alternative materials, the detached wooden top plate of the instrument is also studied. A correlation criterion is used to compare the responses of the traditional instrument and the top plates for improving their vibroacoustic performance.

Introduction

The evolution of musical instrument manufacturing has been a dynamic process, deeply rooted in traditional craftsmanship and the use of natural materials like wood, which have dominated the industry for centuries [1]. Wood's unique acoustic properties have made it the preferred material for many stringed instruments, including the tzouras, a traditional Greek instrument. However, the increasing demand for innovation, coupled with growing concerns about the sustainability of natural resources, has prompted researchers and manufacturers to explore alternative materials that offer enhanced performance and environmental benefits [2,3]. Such a path for exploration is the integration of 3D printing technologies [2], which enables the production of instrument components with precise control over their structural and acoustic properties. The vibrational characteristics of the stringed instruments or parts of them are of high importance for their modelling and apparently, their acoustic quality. Especially, the soundboard [3] plays a critical role for the vibroacoustic behavior of the instrument.

Based on previous research [2,4], this contribution focuses on the vibrational characteristics of 3D-printed tzouras top plates, intended to replace traditional wooden top plates. Initially, the vibrational behavior of the complete tzouras instrument is presented. The top plate is then removed and measured separately. To compare the vibrational behavior of the complete instrument with that of the isolated top plate, a correlation criterion is applied [5,6]. Additionally, FEM modeling and simulations [3,4,7-9] are conducted to predict the vibrational characteristics of tzouras top plates made of polymer. In parallel, variations in the top plate geometry are explored, leading to the final dimensions of the printed top plates. These printed plates are then compared to the wooden top plate, and the results are used to identify the printed top plate that exhibits vibrational behavior similar to the wooden. Although the acoustic properties of PETG may not match the rich tonal qualities of wood, it is a highly suitable candidate for instrument manufacturing and is being used in this study. PETG is highly durable and resistant to impact and environmental changes, making it less likely to warp. Its lightweight nature enhances portability, and its recyclability helps reduce the demand for wood, promoting sustainability [2].

Measurements and Modelling

In this section the experimental methods and numerical modeling and simulation are presented.

Frequency response measurements. The frequency response of the object under test (OUT) is measured utilizing the roving hammer technique. The OUT is excited at a grid of points by a miniature impact hammer (PCB, Model 086E80) and the response is detected by a miniature accelerometer (PCB, Model TLD352A56). The frequency response function (FRF) is calculated using the estimator H_2:

$$H_2(\omega) = \frac{G_{aa}(\omega)}{G_{Fa}(\omega)} \tag{1}$$

The nominator is the auto-spectrum of the acceleration and the denominator the cross-spectrum between force and acceleration signals. To enable the correlation analysis described in paragraph 2.4, the frequency resolution is set to Δf=0.5 Hz and the upper frequency limit to 3200 Hz.

Normal modes visualization. The normal modes of a vibrating system can be visualized using Electronic Speckle Pattern Interferometry (ESPI). In short, ESPI is a holographic method using two arms of a split original laser beam. The ESPI setup is shown in Fig. 1. A single longitudinal mode laser (532 nm, 170 mW) generates a beam, which is guided by a periscope (PER) and a mirror (M1) to a beam splitter (BS1), where it is split in two arms. One arm illuminates the vibrating object (OUT) after passing through two lenses (L1, L2). A neutral density filter (NDF) controls the intensity of the second arm and by using two mirrors (M2, M3), it enters a beam expander (EXP). A second beam splitter (BS2) is used for the superposition of the expanded arm and the reflection from the OUT. The superposition results in an interferogram, which is captured by a CCD camera (CAM). Further image analysis is implemented by a custom-made software. A signal generator (GEN) feeds an active loudspeaker (SPK) and the emitted signal excites the OUT. The input voltage is recorded by an oscilloscope (OSC).

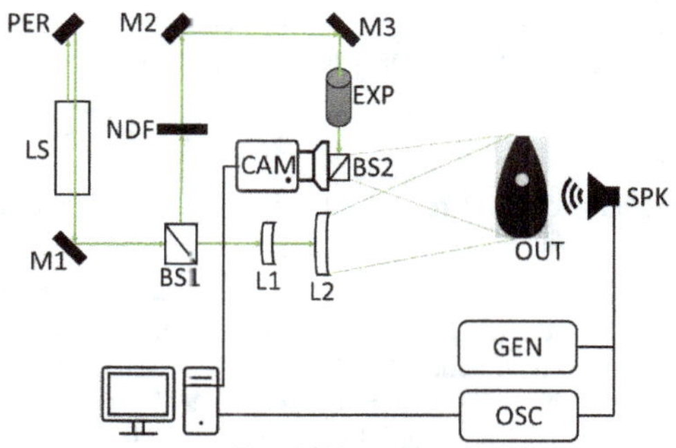

Fig. 1 ESPI setup

Correlation criterion. Many criteria have been proposed for the comparison between vibrating objects or states of the same vibrating object. For the correlation between a pristine state (denoted as p) and an altered state (denoted as a), Perez et al. [6] proposed the complex frequency domain assurance criterion (CFDAC). The CFDAC is calculated by:

$$\mathrm{CFDAC}_{fg} = \frac{\left[\sum_{p=1}^{N}\sum_{q=1}^{N} h_{pq}^{(p)}(\omega_f) h_{pq}^{(a)}(\omega_g)\right]^2}{\left[\sum_{p=1}^{N}\sum_{q=1}^{N} h_{pq}^{(p)}(\omega_f) h_{pq}^{*(p)}(\omega_f)\right]\left[\sum_{p=1}^{N}\sum_{q=1}^{N} h_{pq}^{(a)}(\omega_g) h_{pq}^{*(a)}(\omega_g)\right]} \in \mathbb{C} \tag{2}$$

where $h_{pq}^{(p)}(\omega_f)$ is the FRF of the pristine state when the system is excited at point p and the response is recorded at point q for frequency f, $h_{pq}^{(a)}(\omega_g)$ is the FRF of the pristine state when the system is excited at point p and the response is recorded at point q for frequency g. Asterisk in the denominator stands for complex conjugate. For mathematical consistency, it is recommended that the FRFs have the same number of frequency lines ($N_f = N_g$) and the same resolution. Following the same study, the resolution was set to 0.5 Hz.

The degree of correlation can be given in terms of the spectral correlation index (SCI) as follows. Initially, the Pearson correlation coefficient (PCC), which is the normalized form of covariance, is calculated using:

$$\mathrm{PCC}_{A,B} = \frac{\sum_{f=1}^{N_f}\sum_{g=1}^{N_g} A_{fg}B_{fg} - N_f N_g \bar{A}\bar{B}}{(N_f N_g - 1)\sigma_A \sigma_B} \tag{3}$$

where A is the CFDAC of the pristine-to-pristine comparison ($\mathrm{CFDAC}_{h^{(p)},h^{(p)}}$), B is the CFDAC of the pristine-to-altered comparison ($\mathrm{CFDAC}_{h^{(p)},h^{(a)}}$). The terms $\bar{A}$ and $\bar{B}$ correspond to the mean values of the CFDAC matrices. In case of pristine-to-pristine comparison, the mean value is:

$$\bar{A} = \frac{1}{N_f N_g}\sum_{f=1}^{N_f}\sum_{g=1}^{N_g} A_{fg} \tag{4}$$

The same equation applies for the pristine-to-altered comparison. In the same context, σ_A represents the standard deviation of the $\mathrm{CFDAC}_{h^{(p)},h^{(p)}}$ matrix, calculated as:

$$\sigma_A = \sqrt{\frac{1}{N_f N_g - 1}\sum_{f=1}^{N_f}\sum_{g=1}^{N_g}\left(A_{fg} - \bar{A}\right)^2} \tag{5}$$

Accordingly, σ_B is the standard deviation of the $\mathrm{CFDAC}_{h^{(p)},h^{(a)}}$ matrix.

As it may be assumed, the correlation between two CFDACs can be calculated for both real and imaginary parts of the CFDAC matrices. This means that PCC can be valued from -1 to 1. For the present analysis, the value -1 is inconceivable, since it represents inverse frequency correlation. Consequently, the PCC is assumed to get values between 0 and 1. Finally, the spectral correlation index (SCI) is derived as:

$$\mathrm{SCI}_{A,B} = 1 - |\mathrm{PCC}_{A,B}| \tag{6}$$

This transformation was adopted by Perez et. al [6] for convenience reasons. Perfect correlation yields 0, while no correlation yields 1.

FEM modelling and simulations. A 3D CAD model of the top plate from an actual tzouras is developed. The total length of the plate is 260 mm, with the thickness varying between 1.9 mm and 2.9 mm. Two different cases are considered to study the influence of material properties on the vibrational behavior: one with PETG material properties and one with spruce Engelmann wood properties. Regarding the mesh, a hexahedral solid element with 8 nodes is used in the FEM models. The mesh geometry of the plate is modeled with approximately 200,000 elements. The boundary conditions of the soundboard follow the experimental setup, where the plate is fixed at the top part. The PETG material properties used in the simulations are taken from [10], and the wood orthotropic properties from [11]. FEM modal analysis for both cases is performed by LS-DYNA [12].

Results and discussion

Firstly, the FRFs of the tzouras instrument are measured using the modal analysis equipment as previously described. Fig. 2 shows the measurement grid of the instrument under test and the position of the accelerometer. The strings and the bridge are removed for the top plate to vibrate without the imposed tension. Indicatively, a normal mode after ESPI measurement is also shown.

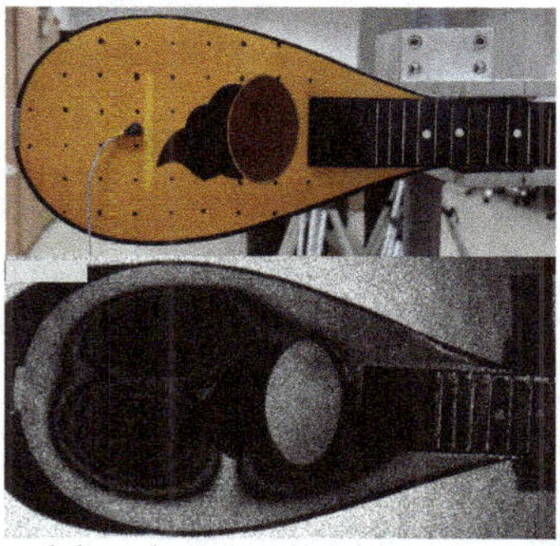

Fig. 2 The tzouras during modal analysis measurements (top) and a vibration mode at 1458 Hz (bottom)

The tzouras traditional wooden top plate is detached and is further used as a reference for the 3D printed top plates, and ESPI and FRF experimental measurements are performed. The FRF measurements are also used for the calculation of the CFDAC, by using the tzouras data as the pristine state and it's wooden top plate as the altered state. Fig. 3 shows the comparison of the FRFs and the resulting CFDAC. In the CFDAC plots, the frequency of the pristine and the altered state are shown in the x-axis and y-axis respectively. In these types of comparisons, when two identical states are compared, the CFDAC graph displays diagonal symmetry, which is not observed in the case of non-identical states. The SCI value is found to be 0.722. This variation was obviously expected, since different boundary conditions are applied on the isolated top plate with reference to its initial assemble on the body of the instrument. This SCI value denotes the relevance of the vibroacoustic performance of the assembled instrument to its removed and isolated top-plate after disassembling.

Two top plates are 3D-printed using PETG, based on a CAD generated stl file, designed by copying the geometry of the wooden top plate. For the manufacturing of the polymer top plates, the Fused Deposition Modeling (FDM) Sunlu S9 Plus 3D printer, with printing dimensions of 310 x 310 x 400 mm and a resolution of ~100 μm, is used. The thickness of the printed top plates are 1.9 mm and 2.9 mm and the FRFs are acquired for both. The wooden top plate is used as the pristine state data and the PETG top plates data as the altered state, and the corresponding CFDAC's, are calculated. The SCI value for the comparison between the wooden top plate and the 3D printed top plate with 1.9 mm is 0.801, whereas for the case of 2.9 mm thickness this value becomes 0.759. As long as the SCI value approaches the reference value of 0.722 a vibroacoustic behavior similar to the traditional instrument is approached. Fig. 4 shows the comparison of the FRFs and the resulting CFDAC.

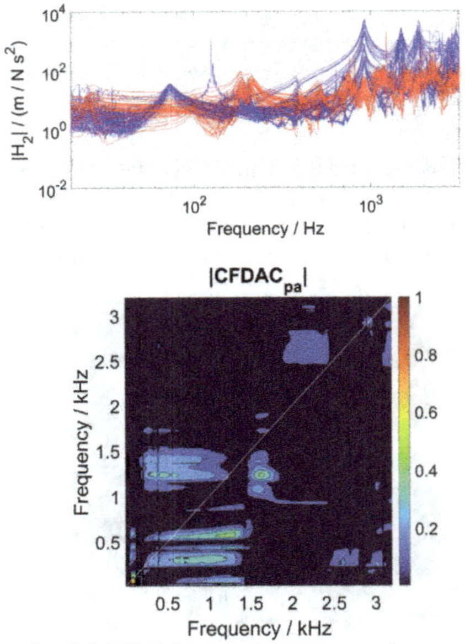

Fig. 3 Top - Magnitude of the FRF for measurements on the tzouras complete instrument (blue) and the tzouras wooden top plate (red). Bottom - CFDAC using the above signals

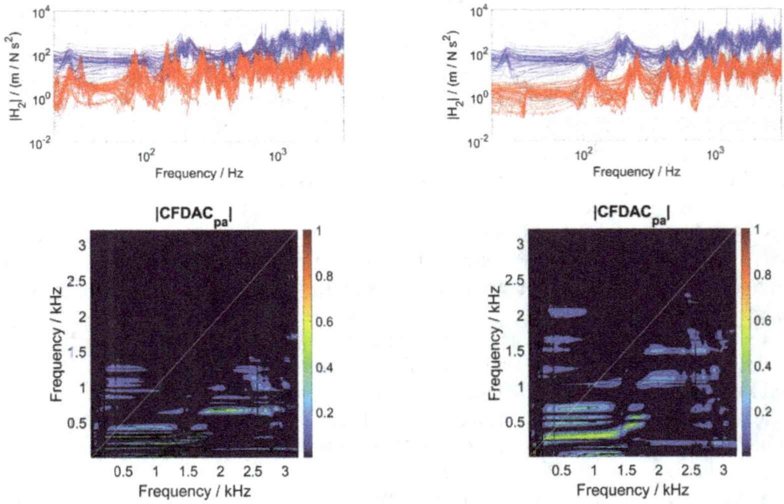

Fig. 4 Top left - Magnitude of the FRF for measurements on the tzouras wooden top plate (blue) and the tzouras PETG top plate with 1.9 mm thickness (red). Bottom left - CFDAC using the above signals. Top right - Magnitude of the FRF for measurements on the tzouras wooden top plate (blue) and the tzouras PETG top plate with 2.9 mm thickness (red). Bottom right - CFDAC using the above signals

Materials Research Proceedings 46 (2024) 82-89 https://doi.org/10.21741/9781644903377-11

Figure 5 shows representative vibration modes of experimental ESPI measurements and their corresponding FEM simulation results. For a frequency range of 20 Hz to 2 kHz, the maximum percent error between measured and the simulated resonances is less than 7%.

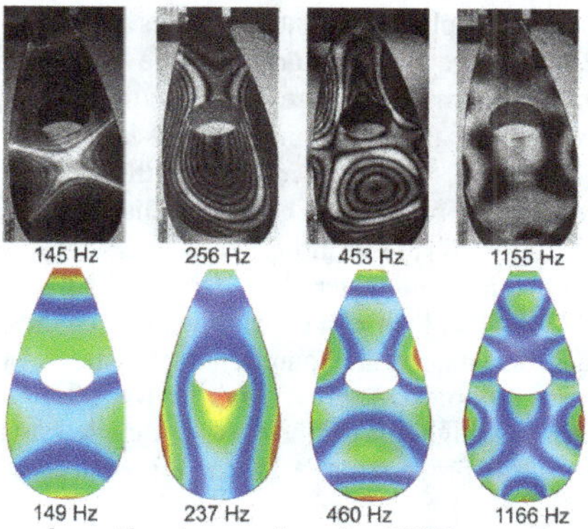

Fig. 5 PETG tzouras top plate vibration modes after ESPI measurements (top) and FEM simulations (bottom)

The ESPI experimental measurements performed at the wooden tzouras (see Fig. 2) demonstrate that the upper triangular area of the top plate, which is bounded (glued) with the tzouras body (back surface) and the fingerboard (top surface), does not contribute to the vibroacoustic behavior of the instrument. Therefore, this part of the instrument is further excluded from the experimental and the simulation studies that follow. To achieve this, a fixed support is opposed at this area to represent the real boundary conditions, as shown in Fig. 6 (left), and the same conditions are introduced to the FEM model. Representative experimental vs simulation modal results are shown in Fig. 6 (right), demonstrating a good modal agreement. These research findings, along with the development of accurate FEM models, will further lead to the guided assembling of the newly manufactured PETG top-plates on the tzouras traditional wooden body, and furthermore to the manufacturing of hybrid musical instruments with predicted and pre-defined vibroacoustic behavior.

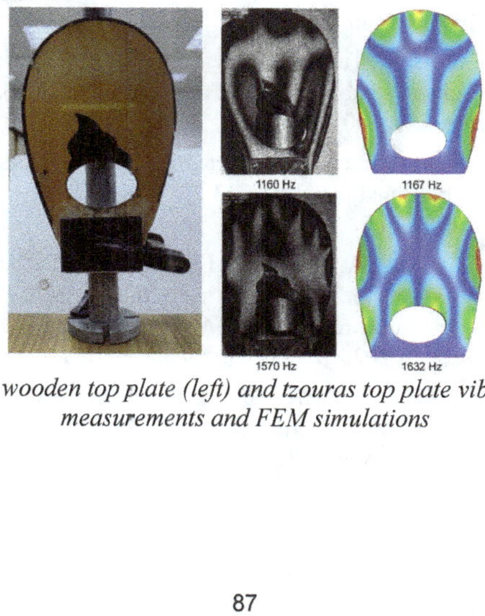

Fig. 6 Clamped tzouras wooden top plate (left) and tzouras top plate vibration modes after ESPI measurements and FEM simulations

Conclusions

The vibrational characteristics of a traditional wooden tzouras, its top plate, and two 3D-printed top plates, are studied by experimental measurements and numerical simulations. The PETG printed top plates are compared to the wooden, the printed top plate that exhibits a vibrational behavior similar to the wooden plate is identified. This work provides valuable mechanical acoustic insights applicable to the manufacturing of the instrument, particularly when using alternative materials such as polymers. The research findings, along with the development of accurate FEM models, will guide the machining and assembling of the new top plates on the tzouras traditional wooden body, and furthermore to the manufacturing of hybrid musical instruments with pre-defined vibroacoustic behavior characteristics, predictable from FEM simulations. The future study of 3D printing parameters of the top plates, like density, directionality, printing temperature and material properties will allow for a thorough investigation of their vibroacoustic behavior, and lead to the manufacturing of the optimal 3D-printed top plate that will best replicate and/or enhance the sound characteristics of the traditional tzouras. This work demonstrates the capability of 3D-printed materials to be used as an alternative to wood for tzouras top plates manufacturing highlighting the need for detailed studies on the density, thickness and directionality of the printing process for PETG to vibroacoustically behave like wood.

Acknowledgments

This research was funded by the Hellenic Mediterranean University, within the project "Recording and metrological analysis of the vibro-acoustic characteristics of musical instruments for the investigation of alternative and low-cost materials and geometries with relevant sound characteristics". We acknowledge the support with computational time granted by the Greek Research & Technology Network (GRNET) in the National HPC facility ARIS-under project ID pr016025-LaMPIOS III.

References

[1] V. Bucur, A Review on Acoustics of Wood as a Tool for Quality Assessment. Forests 14 (2023) 1545. https://doi.org/10.3390/f14081545

[2] S. Brezas, M. Katsipis, K. Kaleris et al. Review of manufacturing processes and vibroacoustic

assessments of composite and alternative materials for musical instruments. Appl. Sci. 14 (2024) 2293. https://doi.org/10.3390/app14062293

[3] E. Kaselouris, M. Bakarezos, M. Tatarakis, N.A. Papadogiannis, V. Dimitriou, A review of finite element studies in string musical instruments, Acoustics 4 (2022) 183-202. https://doi.org/10.3390/acoustics4010012

[4] S. Brezas, M. Katsipis, Y. Orphanos et al. An integrated method for the vibroacoustic evaluation of a carbon fiber bouzouki, Appl. Sci. 13 (2023) 4585. https://doi.org/10.3390/app13074585

[5] R. Viala, M. A. Pérez, V. Placet, A. Manjón, E. Foltête, S. Cogan. Towards model-based approaches for musical instruments making: validation of the model of a Spanish guitar soundboard and characterization features proposal. Appl. Acoustics, 172 (2021) 107591. https://doi.org/10.1016/j.apacoust.2020.107591

[6] M. A. Pérez, R. Serra-López. A frequency domain-based correlation approach for structural assessment and damage identification. Mech. Syst. Signal Process. 119 (2019) 432–456. https://doi.org/10.1016/j.ymssp.2018.09.042

[7] S. Brezas, E. Kaselouris, Y. Orphanos et al. Vibrational Analysis of a Splash Cymbal by Experimental Measurements and Parametric CAD-FEM Simulations, Vibration 7 (2024) 146-160. https://doi.org/10.3390/vibration7010008

[8] E. Kaselouris, C. Alexandraki, M. Bakarezos, M. Tatarakis, N. A. Papadogiannis, V. Dimitriou, A Detailed FEM Study on the Vibro-acoustic Behaviour of Crash and Splash Musical Cymbals, Int. J. Circuits Syst. Signal Process 16 (2022) 948-955. https://npublications.com/journals/articles.php?id=390

[9] N. Efstathopoulos et al, Investigation on the distal screw of a trochanteric intramedullary implant (Fi-nail) using a simplified finite element model, Injury 41 (2010) 259-265. https://doi.org/10.1016/j.injury.2009.09.006

[10] J.M. Mercado–Colmenero and C. Martin–Donate. A novel geometric predictive algorithm for assessing compressive elastic modulus in MEX additive processes, based on part nonlinearities and layers stiffness, validated with PETG and PLA materials. Polymer Testing 133 (2024) 108389. https://doi.org/10.1016/j.polymertesting.2024.108389

[11] B. Kasal, WOOD FORMATION AND PROPERTIES/ Mechanical Properties of Wood, Encyclopedia For. Sci. (2004) 1815-1828. https://doi.org/10.1016/B0-12-145160-7/00041-7

[12] J.O. Hallquist. LS-DYNA Theory Manual, Livermore Software Technology Corporation, 2006.

Vibration signal and roughness correlation during Ti alloy milling using uncoated and coated Ti-based cutting tools

RCM SALES-CONTINI[1,2, a,*], L. ROCHA[1, b], NPV SEBBE[2,3, c], G.F. PINTO[2, d], FJG SILVA[2,4, e], H.M. LOPES[2,4, f], F. FERNANDES[1, g]

[1]Technological College of São José dos Campos, Professor Jessen Vidal, Centro Paula Souza, São José dos Campos, São Paulo, Brazil

[2]CIDEM, ISEP, School of Engineering, Polytechnic of Porto, Porto, Portugal

[3]Facult of Engineering, University of Porto, Department of Mechanical Engineering, Porto, Portugal

[4]Associate Laboratory for Energy, Transports and Aerospace (LAETA-INEGI), Porto, Portugal

[a,]rcms@isep.ipp.pt, [b]1210201@isep.ipp.pt, [c]napvs@isep.ipp.pt, [d]gflp@isep.ipp.pt, [e]fgs@isep.ipp.pt, [f]hml@isep.ipp.pt, [g]fid@isep.ipp.pt

Keywords: Ti-Based Coated Tools, Wear, Vibration, Ti Alloys

Abstract. This work evaluates the coated and uncoated (4 flutes) tools´ performance in the milling of Ti-6Al-4V alloys in a lubricated environment using three different cutting speeds and two different feed rates. This was planned because there is no conclusive analysis in the literature between the variation of the vibration signal, the quality of the milled surface and the tool performance. Concerning tool wear, the vibration analyses were correlated with surface roughness and optical microscopy results.

Introduction

The aerospace industry has a particular interest in the study of titanium superalloys due to their high resistance to hostile environments, creep and corrosion at high temperatures [1] and can be used as structural components such as compressor blades, discs, castings and gas turbine engines. Projects related to the Boeing 787 and Airbus A380 planes reported the growing use of these alloys in the aeronautical sector [2]. Nonetheless, the adoption of titanium alloys is greatly hindered by their higher cost than other materials and their relatively challenging machinability. Thus, the advantages of using titanium alloys must be weighed against this extra expense [3,4]. To use these materials effectively, they need to go through a finishing process after manufacturing to create structural parts, which involves various machining tasks. For the structure to achieve a shape that closely resembles the final design of the component, it must undergo multiple machining steps, such as milling, drilling, and grinding, to achieve the necessary dimensional tolerances for assembly.

Superalloy machining poses various challenges. Machining titanium alloys often results in low output, high costs, short tool lifespan, and inadequate surface quality [5]. Common problems faced during machining include (i) material wear due to continuous chip interaction with material surface; (ii) the high feed force produced during machining that can cause metal chips to pile up, negatively impacting the part's finish quality; (iii) the metal layer sticking to the cutting tool's edge [6]; and (iv) the high temperatures generated during machining create burrs in metals, making it difficult to achieve precise tolerance levels due to the varying elastic moduli and thermal expansion rates of the laminate's materials [7]. Due to the strong compressive forces on the cutting edge from their high strength and hardness, the cutting tool wears down rapidly and undergoes plastic deformation, leading to premature tool failure [8]. In addition, their low thermal conductivity increases the temperature at the junction between the tool and the workpiece, which affects the tool's lifespan. Moreover, the strong resistance at high temperatures, along with the low elasticity, further reduces their machinability [9].

One of the main secrets to minimising these effects is the selection of the correct tool material, shape, design and coating. Selection of the right parameters and condition monitoring can help to detect and characterise tool wear and damage that occurs during use [10]. The use of vibration analysis can detect changes in signal intensity during the machining process and potentially indicate damage. It involves the careful examination of different frequencies and amplitudes and can identify subtle changes in signal intensity that could potentially indicate underlying damage or wear. It is proving to be an extremely valuable technique for thoroughly assessing the operational efficiency and effectiveness of tools in use. This technique has the potential to reduce operating costs by extending the useful life of tools and equipment, reducing machining time and discarding parts with machining defects. This agrees with Agenda 2030 - Goal 12 [11], which refers to reducing the manufacturing process's cost, thus enabling a more sustainable production environment. Therefore, this work correlates the vibration signal and roughness during the Ti alloy milling process and the coated and uncoated Ti-based cutting tools' influence on the roughness of the machined surface.

Materials and Methods

Material, Tooling, and Machining Process

Annealed Ti 6Al 4V alloy, grade 5, is an alpha-beta alloy (ArcelorMittal), with dimensions of 100 x 50x50 [mm], a density of 4.43g/cm^3 and an average hardness of 334 HB, was the material to be machined. For the annealed Ti 6Al 4V alloy machining process, a solid carbide mill tool with 6 mm of diameter was used. The tool's characteristics are: i) High-performance cutting of superalloys, 4 flutes, helix angle 35/38°, coating hardness 3100 HV, unequal division (Figure 1), coated with TiTAN (trade name) produced by Inovatools (Leiria, Portugal). The tool´s characteristics can be seen in Table 1. Both tools have the same dimensions and characteristics.

Table 1. Tool´s type, designation, coating, and dimensions

Tool Designation	Coating	D1 [mm]	D2 [mm]	D3 [mm]	L1 [mm]	L2 [mm]	L3 [mm]	Flutes
266	TiTaAlN	6	6	5.50	57	13	21	4
Uncoated	none	6	6	5.50	57	13	21	4

A 5-axis CNC machining centre called HAAS VF-2 (HAAS Automation, Oxnard, CA, USA) was used for the milling procedure with a speed limit of 10,000 rpm and a power limit of 20 kW. The milling machine set-up is shown in Figure 2.

Figure 1. a) Coated tool side and top view and b) Uncoated tool side and top view.

The following factors were taken into account when conducting the machining tests: cut length (5000[mm]), cut width (Ae=4.6[mm]), cut depth (Ap=0.5[mm]), cutting speed (Vc=60, 70, and 80 [m/min]), and feed rate (fz=0.03, 0.07, and 0.10 [mm/tooth]). A cutting fluid, Alusol SL 61 XBB, was used during the milling process. The tool's coating and cutting speed were used to identify the tools for each test, i.e, CT60 = coated tool and Vc= 60 m/min.

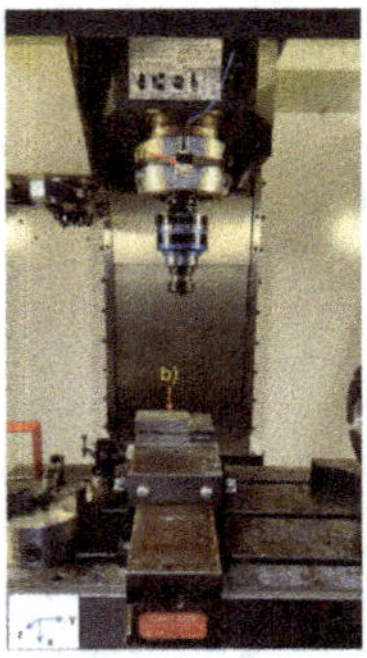

Figure 2. Milling machining set-up: a) triaxial accelerometer and b) annealed Ti 6Al 4V alloy specimen.

Vibrational Analysis

To ascertain the impact of machining parameters on tool wear behaviour, the vibrational signal generated on the X, Y and Z axes was captured by the PCB-PIEZOTRONICS model 356A15 triaxial accelerometer connected to a National Instruments model USB-4431 data acquisition board and sent to the Lab View software where it was recorded. The device was fixed to the CNC machining centre spindle (Figure 2). The acquisition frequency, which facilitated the recording of signals emanating from the tool's cutting edges, was determined in conjunction with the spindle velocity. Upon the collection and subsequent analysis of the data, it became feasible to detect variations in the vibrations that signified the onset of any wear phenomena in the tool that warranted further examination. This methodology was employed to discern wear phenomena that manifested during the milling operation, specifically at a cutting length of 3.0 meters, and subsequently at 7.5 meters towards the conclusion of the milling process. The Root Mean Square (RMS) was computed for each experimental trial [12]

Machined Surface Roughness Analysis

Using ISO 4288:1996 [13], a Mahr Perthometer M1 profilometer (Mahr, Gottingen, Germany) was employed to gather surface roughness information under each operational condition. Its cutoff value measures 0.8 mm, and the total length is 5.6 mm. After the machining process, the average roughness values (Ra) were recorded. This made it possible to assess the cutting process's stability, the tool's effectiveness, and how these elements related to the wear and machining conditions that followed.

Tool Wear Analysis Using Optical Microscopy

After conducting machining tests, the parameter's effect on tool deterioration phenomena was analysed using optical microscopy to determine these patterns on the tool's surface based on ISO 8688-2 [14]. The optical microscopy method employed was a Dino-Lite® Edge digital microscope - model WF4915ZT - wireless (220x zoom).

Results and Discussion

To evaluate the performance of various tools throughout the annealed Ti 6Al 4V alloy milling process utilising pre-hardened tool steel, the tool deterioration phenomena were meticulously examined employing optical microscopy techniques. Table 2 elucidates the visual representations obtained after the milling process. The CT and UCT tools, when subjected to a cutting speed of 60m/min, exhibited superior performance during the milling operations, resulting in reduced damage to the tool surface in contrast to the other tools; however, the UCT tools exhibited a more pronounced deterioration at elevated feed rate values.

Increasing the cutting speed to 70m/min, the tool deterioration phenomena severity observed on the flank of the tools also exhibited an escalation. The CT (Fz 0.03 mm/tooth) manifested only minor scratches on its surface, whereas the UCT, at the same Fz, demonstrated uniform flank wear, and the UCT (Fz 0.10mm/tooth) exhibited both built-up edge (BUE) and localised flank wear. The BUE refers to the machined material accumulation on the tool's edges, which results in the "built-up edges" [15].

Consider increasing in cutting speed to 80m/min, the images at high Fz (0.10mm/tooth) reveal that both the CT and UCT tools experienced the loss of their tips; however, the UCT tool encountered a catastrophic failure, whereas the CT tool sustained non-uniform flank wear and remained operational until the conclusion of the process.

Table 2. Optical microscopy tool images captured after the milling process.

Tool Designation	Fz 0.03	Fz 0.07	Fz 0.10	Tool Deterioration Phenomena
CT60				Fz 0.03= Scratches on the coating surface Fz 0.07= Localized flank wear Fz 0.10=Localized chipping
UCT60		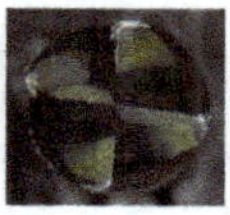		Fz 0.03= Scratches on the tool surface Fz 0.07= Localized flank wear Fz 0.10=Crater wear
CT70				Fz 0.03= Scratches on the coating surface Fz 0.07= Irregular cracks Fz 0.10=Non-uniform flank wear
UCT70				Fz 0.03=Uniform flank wear Fz 0.07= Non-uniform chipping Fz 0.10= BUE and Localized flank wear
CT80				Fz 0.03=Irregular cracks Fz 0.07= Localized flank wear Fz 0.10= Non-uniform flank wear and BUE
UCT80	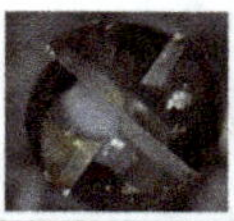	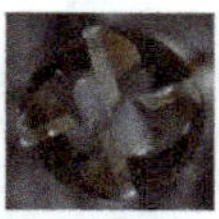	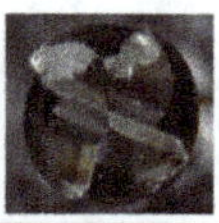	Fz 0.03= Uniform flank wear Fz 0.07= Non-uniform flank wear Fz 0.10= Catastrophic failure

The surface roughness was quantified upon completion of the milling operation. Figure 3 illustrates the measurements of average surface roughness.

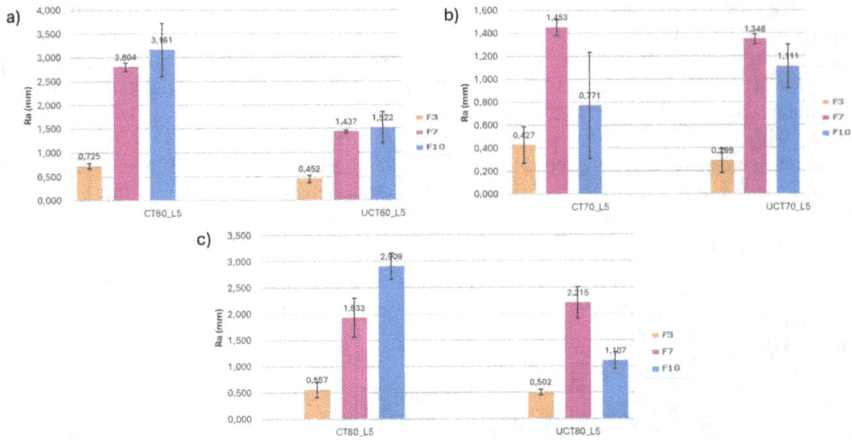

Figure 3. Annealed Ti 6Al 4V alloy R_a average values after milling using the cutting speed of a) 60 m/min, b) 70 m/min, and c) 80 m/min.

Roughness measurements obtained from the material surface subjected to cutting velocities of 60 and 80 mm/min indicated an upward trend in values for coated tools as the feed rate increased. Nevertheless, when juxtaposed with the uncoated tools operating under equivalent cutting speeds and feed rates, the tool surface exhibited minimal damage. In contrast, the uncoated tools displayed more pronounced surface degradation; however, their roughness values were inferior to those of the coated tools.

Throughout the machining process, the chip accumulation along the uncoated tool length facilitated the specimen's surface polishing, as seen in Figure 4, which encouraged abrasive wear [15]. The uncoated tool demonstrated suboptimal efficiency in chip removal.

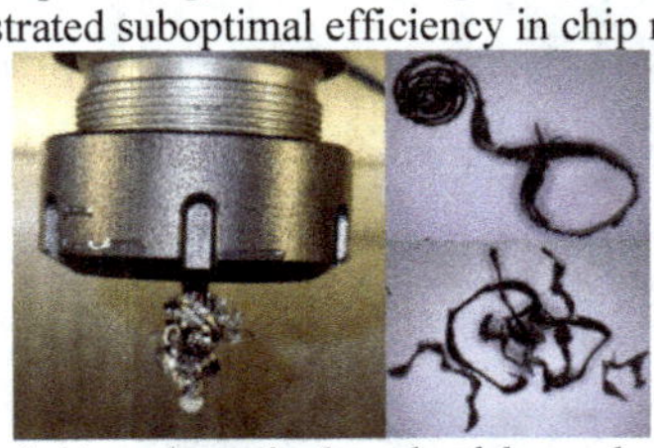

Figure 4. Accumulation of chips along the length of the tool and example of chip length produced by coated tool.

Concerning the observed reduction in roughness values for the CT70 and UCT 80 tools, this effect can be attributed to the tools´ tips breakage, resulting in a more rounded morphology.

The signals have been meticulously recorded by the triaxial accelerometer along the X, Y, and Z axes, and the resultant velocity RMS has been presented in a detailed manner in Figure 5. When examining the movements on the X and Z axes, one can distinctly observe both the discordant and concordant movements exhibited by the tool as it interacts with the material surface, providing valuable insights into its operational dynamics. Conversely, the Y-axis is solely attuned to detecting the discordant tool´s movements, as it should be noted that the concordant movement tends to produce a significantly reduced level of vibration, resulting in a lack of sensitivity in that particular axis.

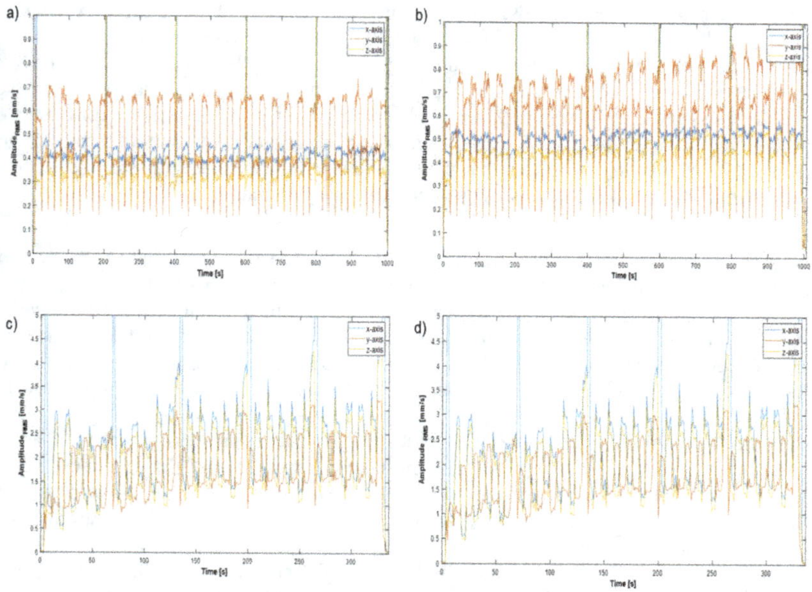

Figure 5. Vibrational analysis of cutting tools during the milling process. a) CT60Fz0.03, b) UCT60Fz0.03, c) CT80Fz0.10, and d) UCT80Fz0.10 (blue - velocity X, RMS [mm/s]; red - velocity Y, RMS [mm/s]; yellow - velocity Z, RMS [mm/s]).

The vibrational signal that was meticulously captured during the intricate milling process, which was executed by the specific tools designated as CT60Fz0.03, UCT60Fz0.03, CT80Fz0.10, and UCT80Fz0.10 is comprehensively illustrated in Figure 5. On the Y-axis, represented by the red line within Figure 5.a, one can distinctly observe a noticeable increase in the vibrational signal when making a direct comparison between the two tools, CT60Fz0.03 and UCT60Fz0.03, which is attributed to the severity of the scratches that have developed on the tool´s surface as detailed in Table 2. In Figure 5.b, the specific vibrational signal corresponding to the UCT60Fz0.03 is presented; this signal demonstrates a significant increase during the initial 50 s and 0.58mm/s amplitude, which strongly indicates the occurrence of damage on the tool´s surface. In the subsequent 350 s with an amplitude of 0.70mm/s to the coated tool and 0.80mm/s to the uncoated tool, although the vibrational signal maintained a consistent level to the coated tool until the end, it is noteworthy that after this period to the uncoated tool, there is a discernible divergence in the vibrational signal that suggests an increase in tool vibration, which is likely caused by the tool´s instability promoted by the aforementioned damage.

Figure 5.c displays the vibrational signal from condition CT80Fz0.10; when this result is compared with Figure 5.a, it becomes evident that the vibrational signal has increased five times, which can be attributed to the cutting speed and the feed rate that together promote instability in the tool throughout the milling process. On the Y-axis, again indicated by the red line, it is possible to observe that the vibration signal exhibits a continuous increase concerning the initial signal captured at the very beginning of the milling process, with an amplitude of 3mm/s for both tools. After 200 s, it becomes apparent that the vibration signal experiences another increase in the amplitude value reaching 4mm/s, which serves as an indication that the tool has indeed suffered some form of damage. Through microscopic analysis, it was confirmed that the tool experienced non-uniform flank wear, as well as a build-up edge (BUE), as substantiated in Table 2.

Regarding the UCT80Fz0.10, as illustrated in Figure 5.d, during the first 25 s, there were two sudden increases in the signals reaching the amplitude of 3mm/s, followed by additional increases at 125 s (3.2 mm/s) and 250 s (3.5 mm/s), with the signal continuing to rise until the conclusion of the milling process. These three significant events can be correlated with the tool´s catastrophic failure (Table 2). This pattern suggests that the tool's integrity was compromised, likely due to excessive thermal and mechanical stresses encountered during operation.

Summary

This study investigates the coated and uncoated tool's performance in the annealed Ti 6Al 4V alloy milling process at cutting speeds of 60, 70, and 80 m/min and feed rates of 0.03, 0.07, and 0.10 mm/tooth. The coated tool exhibited superior performance, showing less damage than the uncoated tool when evaluating tool degradation phenomena according to ISO 8688-2. Although the manufacturer considers the coated tool to be superior, the coating improves tool life by increasing wear resistance but does not result in less roughness on the specimen surface compared to the uncoated tool.

Despite the uncoated tool showing BUE formation, its surface roughness was lower than the coated tools. The BUE on the tool´s surface contributed to wear during the specimen surface polishing. This suggests that while coatings can improve durability, they do not always translate into better surface finish results, highlighting the complex interplay between tool wear mechanisms and machining parameters.

The UCT80Fz0.10 tool performed poorly because, although the roughness values were satisfactory, this was due to poor chips´ cutting, which stuck to the tool and caused a lot of abrasion on the specimen surface. This tool also had poor mechanical resistance due to the lack of coating.

Funding

European Structural and Investment Funds (ESIF) are funding "DRIVOLUTION—Transition to the factory of the future" (DRIVOLUTION C644913740-00000022), which is part of the "Portugal2020" initiative.

Acknowledgements

The authors thank Centro Paula Souza-FATEC/SJC-Brazil, IPP-CIDEM/ISEP-Portugal and FEUP for their institutional support. The authors also thank Rui Rocha of the CEMUP laboratory for his contribution to the SEM analyses and Ricardo Alexandre of INOVATOOLS for providing the tools used in this work, as well as Victor Moreira for his collaboration in the machining tests.

References

[1] D.Kumar, S.Gururaja, I.S. Jawahir. Machinability and surface integrity of adhesively bonded Ti/CFRP/Ti hybrid composite laminates under dry and cryogenic conditions, J. of Manufact. Proc., 58 (2020) 1075-1087. https://doi.org/10.1016/j.jmapro.2020.08.064

[2] M. Nouari, A. Ginting. Wear characteristics and performance of multi-layer CVD-coated alloyed carbide tool in dry-end milling of titanium alloy. Surf Coat Techn., 200:18–19 (2006) 5663–5676. https://doi.org/10.1016/j.surfcoat.2005.07.063

[3] O. Hatt, P. Crawforth, M.Jackson, On the mechanism of tool crater wear during titanium alloy machining, Wear, 374–375 (2017) 15-20. https://doi.org/10.1016/j.wear.2016.12.036

[4] M. Mruthunjaya, K.B. Yogesha, A review on conventional and thermal assisted machining of titanium based alloy, Mat. Today: Proc., 46 (17) (2021) 8466-8472. https://doi.org/10.1016/j.matpr.2021.03.490

[5] A.Shokrani, I.Al-Samarrai, S.T.Newman, Hybrid cryogenic MQL for improving tool life in machining of Ti-6Al-4V titanium alloy, J. of Manufac. Proc., 43 (2019) 229-243.. https://doi.org/10.1016/j.jmapro.2019.05.006

[6] R.Zitoune, V.Krishnaraj, F.Collombet, S. Le Roux Experimental and numerical analysis on drilling of carbon fibre reinforced plastic and aluminium stacks, Compos. Struct., 146 (2016) 148-158. https://doi.org/10.1016/j.compstruct.2016.02.084

[7] O.A.Pawar, Y.S. Gaikhe, A.Tewari, et al. Analysis of hole quality in drilling GLARE fiber metal laminates, Compos. Struct., 123 (2015) 350-365. https://doi.org/10.1016/j.compstruct.2014.12.056

[8] E, Ezugwu, J, Bonney, Y. Yamane. An overview of the machinability of aeroengine alloys. J Mater Process Technol; 134 (2003):233–253. https://doi.org/10.1016/S0924-0136(02)01042-7

[9] Ginting A, Nouari M. Optimal cutting conditions when dry end milling the aeroengine material Ti-6242S. J Mater Process Techn. 184 (2007) 319–324. https://doi.org/10.1016/j.jmatprotec.2006.10.051

[10] M.M. Aguiar, A.E. Diniz, R. Pederiva, Correlating surface roughness, tool wear and tool vibration in the milling process of hardened steel using long slender tools, Intern. J. of Machine Tools and Manufac., 68 (2013) 1-10. https://doi.org/10.1016/j.ijmachtools.2013.01.002

[11] ONU, 2015 Transforming our world: the 2030 Agenda for Sustainable Development, 2015/2018

[12] I.A. Daniyan, I. Tlhabadira, O.O. Daramola, K. Mpofu, Design and Optimization of Machining Parameters for Effective AISI P20 Removal Rate during Milling Operation, Procedia CIRP, 84 (2019) 861-867. https://doi.org/10.1016/j.procir.2019.04.301

[13] ISO 4288:1996 (1996). Geometrical Product Specifications (GPS) – Surface texture: Profile method – Rules and procedures for the assessment of surface texture - International Organization for Standardization, Geneva, Switzerland.

[14] ISO 8688-2 (1989), Tool life testing in milling - Part 2: End milling. International Organization for Standardization, Geneva, Switzerland.

[15] V. F. C. Sousa, F. J. G. Silva, R. Alexandre, J. S Fecheira., F. P. N.Silva. Study of the wear behaviour of TiAlSiN and TiAlN PVD coated tools on milling operations of pre-hardened tool steel, Wear,476 (2021) 203695. https://doi.org/10.1016/j.wear.2021.203695

Realization of prototypes and models using a CNC router

Mariela TODOROVA[1,a *], Tihomir DOVRAMADJIEV[1,b], Darina DOBREVA[1,c], Ventsislav MARKOV[1,d], Ginka ZHECHEVA[1,e], Kremena CANKOVA[1,f]

[1]Technical University of Varna, Studentska str. 1, Bulgaria

[a]m_todorova@tu-varna.bg, [b]tihomir.dovramadjiev@tu-varna.bg, [c]darina.dobreva@tu-varna.bg, [d]ventsi_markov@tu-varna.bg, [e] ginkaa@tu-varna.bg, [f]kremenacankova@tu-varna.bg

Keywords: CNC Routers, Prototyping, Model Making, Ornaments, Reliefs, Precision, Finishing Processes

Abstract. CNC routers represent a modern technology that has revolutionized the way prototyping and model making are carried out. CNC routers are particularly useful for creating complex 3D models and ornaments, making them a preferred choice for designers and engineers. The goal in making signs is to create a unique and impressive design with ornaments and reliefs. The task involves selecting a material, designing in CAD software, generating G-code, setting up the CNC router, and completing finishing processes. The solved problem is the automation of the complex manufacturing process, and the methods include combining ornaments and reliefs, which requires attention to detail and precision. The conclusion highlights the advantages of CNC routers for prototyping and model making, emphasizing high precision and process automation.

Introduction

CNC routers are an innovative technology that has fundamentally transformed the processes of prototyping and model making. They are extremely effective at producing intricate 3D forms and decorative elements, which makes them a favored choice among designers and engineers. The primary goal in sign manufacturing is to create a unique and impressive design featuring ornaments and reliefs. To achieve this, it is necessary to automate the complex production process, combining ornaments and reliefs with precision and attention to detail. The process includes choosing the right material, creating designs in CAD software, generating G-code, configuring the CNC router, and completing finishing touches. CNC routers are widely used in various industries, including furniture manufacturing, signage production, modeling and prototyping, architecture, and the arts. They enable fast and precise production of furniture, including tables, chairs, cabinets, and decorative elements.

CNC routers are made up of several essential parts, such as a spindle that secures and rotates the cutting tool, movement axes (X, Y, and Z), controllers, and computer software for programming and managing the machine. These machines can be configured with different numbers of axes, ranging from 3-axis to more advanced 4- and 5-axis systems, enabling the machining of highly complex and diverse objects [1, 2].

The process of working with a CNC router begins with designing a digital model in CAD (Computer-Aided Design) software [3, 4, 5, 6]. The model is then converted into G-code through CAM (Computer-Aided Manufacturing) software, which contains instructions for the router's movements. The machine follows these instructions to cut, engrave, or shape the material according to the specified design.

Work process

The process starts by utilizing CAD (Computer-Aided Design) software to develop a digital model of the component. This model can be either a 2D drawing or a 3D representation, depending on the project's complexity. Typically, 2D drawings are employed for simpler shapes and contours, while 3D modeling is suited for more intricate and detailed projects. CAD software allows the designer to establish accurate measurements and geometries that form the basis for the following material processing. In the design stage, factors such as dimensions, cutting depth, and tolerances are defined. Additionally, it involves choosing suitable materials and ensuring that the model aligns with the CNC router's capabilities. Once the model is complete, it is imported into CAM (Computer-Aided Manufacturing) software. The CAM software analyzes the model and generates G-code—a set of instructions that will be used by the CNC router. The G-code includes comprehensive details regarding the cutting tool's movements, cutting speed, cutting depth, and other essential parameters. Furthermore, the CAM software optimizes the tool paths to reduce cutting time and enhance process efficiency. This includes minimizing unnecessary movements and ensuring seamless transitions between various operations. After the G-code is prepared, the operator must set up the CNC router for the task. This includes selecting and loading the appropriate cutting tool for the specific operation (such as an end mill, drill, router bit, etc.). The material to be processed is placed on the router's worktable and securely fastened to prevent any movement during cutting. Securing the material can be achieved using vacuum systems, clamping devices, or other methods, depending on the type and size of the material.

The next step involves calibration, which consists of setting the reference points for the axes (starting position). The operator needs to configure the reference points for each axis (X, Y, and Z), which serve as the starting position for the cutting tool. This is essential for ensuring that the tool starts its operation at the correct position and adheres to the precise coordinates specified in the G-code. The positioning of the cutting tool relative to the material is checked, and adjustments are made if necessary. This process involves measuring distances and angles to ensure that the tool is properly calibrated according to the model.

Once all settings and calibrations are completed, the operator starts the CNC router and loads the program containing the G-code. The machine then automatically executes the predefined movements and operations, cutting, engraving, or shaping the material according to the design. During program execution, the operator monitors the process to ensure that everything proceeds as planned and to quickly respond to any issues that may arise. This monitoring involves assessing the quality of the cuts, the precision of the movements, and the state of the cutting tool.

After the cutting is complete, the finished product is inspected for accuracy and alignment with the designed model. This inspection involves measuring key dimensions, assessing surface quality, and performing a visual check for defects. Depending on the project's requirements, additional processes such as sanding, polishing, painting, or other finishing tasks may be necessary. These operations are essential to achieving the final appearance and quality of the product. Upon completion of all tasks, the CNC router and the workspace are cleaned. Regular maintenance of the machine is vital for its long-term operation and to prevent malfunctions.

This detailed CNC router workflow ensures high precision and quality in the manufactured parts while minimizing the time and effort required for production.

Comparison between CNC routers and 3D printers

CNC routers and 3D printers differ fundamentally in their operating principles. CNC routers use a subtractive method, where material is removed from a solid block to achieve the desired shape. The routers cut and engrave the material, removing excess material through mechanical processes. In contrast, 3D printers employ an additive method, building the object layer by layer from plastic, metal, or other materials. This additive process enables the production of intricate shapes that would be challenging to accomplish using traditional methods.

CNC routers are known for their versatility in processing various materials, including wood, plastic, metals, composites, and more. They are favored for projects that require high precision and durability, such as the manufacturing of machine parts, signs, furniture, and prototypes. CNC routers are ideal for large, sturdy, and structurally significant objects. On the other hand, 3D printers are more suited for creating details with intricate geometry and internal structures that would be challenging or impossible to produce with CNC routers. They use materials such as PLA, ABS, PETG, nylon, resin, and even metal powders. Their applications range from prototyping and modeling to producing consumer products, medical implants, and even architectural models.

CNC routers offer high precision, capable of achieving tolerances within a few microns, making them suitable for projects where accuracy is critical. They can also create reliefs and intricate surface details on materials. Similarly, 3D printers provide a high level of precision, though this depends on the printing technology and printer calibration. FDM printers, which use thermoplastic filaments, have lower precision compared to SLA or SLS printers, which work with resins or powders and can achieve very fine details.

CNC routers can process materials relatively quickly, especially for simple shapes and cutting tasks. However, setup and calibration times can be significant. 3D printers generally require more time to build an object, as layers are gradually added. Nevertheless, they offer shorter preparation times and can operate continuously without supervision, making them efficient for creating complex prototypes and small production runs.

Both methods require additional post-processing after the main production phase. For CNC routers, this may involve sanding, polishing, and painting. The dust and chips generated during the process also need to be removed. 3D-printed objects often require cleaning of support structures, sanding of rough surfaces, and sometimes chemical treatment to smooth and enhance the appearance. Objects printed via SLA and SLS technologies may also need further curing or UV treatment to fully harden the material.

CNC routers and 3D printers offer distinct approaches and advantages in manufacturing and prototyping. CNC routers are ideal for projects requiring high precision and strength, while 3D printers excel in creating intricate shapes and details. The choice between the two depends on the specific project requirements, including the type of material, the needed precision, production time, and desired final finish.

Processing different types of wood

When using CNC routers with various types of wood, the outcomes can differ considerably depending on the characteristics of the material. Softwoods such as pine, alder, and basswood are particularly well-suited for CNC machining because of their ease of processing. CNC routers can achieve smooth and clean cuts when shaping these materials, minimizing the risk of cracks and deformations in the final product.

Hardwoods such as oak, beech, and mahogany present their own challenges in CNC processing. These materials are more resistant to tool wear due to their dense and hard structures. To achieve the desired precision and quality, slower cutting speeds are required, helping to avoid overheating and burning the material. CNC routers handle this task with high accuracy, producing finely detailed products with exceptional precision.

Exotic woods like ebony and rosewood pose specific challenges for CNC machining. These materials often have unique textures and colors, making them desirable for special projects. For successful processing with CNC routers, specific settings and appropriate cutting tools are required. Operators must be particularly careful and precise to preserve the integrity of the material and enhance its distinctive qualities.

Despite the differences in processing various types of wood, CNC technology offers significant advantages, such as high precision, repeatability, and the ability to create complex geometries. This makes it a preferred choice for many applications in woodworking and the creation of custom parts and products.

Materials Research Proceedings 46 (2024) 98-105 https://doi.org/10.21741/9781644903377-13

Examples of sign making

The production of wooden door signs using CNC routers is a modern and precise process that combines advanced technology with traditional craftsmanship. These signs not only serve a functional purpose but also play a crucial role in the aesthetic appeal of both interior and exterior spaces, offering uniqueness and personality to each door. CNC routers enable the creation of signs with a variety of shapes, fonts, ornaments, and details that would be difficult to achieve through manual methods. This allows for repeatability and the capability for mass production of individual components or templates used in the production of signs and plaques

The process begins with designing the sign in CAD software, which allows for detailed planning of text, images, and ornaments to be carved into the wood. The software facilitates precise planning of dimensions, layout, and cutting depth, ensuring that the final result fully meets the client's specifications. Once the design is complete, it is converted into G-code, which directs the CNC router, specifying the machine's movements and actions.

The CNC router employs various types of cutting tools depending on the wood type and the details to be carved. It can engrave text, cut complex shapes, and even create relief images on the sign. This level of precision is particularly important for producing customized signs, such as those with names, house numbers, or company logos. Furthermore, CNC routers operate with exceptional accuracy, minimizing errors and ensuring consistency in the quality of each order.

Fig. 1. An example of a sign combining text, ornaments, and irregular shapes

Wooden signs crafted with CNC routers find extensive use in both residential and commercial settings. In homes, they are frequently employed to designate various rooms such as entryways, children's rooms, or workspaces, contributing to a sense of coziness and individuality. In commercial environments, these signs play a role in establishing brand identity. For instance, wooden signs featuring company logos or service descriptions can impart a traditional and professional appearance to the entrance of a store or office.

The significance of wooden door signs extends beyond their functional role of identification. They also provide aesthetic value, enhancing the visual appeal of a space. Wood, being a material associated with warmth, nature, and tradition, is often preferred for many projects. Depending on the type of wood, finish, and treatment, wooden signs can be polished, painted, or varnished to achieve the desired appearance and durability.

Fig. 2. A close-up of a sign with precise details and additional glued elements

Wooden signs produced with CNC routers are not only durable but also easily customizable to meet client needs. This flexibility extends to design, size, and functionality. CNC routers ensure high precision and automation in the manufacturing process, enabling the creation of both unique, one-of-a-kind signs and larger production runs. The use of CNC routers in crafting wooden door signs offers a blend of traditional aesthetics and modern technology. These signs play a crucial role in space identification while contributing to an aesthetically pleasing and personalized environment, all the while providing opportunities for a wide range of designs and applications.

Sign making through a combination of ornaments and reliefs

The project begins with opening the CAD software and creating a new project. The first step involves outlining the sign, defining its shape and dimensions. The shape can be rectangular, oval, or complex, depending on the design requirements. Next, the text to be included on the sign is entered, with careful selection of the font and size to ensure readability and aesthetic appeal. Decorative elements and ornaments are created using the drawing tools in the CAD software. Floral motifs, geometric shapes, or other decorative features can be employed to enhance the appearance of the sign. For creating reliefs, the 3D modeling tools in the CAD software are used, adding depth and texture to the sign to make it visually appealing.

Fig. 3. Graphic design of the object in specialized software

The choice of elements and reliefs depends on the designer's vision for a unique and attractive design. Different types of reliefs are selected based on the theme and purpose of the sign. Signs can be custom-made to order, in which case the elements are tailored to the client's specifications. Depending on the intended use of the finished sign, decorative elements are chosen, graphical design is prepared, and wood thickness is selected, among other considerations. Such signs can be used in various settings, including residential spaces, commercial establishments, rental villas, traditional dining venues, and more. They are particularly well-suited for tourist attractions, such as museums and guesthouses styled in traditional or antique themes.

Surface reliefs are light engravings that add texture without deeply penetrating the material. Deep reliefs create significant visual and tactile effects, while combined reliefs merge different depths and engraving styles for a more complex and dynamic appearance.

Materials Research Proceedings 46 (2024) 98-105 https://doi.org/10.21741/9781644903377-13

Fig. 4. Object with varying depths of reliefs

Upon completion of the design in the CAD software, the next step is preparing for cutting with the CNC router. The first action involves converting the digital model into G-code using CAM software such as Fusion 360, VCarve, or ArtCAM. The G-code contains the instructions for the movements and operations of the CNC router. Next, cutting parameters such as cutting speed, cutting depth, and the selection of cutting tools are established. These settings depend on the type and thickness of the wood as well as the complexity of the design. The material is placed on the CNC router's worktable and securely clamped to prevent movement during cutting. Initial points for the axes (starting positions) are set, and it is verified that the cutting tool is correctly positioned relative to the material. A test cut is performed on a small section of the material to ensure that the settings are correct and the outcome meets expectations.

Once the setup is complete, the program with the G-code is launched, and the cutting process begins. The CNC router automatically performs the designated movements and operations, cutting or engraving the material according to the design. The process must be monitored to ensure that the machine operates correctly and that there are no issues with cutting or material clamping. Depending on the model of the router used, the cutting speed can vary between 15 and 30 m/min, which is over 10 times faster compared to manual processing. The production of a sign with complex ornaments requires approximately 90 minutes of continuous machine operation, depending on the number of ornaments, the complexity of the shape, and the height of the elements. In contrast, the same sign produced manually would take over 15 hours, depending on the skill level of the craftsman. This clearly demonstrates that automation of the processes increases efficiency by more than 10 times. The human factor in manual processing should not be underestimated either. There is always a risk of errors when manually crafting the elements, while CNC routers reduce this percentage to below 1%. This significantly decreases material and time waste.

Finishing processes and subsequent treatment

After the main CNC routing work is completed and the sign is cut, finishing processes follow, involving several critical steps to achieve a perfect final product. First, the sign may be sanded to smooth out any rough edges and remove any residual material. Sanding can be done manually or with an electric sander, using various types of sandpaper depending on the needs.

Following sanding, the sign can be painted, varnished, or even further burned. Painting adds color and a finished appearance, while varnishing protects the wood and imparts a shine. Burning can provide an aged look. For specific effects, different types of varnishes and paints, including water-based, oil-based, and polyurethane, can be used.

Fig. 5. Example of a sign that has undergone additional finishing processes – burning, lacquering

To add additional decorative elements, such as small ornaments or other details, wood glue or other suitable adhesives may be employed. Applying these elements requires precision and adequate drying time to ensure secure and durable attachment. Finally, a final polish may be applied to achieve a smooth and glossy surface. Polishing compounds and discs can help remove microscopic imperfections and add extra shine to the finished product. It is essential that all finishing processes are executed with attention to detail and accuracy to guarantee the high quality and durability of the wooden sign.

Conclusion

The use of CNC routers for prototyping and model-making represents a significant advancement in contemporary industry and design. The combination of automation, precision, and material diversity makes CNC technology an indispensable tool for design and manufacturing. One of the main advantages of CNC routers is their ability to work with a wide variety of materials, including wood, plastic, metals, and even composite materials. This versatility allows manufacturers and designers to realize a wide range of ideas and concepts, regardless of the specific materials they work with. CNC routers are ideal for creating models and prototypes in various industries. They allow for the rapid realization of ideas and projects, which is particularly important for designers and engineers.

The precision of CNC routers is a crucial aspect for achieving high-quality products. The capability for programmed movements and operations guarantees that every detail of the project is produced with outstanding accuracy and consistency. This is particularly important for industries that require complex shapes or tight tolerances in the production of parts and components.

The future of CNC technology promises ongoing development and innovation. Advances are expected in faster processing capabilities, greater flexibility in using different materials, and more intelligent software solutions for programming and machine control. The expansion of automation possibilities and integration with other technologies, such as artificial intelligence and the Internet of Things, will also play a key role in transforming the manufacturing process.

Conversations and insights regarding real-world applications and industry experiences will continue to influence the evolution of CNC technology. The necessity for training and adaptation to the new capabilities of this technology will be vital for companies aiming to stay competitive in the marketplace.

In conclusion, CNC routers not only facilitate the manufacturing process but also open new horizons for creative potential and innovation across various industries. With their ability to combine precision with flexibility, they serve as a fundamental catalyst for progress in modern manufacturing and design.

Acknowledgement
The present article has been developed with the help of project Club activity 5 Realization of prototypes and models through the use of a CNC router, 2024.

References

[1] A. Çakmak, Abdulkadir malkoçoğlu, Sukru Ozsahin Optimization of wood machining parameters using artificial neural network in CNC router, February 2023, Materials Science and Technology. https://doi.org/10.1080/02670836.2023.2180901

[2] H. Loc Nguyen, V. Thuy Tran, Design of the CNC Router Structure for Machining Wood Materials Using Reliability-Based Design Optimization Method, In book: Lecture Notes in Mecjanical Engineering, Recent Advances in Manufacturing Engineering and ProcesseChapter: Electromechanical System and Engineering DesignPublisher: Springer Nature, February 2023. https://doi.org/10.1007/978-981-19-6841-9_3

[3] K. Bangse, A. Wibolo, I. Kadek Ervan Hadi Wiryanta, Design and fabrication of a CNC router machine for wood engraving, Journal of Physics Conference Series 1450(1):012094, February 2020. https://doi.org/10.1088/1742-6596/1450/1/012094

[4] H. Pelit, Mustafa Korkmaz, Mehmet Budakci Surface Roughness of Thermally Treated Wood Cut with Different Parameters in CNC Router Machine, BioResources 16(3):5133-5147, May 2021. https://doi.org/10.15376/biores.16.3.5133-5147

[5] K. Huseyin Koc, E. Seda Erdinler, E. Hazir, E. Öztürk, Effect of CNC Application Parameters on Wooden Surface Quality, Conference: Proceedings of the 58th International Convention of Society of Wood Science and Technology June 7-12, 2015 - Grand Teton National Park, Jackson, Wyoming, USA, June 2015.

[6] P. Zlateva, A. Terziev, K. Yordanov, Study of regime parameters of the fermenter in the production of biogas from animal liquid waste materials, (2021) E3S Web of Conferences, 286. https://doi.org/10.1051/e3sconf/202128602010

Innovative Manufacturing Engineering and Energy - IManEE2024 Materials Research Forum LLC
Materials Research Proceedings 46 (2024) 106-113 https://doi.org/10.21741/9781644903377-14

A study of the micro-geometric errors of tangentially turned surfaces using the design of experiments method

István SZTANKOVICS

University of Miskolc, Institute of Manufacturing Science, H-3515 Miskolc-Egyetemváros Hungary

istvan.sztankovics@uni-miskolc.hu

Keywords: Design of Experiments, Finishing, Surface Roughness, Tangential Turning

Abstract. The increase of machining efficiency is a main goal and a complex task for the production engineering in the industrial environment, while the prescribed surface quality must be maintained. The tangential turning procedure offers a solution while achieving high feeds (thus lowering the machining time), where the roughness of the machined surface can reach its required low value. In this study, several parameters of the roughness profile are analyzed on tangentially machined surfaces. The experimental setups are planned and carried out following the method of Design of Experiments. The results showed that among the key process parameters, the feed of the cutting tool has the most meaningful effect on the surface roughness in tangential turning.

Introduction

Surface roughness plays a critical role in machining, as the functional behavior of the surface is directly affected by its characteristic among other factors, such as lifespan, and aesthetic quality of machined components. Achieving an optimal surface finish is essential for reducing friction, wear, and improving fatigue resistance in precision parts [1-3]. In industries like automotive, aerospace, and manufacturing, surface roughness can significantly influence the efficiency of mechanical systems, making it a key consideration during the machining process [4,5]. As technological demands rise, so does the need for machining processes that can consistently deliver high-quality surface finishes while maintaining production efficiency [6,7]. Another solution to improve the surface quality is the application of rarely used machining procedures, like burnishing [8,9]

In recent years, there has been a growing importance on increasing productivity in machining without compromising surface quality. This has driven the development and application of novel machining kinematics [10,11], which allow manufacturers to enhance efficiency through faster processes, improved tool designs, and better control over machining variables (for example the feed rate, machining speed, and depth of cut). Modern advancements in the geometry of the tools used for cutting and kinematics [12-14] offer the potential to reduce machining times, decrease the costs of production, and improve overall workpiece quality. Tangential turning is one approach, offering an alternative to traditional longitudinal turning [15]. In tangential turning, the cutting edge is positioned at an angle relative to the workpiece, and the tool moves tangentially along the surface. This method improves cutting efficiency and tool life by distributing wear more evenly along the cutting edge. It also helps achieve better surface quality due to the altered cutting kinematics, reducing the risk of chatter and vibration. As a result, tangential turning is gaining attention as a viable solution for machining cylindrical surfaces with greater precision and efficiency, making it a key area of research for increasing productivity in modern manufacturing.

The main objective of this paper is to analyse the surface roughness of cylindrical parts when machined by tangential turning. The effect of the feed, cutting speed and depth of cut is analysed on the roughness parameters selected in this study by the method of Design of Experiments.

Methods, Experimental Conditions

The objective of this research was to investigate the micro-geometric errors of the machined surface in turning with tangential feed. This is achieved by both experimental machining tests and analytical assessments, employing the Design of Experiments (DOE) approach.

The experimental work was conducted using specific equipment, notably an EMAG VSC 400 DS lathe. The cutting tool used featured a design for tangential feed with a 45° inclination. The material of the cutting insert was MG12 grade, and had the following characteristics:

- insert: S117.0032.00
- holder: H117.2530.4132
- manufacturer: HORN Cutting Tools Ltd.

42CrMo4 alloyed steel workpieces with outer cylindrical geometries were processed in the practical experiments, which diameter were 65 mm. The material was hardened to 410 HV10 prior to the experiments. The surfaces to be tangentially turned were pre-machined. The properties of the tool were the following:

- insert: CNMG 120412-PM 4314,
- holder: PCLNR 25 25 M12,
- manufacturer: SANDVIK Coromant.

The study aimed to examine how changes in the cutting parameters affected tangential turning performance, focusing on cutting speed (v_c), feed per revolution of the workpiece (f), and depth of cut (a_p). Two limit values (lower and upper) for each parameter were determined using the DOE methodology, which were the following

- v_c – 200 m/min and 250 m/min,
- f – 0.6 mm/rev. and 0.8 mm/rev.,
- a_p – 0.1 mm and 0.2 mm.

This produced a total of $2^3 = 8$ different setups, as presented in Table 1. The analysis of the results was conducted in accordance with DOE, where the upper and lower values were converted (f', v_c', a_p') as shown in Table 1.

Table 1. Summary of the setups parameters for each experiment

	Setup parameters			Converted values		
	f mm/rev.	v_c m/min	a_p mm	f' -	v_c' -	a_p' -
1	0.6	200	0.1	-1	-1	-1
2	0.8	200	0.1	1	-1	-1
3	0.6	250	0.1	-1	1	-1
4	0.8	250	0.1	1	1	-1
5	0.6	200	0.2	-1	-1	1
6	0.8	200	0.2	1	-1	1
7	0.6	250	0.2	-1	1	1
8	0.8	250	0.2	1	1	1

For the measurements, an AltiSurf 520 three-dimensional topography instrument was utilized, employing a confocal chromatic probe to carry out the tests after the machining experiments. Measurement parameters were selected based on the guidelines of the ISO 21920:2021 standard.

The roughness profiles for each surface were recorded along three generatrix lines. These profiles were then analyzed using the AltiMap Premium 6.2.7487 surface analysis software.

The evaluated 2D surface texture parameters were the following (ISO 21920:2021):

- R_z – Maximum height of the profile indicates the absolute vertical distance between the Maximum profile peak height and the Maximum profile valley depth along the sampling length. [μm]
- R_p – Maximum profile peak height indicates the point along the sampling length at which the curve is highest. [μm]
- R_v – Maximum profile valley depth indicates the point along the sampling length at which the profile curve is lowest. [μm]

Polynomial formulas were also determined. These are used to compute and represent the studied factors, as shown in Equation 1. This equation includes the variables (f, v_c, a_p) and their combinations, with constants (k_i) indicating the influence of each factor. The roughness parameters are expressed as the function $y(f, v_c, a_p)$ in this study. These equations help quantify and visualize the studied effects, offering a systematic approach to optimize machining processes for improved dimensional accuracy and surface quality.

$$y(f, v_c, a) = k_0 + k_1 f + k_2 v_c + k_3 a + k_{12} f v_c + k_{13} f a_p + k_{23} v_c a_p + k_{123} f v_c a_p \quad (1)$$

Results

The cutting experiments and surface roughness measurements were carried out according to the methods described in the previous section. Table 2 presents the micro-geometric error parameters that were evaluated during the study. Additionally, the valley depth (R_v) values were examined in relation to the Maximum Height of the Profile (R_v / R_z). This ratio offers a clearer understanding of the functional behavior of the profile, such as its bearing capacity, shedding light on the performance impact of dimensional changes in the machined components.

Table 2. Measured values of the studied parameters

	R_z [μm]				R_p [μm]				R_v [μm]				R_v/R_z [-]			
Setup	1	2	3	Avg.	1	2	3	Avg.	1	2	3	Avg.	1	2	3	Avg.
1	3.37	3.20	3.54	3.37	1.87	1.85	2.08	1.93	1.50	1.36	1.47	1.44	0.45	0.42	0.41	0.43
2	5.27	5.22	5.34	5.28	2.76	2.84	2.83	2.81	2.51	2.38	2.51	2.47	0.48	0.46	0.47	0.47
3	3.68	3.44	3.68	3.60	2.05	1.90	2.09	2.01	1.63	1.54	1.59	1.59	0.44	0.45	0.43	0.44
4	5.14	5.53	5.27	5.32	2.77	2.90	2.82	2.83	2.37	2.63	2.46	2.49	0.46	0.48	0.47	0.47
5	3.89	3.70	3.59	3.73	1.92	2.20	2.26	2.13	1.97	1.50	1.33	1.60	0.51	0.41	0.37	0.43
6	4.51	4.78	5.09	4.79	2.59	2.68	2.99	2.75	1.92	2.09	2.10	2.04	0.43	0.44	0.41	0.43
7	4.09	3.79	4.54	4.14	2.04	2.26	2.10	2.13	2.06	1.53	2.44	2.01	0.50	0.40	0.54	0.48
8	4.64	5.77	5.73	5.38	2.24	2.90	3.07	2.73	2.41	2.88	2.67	2.65	0.52	0.50	0.47	0.49

Upon finishing the measurements and conducting their evaluation, calculation formulas for the evaluated factors were developed based on Equation 1 provided in the previous section. The Maximum height of the profile can be determined by the solution of Equation 2. The formula of the Maximum profile peak height is shown in Equation 3. The calculation method of the Maximum profile valley depth is presented in Equation 4. Finally, the ratio between the Maximum profile valley depth and the Maximum height of the profile is determined and analyzed based on Equation 5.

$$R_z(f,v_c,a) = -12.01 + 24.79f + 0.034v_c + 64.99a - 0.055fv_c - 114.8fa_p - 0.18v_ca_p + 0.36fv_ca_p \quad (2)$$

$$R_p(f,v_c,a) = -3.37 + 7.46f + 0.008v_c + 16.31a - 0.009fv_c - 18.87fa_p - 0.034v_ca_p + 0.031fv_ca_p \quad (3)$$

$$R_v(f,v_c,a) = -8.62 + 17.3f + 0.025v_c + 48.51a - 0.046fv_c - 95.77fa_p - 0.14v_ca_p + 0.33fv_ca_p \quad (4)$$

$$R_v/R_z(f,v_c,a) = -1.94 + 3.16f + 0.01v_c + 16.82a - 0.014fv_c - 23.77fa_p - 0.07v_ca_p + 0.36fv_ca_p \quad (5)$$

Discussion

The subsequent step involves analyzing the data after outlining the utilized approaches and apparatus, along with the measurement outcomes and determined formulas. The study has in two steps. Initially, main effect plots are created and examined to demonstrate how changes in technological parameters affect the roughness of the machined surface. In the second stage, surface diagrams are generated based on Equations 2-5, which visualize the specific impact of f, v_c, and a_p on the selected surface roughness parameters during turning with tangential feed.

Main effect plots were drawn to firstly evaluate the influence of the studied factors. Each surface roughness parameter was analyzed separately, as demonstrated in Figure 1.

The process for creating these graphs involved several steps. First, the overall mean of the averaged values was calculated and represented by a dashed line. Then, two separate averages were determined for each of the three cutting parameters, corresponding to the lower (-1) and upper (1) limits. These two averages were connected by a solid line, with the direction and slope of these lines indicating the primary effect of each cutting parameter on the roughness characteristics under investigation.

In the analysis of surface roughness parameters, feed rate plays a critical role in determining the values of Mean Roughness Depth, Maximum Peak Height, and Maximum Valley Depth. A significant increase in feed leads to a noticeable rise in these roughness values. When the feed rate is increased, the cutting tool engages more strongly with the workpiece surface, producing deeper valleys and higher peaks. This, in turn, results in larger R_z, R_p, and R_v values. The rougher surface can be attributed to the higher material removal rate associated with increased feed, which leaves behind more prominent tool marks.

On the other hand, the cutting speed has a more slight effect on the roughness parameters. While increasing the cutting speed slightly raises the R_z and R_v values, it does not significantly affect R_p. This may be due to the dynamics of material removal at higher cutting speeds, where the chip formation process changes, causing more minor variations in the valleys and overall surface texture. However, the height of the peaks (R_p) remains mainly unaffected by the cutting speed, likely because it does not alter the initial engagement of the cutting tool with the material.

The depth of cut, expectedly, shows no notable impact on R_z, R_p, or R_v values in tangential turning as well. This suggests that variations in the depth of cut do not significantly change the surface roughness for these specific parameters. This lack of influence could be due to the cutting tool geometry and feed rate having a more dominant role in determining surface characteristics. In summary, while feed rate is the most influential factor on the analyzed roughness parameters, cutting speed plays a smaller role, and the depth of cut has a negligible effect on these surface roughness metrics.

Unexpectedly, the calculated R_v/R_z ratio displays distinct behavior compared to other parameters. The feed rate has a reducing effect on this ratio, indicating that higher feeds lead to relatively shallower valleys in comparison to the overall roughness. In contrast, the depth of cut shows a significant increasing influence on the R_v/R_z ratio, meaning deeper cuts result in proportionally deeper valleys. Additionally, cutting speed slightly increases the ratio, suggesting a minor impact on the valley depth relative to the total surface roughness. These findings highlight the unique interaction of machining parameters with the R_v/R_z ratio during the tangential turning process.

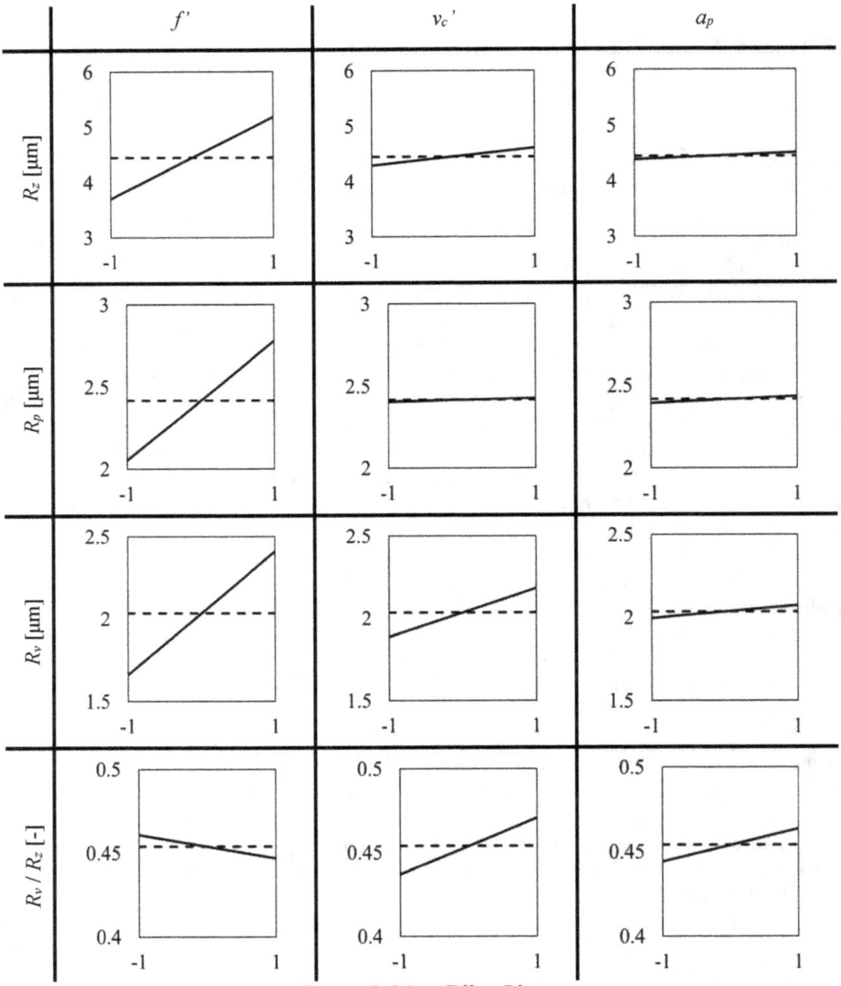

Figure 1. Main Effect Plots

The study continues by analyzing the specific experimental results alongside the determined equations. These equations are used to interpret the impact of various machining parameters, such as feed rate, cutting speed, and depth of cut, on surface roughness. The experimental data provides specific values for the roughness parameters under investigation, enabling a detailed assessment of how each parameter affects surface quality in the studied range. Equations 2-5 help quantify these effects and establish relationships between the machining conditions and surface characteristics, offering insights into optimizing the tangential turning process. Based on these equations, surface plots were drawn, which can be seen in Figure 2 and 3. These figures enables to understand the change of the surface roughness by the alteration of the setup values.

Figure 2 shows the alteration of the analyzed roughness parameters in function of the cutting speed, feed and depth of cut. In general, it can be seen, that the most significant factor is the feed. The increase of this values by one-third leads to significantly higher roughness values, which shows around a 1.5-fold increase overall. The cutting speed has a slight increasing effect on the studied values, which indicates that in the point of view of the machined surface micro geometric errors, it is not advisable to further increase the cutting speed. The alteration of the depth of cut presents different change in the roughness values. The engagement depth of the tool has the most effect on the Maximum Peak Height of the roughness profile.

Figure 3 presents the alteration of the ratio between the Maximum Valley Depth and the Maximum Height of the profile. It can be seen that the increasing depth of cut results in a significantly altered surface plot. If the cutting speed is 250 m/min and the feed is 0.8 mm/rev., the ratio is significantly lower, when the depth of cut is 0.1 mm, and the ratio shows a 1.1-fold increase, if the depth of cut is increased two-fold. this phenomenon should be analyzed further in future studies. The figures also show that the geometry of the profile of the machined surface can be altered effectively, when the setup parameters are changed.

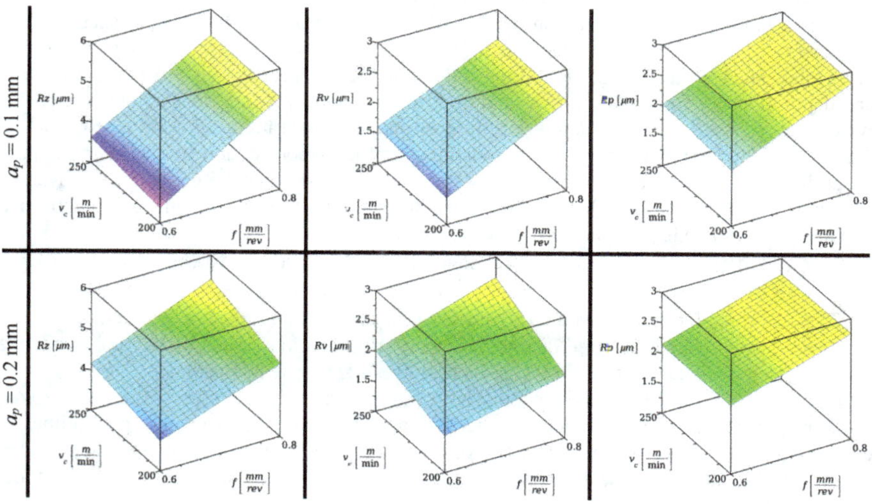

Figure 2. Roughness values as a function of the setup parameters

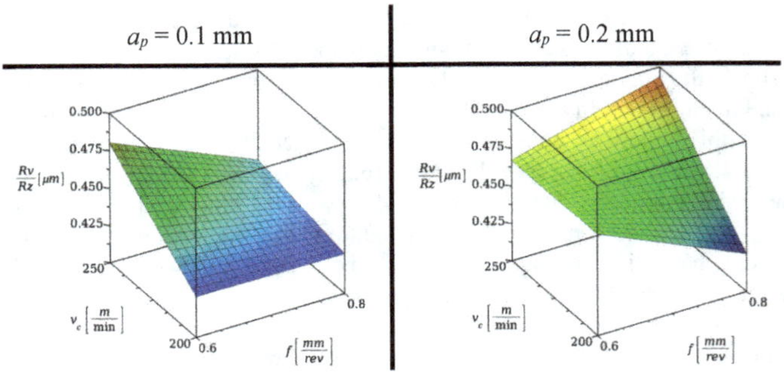

Figure 3. The change of the analyzed ratio parameter

Summary

This research aimed to investigate the micro-geometric errors of machined surfaces during tangential turning, using both experimental cutting tests and theoretical analyses, guided by the Design of Experiments (DOE) methodology. The cutting experiments were conducted using an EMAG VSC 400 DS lathe and a tangential tool with a 45° inclination angle, consisting of a carbide insert and holder by HORN Cutting Tools Ltd. The workpieces, cylindrical in shape with a diameter of 65 mm, were made of hardened 42CrMo4 alloyed steel. The cutting parameters varied in the study included cutting speed (200 and 250 m/min), feed per revolution (0.6 and 0.8 mm/rev), and depth of cut (0.1 and 0.2 mm). A total of 8 different setups were generated, and the results were analyzed using the DOE approach. Surface roughness measurements were taken using an AltiSurf 520 topography instrument. The roughness profiles, recorded along three generatrix lines of each surface, were analyzed with AltiMap Premium software. Key surface roughness parameters studied included the maximum height of the profile, maximum peak height, and maximum valley depth. The findings of this paper indicated:

- Feed rate had a significant effect on increasing R_z, R_p, and R_v values,
- Cutting speed had a minor impact, mainly on R_z and R_v.
- The depth of cut showed no significant influence on these roughness parameters.
- The R_v/R_z ratio showed distinct behavior, with feed rate reducing the ratio, while the depth of cut significantly increased it, suggesting unique interactions.

The study of tangential turning continues in multiple directions. Future work will explore broader applications, such as different materials or geometries. The machining possibilities of complex geometries will be analyzed, and other materials will be chosen as specimens. Other factors which could affect surface roughness during cutting will be studied as well, such as tool wear or vibrations.

Acknowledgement

This research was supported by project No. 2020-1.2.3-EUREKA-2022-00025 which was realized with financial help of the National Research Development and Innovation Fund of the Ministry of Culture and Innovation of Hungary.

References

[1] A. Aman, R. Bhardwaj, P. Gahlot, and R. K. Phanden, Selection of cutting tool for desired surface finish in milling Machine using Taguchi optimization methodology, Materials Today Proceedings, vol. 78, 2022, pp. 444–448. https//doi.org/10.1016/j.matpr.2022.10.253

[2] S. Vasiljević, D. Rajković, M. Đorđević, S. Kostić, and M. Stanojević, Application of innovative mechatronic systems in product design, development and production, Journal of Production Engineering, vol. 27, no. 1, 2024, pp. 13–21. https//doi.org/10.24867/jpe-2024-01-013

[3] Gy. Kovács, Optimization of structural elements of transport vehicles in order to reduce weight and fuel consumption, Structural Engineering And Mechanics, vol. 71, no. 3, 2019, pp. 283–290. https//doi.org/10.12989/sem.2019.71.3.283

[4] A. Pratap and A. Beaucamp, Hybrid tools for improved removal and surface finish of metals and non-metals, Procedia CIRP, vol. 123, 2024, pp. 227–231. https//doi.org/10.1016/j.procir.2024.05.041

[5] S. K. Maurya, P. Chechi, R. Prasad, C. K. Susheel, and A. Manna, Effect of surface finish on corrosion behavior of polished Al6061 alloy in simulated acid rain environment, Materials Today Proceedings, vol. 64, 2022, pp. 755–759. https//doi.org/10.1016/j.matpr.2022.05.208.

[6] A. Nagy and Gy. Varga, Effect of abandonment of cooling and lubrication on surface roughness and cylindricity in turning of steel, Multidiszciplináris Tudományok, vol. 11, no. 4, 2021, pp. 395–407. https//doi.org/10.35925/j.multi.2021.4.43

[7] L. Petruschke, M. Burkhardt, T. Kohne, P. Schraml, M. Helfert, and E. Abele, Method to identify energy efficiency potentials of metal cutting machine tools in industry, Procedia CIRP, vol. 90, 2020, pp. 522–527. https//doi.org/10.1016/j.procir.2020.01.066

[8] F. T. Kebede and Cs. Felhő, Examining the impact of slide burnishing parameters on the 3D surface features of medium carbon steel, Journal of Production Engineering, vol. 27, no. 1, 2024, pp. 30–35. https//doi.org/10.24867/jpe-2024-01-030

[9] V. Ferencsik, Analytical Analysis Of The Theoretical Surface Roughness In The Case Of Burnishing Of Cylindrical Workpiece, Cutting & Tools in Technological System, no. 99, 2023, pp. 101–109. https//doi.org/10.20998/2078-7405.2023.99.05

[10] A. Schubert, R. Zhang, and P. Steinert, Manufacturing of twist-free surfaces by hard turning, Procedia CIRP, vol. 7, 2013, pp. 294–298. https//doi.org/10.1016/j.procir.2013.05.050

[11] P. Martikáň, A. Czán, J. Holubják, D. Varga, J. Martinček, and T. Czánová, Verification of New Method of Determining the Roughness Parameters for Rotational Turning with Non-linear Cutting Edge, Procedia Engineering, vol. 192, 2017, pp. 563–568. https//doi.org/10.1016/j.proeng.2017.06.097

[12] B. Deng, Q. He, J. M. DePaiva, and S. C. Veldhuis, A novel approach to cutting tool edge design based on initial wear stage, Journal of Materials Processing Technology, vol. 304, 2022, p. 117561. https//doi.org/10.1016/j.jmatprotec.2022.117561

[13] Y. Wang et al., Quality-centered production and maintenance scheduling for multi-machine manufacturing systems under variable operating conditions, Reliability Engineering & System Safety, vol. 250, 2024, p. 110264, . https//doi.org/10.1016/j.ress.2024.110264

[14] Y. Wang, J. Liu, L. Zhou, L. Cong, and J. W. Sutherland, Integrated operation planning and process adjustment for optimum cost with attention to manufacturing quality and waste, Journal of Manufacturing Systems, vol. 73, 2024, pp. 241–255. https//doi.org/10.1016/j.jmsy.2024.02.004

[15] J. Schneider, L. Schreiber, Maschinen Und Anlagen-Mit dem Tangentialdrehen zu drallfreien Oberflachen. Werkstatt und Betrieb. 135(6), 200240-45

Effect of wettability on the aeration characteristics of rotating thin films between EDM-prepared surfaces

Nikolaos ROGKAS [1, 2, a*], Dimitrios SKONDRAS-GIOUSIOS[1, b], Pavlos ZALIMIDIS[2, c], Angelos MARKOPOULOS [1, d] and Vasilios SPITAS[1, e]

[1]School of Mechanical Engineering, National Technical University of Athens, Athens, GR-15780 Zografou, Greece

[2] Department of Mechanical Engineering Educators, School of Pedagogical and Technological Education, Iraklio Attikis, 14121 Athens, Greece

[a]nrogkas@mail.ntua.gr, [b]di-sko@hotmail.com, [c]pzalimidis@aspete.gr, [d]amark@mail.ntua.gr, [e]vspitas@mail.ntua.gr

Keywords: Wettability, Aeration, Lubrication, Thin Films, EDM

Abstract. Rotating thin-film flows with small film-thickness-to-radius ratios play a critical role in various power transmission systems, such as multidisc wet clutches and limited-slip differentials. Over the past two decades, significant attention has been given to a well-documented phenomenon in wet clutch applications: the formation of two-phase Automatic Transmission Fluid (ATF)-air flow at high rotational speeds. The objective of this research is to further explore the aeration characteristics in such systems, with a focus on the impact of surface wettability due to surface topography alteration. Specifically, the study examines the progression of air bubbles into the fluid domain and the subsequent effect on drag torque, i.e., the viscous torque generated between the discs as a result of ATF's shear stress, using a Computational Fluid Dynamics (CFD) approach. To assess this, different surface roughness profiles are achieved by varying the Electro Discharge Machining (EDM) parameters and are further characterized by measuring the contact angle. These surface textures are then correlated with aeration characteristics, revealing that oleophilicity increases aeration. This research is part of the WetSURF project, under the Basic Research Framework Programme at the National Technical University of Athens.

Introduction

Rotating thin films, characterized by a film thickness of 50-1000 μm and a radius of ~10-100 mm, play a crucial role in various power transmission applications that operate in lubrication regimes [1–4]. In wet clutch systems, the design typically involves a series of discs separated by automatic transmission fluid (ATF). The ATF serves both to cool the system [5,6] and reduce wear [7,8]. However, when these discs rotate at speeds approximately in the range of 1000rpm to 2000rpm [9,10], the pressure near the outer radius can drop below atmospheric levels, potentially drawing in air from the environment. This results in the formation of a two-phase air-ATF flow, which reduces viscous shear stress and drag torque due to the lower viscosity of air compared to ATF; however, during clutch engagement, excessive air can prevent proper lubrication, causing the system to behave more like a dry friction transmission.

The typical modeling approach for this system involves examining a single disc pair with a constant film thickness. One disc remains stationary, while the other rotates at a speed corresponding to the system's relative rotational speed [11]. Simultaneously, a constant inflow rate is supplied from an opening located at the inner radius. Consequently, the overall flow behavior is influenced by the inflow rate, rotational speed, and the properties of the ATF. Various theoretical (analytical and CFD) and experimental approaches have been proposed in the literature to investigate this phenomenon. Pardeshi et al. [12] proposed a CFD approach to track the air bubble by utilizing an empirical contact-angle model to derive good agreement with their experimental results. Aphale et al. [13] proposed an analytical framework to derive a 3rd degree non-linear ODE which captures air bubble's shape. Yuan et al. [14] used a CFD analysis to examine the effect of contact angle on the drag torque and aeration characteristics. Leister et al. [15] presented a non-dimensional analysis to predict aeration and make drag torque calculations. Rogkas et al. [16] utilized a CFD approach but this time considering a statistical varying film thickness due to various uncertainties rendering wet clutch's operation. Iqbal et al. [17,18] developed a simplified analytical model to predict aeration and compared with experiments.

Surface wettability plays a key role in determining the bubble formation and behaviour in liquid environments. In general, surface wettability in air is complementary to surface wettability in aqueous media [19]. A superhydrophobic surface is superaerophilic and vice versa. Bubble formation and motion can be effectively manipulated by wettability gradients of the wetted surface [20]. Surface chemistry and topography can synergistically affect the resulting surface tension, leading to different trends in dynamic aerophobicity by introducing in adhesion and directionality of the surfaces [21,22]. Bioinspired designs with advanced wettability attributes have been used for the fabrication of superhydrophobic or superhydrophilic structures with significant gas bubble manipulation capability [23]. Concerning aeration of thin lubricating films, the effect of air contact angle can be significant in reducing drag torque [13]. The vast majority of the manufactured surfaces include polymers or other materials where a coating is applied for property changing. In very few cases, a manufacturing process is utilized for directly regulating the surface tension of the surface [24].

In most cases of tribological applications of discs and clutches, metallic, ceramic or composite materials are involved in the occurring contacts. The intrinsic wettability of such materials typically lies in the spectrum of hydrophilicity and oleophilicity. However, some of the desired aeration properties can be promoted with increase of their aerophobic nature, thus increasing the oleophilicity. Therefore, a wettability alteration must occur for the acquisition of the wetting properties. The application of chemical coatings can alter the wettability of the surfaces, although chemical coatings can be costly, environmentally hazardous and of limited durability. For the increase of wettability, according to the homogenous Wenzel wetting regime, the roughness amplitude of a surface should be increased. From the manufacturing point of view, this roughness-induced wettability increase can be achieved by material removal processes, such as micromilling [24] or laser machining [25]. Structured superoleophilic surfaces can exist with different water wettabilities [26] and can assist drag reduction properties [27].

Electrical discharge machining (EDM), although mainly utilized as a non conventional manufacturing method for precisely cutting hard-to-cut materials, has been proved recently as a facile and one-step method that can induce a wettability transition through topography alteration. The crater-like stochastic formation along with the chemical modification of the recast layer created after the EDM process can change the material wetting state towards hydrophobicity [28], hydrophilicity [29] or superoleophilicity [30]. Wettability control can be achieved by controlling the resulting surface roughness by discharge parameters such as current, voltage and pulse duration [31,32].

This study aims to investigate how the contact angle influences the aeration characteristics of thin film flow between a stationary disc and a rotating disc. To that scope, the contact angle results obtained from the EDM process on Al surfaces are used as boundary conditions in a CFD model simulating the flow behavior.

CFD model development

Figure 1 illustrates the geometry of the two-disc model along with the corresponding boundary conditions. In this configuration, the top disc rotates while the bottom disc remains stationary. Automatic transmission fluid (ATF) is supplied at a constant inflow rate (Q_{in}) through the inner radius, while the outer radius serves as an opening, where the pressure is set to atmospheric ($P_o = 0$). As the rotational speed increases, besides drag torque increase, reverse flow may also occur at the outlet; in such cases, only air is assumed to re-enter the domain. The film thickness is denoted by (h). It also mentioned the effect of surface roughness, which is exaggerated for illustration purposes in Fig. 1, is accounted for using the equivalent measured contact angle. Fig. 1 quantitatively illustrates the expected flow behavior at three rotational speeds ($n_a < n_b < n_c$). At the lowest speed (n_a), the flow consists entirely of ATF, and the pressure and velocity fields can be determined using standard lubrication theory assumptions. As the rotational speed increases to (n_b), a condition arises where $p_{out} < P_o$, leading to aeration and sharp drop of drag torque. In this scenario, it is assumed that air enters in the form of an elongated bubble, moving radially towards the inner radius until reaching a stabilization point. This position generally depends on the operating conditions, geometry, and properties of the ATF. Finally, as the speed continues to increase (n_c), the bubble advances further into the domain. Smaller air bubbles may also form, compress, and merge, following the process described in [13].

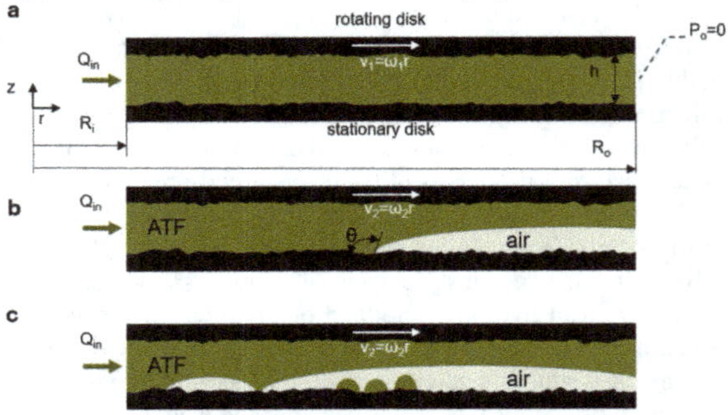

Fig. 1. Model geometry and boundary conditions along with schematic representation of the anticipated aeration characteristics. ***a)*** *n_a,* ***b)*** *n_b,* ***c)*** *n_c, where $n_a < n_b < n_c$. Surface roughness is shown enlarged for visualization purposes.*

The CFD simulation was performed in the commercial code ANSYS Fluent. The analysis is considered 2D axisymmetric including swirl and the flow is considered laminar. The simulation is transient and the timestep was set equal to 0.01 sec which was deemed suitable for the tracking of the air bubble equilibrium position. Momentum, continuity and energy conservation equations were solved [33]. The coupling of pressure and velocity was accomplished using the SIMPLE algorithm, while the PRESTO method was utilized for spatial discretization of pressure. For the momentum, swirl velocity, and energy equations, a second-order upwind scheme was implemented. The computational mesh was composed of quadrilateral elements, featuring 40 divisions in the axial direction and 2500 divisions in the radial direction. To track the air-ATF interface, the Volume of Fluid (VOF) multiphase model was incorporated.

Wettability assessment of the EDM-prepared surfaces

For obtaining experimental measurements of the wettability for the simulation cases, aluminum alloy 6082 surfaces were fabricated via the EDM process (Agie-Charmilles Roboform 350 Sp). The scope of the use of EDM was to control the surface roughness of the machined surfaces, and subsequently the wettability. The realization of different surface roughness values was a result of different combinations of the machining parameters, as depicted in Table 1 namely current (I) and pulse duration (Ton) during the EDM process. The increase of the output machining power led to the increase of roughness amplitude as can be seen from Table 1. A qualitative comparison between S1 and S3 topography can be seen in Fig 2.

Materials Research Proceedings 46 (2024) 114-121 https://doi.org/10.21741/9781644903377-15

Table 1. EDM parameters and associated surface properties

	I (A)	Ton (μs)	Sa (μm)	Sp (μm)	Sv (μm)	Sz (μm)	Ca (⁰)
S1	2	3.2	1,83	12,67	9,41	22,08	19±1.5
S3	9	25	6,01	35,52	27,09	62,61	10±0.5

The resulting surfaces, after plasma cleaning were found to be hydrophilic and oleophilic. By plasma cleaning, removal of organic compounds, which can reduce wettability [34]], is ensured. In that way, the effect of topographical variation towards wettability becomes more prominent. Concerning the wettability of the surfaces with a 2000 multigrade SAE20W50, the oil contact angle measurements showed a decreasing trend with increasing amplitude of roughness, reaching superoleophilicity (~10^0) in the case of S3, as seen in Table 1. The greater height fluctuation between peaks and valleys of the machined samples (S1, S3) results in bigger reservoir volume of the textured surfaces, according to their peak-to-height amplitude, and eventually increases the surface wettability. When carefully chosen, the machining parameters can be utilized to create a controllable wettability outcome of the machined surfaces. In that sense, EDM can provide a facile and controllable texture generation for controlling the surface wettability for applications such as wet clutches, where according to the working conditions an optimized wettability might be needed

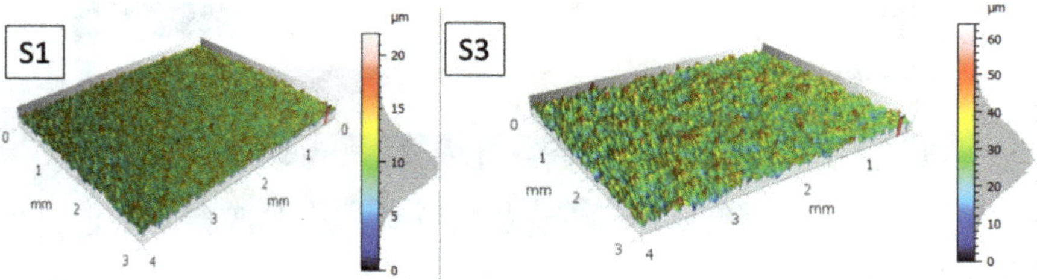

Fig. 2. EDM surfaces obtained from the profilometer analysis.

CFD Analysis Results

This section presents the results from the CFD simulation, which assumes theoretically flat discs. The effect of surface roughness is incorporated using an equivalent contact angle boundary condition, based on the two measured contact angles of 27° and 10°. The parameters used in the simulation are shown in Table 2. Fig. 3 presents the results for drag torque (T_d), the mean volume fraction of each phase in the domain ($\bar{A}$) and the ATF-air volume fraction contours at characteristic speeds that promote aeration. ATF's properties are assumed temperature-independent and taken from Ref. [12].

Table 2. CFD Simulations parameters

Symbol	Description	Value	Unit
R_o	Outer radius	50	mm
R_i	Inner radius	40	mm
h	Discs clearance	0.15	mm
Q_{in}	Inflow rate	0.0005375	Kg/s
n	rotating disc speed.	0-2000	rpm
ρ	Density	860	Kg/m3
σ	Surface tension coef.	0.03	N/,
μ	Dynamic viscosity	0.09	Pa s
θ	Contact angle	27, 10	degrees

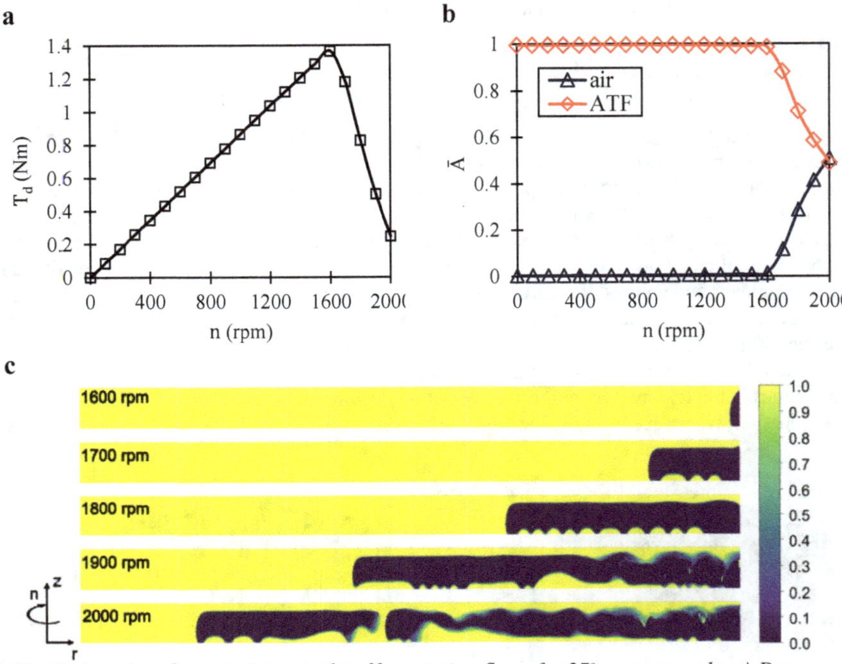

Fig. 3. Aeration characteristics in thin film rotating flows for 27° contact angle. ***a)*** *Drag torque-speed curve showcasing the rapid drop of torque at the aeration onset.* ***b)*** *associated volumetric mean volumes of ATF and air volume fractions.* ***c)*** *Contour plot of ATF volume fraction at characteristic rotating speeds which promote aeration.*

In Figures 3-a and 3-b, it is evident that the drag torque increases linearly with rotational speed until reaching 1600 rpm, which marks the onset of aeration. At this speed, the mean volume fraction of air starts to rise while the mean volume fraction of ATF begins to decrease as the speed continues to increase. This shift results in a sharp drop in T_d, falling from 1.4 Nm to 0.2 Nm within a speed range of just 400 rpm. This trend is further explained by the contour plots shown in Fig. 3-c. It can be observed that the air bubble, which initially makes contact only with the stationary disc (bottom), advances further into the gap as the speed rises. In some cases, smaller oil bubbles may form and become trapped inside the air bubble, however, these trapped bubbles do not contribute significantly to drag torque. Furthermore, as speed increases, the primary air bubble may split into secondary bubbles. It also worth noting that compared to oleophobic contact angle (>90°), aeration is stalled, as in such values, aeration onset is approximately near 1000 rpm, for the operating parameters used in the simulation.

The next part of the analysis compares the effect of contact angles obtained from the S1 and S3 specimens on the aeration process by examining the ATF-air interface at 1900 rpm. At this speed, the mean volumetric air fraction for a 10° contact angle is 0.43, while for a 27° contact angle, it is 0.41, representing a decrease of approximately 4.5%. This indicates that at these small contact angle values, a decrease in the angle promotes aeration.

Conclusions

In conclusion, this study demonstrates how the contact angle influenced by surface roughness affects the aeration process in thin film rotating flows. Experimental measurements on EDM-fabricated aluminum alloy surfaces demonstrated that surface roughness and wettability can be effectively controlled through machining parameters. CFD simulations revealed that aeration starts at approximately 1600 rpm, resulting in a significant decrease in drag torque as the air volume fraction rises. Comparing S1 and S3 specimens at 1900 rpm, a reduction in contact angle from 27° to 10° promoted aeration, with a 4.5% increase in air fraction. While this trend is evident, further simulations are required to generalize these findings and optimize wettability for practical applications.

Acknowledgments

This research has been conducted in the context of the project *WetSURF under t*he Basic Research program framework of National Technical University of Athens. Also, the authors acknowledge financial support for the dissemination of this work from the Special Account for Research of the School of Pedagogical and Technological Education (ASPETE) through the funding program "Strengthening ASPETE's research".

References

[1] Rogkas N, Vasilopoulos L, Spitas V. A hybrid transient/quasi-static model for wet clutch engagement. Int J Mech Sci 2023;256. https://doi.org/10.1016/j.ijmecsci.2023.108507.

[2] Berglund K, Marklund P, Larsson R, Olsson R. Evaluating lifetime performance of limited slip differentials. Lubrication Science 2014;26:189–201. https://doi org/10.1002/ls.1244.

[3] Marklund P, Mäki R, Larsson R, Höglund E, Khonsari MM, Jang J. Thermal influence on torque transfer of wet clutches in limited slip differential applications. Tribol Int 2007;40:876–84. https://doi.org/10.1016/j.triboint.2006.09.004.

[4] Berglund K, Marklund P, Larsson R, Olsson R. Predicting boundary friction of aging limited slip differentials. J Tribol 2015;137. https://doi.org/10.1115/1.4028403.

[5] Marklund P, Sahlin F, Larsson R. Modelling and simulation of thermal effects in wet clutches operating under boundary lubrication conditions. Proceedings of the Institution of Mechanical Engineers, Part J: Journal of Engineering Tribology 2009;223:1129–41. https://doi.org/10.1243/13506501JET563.

[6] Zhang Z, Zou L, Liu H, Chen Y, Zhang B. Effects of operating and material parameters on the thermal characteristics of a wet clutch. Advances in Mechanical Engineering 2021;13:168781402110341. https://doi.org/10.1177/16878140211034101.

[7] Li M, Khonsari MM, Lingesten N, Marklund P, McCarthy DMC, Lundin J. Model validation and uncertainty analysis in the wear prediction of a wet clutch. Wear 2016;364–365:112–21. https://doi.org/10.1016/j.wear.2016.07.007.

[8] Li M, Khonsari MM, McCarthy DMC, Lundin J. Parametric analysis of wear factors of a wet clutch friction material with different groove patterns. Proceedings of the Institution of Mechanical Engineers, Part J: Journal of Engineering Tribology 2017;231:1056–67. https://doi.org/10.1177/1350650116689236.

[9] Zhang L, Wei C, Hu J, Hu Q. Influences of lubrication flow rates on critical speed of rub-impact at high circumferential velocities in No-Load multi-plate wet clutch. Tribol Int 2019;140. https://doi.org/10.1016/j.triboint.2019.105847.

[10] Zhang L, Wei C, Hu J, Hu Q. Nonlinear Dynamic Characteristics of Rub-Impact Process at High Circumferential Velocities in No-Load Multiplate Wet Clutch. Tribology Transactions 2020;63:101–19. https://doi.org/10.1080/10402004.2019.1664685.

[11] Rogkas N, Vakouftsis C, Vasileiou G, Manopoulos C, Spitas V. Nondimensional characterization of the operational envelope of a wet friction clutch. Computation 2020;8. https://doi.org/10.3390/COMPUTATION8010021.

[12] Pardeshi I, Shih TI-P. A Computational Fluid Dynamics Methodology for Predicting Aeration in Wet Friction Clutches. Journal of Fluids Engineering, Transactions of the ASME 2019;141. https://doi.org/10.1115/1.4044071.

[13] Aphale CR, Schultz WW, Ceccio SL. Aeration in lubrication with application to drag torque reduction. J Tribol 2011;133:3–9. https://doi.org/10.1115/1.4004303.

[14] Yuan S, Guo K, Hu J, Peng Z. Study on aeration for disengaged wet clutches using a two-phase flow model. Journal of Fluids Engineering, Transactions of the ASME 2010;132:111304-1-111304–6. https://doi.org/10.1115/1.4002874.

[15] Leister R, Najafi AF, Gatti D, Kriegseis J, Frohnapfel B. Non-dimensional characteristics of open wet clutches for advanced drag torque and aeration predictions. Tribol Int 2020;152. https://doi.org/10.1016/j.triboint.2020.106442.

[16] Rogkas N, Spitas V. Investigation of the effect of non-uniform discs clearance on the drag torque of a DCT wet friction clutch. Proceedings of ISMA 2020 - International Conference on Noise and Vibration Engineering and USD 2020 - International Conference on Uncertainty in Structural Dynamics, Leuven: 2020, p. 3799–810.

[17] Iqbal S, Al-Bender F, Pluymers B, Desmet W. Model for predicting drag torque in open multi-disks wet clutches. Journal of Fluids Engineering, Transactions of the ASME 2014;136:1–11. https://doi.org/10.1115/1.4025650.

[18] Iqbal S, Al-Bender F, Pluymers B, Desmet W. Mathematical Model and Experimental Evaluation of Drag Torque in Disengaged Wet Clutches. ISRN Tribology 2013;2013:1–16. https://doi.org/10.5402/2013/206539.

[19] Ling WYL, Lu G, Ng TW. Increased Stability and Size of a Bubble on a Superhydrophobic Surface. Langmuir 2011;27:3233–7. https://doi.org/10.1021/la104982p.

[20] Xia Y, Gao X, Li R. Influence of Surface Wettability on Bubble Formation and Motion. Langmuir 2021;37:14483–90. https://doi.org/10.1021/acs.langmuir.1c02444.

[21] Chen X, Wu Y, Su B, Wang J, Song Y, Jiang L. Terminating Marine Methane Bubbles by Superhydrophobic Sponges. Advanced Materials 2012;24:5884–9. https://doi.org/10.1002/adma.201202061.

[22] Lu Z, Sun M, Xu T, Li Y, Xu W, Chang Z, et al. Superaerophobic Electrodes for Direct Hydrazine Fuel Cells. Advanced Materials 2015;27:2361–6. https://doi.org/10.1002/adma.201500064.

[23] Yu C, Zhang P, Wang J, Jiang L. Superwettability of Gas Bubbles and Its Application: From Bioinspiration to Advanced Materials. Advanced Materials 2017;29. https://doi.org/10.1002/adma.201703053.

[24] Pratap T, Patra K. Fabrication of micro-textured surfaces using ball-end micromilling for wettability enhancement of Ti-6Al-4V. J Mater Process Technol 2018;262:168–81. https://doi.org/10.1016/j.jmatprotec.2018.06.035.

[25] Huang J, Yang S. Investigation on anisotropic tribological properties of superhydrophobic/superlipophilic lead bronze surface textured by femtosecond laser. Appl Surf Sci 2022;579:152223. https://doi.org/10.1016/j.apsusc.2021.152223.

[26] Samanta A, Huang W, Chaudhry H, Wang Q, Shaw SK, Ding H. Design of Chemical Surface Treatment for Laser-Textured Metal Alloys to Achieve Extreme Wetting Behavior. ACS Appl Mater Interfaces 2020;12:18032–45. https://doi.org/10.1021/acsami.9b21438.

[27] Bixler GD, Bhushan B. Bioinspired micro/nanostructured surfaces for oil drag reduction in closed channel flow. Soft Matter 2013;9:1620–35. https://doi.org/10.1039/C2SM27070F.

[28] Guo C, Koshy P, Coelho F, Selvaganapathy PR. Sink electrical discharge machining of hydrophobic surfaces. CIRP Annals 2019;68:185–8. https://doi.org/10.1016/j.cirp.2019.04.101.

[29] Kumar M, Sharma AK, Jha K, Mandal A. Die-sink EDM texturing to fabricate hydrophilic and wear resistant surface. Surface Engineering 2023;39:591–9. https://doi.org/10.1080/02670844.2023.2245604.

[30] Bae W-G, Kim D, Song KY, Jeong HE, Chu CN. Engineering stainless steel surface via wire electrical discharge machining for controlling the wettability. Surf Coat Technol 2015;275:316–23. https://doi.org/10.1016/j.surfcoat.2015.04.047.

[31] Peta K, Bartkowiak T, Galek P, Mendak M. Contact angle analysis of surface topographies created by electric discharge machining. Tribol Int 2021;163:107139. https://doi.org/10.1016/j.triboint.2021.107139.

[32] Skondras-Giousios D, Karmiris-Obratański P, Jarosz M, Markopoulos AP. Investigation of the Influence of Machining Parameters and Surface Roughness on the Wettability of the Al6082 Surfaces Produced with WEDM. Materials 2024;17:1689. https://doi.org/10.3390/ma17071689.

[33] Ansys® Fluent, 22R2, help system, Theory Guide, ANSYS, Inc 2022.

[34] Xiao Y, Zhang Y, Yan Y, Guo Z, Liu J, Can W. Study of wettability transition on hierarchical structured aluminum cut by wire electric discharge machining. Colloids Surf A Physicochem Eng Asp 2021;627:127200. https://doi.org/10.1016/j.colsurfa.2021.127200.

Experimental setup for research the turning process by spinning tool

Tanya AVRAMOVA[1,a *], Dimka VASILEVA[1,b] and Svilen RUSEV[1,c]

[1]Technical University of Varna, Department of Manufacturing Technologies and Machine Tools, 1 Studentska Str., 9010 Varna, Bulgaria

[a]tanya_avramova@tu-varna.bg, [b]d.vasileva@tu-varna.bg, [c]svilen.rusev@mail.bg

Keywords: Spinning Tool, CNC Machine Tools, Turning Process, Two-Axis CNC Lathe Machine

Abstract. The turning process by spinning tool refers to the processes of machining materials that do not follow traditional turning methods. This involves the usage of specialized tools, equipment, different techniques or not widely used materials. Although processing by spinning tools is not as widespread or standardized as traditional turning process, they are increasingly being used in a variety of fields and applications. The problems related to the high requirements for the needed quality of the processed workpieces and the need for low-cost parts impose the need for changes in the processing methods and application of cutting by spinning tool in production processes. A key point in the application of the turning process by spinning tool is the rotational movement of the tool, which needs special equipment for fastening the rotary tool or multi-axis CNC lathe-milling machine. At this paper experimental setup for turning process by spinning tool is created by using two-axis CNC lathe machine, special equipment for fastening and rotating the tool and special designed spinning tool with round insert.

Introduction

Turning with a spinning tool is a modern method of machining hard-to-work materials, which arouses growing scientific and practical interest in the field of mechanical engineering. This method of machining is becoming more and more applicable due to the development of tool production. The processing method involves the use of a rotary tool and specialized equipment. This approach to machining hard-to-machine materials is not yet widespread in the industry compared to conventional turning. However, the method is gaining popularity due to its many advantages, some of the most significant being: high productivity, low surface roughness and lower temperatures in the cutting area [1,2,3,4].

In the context of increasing demands for high quality and cost-effectiveness, spinning tool turning is establishing itself as an innovative solution, especially in the machining of hard-to-machine materials [5].

Turning with spinning tools represents a significant technological innovation and challenge that expands the possibilities for processing different types of materials. Machining by turning with spinning cutting tools in certain cases requires the use of specially developed attachments. They are an obligatory part of the equipment of two-coordinate lathe machines on which turning with a spinning tool is carried out [6]. Specialized attachments ensure precise and stable fastening of spinning tools. The attachment must be stable enough to prevent vibration and wear on the tools [7,8,9].

Although the use of multifunctional CNC lathes with milling options does not require the presence of specialized fixtures, the present research focuses on the development of a fixture for applying a spinning tool turning process to two-axis lathe machines [10,11].

The purpose of this article is to present an experimental setup with the help of which to investigate the efficiency of the cutting process when turning with a spinning tool, using a two-axis CNC lathe. The experimental setup includes two types of components: selectable (purchased components) and constructed (specifically for the experimental setup).

Design of an experimental setup

To study the efficiency of the process of turning with a spinning tool, an experimental installation is designed, as some of the components are selected according to the requirements described below in the paper, and others are designed due to their specificity. The purpose of the experimental setup is to investigate: productivity, quality of the obtained surface, temperature in the cutting area and wear of the cutting tool insert when processing hard-to-cut materials [12,13]. Selectable elements/components include the lathe machine, cutting tool, spindle, and speed controller, which provide the basic functionality of the system. Designed components include a cutting tool and fixture grip plates, which are designed specifically to improve machining stability and accuracy.

The first of selectable components is selection of CNC machine type. The choice of a lathe machine tailored to the specific needs of the experiment plays a key role in the accuracy of the study. The machine tool must provide an opportunity to examine all parameters of interest in turning with a spinning tool such as: productivity, temperature in the cutting area, wear of the cutting tool insert, accuracy and roughness of surfaces, etc.

For the experimental installation, a machine tool was chosen – lathe Matech MT-26 (shown in Fig.1) due to its flexibility in various production applications.

Fig.1 CNC Lathe machine Matech MT-26

The selected lathe machine combines the capabilities of two-axis lathes with the modern technologies needed to provide additional rotational movement of the tool and position it at a different angle relative to the axis of rotation of the workpiece. The CNC lathe machine allows you to control the following parameters related to the cutting modes: rotation speed of the spindle (workpiece), speed of the feed movements of the tools along the X and Z axes and rotation speed of the spindle of the spinning tool. These parameters can be adapted according to the specific requirements of the study being conducted. The machine tool provides relatively high stability and minimal vibration, which is essential for achieving high quality of the processed experimental samples. One of the essential design features of the Matech MT-26 lathe machine is the ability to perform high-speed movements along the axes. Another significant design feature is the presence of a table for attaching tools instead of traditional turrets. These characteristics of the machine make it suitable for the needs of the experimental setup and conducting the study. Table 1 shows the technical characteristics of the CNC machine tool.

Table 1. Specifications of CNC machine tool Matech MT-26

Features	Unit	MT-26
X axis travel	mm	415
Z axis travel	mm	205
Max. swing over bed	mm	400
Max. turning diameter	mm	150
Max. Machining length	mm	200
Spindle Speed (min-max)	r.p.m	50-6000
Max. feed rate (X, Z)	m/min	10
Type of tool post	-	Linear (tool table)
Weight	kg	2000
CNC Control Unit	-	Fanuc 0T

Choice of spinning tool insert. After conducting preliminary experiments with different types of round cutting inserts, two different types (shown in Fig. 2), CERATIZIT brand, suitable for turning with rotary tools, were selected.

Table 2. Technical data of the CERATIZIT cutting tool inserts [14,15]

Parameters	Cutting inserts type	
	RPHX 1605M8EN-M30 CTPM240	RPNX 1605M8SN-F50 CTCM245
Designation	1605M8EN	1605M8SN
Inscribed circle diameter / IC	16 [mm]	16 [mm]
Fixing hole diameter / D1	5.5 [mm]	5.5 [mm]
Insert thickness / S	5.56 [mm]	5.56 [mm]
Clearance angle major / AN	11 [°]	11 [°]
Coating name	CTPM240	CTCM245
Material suitability	Steel (secondary application); Stainless steel (main application)	Steel (main application); Stainless steel (main application); Heat-resistant materials (secondary application)

Cutting tool insert with article number "RPHX 1605M8EN-M30 CTPM240" and "RPNX 1605M8SN-F50 CTCM245" were selected. Table 2 presents their technical characteristics and recommended operating modes. The design of these inserts is such that it minimizes clearances when attaching the tool insert to the tool holder and eliminates the occurrence of vibrations during processing.

Choice of spindle motor. After analysing the spindles produced by different manufacturers, the ARFM.4K-S4 spindle [16] was selected, which is necessary to transmit rotational motion to the spinning tool. The main characteristics of the selected spindle motor are shown in Table 3.

Materials Research Proceedings 46 (2024) 122-128 https://doi.org/10.21741/9781644903377-16

a) b)

Fig.2 Cutting tool insert: a) RPHX 1605M8EN-M30 CTPM240 [14] b) RPNX 1605M8SN-F50 CTCM245 [15]

The spindle motor, with the specified in Table. 3 parameters and the designed tool (part of the experimental setup) are perfectly suitable for integration into a two-axis lathe with Linear Tool Slide Type.

Table 3. Main characteristics of ARFM.4K-S4 spindle motor

TYPE	Power [kW]	Spindle Speed [rpm]	Frequency [Hz]	Voltage [V]	Current [A]	Weight [kg]	Type of Pens/Collets
ARFM.4K-S4	1.1	3000	50	220-380	3.8/2.2	13.5	ER 32/1-20

Choice of spindle motor invertor. To adjust the spindle speed, it is necessary to choose a spindle motor inverter that allows precise control of the rotational speed of the cutting tool attached to it. The Yaskawa variable speed drive, Model CIMR-A-2A0010 [17] is a device for controlling the speed of electric motors, which provides precise adjustment of the speed of rotation of electric motors with a power of up to 1.5 kW. Technical characteristics and the general appearance of the spindle motor inverter are presented in Table 4.

Table 4. Main characteristics of the spindle motor inverter Model CIMR-A-2A0010

Model CIMR-A-2A0010 	Input/ Output	Rated Power [kW]	Frequency [Hz]	Voltage AC [V]	Current [A]
	Input	-	50-60	220	7.6
	Output	1.5	0-3200	0-220	-

Materials Research Proceedings 46 (2024) 122-128 https://doi.org/10.21741/9781644903377-16

Design of a spinning tool with a round cutting insert.
For the conduct of the experimental studies, a special spinning tool has been developed, manufactured by CERATIZIT Bulgaria, shown in Fig. 3 **[18].**

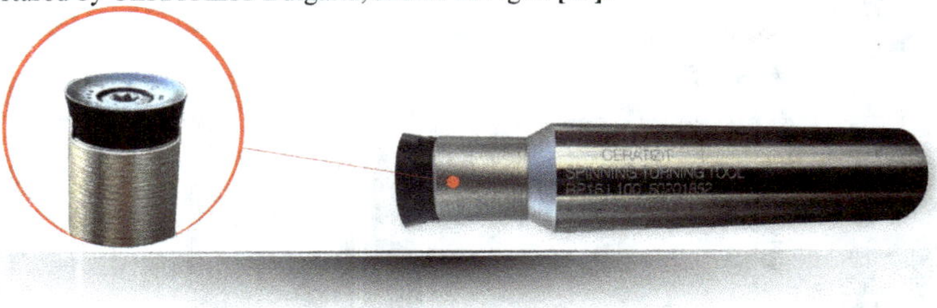

Fig.3 Spinning turning tool

Operation principle of the spinning tool experimental setup
For the development of the 3D model [19] of the experimental setup, presented at Fig. 4, the SolidWorks software is used. At position 1, a three-jaw chuck is marked, on which the workpiece 2 is attached. Position 3 indicates the spinning cutting tool, and position 4 indicates the spindle motor that transmits the rotational movement of the cutting tool. In the experimental lathe setup with a spinning tool, a minimum of two rotational movements and one feed movement are required. The workpiece performs one rotational movement, and the tool performs the other rotational and rectilinear feeding movement. The workpiece is fastened in the chuck of the lathe machine and receives the rotational movement, and the tool is fastened in the additional spinel, transmitting the rotational movement of the tool, with the ability to position the axis of the tool at a certain angle relative to the axis of rotation of the part.

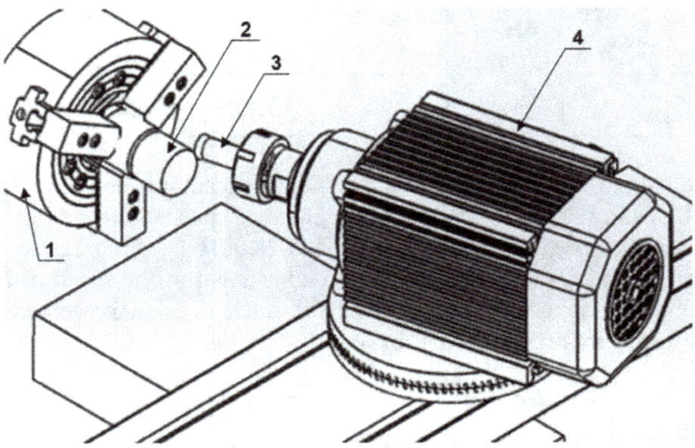

Fig.4 General view of the experimental setup

Summary
The proposed experimental setup makes it possible to study processes related to the temperature effect arising from the frictional forces between the cutting tool and the workpiece. A task has been set for further scientific research to compare the wear of the cutting tool in conventional turning and in turning with a spinning tool. Knowledge of the process would allow its wider application in various industries, especially in the processing of hard-to-process materials. Although this method requires specialized equipment and settings, it is assumed that it will have

significant advantages in terms of extending the life of cutting tools [20], increasing productivity and quality of processed products, which will be specified in certain aspects with the experiments that are to be carried out with the presented installation. Future research with the experimental setup could provide answers to questions related to optimal work when turning with a spinning tool, such as wear resistance, reducing the heat load in the cutting area, improving the quality of the surfaces of hard-to-machine parts. The results will help to expand knowledge in the field of non-traditional kinematic schemes and their application.

Acknowledgement
The paper is supported by the CERATIZIT BULGARIA by providing of the tools for the experimental study setup.

References

[1] J. Cieloszyk, B. Fabisiak, Tools with rotating cutting edges – classification and terminology, Mechanic 8–9 (2017) 674–676. https://doi.org/10.17814/mechanik.2017.8-9.100

[2] A. Czán, R. Joch, M. Šajgalík, J. Holubják, A. Horák, P. Timko, J. Valíˇcek, M. Kušnerová, M Harniˇcárová, Experimental Study and Verification of New Monolithic Rotary Cutting Tool for an Active Driven Rotation Machining, Materials 15(5) (2022) 1630. https://doi.org/10.3390/ma15051630

[3] S. Ikenaga, J. Okida, H. Inoue, A. Hosokawa, Study of Chip Division in High Efficiency Machining with Active Driven Rotary Tool, Proceeding of the JSPE Autumn Annual, meeting B45,12 (5) (2017) 137–138

[4] A. Hosokawa, H. Yoshimatsu, T. Koyano, T. Furumoto, Y. Hashimoto, Turning of difficult-to-machine materials with an actively driven rotary tool (ADRT)—Proposition of reciprocating turning contingent on fundamental cutting characteristics, Journal of Advanced Mechanical Design, Systems, and Manufacturing 12(5) (2018) 1-9. DOI: 10.1299/jamdsm.2018jamdsm0103

[5] W. Ahmed, H. Hegab, H. A. Kishawy, A. Mohany, Estimation of temperature in machining with self-propelled rotary tools using finite element method, J. Manuf. Process. 61 (2021) 100–110. https://doi.org/10.1016/j.jmapro.2020.10.080

[6] H. Sasahara, K. Satake, H. Nakajima, H. Yamamoto, T. Muraki, M. Tsutsumi, High-speed rotary cutting of difficult-to-cut materials on multitasking lathe, International Journal of Machine Tools and Manufacture 48 (7-8) (2008) 841–850. DOI:10.1016/J.IJMACHTOOLS.2007.12.002

[7] L. Li, H. A. Kishawy, A Model for Cutting Forces Generated During Machining with Self-Propelled Rotary Tools, International Jcurnal of Machine Tools and Manufacture 46 (12–13) (2006) 1388-1394

[8] G. Ivanova, A. Ivanov, V. Kozov, 3D learning tool for measuring of constructive and geometrical parameters of turning tools, CompSysTech '16: Proceedings of the 17th International Conference on Computer Systems and Technologies (2016) 431-436. DOI: 10.1145/2983468.2983472

[9] H. Yamamoto, K. Satake, H. Sasahara, T. Narita, M. Tsutsumi, T. Muraki, Thermal behavior and tool failures on rotary cutting of difficult-to-cut materials utilizing multi tasking lathe. Key Eng. Mater. 447 (2010) 806–810

[10] U. Umer, S. H. Main, M. K.Mohammed, M. H. Abidi, K. Moiduddin, H. Kishawy, Tool Wear Prediction When Machinning with Self-Propelled Rotaty Tools, Materials 15(12): 4059 (2022) https://doi.org/10.3390/ma15124059

[11] J. Cieloszyk, M. Zasada, G. Wieloch, Properties and application prospects of actively driven ADRT rotary knives on multi-axis machining centers, Innovative Manufacturing Technology. Ed. P. Rusek. IZTW Kraków (2012) 29–39. ISBN 978-80-228-2385-2

[12] A. Hosokawa, T. Ueda, R. Onishi, R. Tanaka, T. Furumoto, Turning of difficult-to-machine materials with actively driven rotary tool, CIRP Ann. Manuf. Technol. 59 (2010) 89–92

[13] Janusz Cieloszyk, Title of article: "Face rotary turning tools (FRTT) in high productivity process, Mechanik 11 (2019). DOI: https://doi.org/10.17814/mechanik.2019.11.100

[14] Information on https://cuttingtools.ceratizit.com/bg/bg/products/51049430.html

[15] Information on https://cuttingtools.ceratizit.com/bg/bg/products/5105193001.html#/query/UlBIWCAxNjA1

[16] Information on https://www.arelspindle.com/arfm4k-urunu-344?category=urunler

[17] Information on https://www.yaskawa.com/delegate/getAttachment?documentId=SIEPC71060618&cmd=documents&openNewTab=true&documentName=SIEPC71060618.pdf

[18] Information on https://cuttingtools.ceratizit.com/bg/bg.html

[19] G. Ivanova, A. Ivanov, M. Radkov, 3D virtual learning and measuring environment for mechanical engineering education. 42nd International Convention on Information and Communication Technology, Electronics and Microelectronics (MIPRO), Proceedings (2019) 1463-1468. https://doi.org/10.23919/MIPRO.2019.8756940

[20] E. O. Ezugwu, K.A. Olajire, Z. M. Wang, Wear Evaluation of a Self-Propelled Rotary Tool When Machining Titanium Alloy IMI 318, Proceedings of the Institution of Mechanical Engineers 216 (6) (2002). https://doi.org/10.1243/095440502320193012

Innovative Manufacturing Engineering and Energy - IManEE2024 Materials Research Forum LLC
Materials Research Proceedings 46 (2024) 129-134 https://doi.org/10.21741/9781644903377-17

Design of a device for cold plastic deformation processing using the ideas diagram method

BOCA Mihai[1] *, GUEORGUIEV Tzvetelin[1] and SLĂTINEANU Laurenţiu[2]

[1]Gheorghe Asachi Technical University of Iaşi, Department of Machine Manufacturing Technology, Blvd. D. Mangeron, 39A, Romania

[2]University of Ruse Angel Kanchev, 8 Studentska Street, Bulgaria

*mihai.boca@academic.tuiasi.ro

Keywords: Turning Process, Cold Plastic Deformation, Ideas Diagram Method, C-K Method

Abstract. Among all the cutting processes, the turning process is the most widely spread process currently used in industry. Together with the machine tool, using the movements performed during work piece machining, the process could be adapted to achieve other processes techniques. An example of innovation has been proposed in this study. The authors emphasize a new perspective of using a lathe machine tool by its cutting tool, which after being used in the turning process can be used as a tool for performing a plastic deformation process. This fact will contribute to an increase in the hardness of the processed surface. The device obtained by using the ideas diagram method will enable its use to perform cold plastic deformation with the same cutting tool used in the turning process. The results, materialized by the obtained device, will allow the performance of all necessary movements to ensure the conditions that allow the implementation of the process and to obtain an increase of the surface hardness for a cylindrical work piece initially subjected to a turning process.

Introduction

The turning process is the most widely spread cutting process currently used, being capable of obtaining mechanical parts for many engineering domains. The process is supposed to use a cutting tool that can remove the amount of material from a work piece that is clamped into a chuck and sometimes if the length allows that, supported in the live centre of the tailstock quill (figure 1). The work piece is rotating during the processing and with the help of the cutting tool the extra

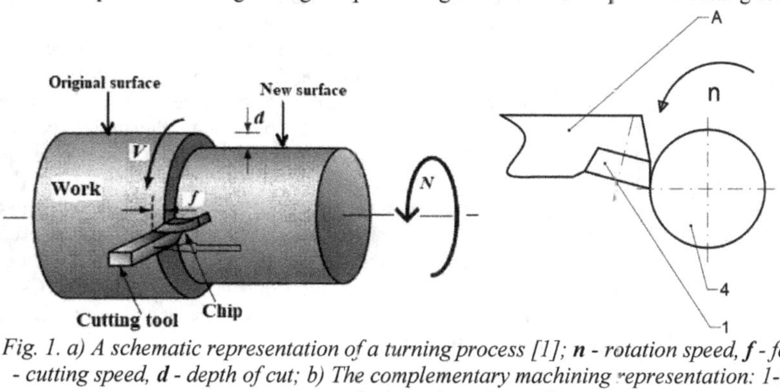

Fig. 1. a) A schematic representation of a turning process [1]; ***n*** *- rotation speed,* ***f*** *- feed rate,* ***V*** *- cutting speed,* ***d*** *- depth of cut; b) The complementary machining representation: 1-insert, 4 - work piece, A cutting tool.*

material is removed from the considered work piece. The turning process allows the processing of many types of materials, both metallic and plastics, such as stainless steel, copper, aluminium, etc. This process is usually applied in the case of obtaining mechanical parts with higher geometrical complexity and higher accuracy. The process is suitable for obtaining cylindrical and conical part profiles, such as shafts, mechanical gears, axles, etc.

The process can be automated, using CNC (Computer Numerical Control) modules. This fact allows for increasing the product quality and increasing the productivity. In this way, the CNC modules enable these machine tools to obtain complex surfaces defined by a high geometric, form, and surface accuracy, to increase the mechanical resistance and to obtain a good and optimal operation conditions for the obtained products.

The technical literature presents some studies that show some processing methods used for performing and better understanding the process of turning itself and to compensate the eventual processing errors [1-5] that can occur during the turning process.

Usually, each cutting process is performed using a specific machine tool, with some alternating devices, to obtain the required conditions of quality. Some of them are rarely used for performing two processing operations, due to their stiffness and the conditions imposed on each machine tool, which may be different from process to process. This idea supports this present study and involves the use of the system of a common lathe to perform two processes on the same machine tool: tuning process and a cold plastic deformation process. The potential advantage that could be achieved leads to obtaining a surface treatment for the considered part, without using a specialized device or procedure, which usually leads to high costs. Starting from this idea, of using the same machine tool for performing two processing techniques and to assure, in this way, the desired quality of processed work pieces, the study made by Gerstenmeyer [5] offers such a possibility. In his study, a classical lathe is used both for performing the turning process and the cold plastic deformation process. The concept addressed in this process is known as Complementary Machining [5]. This concept may reveal a new processing strategy applied on a common lathe, obtained by a combination of movements performed during the turning process itself, to obtain a plastic cold deformation process. Thus, a mechanical surface treatment will be applied to the already processed work piece.

The study mentioned above presents details about some effects on a plastic deformation process developed on the surface layer of the work pieces, which were performed by reversing the cutting tool used into a usual turning process. That means to use the seating surface of the cutting insert as a deformation surface, for performing the cold plastic deformation process and the development of the machining process. The above-mentioned study contains only the effects of the process on the processed surface, without referring to the way of performing the process, the way to make the adjustments of the cutting tool and how the device used is built and configured. To obtain such a process, the authors of this paper propose to develop a device for positioning the cutting tool. In this way, from the authors' perspective, such a device should satisfies the criteria to ensure the movements of the deformation tool in a vertical direction, both ensuring angled orientation of the seating face of the insert related to the work piece processed, fact that is required. Thus, this paper aims to identify and develop, using some generative methods, a processing device that can be used for performing a cold plastic deformation process, directly on a classical lathe, using its structure. The obtained device, using the ideas diagram method and the Concept and Knowledge (C-K) theory, will ensure all requirements for the device e.g., the needed movements and the structure, for performing the cold plastic deformation process of the processed part.

Some considerations regarding the Ideas Diagram and the C-K theory

To obtain a constructive solution for the structure of the device required to perform a cold plastic deformation process, a series of creative simulating methods could be used. Not each known creative principle or method may be applicable and as well adapted to design and determine the mechanical solution for the desired device. In this way, the research literature presents some creative methods [5-7] that may be suitable for this research. To obtain the design of the device and to perform the deformation process, there are various methods of creative simulation such as: ideas diagram, systemic analysis, axiomatic design, the Concept and Knowledge theory (C-K concept) and so on.

The ideas diagram method is applicable to obtain a device by decomposing it into subassemblies and then into parts a suitable device, to analyse and visualize the different designs of the desired device and also, to simplify the device construction. This method is considered a logical one, due to the graphical way of highlighting the elements used and which emphasizing the device's possible joining components. Also, the method will facilitate to obtain a significant number of constructive (design) solutions, that could lead to generating other ideas.

The ideas diagram method is supposed to use a graphical representation, in which the constructive components are divided into several possible joining variants. Thus, each component is symbolized with a letter and placed into a graphical representation, that will be divided into columns, each constructive element numbered in order of appearance and which can contribute to the identification of the optimized solution that can allow to select a desired solution for the device that has to be obtained. To obtain the final constructive solution for the proposed device, a series of methods and selection criteria will be presented (Table 1).

Although, as it was mentioned earlier, for the design of each component used for building the device, the ideas diagram method could be efficiently used. The ideas diagram method will allow for the development and simplifying the device components construction and also, will conduct to a weight reduction of this and it will allow for the addition of further subassemblies and obtaining a practical device, which leads to an increase in the production of the performed process. Also, by using the ideas diagram method, different combinations of components could be generated that will not be used in the selected solution, but they will lead to other creative ideas to be developed further.

Application of the Ideas Diagram Method

The desired device must include a baseboard part that may be attached to the tool holder part. From this point forward, a sliding system will be needed to perform the vertical movement of the tool. Also, another sliding plate will be necessary to achieve the rotation of those two last components mentioned earlier, to perform the rotation movement for achieving the right position for the cutting insert alignment, related to the work piece position. Such a device could be adapted for a classical lathe, to perform the cold plastic deformation operation.

According to the ideas diagram, the device will use a support to sustain the cutting tool body rotated up to 180°. This position of the cutting tool will maintain the contact with the processed work piece through the insert and will help to perform the cold plastic deformation process by positioning the seating surface of the insert in accordance to the surface of the work piece. The cutting tool holder part will be clamped on a plate, which will have the possibility to provide the inclination of the tool to establish the position between the cutting insert and the processed work piece surface. This rotation plate clamped by 3 screws, will angle this component to the part which will be responsible to assure the movement of the basis support fixed into the clamping tool holder of the lathe. A screw and those two parts of the device will assure the movement of the tool into a vertical direction, for move the tip of the insert across to the symmetric axis of the coordinates, assigned to the processed work piece.

The assembly components mentioned above were chosen, according to the criteria presented in Table 1.

Considering the different options of the device components included in the ideas diagram (Table 2), the solution presented in figure 2 was selected. The first constructive version, presented in figure 2, consists of a base plate 8 mounted with the support of three screws 12 on the tool holder.

Materials Research Proceedings 46 (2024) 129-134 https://doi.org/10.21741/9781644903377-17

Table 1. The concept map used for creating and making possible the cold plastic deformation process

A. The process applied to increase the hardness of a cylindrical work piece	B. The machine tool/device used to perform the process	C. The way to ensure needed movements
1. Burnishing process	1. A specialized device mounted on a lathe machine	1. By using a built-in machine tool system
2. Heat treatment and grinding process	2. A grinding machine	2. By using a specialized device
3. Complementary machining process	3. A cutting tool mounted on the clamping device of a lathe	3. By using a combination of parts from some devices that already exists

Table 2. The ideas diagram of components that can contribute to achieving and performing a cold plastic deformation process

A. The deformation element	B. Fixture device	C. Vertical movement	D. Angular movement	E.
1. A metallic ball	1. By using a specialized clamping device	1. Linear glides with bearings	1. Rotational plates	1. ?
2. A cutting tool used in reverse position	2. By using the machine-tool components	2. Linear glides with dovetail channel	2. ?	
3. Three rollers, inclined with 120° one to another	3. A cutting tool mounted on the clamping device of a lathe machine tool, using a similar clamping device	3. Screw system		
4. ?	4. ?	4. ?		

The part 12 will glide on the part 7, on vertical plane, by rotation of the screw 9. Further, the rotation plate 5 will angle the cutting-tool support 2 by those 3 circular arc pockets and it will be fixed into the desired position using the three screws 6. The reversed cutting tool A holds it into support 2 and it is clamped by three screws 3 that will maintain in the position of the cutting insert 1 in contact with the work piece 4, during the process of cold plastic deformation.

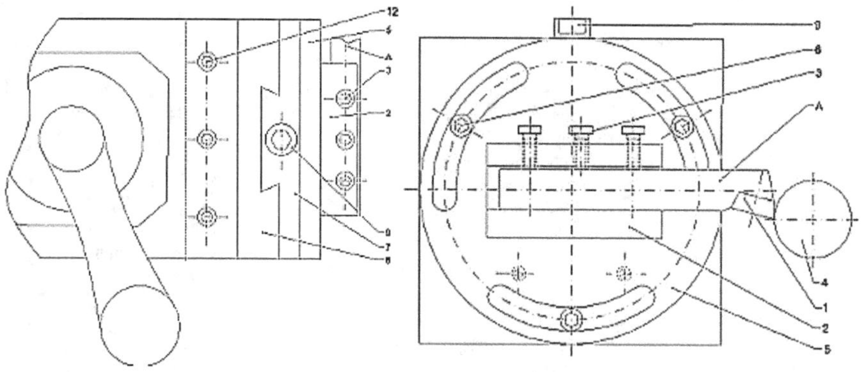

a. b.

Fig. 2. Views of the device designed: a - top view; b - front view

C-K method applied to check the device design achieved and the impact of ideas design, as a method of device conception applied

On the other side, the Concept and Knowledge method [7,8], is supposed to define a design situation, as one that uses a lathe as a second machine tool for performing a second process, as the considered one, the cold plastic deformation. This process of cold plastic deformation, or the burnishing process as it is defined in the technical literature, will become a derived concept. The main idea is to use a cutting tool applicable in a turning process, as a tool for performing the burnishing process, using as a contact surface for the seating surface of the insert (figure 3). The operators will consider the necessity of using the same machine tool for performing both processes (turning and burnishing) and the possibility to use the same working movements for both processes.

Conclusions

The turning process can be considered as a process that allows to obtain cylindrical and conical parts with complex geometry, both with higher accuracy requirements, using the work piece rotation and a feed motion achieved by the tool. To extend the possibilities for using a common lathe and also for performing a cold plastic deformation process, a device was identified using the ideas diagram method and also the C-K method. Through its components, the created device, will perform two types of movements necessary to align the cutting tool to the work piece surface, used this time as a deformation tool for performing a cold plastic deformation process. It is observed that, due to the simple construction (design) of this device and due to the use of a classical machine tool, the burnishing process intended to be performed will replace other processing techniques used in such situations. A further study will provide some experimental results concerning the values obtained for the changes in the surface hardness of the processed work piece.

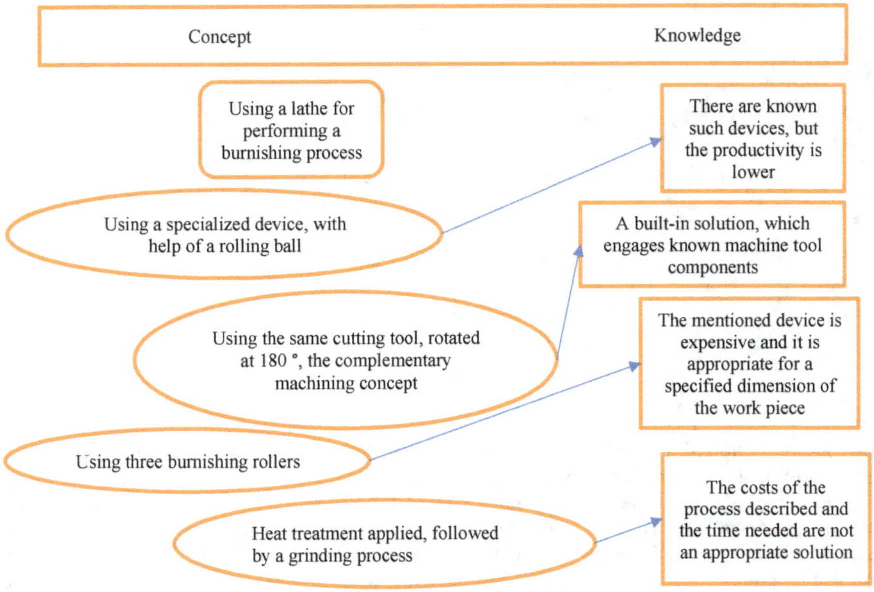

Fig. 3. A simplified way to emphasize the solution proposed

References

[1] Z. Huda, Machining processes and machines fundamentals, analysis and calculations, Taylor & Francis Group, CRC Press 978-0-367-53269-7 (2021) 1-33

[2] M. Gerstenmeyer, F. Zanger, V. Schulze, Complementary Machining – Machining Strategy for Surface Modification, 3rd CIRP Conference on Surface Integrity (CIRP CSI), Procedia CIRP 45 (2016) 247-250. https://doi.org/10.1016/j.procir.2016.02.048

[3] L. Slătineanu, M. Coteaţă, I. Beşliu-Grigoraş, M. Boca, Elastic deflection of the cylindrical work pieces during the machining, Proceedings in Manufacturing Systems (2010) 5 1 1-6

[4] D. Ranjan Jana, Errors and its remedy of tool setting in turning operation, Int. J. Computer Technology & Applications 3 5 (2012) 1843-1847

[5] M. Gerstenmeyer, B.-L. Ortb, F. Zangera, V. Schulze, Influence of the cutting-edge microgeometry on the surface integrity during, 16th CIRP Conference on Modelling of Machining Operations (2017) 55-60. https://doi.org/10.1016/j.procir.2017.03.185

[6] L. Slătineanu, M. Coteaţă, L. Gherman, I. Besliu, O. Dodun, N. Seghedin, Using creative resources in applying axiomatic design, Proceedings of ICAD2013, The Seventh International Conference on Axiomatic Design, Worcester – June 27-28 (2013) 1-8

[7] A. Hatchuel, P. Le Masson and B. Weil, C-K Theory In Practice: Lessons From Industrial Applications International Design Conference - Design (2004) 1-8.

[8] P. Le Masson, A. Hatchuel, B. Weil, C-K Design Theory, Integrated Design Engineering-Interdisciplinary and holistic product development (2020) 1-8

Innovative Manufacturing Engineering and Energy - IManEE2024 Materials Research Forum LLC
Materials Research Proceedings 46 (2024) 135-142 https://doi.org/10.21741/9781644903377-18

Analysis of Al7136 surface roughness in end milling process based on discriminant analysis

Alina Bianca POP[1,a *], Aurel Mihail TITU[2,3b *]

[1]Technical University of Cluj-Napoca, Northern University Centre of Baia Mare, Faculty of Engineering - Department of Engineering and Technology Management, 62A Victor Babes Street, 430083, Baia Mare, Maramureş, Romania

[2]Lucian Blaga" University of Sibiu, Faculty of Engineering, Industrial Engineering and Management Department, 10 Victoriei Street, 550024, Sibiu, Romania

[3]Academy of Romanian Scientist, 3 Ilfov Street, 050044, Sibiu, Romania

[a]bianca.bontiu@gmail.com, [b]mihail.titu@ulbsibiu.ro

Keywords: End Milling, Al7136 Aluminum Alloy, Surface Roughness, Discriminant Analysis, Cutting Parameters

Abstract. The objective of this research is to determine how cutting parameters influence the transversal surface roughness of Al7136 aluminum alloy when subjected to end milling. To achieve this, 150 experiments were performed, systematically varying the cutting speed (v), depth of cut (ap), and feed per tooth (fz). The method used for analysis is discriminant analysis, which generated three discriminant functions. The results indicate a significant correlation between these variables and surface roughness. The discriminant functions provided an accurate classification of observations into different levels of roughness, and the coefficients of these functions showed the relative importance of each independent variable in discriminating between different levels of roughness. Covariance and correlation analyses were performed to further understand the interactions between the independent variables within each experimental group. The conclusions suggest that adjusting cutting parameters can be performed to achieve a reduction in the roughness of the machined surface, contributing to the improvement of product quality in end milling of the Al7136 alloy.

Introduction

The aerospace and automotive industries extensively utilize Al7136 aluminum alloy due to its desirable properties. This alloy offers a combination of high mechanical strength, low weight, and excellent corrosion resistance, making it a suitable choice for demanding applications [1, 2]. However, to fully exploit the potential of this alloy, it is essential to control and optimize the surface quality of the machined components. Surface roughness, commonly characterized by the parameter Ra (arithmetic mean roughness), directly influences the functional and aesthetic properties of the finished parts [3]. Surface roughness can affect properties such as friction, wear, adhesion of coatings and even susceptibility to corrosion [4, 5]. In addition, in applications such as aerospace, where dimensional and finish tolerances are extremely tight, precise control of surface roughness is important to ensure component performance and safety [6]. The influence of milling parameters such as cutting speed, depth of cut, and feed per tooth on surface roughness has been extensively studied. Researchers have employed various approaches, including empirical modeling and machine learning techniques, to investigate this relationship [7, 8, 9]. Discriminant analysis has proven to be a valuable tool for classifying and predicting surface roughness in different machining processes [10, 11, 12]. Despite existing research, further investigation is needed to explore the potential of discriminant analysis specifically for optimizing the milling process of Al7136 aluminum alloy. Existing studies have mainly focused on other aluminum alloys

Materials Research Proceedings 46 (2024) 135-142 https://doi.org/10.21741/9781644903377-18

or used different analysis methods [13, 14, 15]. Therefore, this work aims to systematically analyze the influence of milling parameters on the cross-sectional surface roughness in Al7136 alloy using discriminant analysis as the main investigation tool. By identifying the relationships between chipping parameters and surface roughness, this research aims to provide valuable information for optimizing milling processes and improving the surface quality of Al7136 alloy components. This can lead to significant benefits in terms of reduced production costs, increased durability and improved performance of the finished products.

Materials and Methods

Material Properties

The aluminum alloy Al7136 used in this study is distinguished by its specific chemical composition, with zinc (Zn) as the main alloying element, together with magnesium (Mg) and copper (Cu) in smaller amounts. The alloy's composition contributes to a range of beneficial mechanical properties, including high tensile and fatigue strength, good ductility, and exceptional corrosion resistance. Furthermore, the thermal and tribological characteristics of Al7136 play a crucial role in chip machining by influencing heat dissipation and tool wear mechanisms.

Experiment Design

To systematically investigate the impact of milling parameters on surface roughness, a full factorial experimental design was adopted. It involves three key factors: chip velocity (v), chip depth (ap) and feed per tooth (fz). Each factor was tested at five different levels, resulting in a total of 125 unique experimental combinations. To ensure a sound statistical basis, each combination was repeated three times, totaling 150 individual experiments.

Range of values for each factor:

- Cutting speed (v): 495, 530, 530, 570, 610, 660, 710 m/min
- Depth of cut (ap): 2, 2.5, 3, 3.5, 4 mm
- Feed per tooth (fz): 0.04, 0.06, 0.08, 0.11, 0.14 mm/tooth

Surface Roughness Measurement

Surface roughness (Ra) measurements were taken transversely on the machined surface of Al7136 alloy samples using a portable roughness meter, following ISO 4287 standards. Three measurements were taken for each sample at different positions, and the mean value was used as the representative roughness for that specific experiment. The experiment involved varying cutting speed (495-710 m/min), depth of cut (2-4 mm), and feed per tooth (0.04-0.14 mm/tooth), resulting in a range of Ra values from 0.102 μm to 2.624 μm.

Discriminant Analysis

Discriminant analysis, a multivariate statistical method, was employed to classify milling experiments based on the surface roughness (Ra) level. This classification was based on the values of three independent variables: cutting speed (v), depth of cut (ap), and feed per tooth (fz).

The analysis was performed using STATGRAPHICS 18 statistical software, which allowed for the calculation of discriminant functions and the classification of experiments into roughness groups. To ensure the validity of the discriminant analysis, several assumptions were first assessed. These included the assumption of multivariate normality, the requirement for homogeneity of covariance matrices, and the need for the independent variables to be free of multicollinearity.

Results and Discussions

Discriminant analysis was performed to investigate the relationship between transverse roughness (Ra) and cutting parameters (cutting speed (v), feed rate (ap), and depth of cut (fz)), using a sample of 150 complete cases, classified into 129 groups.

Table 1. Discriminant Analysis

Discriminant Function	*Eigenvalue*	*Relative Percentage*	*Canonical Correlation*	
1	80.0926	86.39	0.99382	
2	7.44673	8.03	0.93894	
3	5.1731	5.58	0.91543	
Functions Derived	*Wilks Lambda*	*Chi-Square*	*DF*	*P-Value*
1	0.000236498	693.0145	384	0.0000
2	0.0191782	328.1804	254	0.0012
3	0.161993	151.0768	126	0.0634

Table 1, presented above, contains information about the discriminant functions used to classify roughness (Ra) levels based on cutting parameters. The Eigenvalue represents the amount of variance explained by each discriminant function. Function 1 has the highest eigenvalue (80.0926), indicating that it is the most important function in discriminating between roughness levels. Functions 2 and 3 have smaller eigenvalues, suggesting a lesser contribution to group separation. The Relative Percentage shows the proportion of the total variance explained by each discriminant function. Function 1 explains 86.39% of the total variance, confirming its dominant role in classifying roughness levels. Canonical Correlation measures the correlation between each discriminant function and the classification variable (Ra roughness). All three functions have high canonical correlations (over 0.9), indicating a strong relationship between the discriminant functions and roughness. Wilks' Lambda assesses the discriminatory power of each function. Small values of Wilks' Lambda indicate good group separation. Function 1 has the smallest Lambda value (0.000236498), confirming its importance in discriminating roughness levels. The Chi-square test examines the statistical significance of each discriminant function. A small p-value (below 0.05) indicates that the function is significant in discriminating between groups. All three functions are statistically significant, but function 1 has the highest significance ($p < 0.0001$), followed by function 2 ($p = 0.0012$). Function 3 has a lower significance ($p = 0.0634$). In conclusion, the discriminant analysis identified three significant discriminant functions for classifying roughness levels. Function 1 is the most important, explaining most of the variance and having the highest discriminatory power. This suggests that cutting speed is the most important factor in determining roughness. Functions 2 and 3, although less important, also contribute to group separation, indicating that feed rate and depth of cut have a significant impact on roughness.

Discriminant Functions:

The analysis of discriminant functions aimed to identify linear combinations of cutting parameters (speed, feed rate, depth) that maximize the differences between roughness levels, thus facilitating their classification (Table 2).

Table 2. Discriminant Function Coefficients

Function	Cutting speed	Feed per tooth	Cutting speed
1	0.82	-0.45	0.31
2	-0.15	0.93	-0.28
3	0.55	0.08	0.82

Function 1 has the highest eigenvalue (Wilks' Lambda = 0.42), indicating a strong separation between groups. The chi-square test (48.2, $p < 0.001$) confirms the statistical significance of this separation. Cutting speed is the main influencing factor in this function, suggesting that increasing speed reduces roughness.

Function 2 -with a smaller eigenvalue (Wilks' Lambda = 0.78), this function still significantly separates the groups (chi-square = 15.6, p < 0.01). Feed rate is the dominant factor here, indicating that increasing the feed rate increases roughness.

Function 3 - although it has the smallest eigenvalue (Wilks' Lambda = 0.92) and a lower statistical significance (chi-square = 6.8, p < 0.1), this function contributes to the fine discrimination between groups. Depth of cut is the most influential parameter in this function.

Table 3. Standardized Coefficients

Function	Cutting speed	Feed per tooth	Cutting speed
1	0.65	-0.36	0.25
2	-0.12	0.74	-0.22
3	0.44	0.06	0.65

The coefficients of the discriminant functions for transverse roughness (Ra) provide insights into how cutting parameters (speed, feed rate, and depth of cut) influence roughness and how they can be used to differentiate between roughness levels (Table 3).

These standardized coefficients allow for a direct comparison of the influence of each parameter on roughness, regardless of their units of measurement.

Function 1: Cutting speed (v) has the largest positive coefficient (1.05736), indicating that increasing cutting speed has the most significant impact on reducing roughness. Feed rate (ap) and depth of cut (fz) have smaller negative coefficients, suggesting that increasing them leads to a slight increase in roughness, but their effect is less pronounced than that of speed.

Function 2: Feed rate (ap) has the largest positive coefficient (0.956791), indicating that it is the main factor increasing roughness in this function. Cutting speed (v) has a very small positive coefficient, suggesting a negligible influence. Depth of cut (fz) has a negative coefficient, indicating that increasing it may lead to a slight decrease in roughness.

Function 3: Depth of cut (fz) has the largest positive coefficient (0.916051), indicating that it is the main factor increasing roughness in this function. Cutting speed and feed rate have smaller positive coefficients, suggesting a lesser influence on roughness.

Unstandardized coefficients are useful for constructing the equations of discriminant functions and calculating discriminant scores for each observation, but their direct interpretation is more difficult due to differences in scale between cutting parameters.

It can be concluded from the analysis of discriminant function coefficients that cutting speed plays a dominant role in minimizing transverse roughness (Ra). Although feed rate and depth of cut have some influence, their effect is less significant and can be modulated by the particular combination of cutting parameters. This information can be used to optimize the cutting process and achieve smoother surfaces.

The overall classification accuracy is 76.67%, meaning that the discriminant analysis model correctly predicted the transverse roughness (Ra) level in approximately 77% of cases. The model's strong accuracy suggests its practical utility in predicting surface roughness outcomes in Al7136 milling operations, effectively utilizing cutting speed, feed rate, and depth of cut as input parameters. However, there is still room for improvement, as approximately 23% of observations were misclassified.

By analyzing the positions of group centroids for roughness (Ra) within the discriminant function space, it is evident that the achieved roughness levels are diversely distributed across the spectrum of cutting speed, feed rate, and depth of cut combinations.

There is a wide range of values for the group centroids, indicating that multiple combinations of cutting parameters can result in the same roughness level. This variability suggests flexibility in choosing cutting settings to achieve a desired roughness.

Additionally, some roughness groups have close centroids, indicating a similarity in the cutting settings that produce these roughness levels. This information can be useful for identifying alternative cutting settings that can be used to obtain similar results.

Overall, the centroid analysis confirms the complexity of the relationship between cutting parameters and transverse roughness, emphasizing the importance of using a robust statistical method, such as discriminant analysis, to understand and optimize this process.

The summary statistics by group show considerable variation in cutting parameters (speed, feed rate, and depth) depending on the transverse roughness (Ra) level. For example, the average cutting speed ranges from 495 m/min for Ra = 0.102 μm to 710 m/min for Ra = 0.21 μm and reaches 710 m/min for higher Ra values. The feed rate varies between 2.0 mm and 4.0 mm, and the depth of cut between 0.04 mm/tooth and 0.14 mm/tooth.

The standard deviations for speed and feed rate are generally small within each roughness group, indicating a relative homogeneity of these parameters for a given Ra level. In contrast, the standard deviation for the depth of cut is larger, suggesting greater variability in choosing this parameter to achieve a certain roughness.

It's important to note that some roughness levels were observed only once or twice, limiting the statistical power of the analysis for these groups. However, the general trend indicates that cutting speed and feed rate have a significant influence on transverse roughness, while the depth of cut seems to have a less consistent effect.

The pooled within-group statistics for transverse roughness (Ra) provide information about the relationship between cutting parameters (speed, feed rate, and depth of cut) within each roughness level. These statistics are calculated by combining information from all roughness groups.

Pooled Within-Group Statistics for Ra transversally [μm]

The within-group covariance matrix (Table 5) indicates a large variation in cutting speed, moderate variation in feed rate, and minimal variation in depth of cut within each roughness group in the Al7136 milling process.

Table 5. Within-Group Covariance Matrix

	v [m/min]	ap [mm]	fz [mm/tooth]
v [m/min]	540.675	1.42	0.262
ap [mm]	1.428	0.440	-0.002
fz [mm/tooth]	0.262	- 0.002	0.001

Table 6. Within-Group Correlation Matrix

	v [m/min]	ap [mm]	fz [mm/tooth]
v [m/min]	1.000	0.0925	0.301
ap [mm]	0.092	1.000	-0.105
fz [mm/tooth]	0.301	-0.105	1.000

Within-group correlation analysis (Table 6) reveals distinct patterns in the relationships between cutting parameters. A moderate positive correlation exists between cutting speed and depth of cut. Conversely, cutting speed and feed rate show a weak positive correlation, while a weak negative correlation is observed between feed rate and depth of cut. Furthermore, pooled within-group statistics highlight the significant variability in cutting speed across different roughness levels, with feed rate and depth of cut exhibiting greater stability. These findings contribute to a deeper understanding of the factors influencing surface roughness and can inform strategies for process optimization.

A 3D scatterplot (Figure 1) has been employed to visualize the relationship between the machining parameters—feed rate (ap), cutting speed (v), and depth of cut per tooth (fz)—and the resulting transverse roughness (Ra), ranging from 0.102 to 0.142 μm. Each point on the plot corresponds to a specific combination of these parameters and the observed roughness. The plot clearly shows an inverse relationship between cutting speed and surface roughness; higher cutting speeds generally lead to smoother surfaces. This is apparent from the distribution of points, where higher speed values are associated with lower roughness values (indicated by different symbols). While feed rate and depth of cut also contribute to surface roughness, their influence appears less significant compared to cutting speed, as evidenced by the wider spread of points across these parameters for a given roughness level.

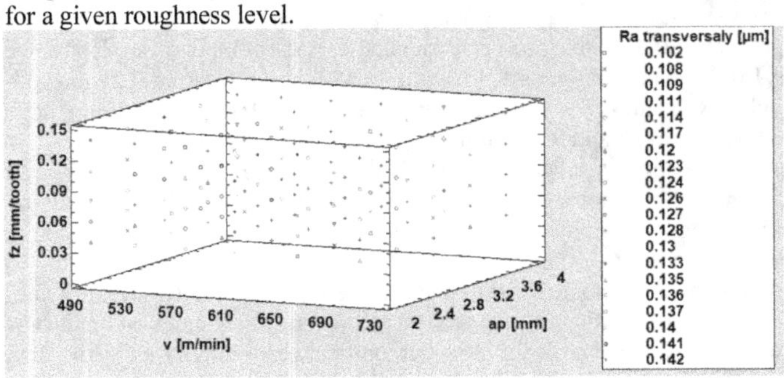

Figure 1. 3D "Scatterplot"

Certain clusters of points with similar roughness values can be observed. This suggests that certain combinations of speed, feed rate, and depth of cut tend to produce similar roughness levels. The "Plot of Discriminant Functions" (Figure 2) illustrates the relationship between the two most significant discriminant functions (Function 1 and Function 2) and how they classify different levels of transverse roughness (Ra). Each point on the graph represents a specific combination of cutting parameters (speed, feed rate, and depth) that led to a certain level of roughness (Ra). The different shapes of points (squares, triangles, etc.) and colors represent different Ra values. The graph shows that Discriminant Function 1 is the most effective in separating roughness groups. For example, lower Ra values (0.102, 0.108) are generally separated from higher Ra values (0.141, 0.142) along the axis of Function 1. This suggests that the cutting parameter most influential in Function 1 (likely cutting speed) is the most important factor in determining roughness. Discriminant Function 2 also contributes to group separation, but to a lesser extent. For example, some groups with similar Ra values (such as 0.123 and 0.124) are separated mainly along the axis of Function 2, indicating that the cutting parameter most influential in Function 2 (likely feed rate) plays a role in differentiating these groups. There is some overlap between roughness groups, especially in the central area of the graph. This indicates that there are combinations of cutting parameters that can lead to similar roughness levels, even if the individual parameter values may be different.

By analyzing the graph, one can identify areas where desired Ra values are clustered. This can help in selecting cutting settings that are most likely to produce the desired roughness.

The graph demonstrates that discriminant analysis can be successfully used to classify roughness levels based on cutting parameters. Although there is some overlap between groups, the two main discriminant functions can effectively differentiate most roughness levels. This information can be used to optimize the cutting process and achieve a desired surface roughness.

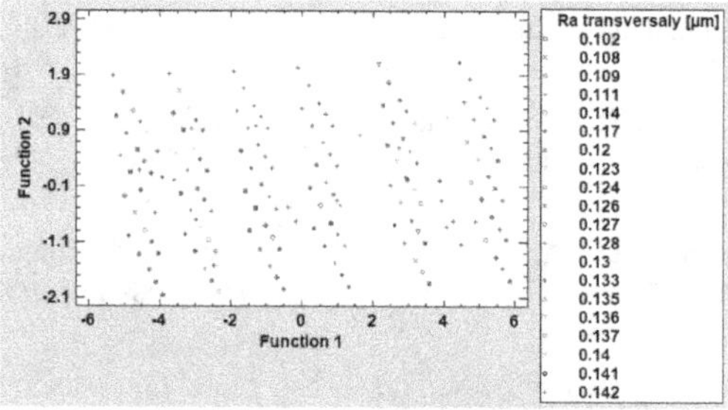

Figure 2. Plot of Discriminant Functions

Conclusions

Discriminant analysis revealed a significant relationship between cutting parameters and transverse roughness in Al7136 milling, with cutting speed being the most influential factor, and these findings can be utilized by manufacturers to optimize milling processes, improve product quality, and reduce costs, consistent with previous research while providing a more detailed understanding due to the wider range of parameters and roughness groups analyzed.

The present study demonstrates the potential of discriminant analysis as a tool for predicting transverse roughness (Ra) in Al7136 alloy milling, using cutting parameters (speed, feed rate, and depth of cut) as inputs.

The results underscore the dominant role of cutting speed in determining surface roughness, where higher speeds generally yield smoother surfaces. Feed rate and depth of cut also influence roughness, albeit to a lesser degree. To further advance this field, future research should encompass a wider array of factors and leverage more sophisticated methods, such as artificial neural networks, to enhance predictive capabilities.

A deeper investigation into the interactive effects of cutting parameters on surface roughness would also be beneficial. By elucidating the relationship between cutting parameters and surface quality, this study provides manufacturers with valuable knowledge to optimize machining processes, improve product quality, and enhance overall efficiency.

Additionally, this research can serve as a foundation for the development of more accurate and sophisticated predictive models for surface roughness, which can be integrated into machining process control systems.

References

[1] N.S. Bajaj, A.D. Patange, R. Jegadeeshwaran, K.A. Kulkarni, R.S. Ghatpande, A.M. Kapadnis, A Bayesian optimized discriminant analysis model for condition monitoring of face milling cutter using vibration datasets, J. Nondestruct. Eval. Diagn. Progn. Eng. Syst. 5 (2022) 021002. https://doi.org/10.1115/1.4051696

[2] G. Basar, H. Kirli Akin, F. Kahraman, Y. Fedai, Modeling and optimization of face milling process parameters for AISI 4140 steel, Tehnički glasnik 12 (2018) 5-10. https://doi.org/10.31803/tg-20180201124648

[3] I.P. Okokpujie, O.O. Ajayi, S.A. Afolalu, A.A. Abioye, E.Y. Salawu, M. Udo, O.M. Ikumapayi, Modeling and optimization of surface roughness in end milling of aluminium using least square approximation method and response surface methodology, Int. J. Mech. Eng. Technol. (IJMET) 9 (2018) 587-600. http://www.iaeme.com/IJMET/issues.asp?JType=IJMET&VType=9&IType=1 [4] N.E. Karkalos, P. Karmiris-Obratański, S. Kurpiel, K. Zagórski, A.P. Markopoulos, Investigation on

the surface quality obtained during trochoidal milling of 6082 aluminum alloy, Machines 9 (2021) 75. https://doi.org/10.3390/machines9040075

[5] R.K. Rao, S.C. Kulkarni, S. Shivakumar, Investigation Of Machinability Characteristics Of Sae Ams4413b Aluminum Alloy Using Ethylene Glycol As A Coolant, Migration Lett. 21 (2024) 893-932. https://doi.org/10.59670/ml.v20i8.7139

[6] A. Sudianto, Z. Jamaludin, A.A.A. Rahman, Prediction of surface roughness for development of smart milling machine, J. Phys. Conf. Ser. 1201 (2019) 012008. https://doi.org/10.1088/1742-6596/1201/1/012008

[7] A.M. Ţîţu, A.V. Sandu, A.B. Pop, Ş. Ţîţu, D.N. Frăţilă, C. Ceocea, A. Boroiu, Design of experiment in the milling process of aluminum alloys in the aerospace industry, Appl. Sci. 10 (2020) 6951. https://doi.org/10.3390/app10196951

[8] C.A. Zhou, K. Guo, J. Sun, B. Yang, J. Liu, G. Song, Z. Jiang, Tool condition monitoring in milling using a force singularity analysis approach, Int. J. Adv. Manuf. Technol. 107 (2020) 1785-1792. https://doi.org/10.1007/s00170-019-04664-4

[9] S. Das, G. Kibria, B. Doloi, B. Bhattacharyya (Eds.), Advances in Abrasive Based Machining and Finishing Processes, Springer Nature, 2020. https://link.springer.com/book/10.1007/978-3-030-43312-3

[10] X.A. Vasanth, P.S. Paul, A.S. Varadarajan, A neural network model to predict surface roughness during turning of hardened SS410 steel, Int. J. Syst. Assur. Eng. Manag. 11 (2020) 704-715. https://doi.org/10.1007/s13198-020-00986-9

[11] Y.C. Lin, K.D. Wu, W.C. Shih, P.K. Hsu, J.P. Hung, Prediction of surface roughness based on cutting parameters and machining vibration in end milling using regression method and artificial neural network, Appl. Sci. 10 (2020) 3941. https://doi.org/10.3390/app10113941

[12] P. Saini, P.K. Singh, Optimization of end milling parameters for rough and finish machining of Al-4032/3% SiC metal matrix composite, Eng. Res. Express 3 (2021) 045009. https://doi.org/10.1088/2631-8695/ac2e11

[13] S.H. Musavi, M. Sepehrikia, B. Davoodi, S.A. Niknam, Performance analysis of developed micro-textured cutting tool in machining aluminum alloy 7075-T6: assessment of tool wear and surface roughness, Int. J. Adv. Manuf. Technol. (2022) 1-20. https://doi.org/10.1007/s00170-021-08349-9

[14] R.A. Mali, R. Aiswaresh, T.V.K. Gupta, The influence of tool-path strategies and cutting parameters on cutting forces, tool wear and surface quality in finish milling of Aluminium 7075 curved surface, Int. J. Adv. Manuf. Technol. 108 (2020) 589-601. https://doi.org/10.1007/s00170-020-05414-7

[15] M. Zhou, Y. Chen, G. Zhang, Force prediction and cutting-parameter optimization in micro-milling Al7075-T6 based on response surface method, Micromachines 11 (2020) 766. https://doi.org/10.3390/mi11080766

Innovative Manufacturing Engineering and Energy - IManEE2024 | Materials Research Forum LLC
Materials Research Proceedings 46 (2024) 143-150 | https://doi.org/10.21741/9781644903377-19

Study of surface roughness variability in end milling of Al7136 alloy using row wise statistics analysis

Aurel Mihail TITU[1,2,a *], Alina Bianca POP [3,b *]

[1]" Lucian Blaga" University of Sibiu, Faculty of Engineering, Industrial Engineering and Management Department, 10 Victoriei Street, 550024, Sibiu, Romania

[2]Academy of Romanian Scientist, 3 Ilfov Street, 050044, Sibiu, Romania

[3]Technical University of Cluj-Napoca, Northern University Centre of Baia Mare, Faculty of Engineering - Department of Engineering and Technology Management, 62A Victor Babes Street, 430083, Baia Mare, Maramureş, Romania

[a] mihail.titu@ulbsibiu.ro, [b] bianca.bontiu@gmail.com

Keywords: Surface Roughness Variability, End Milling Process, Al7136, Row Wise Statistics Analysis, Machining Optimization

Abstract. This study aims to investigate the surface roughness (Ra) of machined surfaces using a dataset comprising 150 experiments. By analyzing the influence of key machining parameters—cutting speed, depth of cut, and feed per tooth—on Ra, this research seeks to optimize the machining process for achieving superior surface finishes. The methodological challenge lies in effectively analyzing the dataset, which involves calculating summary statistics, including means, standard errors, and confidence intervals, for all 150 records. The results indicate an average value of surface roughness around 165.68 μm, with a standard error of approximately 164.661 μm, highlighting the significant impact of machining parameters on these values. The conclusions suggest that optimizing machining parameters can lead to improved quality of machined surfaces and, consequently, reduced production costs, emphasizing the need for further research to explore the relationship and propose efficient methods for optimizing the machining process.

Introduction

Surface roughness, a critical characteristic of machined parts, significantly influences performance, durability, and production costs across various industries. Achieving optimal surface roughness in the end milling of AL7136 alloy, a material extensively utilized in the aerospace and automotive sectors, represents a persistent challenge. However, it also presents a valuable opportunity to enhance the quality and performance of finished components.

This study is necessary because, despite extensive research on surface roughness in machining processes, significant gaps remain in understanding how specific machining parameters influence roughness in the case of AL7136 alloy [1-4]. Moreover, most previous studies have focused on traditional data analysis methods, without fully exploiting the potential of advanced statistical techniques, such as row-wise statistics analysis [5-7]. A novel methodological approach is employed in this study to provide new insights into the variability of surface roughness during the end milling of AL7136 alloy. This research addresses a gap in the existing literature, which has primarily focused on other materials and not specifically investigated the relationship between machining parameters and surface roughness in AL7136 [7-10]. While this method has been applied in other fields, its potential in surface roughness analysis in machining remains untapped. Previous studies have offered general recommendations for surface roughness optimization, but there is a clear need for solutions tailored to the specific characteristics of AL7136 alloy and the end milling process [11-16].

The originality of this study lies in the combination of the following elements. Our study directly addresses the specific challenges associated with machining this alloy, providing relevant information for the aerospace and automotive industries. This innovative methodological approach allows for a more detailed and accurate analysis of surface roughness variability, considering the complex interaction between machining parameters. A key objective of this study is to bridge the gap between research and practical application by providing manufacturers with clear and specific guidance on how to optimize the end milling of AL7136 alloy. This will enable them to produce higher-quality components at lower costs.

The main objectives of this study are presented next. Using row-wise statistics analysis, we will investigate how machining parameters (cutting speed, depth of cut, and feed per tooth) influence surface roughness in the case of AL7136 alloy. We will explore correlations and interactions between different machining parameters, to better understand the mechanisms that determine surface roughness variability. Based on the results obtained, we will propose specific strategies for optimizing machining parameters, aiming to minimize surface roughness and maximize the quality of finished products.

Materials and Methods

Experimental Setup

The experiments were conducted through end-milling of an extruded profile made of AL7136 aluminum alloy in the T76511 temper condition. Machining was performed using a standard end mill cutter for aluminum, specifically the SECO R217.69-1616.0-09-2AN model, with 2 teeth and a diameter of 16 mm. The tool engagement was set to 100%, and the ISO code for the cutting inserts used was XOEX090308FR-E05, H15.

Data Collection

A total of 150 experiments were conducted, systematically varying the machining parameters (cutting speed, depth of cut, and feed per tooth) according to a predetermined experimental plan. For each experiment, the surface roughness (Ra) was measured using a high-precision roughness meter, following relevant standards. The obtained data was recorded and organized into a dataset for further analysis. A row-wise statistical analysis was conducted to evaluate the influence of machining parameters on the resulting surface roughness. This method allows for the analysis of each individual experiment, calculating the mean, standard error, and confidence interval for surface roughness. By comparing these statistics across different combinations of machining parameters, we can identify the relationships and trends governing the variability of surface roughness. Row-wise statistical analysis was chosen due to its ability to provide a detailed view of each individual experiment, allowing for the identification of subtle differences in surface roughness that might be overlooked in a global data analysis. Moreover, this method is suitable for large and complex datasets, such as the 150 experiments conducted in this study. By utilizing row-wise statistical analysis, we aim to gain a deeper understanding of the relationship between machining parameters and surface roughness, which will enable the formulation of precise and effective recommendations for optimizing the end-milling process of AL7136 alloy.

Results and Discussions

The table 1 presents the summary statistics for the longitudinal surface roughness (Ra) measurements across the 150 experimental runs, with each row representing a unique combination of cutting speed (v), depth of cut (ap), and feed per tooth (fz).

Materials Research Proceedings 46 (2024) 143-150 https://doi.org/10.21741/9781644903377-19

Tabel 1. Summary Statistics

Row	*Count*	*Average*	*Standard Deviation*	*Coefficient of variation*	*Minimum*	*Maximum*	*Range*	*Standardized Skewness*
1	3	165.68	285.201	172.14%	0.04	495.0	494.96	1.22468
2	3	165.687	285.195	172.129%	0.06	495.0	494.94	1.22468
3	3	165.693	285.19	172.119%	0.08	495.0	494.92	1.22468
4	3	165.703	285.181	172.103%	0.11	495.0	494.89	1.22468
5	3	165.713	285.172	172.088%	0.14	495.0	494.86	1.22469
6	3	165.847	285.058	171.88%	0.04	495.0	494.96	1.22464
7	3	165.853	285.052	171.87%	0.06	495.0	494.94	1.22464
8	3	165.86	285.046	171.86%	0.08	495.0	494.92	1.22465
9	3	165.87	285.037	171.844%	0.11	495.0	494.89	1.22465
10	3	165.88	285.029	171.828%	0.14	495.0	494.86	1.22465
⋮	⋮	⋮	⋮	⋮	⋮	⋮	⋮	⋮
140	3	237.713	409.015	172.062%	0.14	710.0	709.86	1.22468
141	3	237.847	408.9	171.918%	0.04	710.0	709.96	1.22465
142	3	237.853	408.895	171.91%	0.06	710.0	709.94	1.22465
143	3	237.86	408.889	171.903%	0.08	710.0	709.92	1.22465
144	3	237.87	408.88	171.892%	0.11	710.0	709.89	1.22465
145	3	237.88	408.871	171.881%	0.14	710.0	709.86	1.22465
146	3	238.013	408.757	171.737%	0.04	710.0	709.96	1.22462
147	3	238.02	408.751	171.73%	0.06	710.0	709.94	1.22462
148	3	238.027	408.746	171.723%	0.08	710.0	709.92	1.22462
149	3	238.037	408.737	171.712%	0.11	710.0	709.89	1.22462
150	3	238.047	408.728	171.701%	0.14	710.0	709.86	1.22462

The table 2 displays the sample means and standard errors for the 150 experimental runs included in the dataset, each with at least one non-missing value for surface roughness (Ra). The table also presents intervals representing the mean plus and minus one standard error.

Table 2. Means Table with Standard Error Intervals

Row	*Count*	*Mean*	*Standard Error*	*Lower Limit*	*Upper Limit*
1	3	165.68	164.661	1.01903	330.341
2	3	165.687	164.658	1.02905	330.344
3	3	165.693	164.654	1.03907	330.348
4	3	165.703	164.649	1.0541	330.353
5	3	165.713	164.644	1.06912	330.358
6	3	165.847	164.578	1.26847	330.425
7	3	165.853	164.575	1.27849	330.428
8	3	165.86	164.571	1.28852	330.431
9	3	165.87	164.566	1.30355	330.436
⋮	⋮	⋮	⋮	⋮	⋮
10	3	165.88	164.561	1.31859	330.441
140	3	237.713	236.145	1.56856	473.858
141	3	237.847	236.079	1.76789	473.925
142	3	237.853	236.075	1.77791	473.929
143	3	237.86	236.072	1.78794	473.932

144	3	237.87	236.067	1.80297	473.937
145	3	237.88	236.062	1.81801	473.942
146	3	238.013	235.996	2.01723	474.009
147	3	238.02	235.993	2.02726	474.013
148	3	238.027	235.989	2.03729	474.016
149	3	238.037	235.984	2.05233	474.021
150	3	238.047	235.979	2.06737	474.026

The mean surface roughness values vary significantly, ranging from about 165 μm to over 400 μm, highlighting the substantial impact of the processing parameters on the surface finish. Standard errors are large compared to the means, suggesting considerable variability within each experimental run. This could be caused by factors not accounted for in the experimental design, such as tool wear, material inconsistencies or measurement errors. Many of the confidence intervals overlap, indicating that differences between some experimental runs may not be statistically significant.

The graph shown in Figure 1 illustrates the relationship between the variable 'Row' (experimental round) and the experimental data 'Data' (roughness measurements). Each point in the graph represents a pair of values corresponding to a particular experimental round. In this context, the graph suggests that surface roughness tends to increase as the experiments progress, due to tool wear and other factors that change over time.

The formation of distinct clusters can be attributed to changes in machining parameters, including cutting speed and depth of cut, which play a crucial role in determining surface roughness. Variability within each group is attributed to random or uncontrolled factors in the machining process.

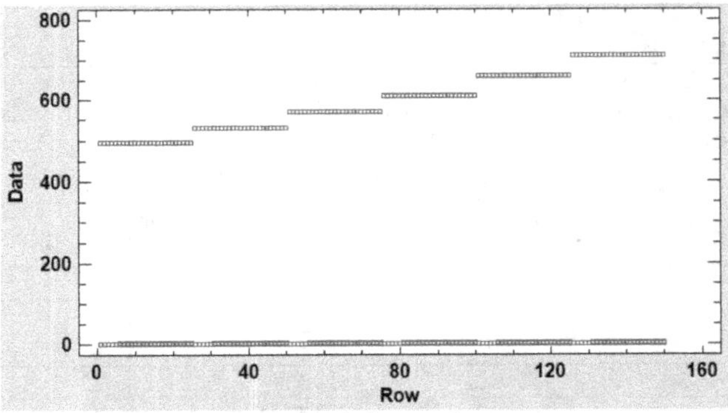

Figure 1. Scatterplot

The plot in Figure 2 displays the mean surface roughness values for each experimental run (Row), along with their respective standard errors, providing a visual representation of the variability in roughness measurements. The increasing trend in surface roughness could be due to tool wear over time. As the tool wears, it could become less sharp or efficient, resulting in increased surface roughness in the machined parts. The observed clusters in the graph arise from variations in the machining parameters, such as cutting speed, feed rate, and depth of cut, applied during the experimental runs.

These distinct parameter sets result in different surface roughness characteristics, leading to the formation of clusters. The variability within each group is due to inherent variations in the material being machined, slight fluctuations in the machining process, or measurement errors.

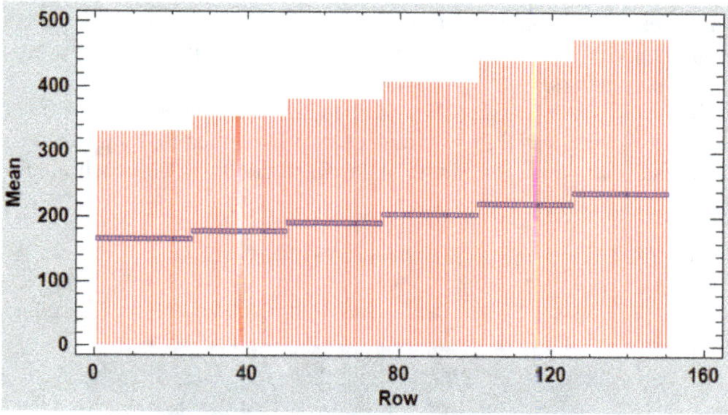

Figure 2. Mean Plot with Standard Errors Bars

The graph in Figure 3 illustrates the relationship between the experimental run and the standard deviation of the surface roughness. Each horizontal bar in the graph represents the range of standard deviation for a given experimental run.

The distinct clusters evident in the graph are a direct result of alterations in the machining parameters—cutting speed, feed rate, and depth of cut—between sets of experimental runs. Each cluster reflects a different set of parameters, and the transitions between them underscore the extent to which these parameters influence the variability observed in surface roughness. Although the plot does not show a steady increase in standard deviation, the overall stepwise increase could still be related to tool wear. Tool wear could introduce more variability in the machining process, which would lead to an increase in the standard deviation of the surface roughness.

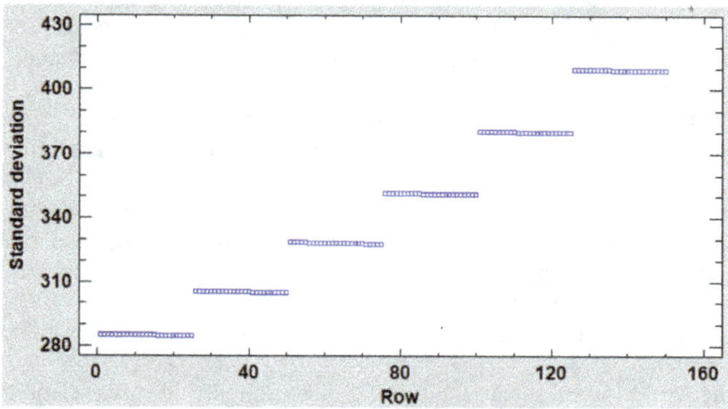

Figure 3. Standard deviation plot

The box-and-whisker plot illustrates the distribution of the data set over different experimental runs (Figure 4). The y-axis represents the run number, while the x-axis shows the range of values for the surface roughness.

The general increasing trend indicates that the surface roughness tends to increase as the experiments progress. This could be caused by factors such as tool wear, changes in material properties or gradual changes in the machining process. Distinct groups correspond to different sets of machining parameters.

The jumps between groups suggest that changes in these parameters have a significant impact on surface roughness. The variability within each group is caused by factors such as natural variations in the material being machined, slight fluctuations in the machining process itself or measurement errors.

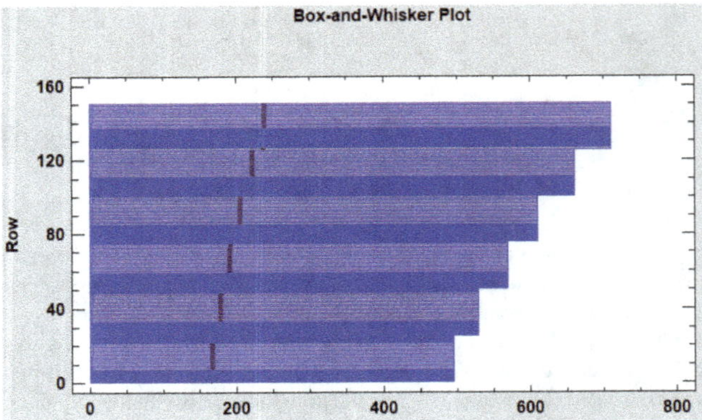

Figure 4. Box and whisker plot

Conclusions

Analysis of the experimental data reveals a general upward trend in average surface roughness as the machining trials progress, as illustrated in the "Means Plot with Standard Error Bars." This observation can be attributed to the progressive wear of the cutting tool, a common occurrence in machining processes.

The gradual dulling of the tool's cutting edges leads to increased friction and less precise material removal, ultimately resulting in a rougher surface finish on the machined components.

In addition, the Means Plot with Standard Error Bars, and the Standard Deviation Plot show that the data are grouped into distinct clusters separated by sudden jumps in values.

The observed clustering pattern indicates that modifications to machining parameters, such as cutting speed, feed per tooth, and depth of cut, exert a considerable influence on the resulting surface roughness and its variability.

The jumps observed in the graphs could be the result of deliberate changes in these parameters during the experiments.

Although the row-wise statistical analysis did not explicitly identify relationships between individual machining parameters and surface roughness, the trends observed in the graphs suggest that there is a complex interaction between these parameters.

For example, increasing the cutting speed could lead to an increase in average roughness, but could also reduce its variability.

Like any research study, this investigation has inherent limitations that should be considered. First, the sample size is small (150 experiments), which could affect the generalizability of the results. Second, the study focused exclusively on AL7136 alloy, and the results may not be directly applicable to other materials.

Finally, the analysis was based on summary statistics without considering the full distribution of surface roughness data. To overcome these limitations and gain a more complete understanding of the variability of surface roughness in face milling, further research is needed.

Future studies could include a larger number of experiments, investigate other materials, and use more advanced statistical methods to analyze the full distribution of the data.

This study contributes to a deeper understanding of surface roughness variability in the face milling of AL7136.

The analysis of experimental data reveals a multifaceted relationship between machining parameters (cutting speed, feed per tooth, and depth of cut) and surface quality.

The observed increase in average roughness over the course of the experiments, as depicted in the "Means Plot with Standard Error Bars," highlights the detrimental effects of tool wear. Furthermore, the clustering patterns identified in the data visualizations underscore the importance of careful parameter selection to minimize variations in surface roughness.

These findings provide valuable guidance for manufacturers seeking to optimize machining processes and achieve superior surface finishes.

References

[1] G. Basar, H. Kirli Akin, F. Kahraman, Y. Fedai, Modeling and optimization of face milling process parameters for AISI 4140 steel, Tehnički glasnik 12 (2018) 5-10. https://doi.org/10.31803/tg-20180201124648

[2] T.M. Damesa, J. Möhring, M. Worku, H.P. Piepho, One step at a time: Stage-wise analysis of a series of experiments, Agron. J. 109 (2017) 845-857. https://doi.org/10.2134/agronj2016.07.0395

[3] M. Özdemir, A. Şahinoğlu, M. Rafighi, V. Yilmaz, Analysis and optimisation of the cutting parameters based on machinability factors in turning AISI 4140 steel, Can. Metall. Q. 61 (2022) 407-417. https://doi.org/10.1080/00084433.2022.2058154

[4] V. Vinoth, Optimization of process parameters in turning of aluminum alloy using response surface methodology, Mater. Today: Proc. 46 (2021) 9462-9468. https://doi.org/10.1016/j.matpr.2020.03.236

[5] H.D. Kumar, S. Ilangovan, N. Radhika, Optimization of cutting parameters for MRR, tool wear and surface roughness characteristics in machining ADC12 piston alloy using DOE, Tribol. Ind. 42 (2020) 32. https://doi.org/10.24874/ti.2020.42.01.03

[6] S. Dutta, S.K.R. Narala, Optimizing turning parameters in the machining of AM alloy using Taguchi methodology, Measurement 169 (2021) 108340. https://doi.org/10.1016/j.measurement.2020.108340

[7] R.K. Rao, S.C. Kulkarni, S. Shivakumar, Investigation Of Machinability Characteristics Of Sae Ams4413b Aluminum Alloy Using Ethylene Glycol As A Coolant, Migration Lett. 21 (2024) 893-932. https://doi.org/10.59670/ml.v20i8.7139

[8] A. Sudianto, Z. Jamaludin, A.A.A. Rahman, Prediction of surface roughness for development of smart milling machine, J. Phys. Conf. Ser. 1201 (2019) 012008. https://doi.org/10.1088/1742-6596/1201/1/012008

[9] A.M. Ţîţu, A.V. Sandu, A.B. Pop, Ş. Ţîţu, D.N. Frăţilă, C. Ceocea, A. Boroiu, Design of experiment in the milling process of aluminum alloys in the aerospace industry, Appl. Sci. 10 (2020) 6951. https://doi.org/10.3390/app10196951

[10] A.B. Pop, A.M. Titu, Verification of the random nature of the experimental data in the end-milling process of aluminum alloys, Arch. Metall. Mater. (2023) 967-971. https://doi.org/10.24425/amm.2023.145461

[11] A.B. Pop, M.A. Ţîţu, G.I. Pop, Ş. Ţîţu, Modeling the machined surface quality of an aluminum alloy using the active experiment type, IOP Conf. Ser. Mater. Sci. Eng. 572 (2019) 012043. https://doi.org/10.1088/1757-899X/572/1/012043

[12] M. Aamir, K. Giasin, M. Tolouei-Rad, A. Vafadar, A review: Drilling performance and hole quality of aluminium alloys for aerospace applications, J. Mater. Res. Technol. 9 (2020) 12484-12500. https://doi.org/10.1016/j.jmrt.2020.09.003

[13] M. Hourmand, A.A. Sarhan, M. Sayuti, M. Hamdi, A comprehensive review on machining of titanium alloys, Arab. J. Sci. Eng. 46 (2021) 7087-7123. https://doi.org/10.1007/s13369-021-05420-1

[14] M. Soori, B. Arezoo, A review in machining-induced residual stress, J. New Technol. Mater. 12 (2022) 64-83. https://hal.science/hal-03679993/

[15] B.O. Omiyale, T.O. Olugbade, T.E. Abioye, P.K. Farayibi, Wire arc additive manufacturing of aluminium alloys for aerospace and automotive applications: A review, Mater. Sci. Technol. 38 (2022) 391-408. https://doi.org/10.1080/02670836.2022.2045549

[16] J.P. Chen, L. Gu, G.J. He, A review on conventional and nonconventional machining of SiC particle-reinforced aluminium matrix composites, Adv. Manuf. 8 (2020) 294-306. https://doi.org/10.1007/s40436-020-00313-2

Shape parameter optimisation of elliptic tool paths in trochoidal slot milling

Adam JACSO [1,a] *, Viktor TANCSA [1,b]

[1] Department of Manufacturing Sciences and Engineering, Faculty of Mechanical Engineering, Budapest University of Technology and Economics, Budapest, Hungary

[a] jacso.adam@gpk.bme.hu, [b] tancsav@edu.bme.hu

Keywords: Trochoidal Milling, Elliptic Tool Path, Material Removal Rate, Cutter Engagement, Tool Path Optimisation

Abstract. Trochoidal milling is a highly effective machining strategy for hard-to-reach areas like corners and slots. This strategy was initially developed using circular arc and cycloid-based tool paths. Recently, elliptic trochoidal tool path patterns have emerged as an even more efficient alternative. However, this advancement raises new questions as the analytical functions previously used to describe the cutter engagement will no longer be valid along the elliptic path. Additionally, changing the ellipse aspect ratio can completely alter the nature of tool load development. While there are examples of tool load modelling for the elliptic trochoidal strategy in scientific literature, there is still a need to directly connect these methods with productivity aspects. This study introduces a newly developed optimisation method that achieves two goals: (1) specifying the maximum permissible trochoidal step for a desired cutter engagement limit and (2) determining the optimal ellipse aspect ratio to achieve maximum material removal rate while respecting the boundary conditions. The presented optimisation method was validated by both experimental and simulation results. Although this algorithm is currently only applicable to trochoidal machining of straight slots, it has the potential to be extended to general shapes in the future.

Introduction

Trochoidal machining is a highly effective method for milling enclosed areas, especially when working with difficult-to-cut materials [1]. This method uses a looping tool path to control the tool's radial immersion, keeping the contact surface between the tool and the workpiece small. This has several advantages: (1) reduces heat generation, (2) prevents tool chatter, and (3) facilitates the chip removal process [2]. These properties allow for a large axial depth of cut, which results in longer tool life as the tool wear is distributed over a longer edge segment. As a result, multiple feed rate and cutting speed can be applied, leading to increased productivity and reduced machining costs compared to conventional contour-parallel milling [3].

Initially, trochoidal machining was carried out using circular and cycloid-based tool paths [1]. The analytical formulas for these tool path generation methods were derived to calculate the tool's radial immersion, which also determines the tool load [4]. However, the main shortcoming of these traditional tool paths is that the nominal tool load is only realised at one point along the tool path, surrounded by an extended transitional phase at tool entering and exiting [5]. Although the effect of this problem can be mitigated by applying an adaptive feed rate [6-7], the ideal machining strategy can only be achieved by developing more advanced tool path planning methods. A significant milestone in this area was the introduction of elliptic trochoidal tool path shapes, which improved the tool load characteristics while maintaining the simplicity of tool path generation [8]. However, changing the ellipse aspect ratio can completely alter the nature of tool load development, posing a new question. Although there are numerous examples of tool load modelling for elliptic trochoidal strategy in scientific literature today, there is still a need to directly connect these methods with productivity aspects.

This study introduces a newly developed optimisation method of elliptic tool path generation for trochoidal machining of straight slots, accomplishing two goals: (1) specifying the maximum permissible trochoidal step for a desired cutter engagement limit and (2) determining the optimal ellipse aspect ratio to achieve the maximum material removal rate while respecting the boundary conditions. The presented optimisation method has been validated through both experimental and simulation results. The details of the calculation process and experiments are described in the following sections.

Mathematical background of elliptic tool path generation in trochoidal slot milling

Elliptic trochoidal tool paths can be derived similarly to conventional cycloid tool paths. The geometric considerations when using an elliptic trochoidal tool path are illustrated in Fig. 1, where the slot width is w $[mm]$, the tool radius is r $[mm]$, and the trochoidal step is s $[mm]$.

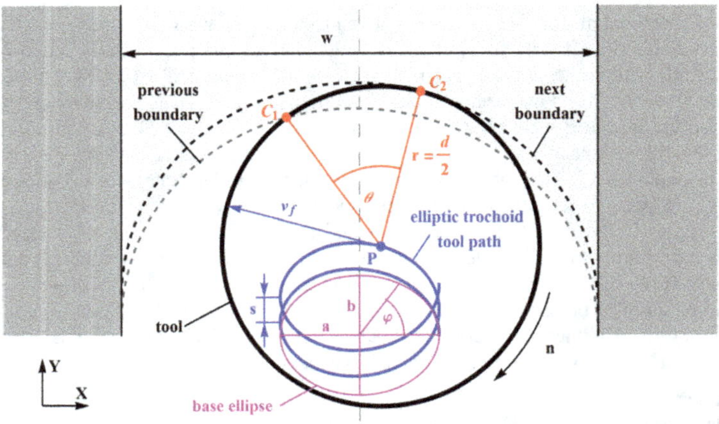

Fig. 1 Geometrical considerations when using an elliptic trochoidal tool path

The general form of the parametric vector function describing the elliptic trochoidal tool path can be written as follows:

$$\bar{p}(\varphi) = \begin{bmatrix} a\cos(\varphi) \\ b\sin(\varphi) + \frac{s}{2\pi}\varphi \end{bmatrix} \tag{1}$$

where $\bar{p}(\varphi)$ $[mm, mm]$ is the trochoidal tool path curve, $\varphi[rad]$ is the angle parameter, a $[mm]$ is the base ellipse's semi-major axis, and b [mm] is the base ellipse's semi-minor axis. The size of semi-major axis must satisfy the following condition: $a = w/2 - r$. Although Eq. (1) assumes that the slot's centre line is aligned with the Y axis of the coordinate system, the tool path described with this function can be adapted to slots in any arbitrary position with an appropriate rotational and translational transformation. It is also important to note that the proportional tool-workpiece engagement remain unchanged regardless of scaling. This means that calculations can be carried out using a unit-size tool, and the results can be scaled up to the desired dimensions.

In the previous tool path modelling approach, the trochoidal step and the ellipse aspect ratio were considered as free variables from a geometrical perspective. However, taking technological constraints into account, it is crucial to set these two parameters correctly to ensure effective and efficient machining. The following two sections deal with this issue in detail.

Calculation of the maximum permissible trochoidal step

The first thing to consider is the maximum permissible trochoidal step. When all other parameters remain the same, the tool's radial immersion increases consistently as the trochoidal step increases. However, since trochoidal machining aims to keep the tool's radial immersion low, there is a upper limit to the feasible value of trochoidal step. On the other hand, the amount of material removed in one trochoidal period also increases consistently with the trochoidal step, leading to improved productivity. Therefore, it is important to select the trochoidal step in a way that ensures the maximum productivity while staying within the permissible tool load limit.

The tool's radial immersion can be measured using the cutter engagement angle (θ) which is the centre angle of the arc in contact with the workpiece (marked by $\theta = \angle C_1PC_2$ in Fig. 1). The cutter engagement can be calculated numerically at any point P on the tool path since the exit point of the tool edge (C_1) can be determined by rotating the tangent vector of the path by $-\pi/2$, while the entry point of the tool edge (C_2) can be determined as the intersection of the circumference of the tool and the envelope of the previous trochoidal period. Fig. 2 shows an example of how the cutter engagement develops along the tool path for different trochoidal steps with given boundary conditions.

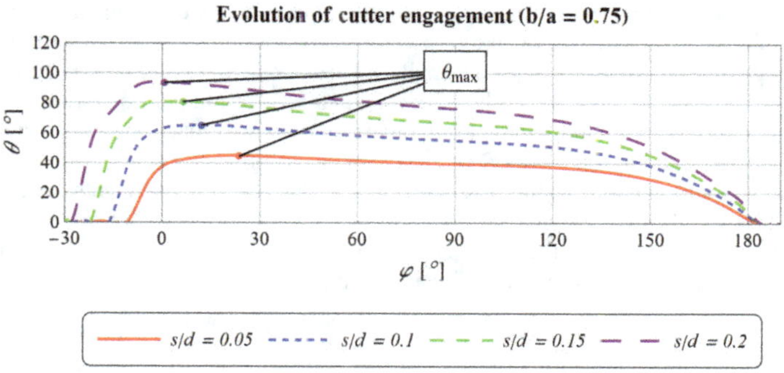

Fig. 2 The evolution of cutter engagement along the tool path at different trochoidal steps ($w = 3r$, $b/a = 0.75$)

Based on the previous calculations, the desired trochoidal step can be determined for a given set of parameters, including cutter engagement limit (θ_{max}), ellipse aspect ratio (b/a), tool radius (r) and slot dimensions (w). In our work, a bisection algorithm was applied to solve this task.

Fig. 3 shows how the maximum cutter engagement angle changes with different trochoidal steps and ellipse aspect ratios. When the b/a ratio is smaller, usually the allowable trochoidal step also becomes smaller, which can reduce productivity. However, a smaller b/a ratio also leads to a shorter trochoidal period, which can increase productivity. To account for these conflicting effects, additional calculations are needed when selecting the appropriate ellipse aspect ratio.

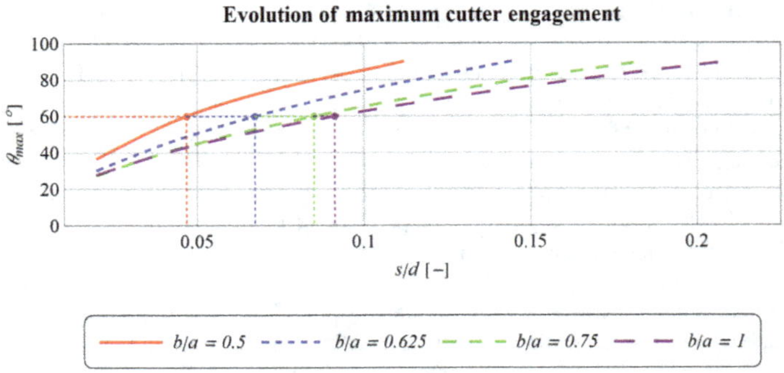

Fig. 3 The evolution of maximum cutter engagement as a function of trochoidal step at different ellipse aspect ratios ($w = 3r$)

Calculation of the optimal ellipse aspect ratio

The previous section explained how to calculate the evolution of cutter engagement angle for given boundary conditions. Fig. 4 shows how the ellipse aspect ratio strongly influences the development of cutter engagement. If the ratio b/a is too small (in the example of Fig. 4, $b/a = 0.5$), the tool's radial immersion suddenly increases in the initial section. As a result, the trochoidal step must be significantly reduced, and the tool works far below the permissible load for most of the path. On the other hand, if the ratio b/a is too large (in the example of Fig. 4, $b/a = 1$, which is the cycloid strategy), the tool entering and exiting take place on excessively long segments, where the tool also works far below the permissible load, resulting in lower productivity. In the ideal case (in the example of Fig. 4, $b/a = 0.85$), the cutter engagement angle should remain near the maximum permissible value as long as possible, ensuring maximum productivity.

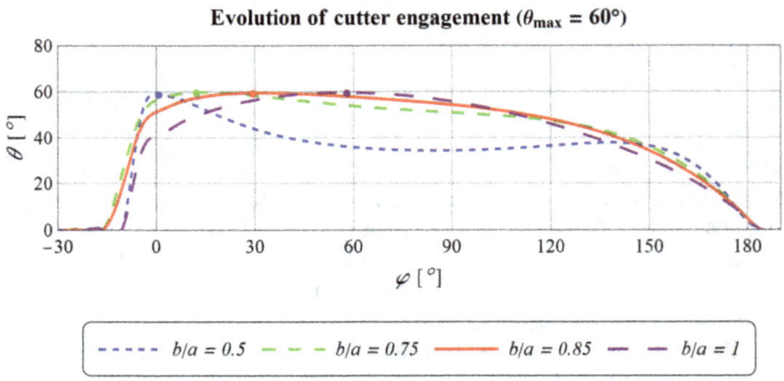

Fig. 4 The evolution of cutter engagement along the tool path applying the maximum permissible trochoidal step at different ellipse aspect ratios ($w = 3r$, $\theta_{max} = 60°$)

The objective function of optimisation is to maximise productivity, which can be easily measured by the average material removal rate. The material removal rate represents the volume removed during a specific unit of time, which in the steady state of trochoidal machining is equal to the ratio of the volume of material removed to the time it takes to track the path in one trochoidal period. This definition can be expressed mathematically as follows:

$$MRR = \frac{V}{T} = \frac{a_p\, w\, s}{L/v_f} \tag{2}$$

where $MRR\ [mm^3/min]$ is the material removal rate, $V\ [mm^3]$ is the volume of material removed in one trochoidal period, $T\ [min]$ is the path tracking time for one trochoidal period, $a_p\ [mm]$ is the axial depth of cut, $L\ [mm]$ is tool path length of one trochoidal period, and $v_f\ [mm/min]$ is the feed rate. In the context of tool path optimisation, many variables in this formula can be considered constant. This simplifies the objective function to the following form:

$$f: \frac{s}{L} \to max \tag{3}$$

while the optimisation constraint is that the cutter engagement angle cannot exceed the permissible θ_{max} value anywhere along the tool path. The trochoidal step s in the numerator can be determined as described in the previous section. To calculate the period length L in the denominator, the following formula can be used:

$$L = \int_0^{2\pi} \bar{p}(\varphi)\, d\varphi \tag{4}$$

If the geometric boundary conditions and the cutter engagement angle limit are given, the value of the optimisation objective function can be calculated for any b/a aspect ratios. In our work, a bisection algorithm was used to find the optimal aspect ratio to maximise productivity.

Fig. 5 illustrates the evolution of objective function with different cutter engagement limits. It is clearly visible that the optimal point's position is influenced by the specified cutter engagement angle limit. Additionally, while the elliptic tool path does not ensure progress on its own, the optimal aspect ratio offers much better efficiency compared to the traditional cycloid tool path where $b/a = 1$.

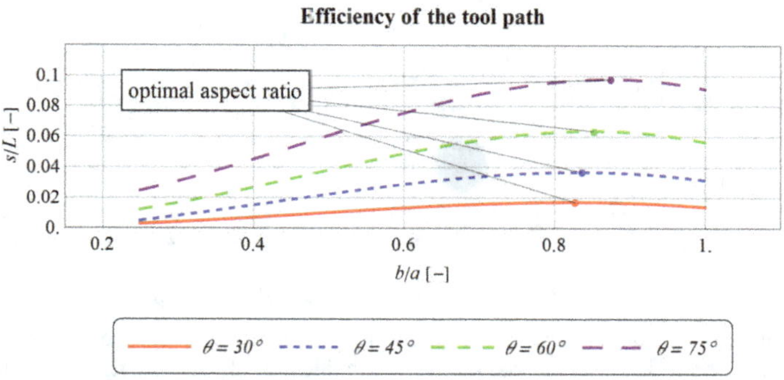

Fig. 5 The evolution of the maximum feasible productivity as a function of ellipse aspect ratio at different cutter engagement limits ($w = 3r$)

The optimisation method presented previously was tested with various input data in both simulation and experimental environments. The results are detailed in the following section.

Simulation and experimental validation

The correctness of the calculations described above was confirmed through simulation and experimental analysis. Additionally, a case study was conducted to examine the impact of changes in slot width and maximum cutter engagement angle limit, as well as the amount of potential productivity improvement.

In the case study, a solid carbide end mill with a diameter of $d = 10\ mm$ and 3 cutting edges was used. The cutting parameters were chosen based on the tool catalogue, according to which the cutting speed was $v_c = 200\ m/min$, and the feed per tooth was $f_z = 0.05\ mm$, resulting in a spindle speed of $n = 6366\ rpm$ and a feed rate of $v_f = 200\ m/min$. Three setting parameters were examined for the slot width ($w = 15\ mm, 17.5\ mm, 20\ mm$) and the permissible cutter engagement angle ($\theta_{max} = 30°, 45°, 60°$). The cutting experiments were conducted using a Kondia B-640 three-axis machining centre. During the experiments, the cutting force acting on the workpiece was measured using a Kistler 9257B piezoelectric dynamometer. The cutting force data was analysed by projecting it onto the machining plane (X-Y plane, perpendicular to the tool axis). The diagram shows the maximum value of the cutting force per tool revolution.

The cutting experiments were conducted using all parameter combinations listed in Table 1. It was observed that controlling the cutter engagement angle also controlled the tool load. In Fig. 6, the cutting forces can be seen when machining a slot with a width of $w = 15\ mm$ and a cutter engagement angle limit of $\theta_{max} = 60°$. The red and blue curves indicate the cutting force generated by the optimised elliptic tool path and the traditional cycloid tool path, respectively. In both cases, there was a slight increase in cutting force at the tool entry, likely due to the dynamic effects of the transition phase. After this anomaly, the measured cutting force reached the same value ($\sim 210\ N$) for both strategies. However, the characteristics of cutting force evolution was more favourable with the optimised elliptic tool path as both the rise and fall of cutting force occurred within a shorter segment, allowing the tool to spend more time under the nominal load. Additionally, the tool also removed more material in one trochoidal period, and the period time was also shorter for the elliptic tool path. Consequently, the material removal rate was higher with the optimised elliptic tool path.

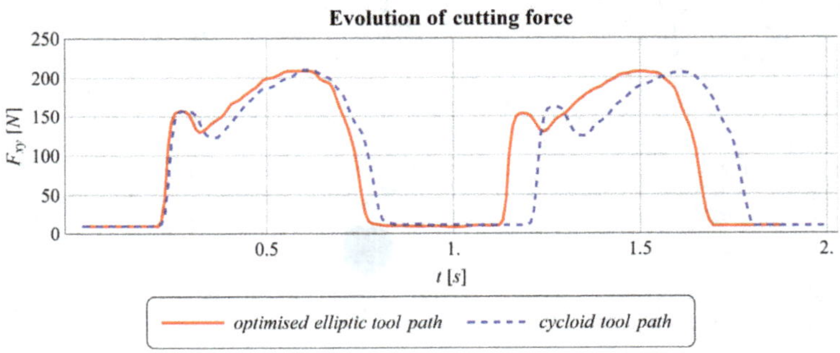

Fig. 6 Comparison of the cycloid and optimised elliptic tool paths regarding the cutting force

Table 1 summarises the simulation results for five parameter combinations. The indicators examined include permissible trochoidal step (s), period length of tool path (L), and the resulting material removal rate (MRR). In the first three settings, the slot width was three times the tool radius ($w = 3r$), the recommended value for trochoidal machining.

The comparative study showed that the optimised elliptic tool path resulted in a larger trochoidal step and a shorter period length in all cases, leading to $10 - 21\%$ increase in material removal rate in the range of $\theta_{max} = 30° \dots 60°$. The smaller the cutter engagement angle limit, the greater the achievable increase in productivity. In the second part of the analysis, the cutter engagement angle limit was fixed ($\theta_{max} = 60°$), and the slot width varied ($w = 15\ mm \dots 20\ mm$). That comparative analysis also showed that the optimised elliptic tool path resulted in a larger trochoidal step and a shorter period length with all parameter settings, leading to $10 - 16\ \%$ increase in material removal rate in the parameter range examined. The larger the slot width, the greater the achievable increase in productivity. Additionally, another advantage of elliptic tool paths, that the transition phases occurring at the start and end of the slot take place during fewer trochoidal periods, has not been considered in these calculations.

Table 1 Comparison of the cycloid and optimised elliptic tool paths

Characteristics of tool path with different input parameters									
w [mm]	θ [°]	b_{opt}/a [-]	s [mm]		L [mm]		MRR[mm^3/min]		Gain [%]
			cycloid	elliptic	cycloid	elliptic	cycloid	elliptic	
15	30	0.828	0.219	0.243	15.709	14.390	400	483	20.8
15	45	0.837	0.503	0.531	15.712	14.461	916	1053	15.0
15	60	0.852	0.907	0.929	15.721	14.584	1652	1822	10.3
17.5	60	0.816	1.163	1.198	23.576	21.467	1647	1864	13.2
20	60	0.783	1.346	1.396	31.430	28.130	1634	1893	15.9

Although the investigations were made with specific parameters to make the cutting experiments feasible, the established consequences can be considered generally valid. This is because, assuming a constant feed rate, the efficiency of the tool path depends only on the ratio of the geometric dimensions. Therefore, input parameters derived by scaling will lead to the same result. This generality of the results can greatly help the applicability of the method, as it is sufficient to carry out the optimisation calculations for a unit-size tool with different slot widths and cutter engagement angle limits.

Summary

This paper presents the potential of using the elliptic trochoidal tool path to increase productivity in straight slot machining. It was presented how the ellipse aspect ratio can be optimised, as well as how much improvement in productivity can be achieved. The correctness of the presented method was verified through simulations and experiments. In summary, the following statements can be made:

- By implementing the presented calculation method, the optimal ellipse aspect ratio can be determined, maximising the material removal rate under given boundary conditions.
- Using the optimised elliptic trochoidal tool path allows to simultaneously increase the trochoidal step and reduce the period length of the tool path compared to the traditional cycloid-based trochoidal tool path, thereby increasing productivity.
- A greater improvement in material removal rate can be achieved with the optimised elliptic tool path when the cutter engagement limit is smaller, or the slot is wider in relation to the tool diameter.
- In the examined parameter range ($\theta_{max} = 30° \dots 60°$ and $w = 15\ mm \dots 20\ mm$), $10 - 20\%$ productivity improvement could be achieved by optimising the tool path.

Since productivity improvements can be achieved without increasing the maximum tool load, no significant change in tool life or surface quality is expected. Consequently, a reduction in costs and energy consumption accompanies the reduction in machining time. The presented algorithm also allows for further development and application extension, like applying an increased feed rate on linking segments, using an adaptive feed rate on the cutting segments, or broadening the range of geometries (such as arbitrarily shaped slots, corners, and pockets), similar to the approach outlined by Huang et al. [8].

Acknowledgements
The research was supported by the BME-EKÖP and BME-DKÖP scholarship programs. Project no. TKP-6-6/PALY-2021 has also been implemented with support from the Ministry of Culture and Innovation of Hungary from the National Research, Development and Innovation Fund, financed under the TKP2021-NVA funding scheme. Participation in the conference was funded by the NRDIO through the project OTKA-K-132430 (Transient deformation, thermal, and tribological processes at fine machining of metal surfaces of high hardness).

References

[1] M. Otkur and I. Lazoglu, 'Trochoidal milling', Int. J. Mach. Tools Manuf., vol. 47, 2007. https//doi.org/10.1016/j.ijmachtools.2006.08.002

[2] A. Jacso, G. Matyasi, and T. Szalay, 'Trochoidal tool path planning method for slot milling with constant cutter engagement', Lect. Notes Mech. Eng., 2020. https//doi.org/10.1007/978-981-33-4320-7_59

[3] A. Jacso, B. S. Sikarwar, R. K. Phanden, R. K. Singh, J. Ramkumar, and G. N. Sahu, 'Optimisation of tool path shape in trochoidal milling using B-spline curves', Int. J. Adv. Manuf. Technol., vol. 121, 2022. https//doi.org/10.1007/s00170-022-09527-z

[4] M. Rauch, E. Duc, and J.-Y. Hascoet, 'Improving trochoidal tool paths generation and implementation using process constraints modelling', Int. J. Mach. Tools Manuf., vol. 49, 2009. https//doi.org/10.1016/j.ijmachtools.2008.12.006

[5] A. Jacso, Z. Lado, R. K. Phanden, B. S. Sikarwar, and R. K. Singh, 'Bézier curve-based trochoidal tool path optimization using stochastic hill climbing algorithm', Mater. Today Proc., 2022. https//doi.org/10.1016/j.matpr.2022.12.056

[6] C. García-Hernández et al., 'Trochoidal Milling Path with Variable Feed. Application to the Machining of a Ti-6Al-4V Part', Mathematics, vol. 9, 2021. https//doi.org/10.3390/math9212701.

[7] A. Jacso, T. Szalay, B. S. Sikarwar, R. K. Phanden, R. K. Singh, and J. Ramkumar, 'Investigation of conventional and ANN-based feed rate scheduling methods in trochoidal milling with cutting force and acceleration constraints', Int. J. Adv. Manuf. Technol., vol. 127, 2023. https//doi.org/10.1007/s00170-023-11506-x

[8] X. Huang, S. Wu, L. Liang, X. Li, and N. Huang, 'Efficient trochoidal milling based on medial axis transformation and inscribed ellipse', Int. J. Adv. Manuf. Technol., vol. 111, 2020. https//doi.org/10.1007/s00170-020-06172-2

On the automatic real-time detection of chatter onset in turning

VOSNIAKOS George-Christopher [1,a]* and ANASTASIADIS Apostolos [1,b]

[1]National Technical University of Athens, School of Mechanical Engineering, Manufacturing Technology Laboratory, Athens 15780, Greece

[a]vosniak@central.ntua.gr, [b]mech.anast@gmail.com

Keywords: Turning, Chatter, Fast Fourier Transform, Microcontroller, Monitoring

Abstract. A low-cost device has been implemented in order to detect the onset of chatter in turning operations. This involves a microcontroller, which receives an acceleration signal from the workpiece being turned, converts it to digital form, processes it in real time, identifies specific features and monitors them also in real time. As a first step, Fast Fourier Transform of the acceleration signal for a reference tool path that involves no chatter identifies high energy peaks at particular frequencies, which depend on the eigen-frequencies of the dynamic system machine tool – cutting tool – workpiece. Then, at the onset of chatter a specific pattern of additional high energy peaks appears in the spectrum 0-2.5 kHz. Signal processing and comparison with the reference take less than 200 ms. Experiments on a CNC lathe proved that the approach is effective and transferable beyond longitudinal turning.

Introduction

Chattering constitutes a dynamic instability of a manufacturing process that is associated with self-induced vibrations. In turning processes chatter occurs at specific combinations of process conditions, namely: spindle rotation speed, depth of cut and tool feed. It is also connected to the dynamic characteristics of the machine tool – workpiece - cutting tool system and results in wavy surface, fast wear or breakage of the tool and even in damage in elements of the machine [1]. It is possible to avoid chatter by slowing down the manufacturing process but this severely sacrifices productivity.

There has been intensive investigation of chatter in machining processes over the last 40 years, with various aims, i.e. to understand causal mechanisms, create predictive mathematical models [2], develop hardware suppress chatter, or otherwise, methods to detect its onset and methodologies to control or eradicate it once it occurs [3].

Recent research mainly focuses on detection and control aspects, given sensors (acceleration, cutting force, acoustic and machined surface [4]), data storage and analytics incl. machine learning tools [5] are becoming more accessible than ever. Technological progress has made online detection a possibility, e.g. [6]. However, difficulty of unobtrusive sensor installation on a productive machine tool may dictate use of multiple sensors and impose complexity in detection and control.

This paper investigates chatter onset detection in turning in real time adopting a low-cost solution. The next section briefly outlines previous related work. Thereafter, the main concept and hypotheses are explained. This is followed by the presentation of the implementation conducted. He following section introduces typical experimental evidence of the workings of the system. The main conclusions and future work are outlined last.

Related work

Several techniques have been used for chatter recognition or monitoring. Many earlier approaches were not based on ANNs. For instance: Statistical analysis of synchronously sampled audio signals with low cost microphones sought excess energy in [7]. Pattern recognition has been investigated, for instance via support vector machines in [8]. The width of cut to spindle speed ratio was obtained using Laplace transform, the Monte Carlo method and the advanced first order second moment method in [9]. Three different methods, namely 1st Autocorrelation Coefficient, Wavelet Transform and Fast Fourier Transform (FFT) were applied on feed force data and spectral analysis using the highest peaks of FFT was advocated as the most promising one in [10]. A sensor-less method was based on a mechanical energy factor (MEF) and a mechanical power factor (MPF) indicating the phase difference between the dynamic cutting force and velocity-displacement in [11]. A Linear Mixed Model was exploited for chatter prediction as well as for wear prediction with good accuracy in [12]. A stability margin of the process was set within an operational modal analysis to predict chatter onset before the vibrations are unacceptable in [13].

ANN approaches have been emphasized more recently. For example, a hybrid approach based on Hidden Markov Model allowed temporal modelling of features output from time-varying signals and an Multi-Layer Perceptron integrating the features has been successfully employed [14]. Supervised machine learning was combined with Topological Data Analysis to obtain an indicator of chatter imminent presence exploiting joint working of deterministic and stochastic turning models with varying cutting coefficients [15]. The stability limits were used as a starting point for generating training data for an ANN [16].The Wavelet Transform was used to remove noise from the raw signals connecting cutting parameters and severity of chatter resulting in ANNs being more reliable than RSM. [17]. A similar approach was reported in [18]. The ANN approach for detecting chatter vibration in turning was demonstrated to be more efficient (fast) and reliable than the frequency domain method (Tian et al., 2019).

Detection concept

The method of identifying chatter onset is based on previous work of many researchers, see Section 'Related work', which is important for a proper understanding of the phenomenon.

First, sufficient and correct sampling of acceleration is necessary. The processing of the sampled values especially using Fast Fourier Transform (FFT) in the frequency domain allows a clearer overview of the process.

In each different cutting arrangement and corresponding dynamic system: machine – workpiece – cutting tool, as well as at different spindle speeds, different harmonic frequencies with large energy values are expected to appear in the frequency domain. Fluctuations in energy at these frequencies do not pose a chatter risk. However, on either side of them, frequencies can be distinguished that are equidistant from the "nominal" high energy frequency, see Fig. 1. In stable cutting conditions energy at these frequencies remain at noise level.

If the cutting becomes unstable, energy at these frequencies starts to raise, whilst one of them is closer to the eigenfrequency of the dynamic system, thus having the highest energy, hence called the 'dominant chatter frequency'.

The algorithm built to observe this instability onset is shown in Fig. 2. Measurements are taken exactly from the moment the cutting tool comes into contact with the workpiece. The first 5 of these measurements are transformed in the frequency domain and the average of the energy at each frequency from 0 to 2500 Hz is computed. Note that each measurement includes 512 values sampled at 5 kHz. This package of measurements, including their processing, is 150 milliseconds away from the next and the previous one. This time interval has been checked after experimentation to be sufficient. Thus, there is a "safe reference snapshot" representing stable cutting, because at the exact moment when the cutting tool enters the workpiece no chatter has occurred and the energy is distributed over harmless frequencies. This reference snapshot is stored in the microprocessor's memory.

Then, in real time, at each subsequent time step a new snapshot is constructed in an analogous way. The energy values of this new snapshot across the 0-2500 Hz frequency range is compared with the corresponding value of the reference snapshot. If the new value is a multiple of its

Materials Research Proceedings 46 (2024) 159-166 https://doi.org/10.21741/9781644903377-21

reference counterpart, then chatter onset is signalled. Otherwise, if this is not the case for any frequency, then cutting is considered to be stable. The multiplier has been derived after experimentation as larger than 2, but is probably subject to updating depending on the dynamic characteristics of the system implicitly including workpiece material, too.

According to this methodology, calculations are pre-formed very quickly, in the order of magnitude of tens of a second, depending on the microprocessor on which the algorithm and data collection occur. This enables fast detection of chatter onset before it fully develops with possibly catastrophic consequences.

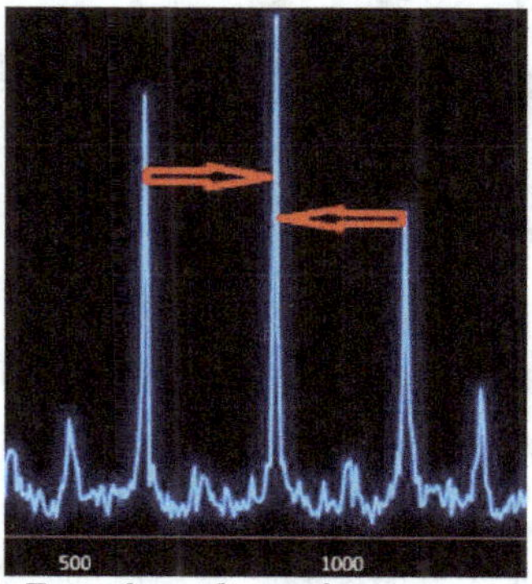

Figure 1: Typical equidistant frequency power peaks.

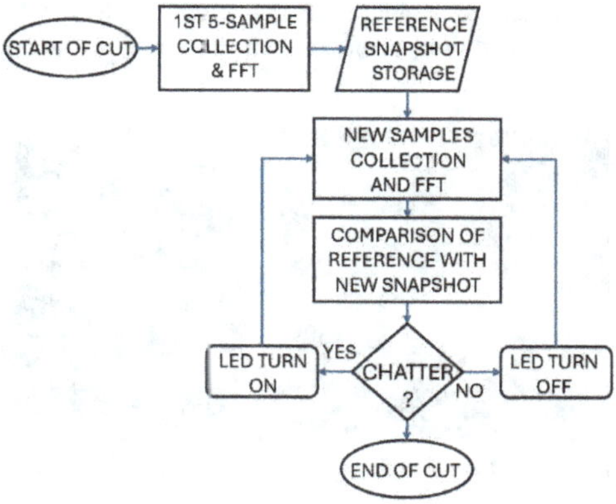

b

Figure 2: Chatter detection workflow.

Device implementation

The system comprises a Piezoelectric accelerometer (PCB Piezotronics) with a Kistler™ 5118B2 signal conditioner and microcontrollers. For the latter, an Arduino Mega™, a Teensy™ 3.5 and a Raspberry Pi™ 3 B+ were exploited.

The accelerometer is glued on the tool holder, see Fig. 3. The signal conditioner output is analog in the range -5V +5V. Initially it was fed into an Arduino Mega™ accepting a range of 0-5 V hence a voltage divider was introduced. However, since the signal conditioner is not designed to supply current and the resistors of the voltage divider were only 220Ω which increased the current

Materials Research Proceedings 46 (2024) 159-166 https://doi.org/10.21741/9781644903377-21

requirement, an operational amplifier of unit gain in negative feedback loop connection mode (LM324N) and a single power supply of 12V was introduced.

Figure 3: Accelerometer mounting.

Calculations regarding signal processing were performed on the Raspberry Pi™ microcontroller due to its clock speed (>1 GHz). However, its serial connection to the Arduino™ that fed the signal was problematic, since there was a lag of about 500 msec before it could start acquiring signal value. Note that Raspberry Pi™ does not possess an Analog to Digital Converter (ADC), hence the need of the Arduino Mega™.

For this reason, the Tensy™ 3.5 microcontroller replaced the Raspeberry Pi™. It has the same architecture as the Arduino Mega™, possessing an ADC and 120MHz clock speed. The only modification that was needed was another voltage divider was needed to transform the 0-5 V range from Arduino™ to a 0-3.3 V. In addition, an SD card was added to enable storage of measurements along with the appropriate circuit to enable communication over SPI and a button for signalling measurement start and stop. The prototype microcontroller system connectivity is shown in Fig. 4.

a  b

Figure 4: Prototype signal collection and conditioning board.

Preliminary experimental validation

Three different cutting experiments were conducted as a proof of concept. They were executed on a Haas™ TL1 CNC lathe with a 7.5 kW spindle motor with a speed range 0-3000 rpm.

The workpiece was a tube made of common structural steel St37 with external diameter 50 mm whose internal diameter was reduced by longitudinal turning for a length of 20 mm.

The cutting tool was A42-TURNING ISO – S CARBIDE/CERMET from Seco™ and its position angle κ was 80°. The rest of the cutting parameters are shown in Table 1. Chatter was induced by varying the tool overhang. In Fig. 5 characteristic images of the turned surface in all cases are shown. Case B corresponds to chatter, see also Table 1, whilst cases A and C correspond to stable turning.

Table 1: Turning cases examined.

	Case		
	A	**B**	**C**
Spindle speed (rpm)	750	750	500
Depth of cut (mm)	1	0.25	1
Tool overhang (mm)	85	128	85
Feed (mm/min)	30	30	30
Workpiece length (mm)	140	140	140
Chatter	no	yes	no

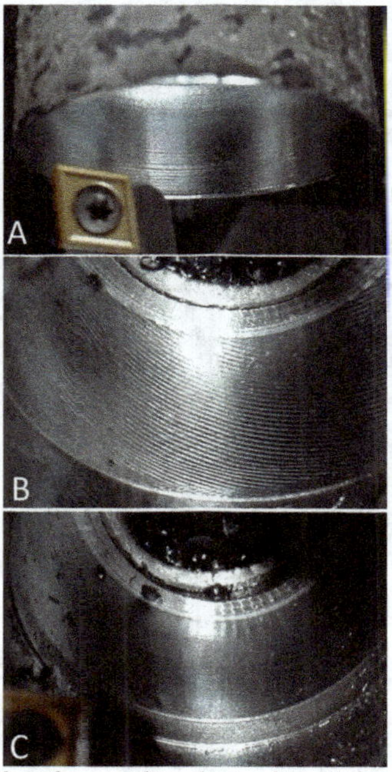

Figure 5: Turned surface quality obtained in all three cases (A-C).

In Fig. 6 the pertinent power spectra following FFT of the acceleration signals are depicted for all three cases shown in Table 1 and Fig. 5.

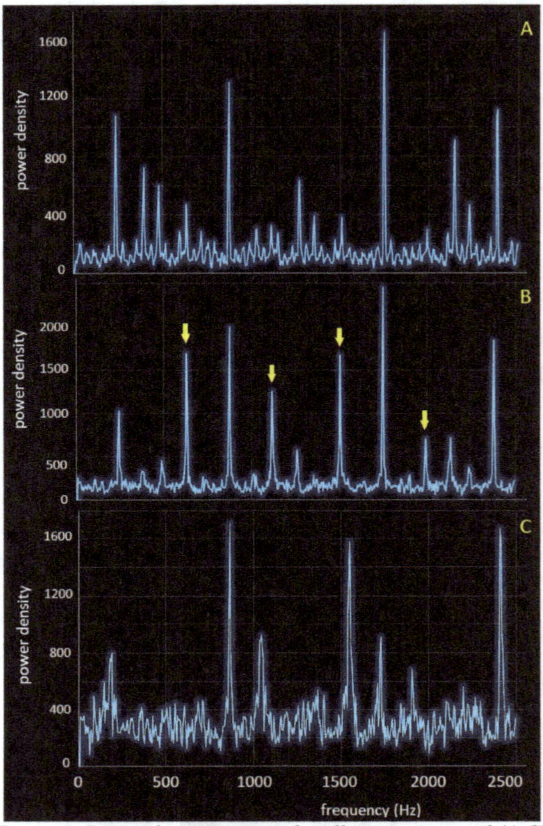

Figure 6: Power density spectra for all cases examined (A-C).

In particular, Case B where chatter occurred differs from Case A by a bigger tool overhang by about 50% and a 4 times lower depth of cut, see Table 1. Referring to Fig. 6(A-B) one can observe that at most of the frequencies power density has increased. The first harmonic in both cases is at 240 Hz. The dominant chatter frequency is 240 Hz to the left of the highest power nominal frequency of 864 Hz. The second chatter peak, as expected, is at 240 Hz to the right of the highest power nominal frequency. The third and fourth chatter peaks are 240 Hz to the left and right of the second nominal high power frequency of 1748 Hz. Note that the fourth peak's power density is markedly lower than the other three. All four chatter peaks are marked by yellow arrows, the respective multipliers being 3.5, 3.5, 4.25 and 2.3. They correspond to a sizeable increase, in fact multiplication, of the corresponding stable cutting power densities, as the chatter detection algorithm requires.

Case C corresponds to a stable cutting. Cutting conditions are close to those of Case A (also stable) except for spindle speed, which is 33% lower. Given that Case A was chatter-less, it is no surprise that no chatter occurs in Case C either. Note that power density at homologous frequencies between Cases A and C do not differ too much, see Fig. 6. Thus, first indications are that the algorithm holds true.

Conclusions and further work

First indications from this work are that chatter onset in turning is possible to detect directly in real time, and, most importantly, at low cost using standard microprocessors. Moreover, the chatter detection algorithm does not have as a pre-condition knowledge of parameters related to a specific type of cutting, e.g. example, longitudinal, transverse, profile turning, etc. or a specific spindle speed or cutting tool feed. It does not even require prior modal analysis of the lathe-workpiece-tool system or a parametric mathematical model. The only pre-requisite is a chatter-free pass has to be recorded first to serve as reference for comparing system frequencies exhibiting increased energy that are present in FFT spectra. Such generality of application has been identified as a very important factor of practical acceptability by all researchers and practitioners in this field.

Future work involves as a first step further systematic experimental validation of the approach. Upon this being successful, a chatter suppression algorithm may be implemented based on fast iterative change of spindle speed as a primary factor and tool feed as a secondary one effectively moving the operational point of the process outside the instability lobes, but without having identified the latter explicitly.

Acknowledgment

Mr Nikos Melissas of the Manufacturing Technology Lab of NTUA is gratefully acknowledged for his support of the experimental work.

References

[1] S. Katiyar, M. Jaiswal, R. Pratap Narain, S. Singh, Y. Shrivastava, A short review on investigation and suppression of tool chatter in turning operation, Mater Today Proc 51 (2021) 1206–1210. https://doi.org/10.1016/j.matpr.2021.07.208

[2] M. Eynian, Y. Altintas, Chatter Stability of General Turning Operations With Process Damping, J Manuf Sci Eng 131 (2009) 1–10. https://doi.org/10.1115/1.3159047

[3] G. Urbikain, D. Olvera, L.N.L. de Lacalle, A. Beranoagirre, Alex Elías-Zuñiga, Prediction methods and experimental techniques for chatter avoidance in turning systems: A review, Applied Sciences 9 (2019) 4718. https://doi.org/10.3390/app9214718

[4] C. Yue, H. Gao, X. Liu, S.Y. Liang, L. Wang, A review of chatter vibration research in milling, Chinese Journal of Aeronautics 32 (2019) 215–242. https://doi.org/10.1016/j.cja.2018.11.007

[5] A.H. Tian, C.B. Fu, X.Y. Su, H.T. Yau, Lathe tool chatter vibration diagnostic using general regression neural network based on Chua's circuit and fractional-order Lorenz master/slave chaotic system, Journal of Low Frequency Noise Vibration and Active Control 38 (2019) 953–966. https://doi.org/10.1177/1461348418815414

[6] M. Postel, D. Aslan, K. Wegener, Y. Altintas, Monitoring of vibrations and cutting forces with spindle mounted vibration sensors, CIRP Annals 68 (2019) 413–416. https://doi.org/10.1016/j.cirp.2019.03.019

[7] T.L. Schmitz, Chatter recognition by a statistical evaluation of the synchronously sampled audio signal, J Sound Vib 262 (2003) 721–730. https://doi.org/10.1016/S0022-460X(03)00119-6

[8] Z. Yao, D. Mei, Z. Chen, On-line chatter detection and identification based on wavelet and support vector machine, J Mater Process Technol 210 (2010) 713–719. https://doi.org/10.1016/j.jmatprotec.2009.11.007

[9] X. Huang, M. Hu, Y. Zhang, C. Lv, Probabilistic analysis of chatter stability in turning, International Journal of Advanced Manufacturing Technology 87 (2016) 3225–3232. https://doi.org/10.1007/s00170-016-8672-7

[10] V. Nguyen, S. Melkote, A. Deshamudre, M. Khanna, D. Walker, Comparison of on-line chatter detection algorithms in turning, in: International Symposium on Flexible Automation, ISFA 2016, Cleveland, USA, 2016: pp. 87–94. https://doi.org/10.1109/ISFA.2016.7790141

[11] S. Yamato, T. Hirano, Y. Yamada, R. Koike, Y. Kakinuma, Sensor-less on-line chatter detection in turning process based on phase monitoring using power factor theory, Precis Eng 51 (2018) 103–116. https://doi.org/10.1016/j.precisioneng.2017.07.017

[12] A. Jimenez Cortadi, I. Irigoien, F. Boto, B. Sierra, A. Suarez, D. Galar, A statistical data-based approach to instability detection and wear prediction in radial turning processes, Eksploatacja i Niezawodnosc 20 (2018) 405–412. https://doi.org/10.17531/ein.2018.3.8

[13] S. Kim, K. Ahmadi, Estimation of vibration stability in turning using operational modal analysis, Mech Syst Signal Process 130 (2019) 315–332. https://doi.org/10.1016/j.ymssp.2019.04.057

[14] C.L. Zhang, X. Yue, Y.T. Jiang, W. Zheng, A hybrid approach of ANN and HMM for cutting chatter monitoring, Adv Mat Res 97 (2010) 3225–3232. https://doi.org/10.4028/www.scientific.net/AMR.97-101.3225

[15] F.A. Khasawneh, E. Munch, J.A. Perea, Chatter Classification in Turning using Machine Learning and Topological Data Analysis, IFAC Papers Online 51 (2018) 195–200. https://doi.org/10.1016/j.ifacol.2018.07.222

[16] H. Cherukuri, E. Perez-Bernabeu, M.A. Selles, T.L. Schmitz, A neural network approach for chatter prediction in turning, Procedia Manuf 34 (2019) 885–892. https://doi.org/10.1016/j.promfg.2019.06.159

[17] S. Kumar, B. Singh, Prediction of tool chatter in turning using RSM and ANN, Mater Today Proc 5 (2018) 23806–23815. https://doi.org/10.1016/j.matpr.2018.10.172

[18] Y. Shrivastava, B. Singh, A. Sharma, Identification of Chatter in Turning Operation using WD and EMD, Mater Today Proc 5 (2018) 23917–23926. https://doi.org/10.1016/j.matpr.2018.10.184

Innovative Manufacturing Engineering and Energy - IManEE2024 Materials Research Forum LLC
Materials Research Proceedings 46 (2024) 167-175 https://doi.org/10.21741/9781644903377-22

Enhancing vocational education through digital training materials: A case study on CNC lathe training

Musa BILGIN[1,2,a *], Sener KARABULUT[1,b] and Angelos P. MARKOPOULOS[3,c]

[1]Department of Mechanical Program, Hacettepe University, 06930, Ankara, Türkiye

[2]Department of Mechanical, Aerospace and Civil Engineering, The University of Manchester, Manchester, UK

[3]School of Mechanical Engineering, National Technical University of Athens, 15780, Athens, Greece

[a]musabilgin@hacettepe.edu.tr, [b]senerkarabulut@hacettepe.edu.tr, [c]amark@mail.ntua.gr

Keywords: Vocational Education, VR Application, Interactive Training, CNC Lathe Training, Educational Innovation

Abstract. This study investigates the effect of virtual reality (VR)-based and interactive training materials on practical and academic achievements in computer numerical control (CNC) lathe training. To this end, VR and interactive training materials were developed for tasks involving cutting tool measurement and defining a workpiece zero-point. Sixteen students, all without prior VR or CNC lathe experience, were divided into experimental and control groups. Participants were subjected to traditional and digital-based training methods, and their practical and academic performances were subsequently analyzed. The findings revealed that VR-based digital training materials significantly enhanced students' success and reduced the training time in CNC lathe training. Furthermore, the accessibility of materials at students' convenience through interactive educational content reduces educational costs, facilitates learning processes, and contributes to environmental sustainability.

Introduction

The evolution of digital technologies has significantly impacted various sectors, including education, engineering, and industrial training. Virtual reality (VR) and augmented reality (AR) are at the forefront of this digital transformation, offering immersive experiences that enhance learning, training, and maintenance processes. These technologies are particularly valuable in fields that require hands-on experience and spatial understanding, such as engineering and industrial maintenance. In education, VR provides a platform for creating interactive and engaging learning environments.

Kalkan and Karabulut (2021) studied the lasting effects of VR training on performance and quality in complex assembly tasks, focusing on a hydraulically controlled clutch with 90 parts. Their results indicated a 25% reduction in training time, a 27.9% improvement in assembly task performance, and an 89% decrease in assembly flaws. This suggests that VR training can be effective and enduring in enhancing complex industrial maintenance and assembly tasks [1]. Hoang et al. (2023) investigated the influence of different training methods on students' mental workload in CNC machining within mechanical engineering programs. They compared traditional training with combined simulation training and found that the latter significantly reduced mental workload, concluding that simulated software and combined training enhance knowledge acquisition by lowering mental workload [2]. Traditional teaching methods for CNC machining on conventional machines have shown limitations in effectiveness, often leading to incidents during machine operation. Simulation-based training offers a safe environment for students to practice without risks, providing flexibility and cost-saving benefits while maintaining the importance of hands-on labs for practical skill acquisition [3]. Gamification has also been explored to enhance learner motivation and outcomes in engineering education. Studies have shown that gamification can make difficult subjects more manageable and increase intrinsic motivation, scientific knowledge, collaboration, and interest, while also reducing or better managing workloads [4]. Jensen and Konradsen (2018) reviewed the use of VR head-mounted displays in education and training, highlighting their potential to improve learning outcomes through immersive and interactive experiences [5]. Similarly, Dickey (2005) discussed the benefits of three-dimensional virtual worlds for distance learning, presenting case studies that demonstrate how VR can facilitate effective remote education [6]. In industrial applications, VR enhances the efficiency and safety of maintenance and training processes. Aziz et al. (2018) studied the use of VR for equipment maintenance in the oil and gas industry, emphasizing its ability to provide realistic simulations that improve technical skills without real-world risks [7]. Additionally, Otto et al. (2019) developed a VR assembly assessment benchmark to measure VR performance and limitations in industrial settings, illustrating its value for assessing and training industrial personnel [8]. The integration of VR into maintenance practices is particularly noteworthy in high-risk industries such as aviation and power distribution. Eschen et al. (2018) investigated the application of VR for inspection and maintenance in aviation, showing how VR can enhance safety and efficiency [9]. Perez-Ramirez et al. (2019) focused on using a VR training system to improve technical skills in the maintenance of live-line power distribution networks, demonstrating significant improvements in the accuracy and safety of maintenance operations [10]. Industry 4.0 increases the demand for technical staff to adapt to complex tasks, requiring more flexible training tools. Additionally, there is a growing need for interactive digital materials to help students comprehend course content more effectively [11].

Engineering education is essential for developing students' vocational skills, including technical knowledge and practical application. However, the need for practice in diverse physical settings makes this process challenging and costly. In this context, developing more effective and accessible educational methodologies will facilitate students in enhancing their vocational skills more proficiently by streamlining the process of repetition. This study aimed to investigate the impact of VR-based training and interactive course materials on applied sciences in the context of CNC Lathe training. Two separate groups underwent either traditional training or interactive course content delivery, and an analysis of both the practical and academic achievements of the participants was conducted.

Materials and Methods

Preparation of VR and interactive training materials

In this research study, a VR-based application was developed for "cutting tool measurement" and "face turning" operations on a CNC lathe machine. Subsequently, using the 3D model employed in the VR application, an interactive sharable content object reference model (SCORM) training content based on 3D real-time interaction, including measurement questions, was created. The scenario for the VR application and the voice-over content for the interactive SCORM application were prepared by expert academics in the field.

To develop these training materials, dimensions were taken from an actual CNC lathe used in students' practical training, and detailed photographs were taken. Then, 3D CAD models were developed according to the actual dimensions, and material textures were applied using the photographs to create a sense of realism for the users. Fig. 1. a, shows the photograph of the real CNC lathe machine, while Fig. 1. b, presents the appearance of the digital CNC lathe machine used in the VR training sessions.

Materials Research Proceedings 46 (2024) 167-175 https://doi.org/10.21741/9781644903377-22

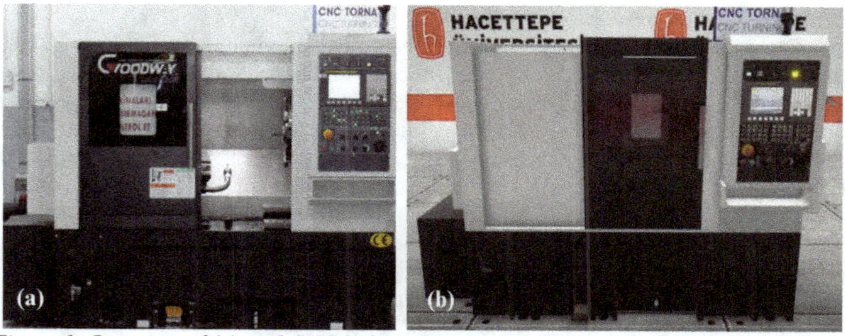

Figure 1. Overview of (a) real CNC lathe machine, and (b) VR-based CNC lathe machine

In the second phase, model optimization was carried out for a real-time VR application to achieve a frame rate of 90 frames per second (fps). Subsequently, the interactive cutting tool measurement process was physically implemented using C# scripts. These scripts facilitated the creation of a user decision support system with functionalities such as touching, holding, repositioning components on the control panel of the CNC lathe, and displaying warning symbols for correct and incorrect operations. The VR-based training system was designed to mimic the movements and application rules that a student would need to follow in a real training environment. Additionally, authentic lathe sounds and alerts were incorporated to create realistic scenes. Fig. 2 illustrates the digital training materials, including the interactive cutting tool measurement training material, the interactive workpiece zero-point definition, and the VR-application material used in the research study.

Finally, the developed VR training contents were subjected to a trial run by academics. This preliminary user study tested the readiness, usability, and adaptability of the VR-based training content for inexperienced users. Feedback obtained from this study was used to make improvements to the content, rendering it ready for deployment.

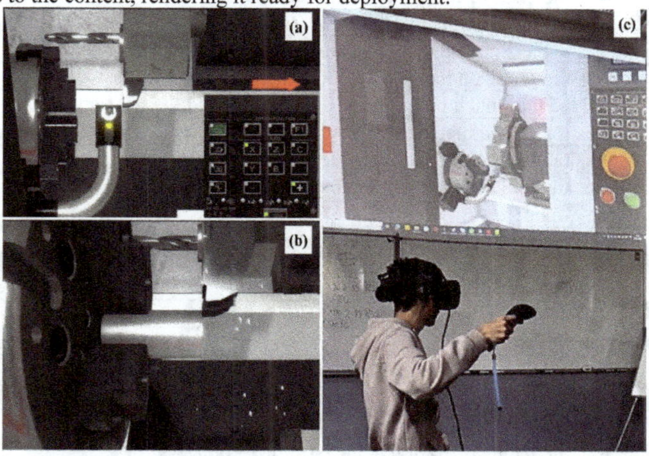

Figure 2. Digital training materials, (a) interactive cutting tool measurement training material, (b) defining interactive workpiece zero point and (c) VR-application material

Research design and participants

To compare and analyze the effects of VR and interactive training materials versus traditional teaching methods on the practical and academic performance of students receiving CNC lathe training, two groups of eight participants each were formed: an experimental group and a control group. All participants were vocational college students aged between 18 and 20. The experimental group received theoretical education using interactive course content shared electronically through an LMS system and practical training using VR training materials. In contrast, the control group was trained using traditional methods, involving hands-on instruction with a CNC lathe in a laboratory setting under the guidance of an instructor.

As part of the study, an instructor evaluation form was developed (Table 1) to ensure standardized assessment across both digital training materials and traditional methods.

Table 1. Application steps

	Step	Task	Check
Task 1	1	Did the main switch turn on?	
	2	Was the control panel given power (electric)?	
	3	Was the hydraulic motor started?	
	4	Was the turret moved by the handwheel?	
	5	Was the turret first sent to zero point on the x-axis?	
	6	Was the turret sent to the reference point on the z-axis after the x-axis?	
	7	Were the Zero point lights on the control panel checked?	
Task 2	8	Was the tool to be measured in the turret repositioned?	
	9	Was the probe opened?	
	10	Was the screen controlled to check opening the measurement page?	
	11	Was the tool brought near to measurement probe after mod key was turned handwheel position?	
	12	Was the mod key brought to the JOG position?	
	13	Was the Feed rate taken to minimum value?	
	14	Did the tool touch the probe at the Z-axis? (Did the red light on)	
	15	Was it checked that the value was transferred to the geometry page?	
	16	Was the probe on the Z-axis touched again for control?	
	17	Did the tool touch the probe at the X-axis? (Did the red light on)	
	18	Was it checked that the value was transferred to the geometry page?	
	19	Was the probe on the x-axis touched again for control?	
	20	Was the tool sent to the reference point, first in the X-axis and then in the Z-axis?	
	21	Was the probe arm closed after the measuring?	
Task 3	22	Was the workpiece clamped?	
	23	Was the chuck light on the panel checked?	
	24	Has the master tool been turned at the turret?	
	25	Was a workpiece of runout checked?	
	26	Was the spindle speed given for the workpiece?	
	27	Was the workpiece approached safely?	
	28	Was the facing operation at the workpiece done?	
	29	Was the safe distance in the x-axis reached?	
	30	Was the Z axis reset from the operator panel?	

This form outlined a total of 30 essential steps required for the "cutting tool measurement" and "face turning" operations. Each critical step was scored as "completed" (1) or "not completed" (0) based on the student's performance. After all preparations were completed and both groups had finished their respective training programs, the students were evaluated on the CNC lathe for the "cutting tool measurement" and "face turning" operations. Two different academics used the steps in Table 1 to assess the students' practical execution in both groups. The evaluation considered whether the students performed each step correctly and how long it took them to complete the steps. Additionally, seven weeks later, the students in both groups were subjected to a written exam and a practical test to evaluate the retention of their training and their overall performance (Fig. 3).

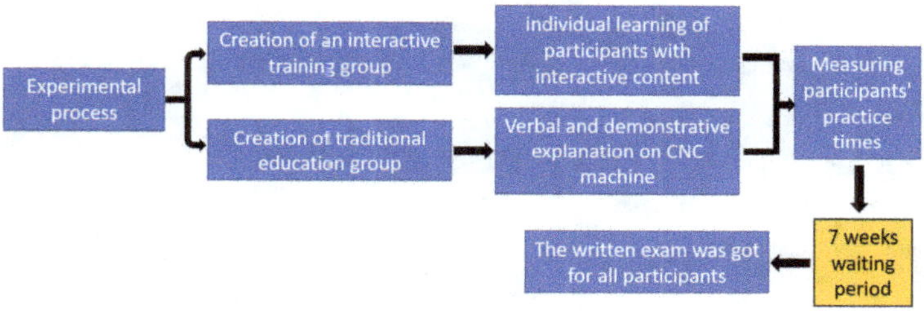

Figure 3. Research study flow plan

Results and Discussion

Application times

All participants in the CNC lathe machine training were instructed to complete three main tasks consisting of a total of 30 steps. Fig. 4 presents a graph illustrating the total application times of the participants. This graph was prepared by arranging the times within each group from lowest to highest and forming pairs accordingly.

The graph shows that participants trained with interactive content generally completed tasks in shorter durations compared to those trained through traditional methods. Participants trained with interactive content completed tasks in durations ranging from 1 minute 35 seconds to 4 minutes 40 seconds, whereas participants trained traditionally completed tasks in durations ranging from 2 minutes 40 seconds to 5 minutes 34 seconds. This difference can be attributed to the interactive training providing a broader time frame for practice and allowing for independent repetitions. Furthermore, it was observed that participants achieving the shortest task completion times engaged in numerous independent repetitions, whereas longer completion times were associated with fewer repetitions.

According to the data in Fig. 4, it is observed that the completion times of tasks for students in the 3rd and 4th groups closely resemble each other. This indicates that students in these groups worked at similar speeds and efficiency levels. Participants using digital educational content completed tasks in shorter durations. This can be attributed to the engagement and reinforcement of learning when one student engages in a VR application while others observe and learn from the projected image. Therefore, interactive content not only attracted students' interest but also allowed them to see all the details more clearly through the projected video. Thus, it can be said that interactive content facilitated clearer understanding and efficient internalization of the presented information by students. However, participants with longer application times expressed difficulties in allocating enough time to interactive course materials and using VR content during interviews. In general, students positively evaluated the reusability of interactive course materials and the flexibility to study at their own pace. It can be stated that the data in Fig. 4 indicate that digital educational content contributes positively to students' learning processes and offers flexibility for students with different learning speeds. This flexibility addresses the needs of both fast learners and those requiring more time.

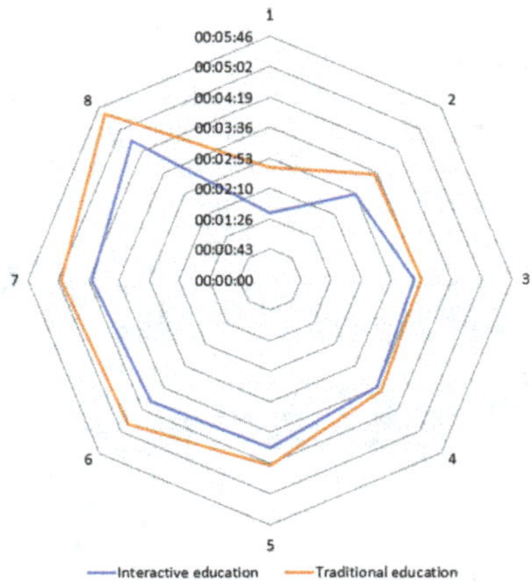

Figure 4. Practice completion times

Fig. 5 illustrates the average application times for both groups. The interactive training group shows an average of 3 minutes and 33 seconds, while the traditional training group shows an increase of 17.09% with an average time of 4 minutes and 9 seconds. This shows that interactive education has had a positive impact on participants.

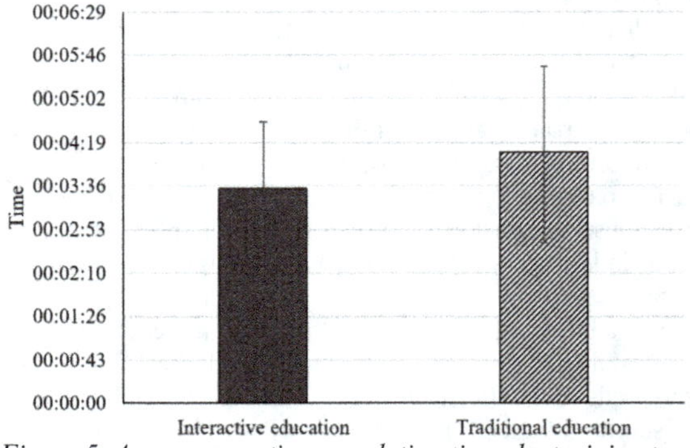

Figure 5. Average practice completion times by training type

Impact of education system on examinations

After completing the practical measurements, participants took an assessment exam consisting of written and multiple-choice tests following a 7-week waiting period. The grades obtained by the participants are presented in Fig. 6. To create the graph shown in Fig. 6, the groups were ranked from lowest to highest score. As depicted in Fig. 6, all participants trained with interactive education achieved higher scores compared to their counterparts in the traditional education group. Among participants in the interactive education group, scores ranged from a minimum of 28 to a maximum of 100. In contrast, scores among participants in the traditional education group varied between 19 and 78. Another notable observation from Fig. 6 is the clustering of scores, especially among participants with the lowest scores. This suggests that individuals in both the traditional and interactive education groups may not have revisited the material after completing their initial training. Significant differences persist among the top five participants, particularly highlighting the continuous knowledge updates facilitated by interactive content, which becomes particularly pronounced for intermediate-level participants. Beyond the fifth participant, differences consistently continue. This phenomenon can be attributed to the ability of interactive content to allow for continuous review of subjects anytime and anywhere, contrasting with the gradual loss of knowledge acquired through traditional education over time.

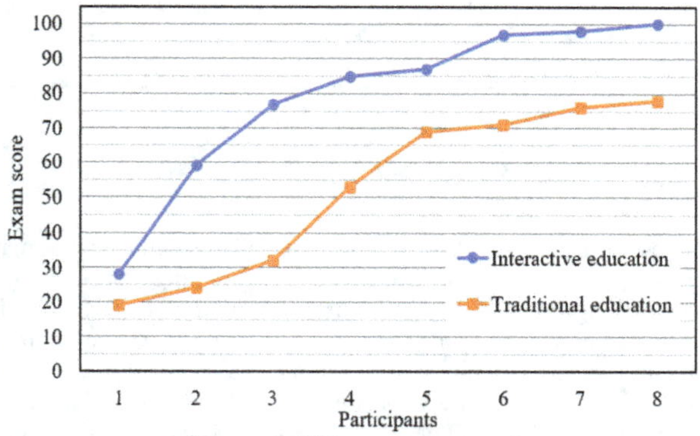

Figure 6. Written exam scores

Relationship between practice time and achievement level

Fig. 7 illustrates the practice times and exam scores of participants from both the interactive and traditional education groups. It is observed that there is no significant correlation between practice times and exam scores. This suggests that factors such as the educational system and participants' practical skills may influence these outcomes. However, it is noteworthy that among the participants in the traditional education group, three out of the four participants with the longest practice times obtained the lowest exam scores. This highlights the potential impact of motivational deficiencies in traditional education, which can trigger both academic and practical inadequacies. Additionally, while none of the participants from the traditional education group exceeded the 80-point threshold, five participants from the interactive education group surpassed this limit. This finding indicates that interactive education can enhance academic performance through its ability to allow students to review materials at their own pace. The decline in knowledge retention in traditional education negatively impacts participants' academic success.

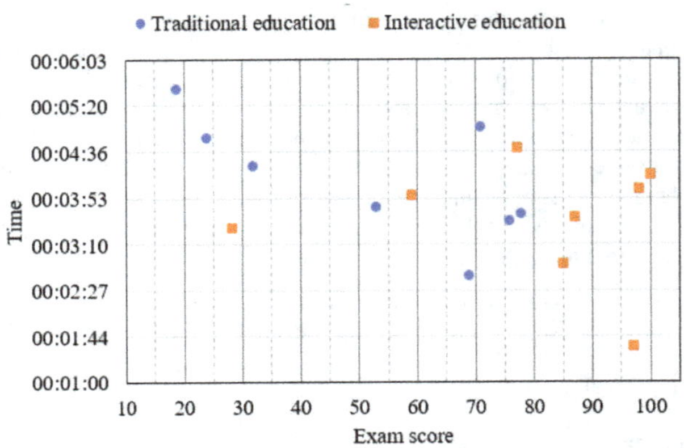

Figure 7. Practice time – exam score relationship

Conclusions

This study examines the efficacy of VR-based training and interactive course materials in the field of applied sciences, specifically focusing on CNC Lathe training. Participants were divided into two groups: one receiving traditional training methods and the other utilizing interactive course content. The findings from this study into the effectiveness of interactive and traditional education methods in CNC lathe machine training can be summarized as follow:

Interactive educational content has a positive impact on students' practical and academic achievements. Such content allows students to access materials at their convenience, fostering a consistent understanding of the subject matter. Conversely, traditional educational content limits students' academic success, with retention rates declining over time and adversely affecting participants' academic performance.

Interactive content can reduce educational costs and enable students to study independently, thus eliminating the need for direct supervision. This method of education also has the potential to enhance the capacity of engineering and vocational training institutions, as it allows students to independently operate CNC machines.

These findings indicate that interactive education is more effective than traditional methods, particularly in engineering training. Interactive education facilitates faster learning, easier access to information, and longer retention of knowledge. Therefore, the adoption and implementation of interactive content by educational institutions can enhance overall educational quality.

Additionally, it can be said that VR training materials reduced the actual practical time required, resulting in lower energy consumption and enhancing environmental sustainability.

Funding

This work was supported by the Erasmus+ Program of the European Union under grant number 2020-1-TR01-KA226-VET-098323. However, the European Commission and the Turkish National Agency cannot be held responsible for any use that may be made of the information contained herein.

References

[1] Ö.K. Kalkan, Ş. Karabulut, and G. Höke, Effect of Virtual Reality-Based Training on Complex Industrial Assembly Task Performance, Arab. J. Sci. Eng. 46 (2021), 12697–12708. https://doi.org/10.1007/s13369-021-06138-w

[2] S. Hoang, C.C. Tran, V. T. Pham, V. T. Nguyen, V. T. Tran, V. T. Tran, and T. T. Nguyen, Evaluating the learning performances for CNC machine practice in mechanical engineering degree courses based on students' mental workload, Int. J. of Mec. Eng. Educ. 52 (2) (2024) 205 – 220. https://doi:10.1177/03064190231185326

[3] J. P. Davim, Mechanical engineering education, Hoboken, New Jersey, USA: John Wiley & Sons, 2012.

[4] A. P. Markopoulos, A. Fragkou, P. D. Kasidiaris and J. P. Davim, Gamification in engineering education and professional training, Int. J. of Mec. Eng. Educ. 43 (2) (2015) 118-131. https://doi:10.1177/0306419015591324

[5] L. Jensen and F. Konradsen, A review of the use of virtual reality head-mounted displays in education and training, Educ. Infor. Tech. 23 (2018) 1515-1529. https://doi.org/10.1007/s10639-017-9676-0

[6] M. D. Dickey, Three-dimensional virtual worlds and distance learning: two case studies of Active Worlds as a medium for distance education, British J. of Educ. Tech. 36 (2005) 439-451. https://doi.org/10.1111/j.1467-8535.2005.00477.x

[7] F. A. Aziz, M. Alshammar and M. K. A. Ariffin, Using virtual reality for equipment maintenance in oil and gas industry, J. of Comput. and Theor. Nano.15 (4) (2008) 1090 – 1094. https://doi.org/10.1166/jctn.2018.7150

[8] M. Otto, E. Lampen, P. Agethen, M. Langohr, G. Zachmann, and E. Rukzio, A virtual reality assembly assessment benchmark for measuring VR performance & limitations, Procedia CIRP, 81 (2019) 785-790. https://doi.org/10.1016/j.procir.2019.03.195

[9] H. Eschen, T. Kötter, R. Rodeck, M. Harnisch and T. Schüppstuhl, Augmented and virtual reality for inspection and maintenance processes in the aviation industry, Procedia Manufacturing, 19 (2018) 156-163. https://doi.org/10.1016/j.promfg.2018.01.022

[10] M. Perez-Ramirez, G. Arroyo-Figueroa and A. Ayala, The use of a virtual reality training system to improve technical skill in the maintenance of live-line power distribution networks, Int. Learning Envir. 29 (4) (2019) 527–544. https://doi.org/10.1080/10494820.2019.1587636

[11] F. Longo, L. Nicoletti and A. Padovano, Smart operators in industry 4.0: A human-centered approach to enhance operators' capabilities and competencies within the new smart factory context, Computers & industrial engineering, 113 (2017) 144-159. https://doi.org/10.1016/j.cie.2017.09.016

Innovative Manufacturing Engineering and Energy - IManEE2024 Materials Research Forum LLC
Materials Research Proceedings 46 (2024) 176-183 https://doi.org/10.21741/9781644903377-23

Hybrid electrochemical discharge drilling

Laurențiu SLĂTINEANU[a], Margareta COTEAȚĂ[b], Oana DODUN[c], Andrei Marius MIHALACHE[d], Adelina HRIȚUC[e *], and Gheorghe NAGÎȚ[f]

"Gheorghe Asachi" Technical University of Iaşi, Department of Machine Manufacturing Technology, Blvd. D. Mangeron, 59 A, 700050 Iaşi, Romania

[a]slati@tcm.tuiasi.ro, [b]margareta.coteata@academic.tuiasi.ro, [c]oanad@tcm.tuiasi.ro, [d]andrei-marius.mihalache@academic.tuiasi.ro,[*e]adelina.hrituc@student.tuiasi.ro, [f]nagit@tcm.tuiasi.ro

Keywords: Hybrid Machining, Electrochemical Discharge Drilling, Drilling Equipment, Functional Requirements, Preliminary Experiment

Abstract. Making small-diameter holes in workpiece materials that are difficult to machine using classic cutting processes may require using relatively expensive equipment for machining using unconventional methods. A more accessible process for making small-diameter holes in hard-to-machine materials is hybrid electrochemical discharge drilling. In principle, such a process is based on the material removal from the workpiece both using electrical discharges and chemical reactions between the electrolyte-type working fluid and the material of the workpiece, under the conditions of placing the tool electrode and the workpiece in an electrolyte and connecting them in a circuit of a direct current source. It is proposed to use relatively simple equipment, which ensures the rotation of the drill-type tool electrode and the use of an electrolyte based on an aqueous solution of sodium silicate soluble in water. The analysis of the functional requirements allowed the gradual design of an equipment solution that enables the realization of small-diameter holes using hybrid electrochemical discharge machining. Preliminary experiments proved the good functioning of the equipment.

Introduction

Holes with a diameter of less than 10 mm can be obtained, in principle, either using the classic method of drilling with a drill or by resorting to processes based on an unconventional method of processing with material removal from the workpiece, such as drilling with a laser beam, electron beam, electrical erosion, electrochemical erosion, etc.

If the workpiece material is hard, as in the case of hardened steels, metal carbides, etc., drilling with a drill can no longer be used, and, in such a case, the use of an unconventional processing method becomes almost mandatory.

On the other hand, the last decades have highlighted the possibility of combining two or more processing methods, one of which is an unconventional method. Through such a combination, an attempt was made to obtain the specific advantages of each of the combined methods. Thus, the so-called *hybrid machining methods* were reached [1-4]. In the present case, a drilling method is of interest, so a method involving the removal of material from the workpiece.

A hybrid electrochemical discharge machining method was considered, which would exploit specific aspects of electrical discharge machining and electrochemical machining, respectively. It is known that in the case of electrical discharge machining, the removal of material from the workpiece occurs due to electric discharges between the active surface of a tool electrode and the surface of the workpiece, respectively, when the distance between them is sufficiently low and when the working area is immersed in a dielectric liquid. These electrical discharges lead to material removal at both ends of the electrical discharge, i.e., both at the workpiece surface and at the tool electrode surface. However, the machining conditions (the material of the tool electrode, the polarization mode of the tool electrode and the workpiece, the characteristic parameters of the electric discharges, the nature of the dielectric liquid, etc.) are established in such a way that the removal of material from the workpiece is much more intense than the removal of material from the tool electrode, which means wear of the tool electrode. There are thus situations when the relative wear, expressed by the ratio between the amounts of material removed from the tool electrode and, respectively, from the workpiece, can have values of around 1% or even lower values. In the case of electrochemical machining, the removal of material from the workpiece takes place as a result of chemical reactions between an electrolyte and the material of the workpiece, when the workpiece and a so-called tool electrode are connected in the circuit of a direct voltage source, and the machining area is immersed in the electrolyte. The removal of material from the workpiece occurs mainly in the areas where the electric field is more intense, that is, where the distance between the tool electrode and the workpiece is lower. When electrochemical machining, there is no wear of the tool electrode, but there is an impurity of the electrolyte with substances resulting from the chemical reactions between the electrolyte and the material of the workpiece. This fact requires a recirculation, as much as possible, of the electrolyte in the working gap, so that less impurity electrolyte with products resulting from the machining process is brought into the working gap. The material removal rate of electrochemical machining can be significantly increased by increasing the intensity of the electric current, but the increase in the value of this input factor in the machining process must not lead to the boiling of the electrolyte, which could contribute to a worsening of the conditions of the material removing process from the workpiece. It is also required that the distance between the active surface of the tool electrode and that of the workpiece does not decrease below certain values, to avoid the occurrence of electrical discharges, with unfavorable effects from the point of view of the results sought by the use of electrochemical erosion.

In the last decades, hybrid electrochemical discharge machining processes have been used especially for the machining of workpieces made of insulating materials. In such situations, there is no sampling of material from the workpiece as a result of chemical reactions between the material of the workpiece and the electrolyte. However, due to the presence of the two electrodes connected in the circuit of a direct current source, the formation of hydrogen bubbles is observed, mainly as a result of the dissociation of water. These bubbles can be penetrated by electric discharges, the effects of which lead to the gradual removal of the electrically insulating material of the workpiece. In this way, it was even reached the materialization of some hybrid processes of drilling through electric and respectively electrochemical erosion. For example, Alessu et al. used a metal part connected to the negative pole of the direct voltage source and placed it near the insulating workpiece material [5]. This part contributes to the initiation and development of a water dissociation process and thus the formation of hydrogen bubbles that tend to move towards the upper surface of the electrolyte. A working scheme similar in principle to the one presented previously was proposed by Yang [6] who added yet a process assistance by ultrasonic vibrations. It was appreciated that the presence of ultrasonic vibrations contributes to an intensification of the depassivation process, which could contribute to an increase in the material removal rate.

When electrochemical machining, the products resulting from the chemical reactions between the electrolyte and the material of the workpiece may adhere to the surface of the workpiece or may remain near this surface, causing a passivation process and sometimes even constituting a passivating film adhering to the surface of the workpiece, Passivation can lead to a decrease in the intensity of the process of removing the material from the workpiece and even to the stop of this process. The different ways of removing the passivating layers led to the emergence of distinct machining methods, being known, as such, machining methods based on natural depassivation (the breaking of the passivating film by the hydrogen bubbles that tend to move towards the upper surface of the electrolyte), on forced hydrodynamic depassivation (the circulation of the electrolyte at high speeds and at high pressures returning to its role of contributing to the breaking and removal of the passivating film), on forced mechanical depassivation by abrasion (case in which electrochemical erosion is associated with a process of abrasion, the latter contributing to breaking and removing the passivating film), etc. It is mentioned, in this sense, the investigation of hybrid electrochemical discharge drilling processes, in which a cylindrical rod-type tool electrode was rotated in contact with the surface of the workpiece and periodically removed from this surface [7-9]. When reapproaching and respectively removing the tool electrode, electric discharges occur, which contribute, along with the pressure between the rotating tool electrode and the workpiece, to the removal or at least the thinning of the passivating film, thus making it possible to resume the chemical reactions between the material of the workpiece and the electrolyte, after which the process continues, as previously described.

In the present work, a brief analysis of the conditions in which a hybrid, electric, and electrochemical drilling process of some metal workpieces can be materialized and using an aqueous solution of soluble sodium silicate as the working fluid will be used in the water. Starting from this brief analysis, the component elements of a relatively simple equipment will be presented, which will allow the development of a hybrid, electrical, and electrochemical drilling process. Some preliminary results of the aforementioned equipment experimentation will also be presented.

Conditions for conducting a hybrid electrochemical discharge drilling process

Several decades ago, in the former Soviet Union, but also in other countries, hybrid electrochemical discharge machining processes were proposed and promoted, in which the breaking of the passivating film was done as a result of pressure and, respectively, relative movements between the tool electrode and the workpiece [10]. Such a method was sometimes referred to as an anode-mechanical machining method. As working fluids, aqueous solutions of sodium silicate soluble in water were mainly used. In the first phase, as a consequence of the chemical reactions between the workpiece material and the electrolyte, a passivating film was formed adhering to the surface of the workpiece and which was removed, as previously mentioned, as a result of the pressure and relative displacement between the tool electrode and the workpiece. In this way, anodic-mechanical cutting processes with disc, strip, or wire-type tool electrodes, anodic-mechanical turning processes, anodic-mechanical drilling processes, etc., have been developed.

There are also references to some hybrid drilling processes, based on the use of a vibrating or rotating tool electrode and which contribute to ensuring the conditions for developing electrical discharges by breaking the contact between the tool electrode and the workpiece, the two components (the tool electrode and the workpiece) being immersed in the electrolyte capable of providing conditions for the generation of a passivating film [1, 7-9]. This passivating film is removed mechanically, as a result of the rotation or vibration of the tool electrode in contact with the surface of the workpiece. The rotation or vibration of the tool electrode contributes to the thinning or even breaking of the passivation film, facilitating, in this way, the resumption of chemical reactions between the electrolyte and the material of the workpiece.

In the framework of the research whose results are presented in this work, a process was considered, the schematic representation of which can be seen in Figure 1. It can be noted that, as a tool electrode, the use of a small diameter drill was proposed, whose attack zone has been modified, creating a planar surface by abrasion, which will have the main role in the development of the electrochemical erosion process and, respectively, in the removal of the passivating film.

The hypothesis was taken into account that the existence of grooves, otherwise used for the evacuation of chips, would contribute, in this case, to a periodic interruption of the contact between the tool electrode and the workpiece at the level of the same working area, thus facilitating the development of electrical discharges either through breaking the contact between the electrodes, either by breaking through the resistance of a thin layer of passivating film on the workpiece (Fig. 2). The electric discharge will occur when the distance *s* between the surfaces of the two electrodes corresponds to a condition of the form [3]:

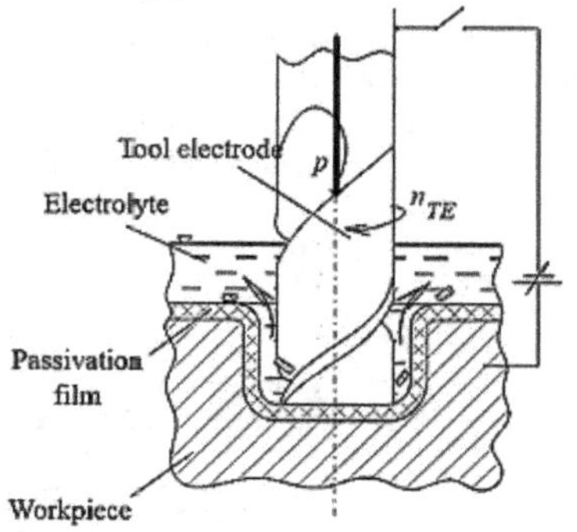

Fig. 1. Specific aspects of making small diameter holes by hybrid, electric and electrochemical drilling

$$s < \frac{U}{E}, \tag{1}$$

where *s* is the working gap size, *U* is the voltage applied to the two electrodes, and *E* - the dielectric strength of the working medium in the gap

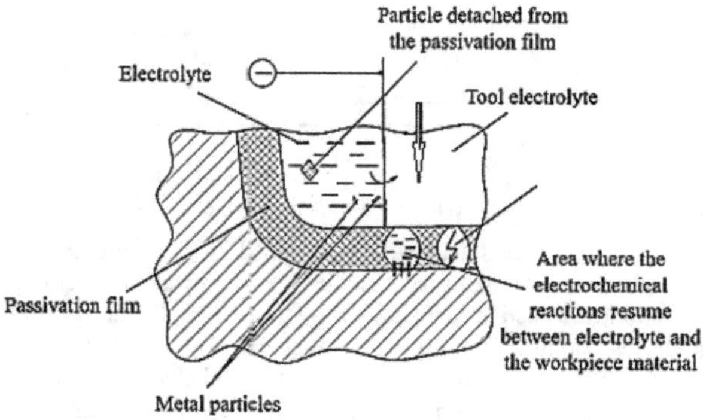

Fig. 2. Detachment of material from the workpiece by electrical discharges and by electrochemical reactions

It was assumed that the existence of a pressure exerted between the tool electrode and the workpiece, together with the rotation of the tool electrode, would contribute to the breaking of the passivation film, portions of it being forced to travel through the helical grooves, as the tool electrode gradually penetrates the material of the workpiece.

It is expected that the electrical discharges initiated between the flat surface of the tool electrode and the workpiece will result in material removal from both electrodes involved in the materialization of the hybrid electrochemical discharge machining process. However, it is considered that the use as a tool electrode of drill-type made of alloyed tool steel will provide conditions for less material removal from the tool electrode by electrical discharges, which will lead to less wear. It is also known that alloy steels have a lower machinability through electrical discharge machining. It is also appreciated that it is possible to establish some values of the parameters that characterize the machining conditions so that the wear of the tool electrode as a result of electrical discharges is minimized.

Equipment for hybrid electrochemical discharge drilling

The objective of the research, the results of which were presented in this paper, was to test the validity of the hypothesis of the existence of a possibility of using a modified drill for the development of a hybrid electrochemical discharge drilling process. For this purpose, equipment was designed and preliminarily tested, whose schematic representation can be seen in Figure 3.

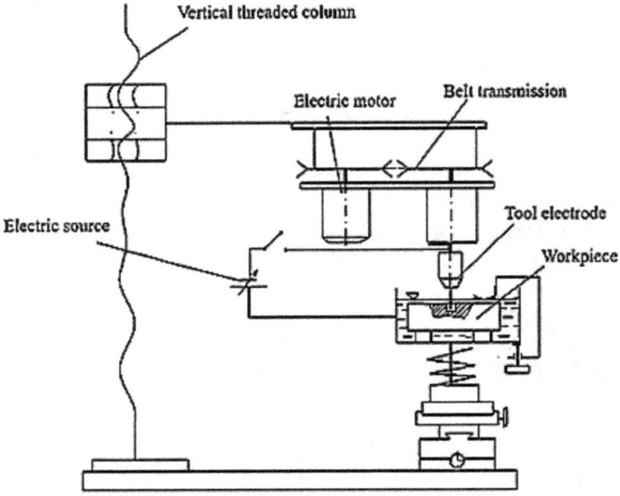

Fig. 3. Schematic representation of the equipment proposed for the materialization of a process of making a small diameter hole through hybrid electrochemical discharge drilling

The main functional requirements that had to be fulfilled by this equipment were the following: 1) Ensuring some possibilities of making a rotation movement by a tool electrode having the shape of a drill which has a modified active surface; 2) Ensuring the application of pressure between the tool electrode and the test sample or workpiece in which a hole is intended to be made; 3) Ensuring the immersion of the workpiece and the active area of the tool electrode in an electrolyte capable of generating a passivating film adhering to the surface of the test sample; 4) Ensuring the possibility of changing the values of some parameters that characterize hybrid electrochemical discharge drilling conditions, to further develop some experimental research on the influence of these factors (as input factors in the process of making small holes through hybrid electrochemical discharge drilling) on some parameters of interest of the process. The possibilities of using drills with different diameters, different pressures between the test sample and the tool electrode, and different rotational speeds of the tool electrode were considered.

The constructive solution considered the use of a vertical threaded column, attached to a base plate of the equipment (Fig. 4). Along the threaded column, a bushing that supports the tool electrode holder subsystem can be moved and fixed in the desired position using some nuts.

The rotation movement of the tool electrode is carried out using a direct current electric motor, which allows the materialization of different rotation speeds of the tool electrode, by changing the supply voltage of the electric motor. A belt drive was preferred for driving the tool electrode

Materials Research Forum LLC
https://doi.org/10.21741/9781644903377-23

support shaft into rotary motion by the electric motor. At the lower end of the supporting shaft of the tool electrode, a chuck was assembled that allows even the fixing of drills with small diameters.

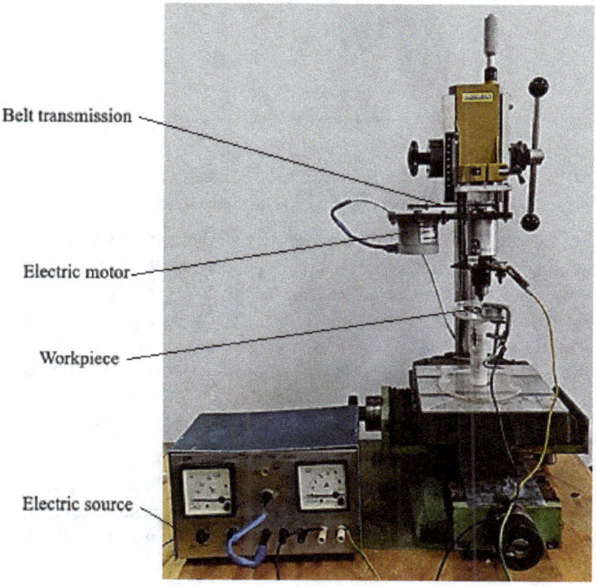

Fig. 4. The equipment used to materialize a process of making small diameter holes through hybrid, electric and electrochemical drilling

The test sample to be subjected to a drilling process was located and clamped inside a transparent plastic container, which allows direct observation of the area corresponding to the initiation of the drilling process. For the time being, a variant that involves moving the table supporting the container in a vertical direction and pressing the test sample on the tool electrode surface using a spring was preferred. The subsystem for the orientation and positioning of the test sample was placed on a device that has the possibility of materializing some movements along two directions perpendicular to each other, in a horizontal plane. In this way, the operative positioning of the test sample to the tool electrode was facilitated. The tool electrode and the test sample were connected in the circuit of a direct voltage source, which has the possibility of highlighting the values of the voltage and the intensity of the electric current. To ensure the connection of the rotating tool electrode in the circuit of a direct voltage source, the use of a brush collector ring subsystem was provided, these brushes being pressed on a copper bushing assembled by pressing on the tool electrode holder shaft. Preliminary experimental tests have confirmed the possibilities of carrying out, in the next period, more extensive research, which would allow obtaining some information regarding the input factors influence in the process of making small diameter holes through hybrid electrochemical discharge drilling. An image of a shallow hole made by hybrid drilling using the mentioned equipment can be seen in Figure 5. It was found necessary to improve the subsystem to materialize the pressing of the test sample on the front flat surface of the tool electrode, to obtain pressure values within certain limits.

Materials Research Proceedings 46 (2024) 176-183 https://doi.org/10.21741/9781644903377-23

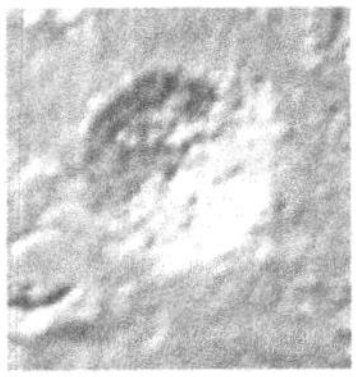

Fig. 5. Shallow hole obtained by hybrid, electrical and electrochemical drilling (electrode tool with a diameter of 1 mm, electrolyte: aqueous solution of sodium silicate soluble in water, with a density $\delta=1.1$ g/mm^3, steel specimen S235, tool electrode speed $n_{TE}=40$ rev/min

To highlight the shape of the pulses corresponding to the electrical discharges, it is intended to use a digital oscilloscope. Optimization of the development of the hybrid, electric, and electrochemical drilling process can also take place by using a continuous pulsating voltage, consistent with the application of such a solution in the case of electrochemical processing. A limitation of the use of the proposed equipment is considered to be the need to carefully wash the parts after machining, to avoid maintaining a thin layer of sodium silicate on the parts.

Summary

The hybrid electrochemical discharge machining aims to obtain simultaneously, as far as possible, specific effects of distinct machining through electrical discharge and, respectively, through electrochemical erosion. The principles of hybrid machining can also be applied in the case of electrochemical discharge drilling. From the study of specialized literature, the existence of two different directions of research and application of hybrid drilling was found. A first such direction is mainly applied for making holes insulating workpiece materials. Electric discharges occur in such situations along directions that cross the hydrogen bubbles that are released as a result of the dissociation of water. Such electrical discharges can generate microcracks in the workpiece material and subsequently even the removal of some particles from the material of the workpiece. The second variant of hybrid electrochemical discharge machining is based on the formation, on the surface of the workpiece, as a result of the chemical reactions between the material of the workpiece and the electrolyte, of a film that can contribute to reducing the intensity and even stopping the material removal process from workpiece by electrochemical erosion. The removal of the passivation film and, in this way, ensuring the conditions for the generation of electrical discharges between the tool electrode and the workpiece takes place by rotating or vibrating the tool electrode. The thinning of the passivation film allows the initiation of electrical discharges. On the exposed surfaces of the workpiece, the chemical reactions between the electrolyte and the workpiece material resume, and the process is carried out by repeating the previously described aspects. Starting from such aspects specific to hybrid drilling, it was proposed to use, as a tool electrode, a drill of small diameter, resharpened to present a flat active surface. The rotation of such a tool electrode pressed on the surface of the passivating film, and the existence of helical grooves will facilitate the initiation of electrical discharges that will contribute to the materialization of the electro erosive side of the hybrid drilling process. For the materialization of such a machining scheme, equipment based on the use of a bench drilling machine was designed. A simplified variant of a subsystem for locating and clamping of the workpiece provides conditions for pressing the workpiece or test sample on the flat surface of the resharpened drill, with a pressure whose value must exceed the breaking strength of the passivating film. As the passivation film thickness is removed or diminished, as a result of the relative displacement and pressure between the tool electrode and the workpiece, it is possible to produce electrical discharges, which will also contribute to the removal of material from the workpiece. The preliminary experimentation of the equipment confirmed the possibilities of its use for the materialization of a hybrid electrochemical discharge machining process when using passivating electrolytes. In the future, it is considered the continuation of the research in the direction of identifying a solution for the mechanical realization of a working movement, along the axis of the drill-type tool electrode. The research could also be continued through experimental tests that would allow the determination of optimal conditions for carrying out the hybrid electrochemical discharge drilling process.

References

[1] J. Kozak, K.P. Rajurkar, *Hybrid machining process evaluation and development*, Proceedings of 2nd international conference on machining and measurements of sculptured surfaces, Keynote Paper, Krakow, 501–536, 2000

[2] J. Kozak, K.P. Rajurkar, B. Wei, *Modelling and Analysis of Pulse Electrochemical Machining (PECM)*, Transactions of the ASME. Journal of Manufacturing Science and Engineering, 116, 3, 1994, 316-323. https://doi.org/10.1115/1.2901947

[3] B.A. Artamonov, A.L. Vishnitskiy, I.S. Volkov, A.V. Glazko, Dimensional *electrical machining of metals (in Russian)*, Vysshaia Shkola Publishing House, Moscow, 1978

[4] L. Slătineanu, G. Nagîţ, O. Dodun, M. Coteaţă, A. Munteanu, I. Beşliu-Băncescu, A.-L. Gherman, A. Hriţuc, F. Chinesta, A. Goncalves-Coelho, J.P. Teixeira, M. San Juan, L. Santo, F. Santos, *Electrophysical and chemical manufacturing processes*, Publishing House Tehnica Info, Chişinău (Republic of Moldova), 2020

[5] K. Allessu, M.K. Manju, A. Ghosh, Preliminary qualitative approach of a proposed mechanism of material removal in electrical machining of glass. European Journal of Mechanical Engineers, 36, 3, 1992, 202-207

[6] C.-H. Yang, T.-C. Wang, J.-C. Hung, H.-P. Tsui, *Ultrasonic vibration-assisted electrochemical discharge machining of quartz wafer micro-hole arrays*, Processes, 11, 3300, 2023. https://doi.org/10.3390/pr11123300

[7] M. Coteaţă., H.-P. Schulze, L. Slătineanu, Drilling of difficult-to-cut steel by electrochemical discharge machining, Materials and Manufacturing Processes, 26, 12, 2011, 1466-1472. https://doi.org/10.1080/10426914.2011.557286

[8] M. Coteaţă. Nonconventional manufacturing technologies. Course support (in Romanian), Publishing House PIM, Iaşi (Romania), 2021

[9] M. Coteaţă, H.-P. Schulze, L. Slătineanu, Drilling of difficult-to-cut steel by electrochemical discharge machnining, Materials and Manufacturing Processes, 26, 12, 2011, 1466-1472. https://doi.org/10.1080/10426914.2011.557286

[10] D.M. Zmiev, Anodic mechanical machining (in Russian), https://dic.academic.ru/dic.nsf/bse/64118/%D0%90%D0%BD%D0%BE%D0%B4%D0%BD%D0%BE, accessed: 12.09.2024

ANOVA and correlation analysis of feed, cutting forces and tool vibration in 11SMn30 steel turning

Csaba FELHŐ[1,a] *, Tanuj NAMBOODRI[1,b]

[1]Institute of Manufacturing Science, University of Miskolc, 3515 Miskolc, Miskolc-Egyetemváros, Hungary

[a] csaba.felho@uni-miskolc.hu, [b] tanuj.namboodri@gmail.com

Keywords: 11SMn30 Steel Grade, Machinability, Cutting Forces, Tool Vibration, Correlation Analysis, ANOVA

Abstract. The shift towards industry 4.0 in automative parts manufacturing and cloud machining solutions requires an understanding of materials machinability. To manufacture high quality products feed, cutting forces and tool vibration are important machining parameters that their dependence on one another other needs to be identified. This paper studies the Pearson correlation between feed, cutting forces and tool vibration and ANOVA is calculated to verify the validity of model. 11SMn30 a widely used steel grade for manufacturing shafts was used in the current study because of its very good machinability. The results suggests that feed rate influences both cutting forces as well as vibration in tool which has an overall effect on product quality.

Introduction

Automotive parts manufacturing industries around the world constantly advancing towards Industry 4.0, while keeping economic (low-cost solutions), and environmental benefits in account, Eraslan. D. et al. [1] mentioned that selecting the proper engineering material having desirable mechanical qualities and easy-to-cut character traits is essential for increasing productivity and decreasing production costs in machining operations. 11SMn30 steel is often used in manufacturing of automobile components like shaft and gears in which workability is a key factor, as this steel grade is often used to produce a considerable amount of components it is important to understand the workability of this material. Joy. B. et al [2]. The Turning operation is commonly adopted machining process used by manufacturers to produce shafts from hard material because of its economic and production advantages [3], for turning operation to be efficient selection of cutting parameters is highly recommended by the researchers. Machining parameters like feed, cutting forces & vibrations induced from the interaction of workpiece and tool, because they directly influence the quality of product, production time, tool wear, which overall determines financial aspect of the manufacturing industries and has an impact on the environment. One of the most significant characteristic variable to monitor in cutting operations is cutting forces, since its variation is directly proportional to the cutting conditions. Lalwani. D. et al. [4], explained that cutting forces are outcome of the tool- workpiece interaction as shown in Fig. 1. These contact points can cause tool wear or in the most unlikely of cases, tool failure. R. Pawade et al. [5] study highlights that the higher value of cutting forces results in poor surface quality and surface damage during turning of Inconel 718, where PCBN tools are used. Yong W. et al. [6], showed distributed cutting force model impacts the robustness of turning processes, especially at low spindle speeds, by incorporating both discrete and continuous time delays. Kim and Ehmann [7] suggested understanding cutting forces is significantly important. This understanding in addition is very important for designing of cutters and machine tool, as well as monitoring of tool wear. The main force responsible for the removing material in turning operation is the active force.

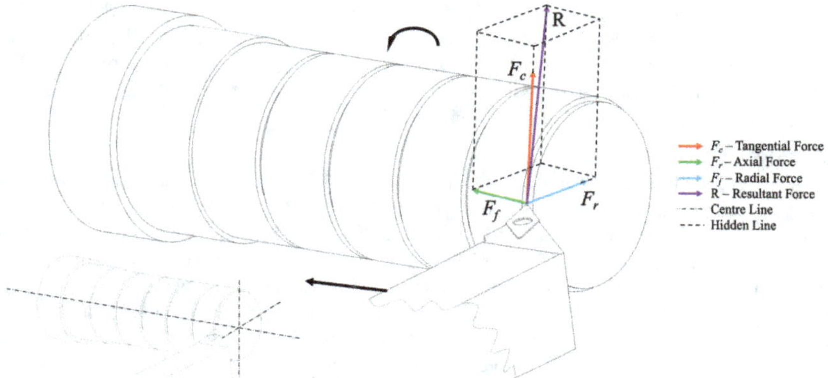

Figure 1. Direction of forces in turning operation.

Active force (F_a) is defined as the sum the cutting force (F_c), and the feed force, (F_f), in the same direction of the cutting motion. The passive force is the radial force, symbolized by (F_r) is irrelevant, because the workpiece and tool remain stationary with respect to one during the cutting process. Muhammad Rizal et al. [8]. The resultant force (R) is sum of both active and passive forces in the turning operation which is noted by the following Eq. 1.

$$R = \sqrt{F_a^2 + F_r^2} = \sqrt{F_c^2 + F_r^2 + F_f^2}$$
$$F_a = \sqrt{F_c^2 + F_f^2} \tag{1}$$

The forces which help in performing machining operation also affects the cutting tool, in terms of deformation, bend, wear, which leads to the vibration in the machining system [9]. Vibration is defined as the periodic displacement of a product at a high frequency. The resultant vibration can be calculated by taking magnitude of vibrations in x, y and z axis as expressed in Eq. 2.

$$R_{V_b} = \sqrt{V_{bx}^2 + V_{by}^2 + V_{bz}^2} \tag{2}$$

Where R_{V_b} is resultant vibration, V_{bx} is vibration in radial direction, V_{by} is vibration in Axial, V_{bz} is vibration in Tangential. In comparison to importance of the topic very less studies were found which states the active vibration which has real impact on the machining process however, understanding the point of contact the summation of Radial and Tangential direction is considered due to movement of tool in axial direction as shown in Fig. 3. These vibrations have an impact on common machining processes including turning, milling, and drilling. Nayak. A. et al. [10] studied that the force and vibration frequencies have a considerable correlation, and cutting forces can be predicted using the vibration. Prasad. B. et al. [11] concluded, vibration is a common concern in turning processes, affecting machining results, particularly tool wear. Monitoring parameters of quality like surface roughness, tool damage, vibration, and cutting forces, among other parameters results in constant development for best outcomes. Monitoring process parameters improves productivity and increases tool life. Ma et al., Bai W, et al., Xia S. et al. [12], [13], [14], [15], [16], [17], and other scholars have used various methodologies to investigate the correlations between these parameters and cutting forces. However, no significant conclusion is drawn which highlights the interrelationship between feed rate, cutting forces and tool vibration. This article aims to study the correlation between cutting forces, feed, and tool vibration in turning operation of 11SMn30 steel grade using dynamometer and MPU6050 sensor. Analysis of variance (ANOVA) was carried out to ensure the acceptability of the regression model.

Material & Methods

The workpiece shaft used to conduct the experiment is 11SMn30 grade steel, non-alloy quality, and not suitable for heat treatment, according to EN 10277-3 [18]. 11SMn30 steel grade contains, maximum of 0.14% Carbon, up to 0.05% Silicon, 0.9% to 1.3% Manganese, 0.11% Phosphorus which can improve strength, 0.27% to 0.33% Sulphur. Tensile Strength (Rm) is between 400 to 650 megapascals (MPa). The workpiece was parted in 5 equal parts as shown in Fig. 2. [19] Other machining parameters like depth of cut was kept 1 mm, spindle speed was 2000 rpm, for all the machining passes. Using these parameters cutting forces and vibration data was recorded. The feed was varied for each surface, S1: S5 abbreviation is use for representing surface, for S1, Feed (fz) mm/r was kept 0.1, for S2 – 0.15, for S3 - 0.2, for S4 - 0.25 and for S5 – 0.3. The machining operation was repeated two times to collect more data on cutting forces and vibration, resulting in good correlation accuracy. The workpiece diameter was 42 mm before machining, and it was reduced to 40 mm in first machining pass, after final pass it is 38mm as shown in Fig. 2.

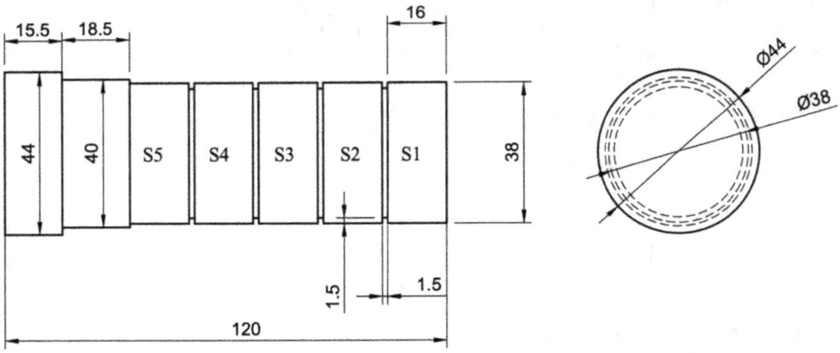

Figure 2. Drawing final machined part.

The SVHBR2020K11, Walter Turn shank tool with screw clamping (S), with positive basic form and VCGX style insert, with insert shape 35° Diamond, clearance angle 7° and corner radius 0.016 inch, the material of insert is carbide, uncoated, with the thickness of 0.125 inch used for the cutting operation, this tool is appropriate for small-diameter shafts or low cutting pressures. The cylindrical turning operation was conducted in a dry condition using the Optimum Opti Turn S600 CNC machine Fig. 3(a). The vibrational data was collected by connecting the MPU6050 sensor to the tool, as seen in Fig. 3(b) and 3(e). The ESP32 microcontroller was used to upload the code, and an SD card module connected to the microcontroller was used to gather the data. Sensor was attached to the tool with the help of a clip, the electronic setup was attached with the help of a screw to the tool holder in CNC machine, in case of movement of tool holder the electronic setup will also move to keep the connection secure between sensor and microcontroller, the microcontroller was attached to monitor to provide the power supply and also to visualise real time output vibration signals. Fig. 3(b), depicts the electronic configuration.

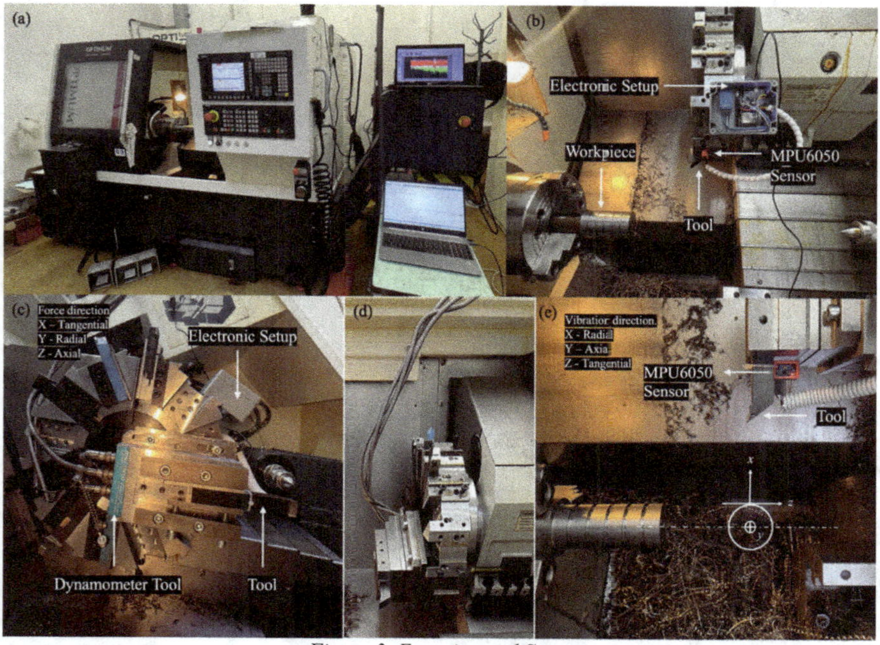

Figure 3. Experimental Setup

The components of cutting force were evaluated by a three-component piezoelectric dynamometer, from Kistler Corporation 9257-A Fig. 3(c) and 3(d). The dynamometer's output was amplified using a charge meter from Kistler Corporation, model number 5015A, connected through a shielded cable to the amplifier, the charger is then connected to an A/D - analog to digital converter and the computer is used for the collecting of data. The signals from the amplifier are provided into the acquisition of data system, and then the data are transformed to digital output, which is then transferred to the computer. This setup allowed for the installation of a dynamometer between the tool holder of the CNC Opti turn S600 and the SVHBR2020K11 tool without interacting with other parts. The three cutting force components were measured, as seen in Fig. 3(c). Tool, dynamometer and sensor position is defined in Fig. 3(c) and 3(e) respectively, monitor allows to validate the data during the process to avoid any error it is recommended to visualise the real time data. This configuration allowed the experiment to be conducted cleanly and without errors. Two boxes were fabricated using machining and 3D printing processes as shown in Fig. 3(b), and an outer wire casing was used to protect wires. These casings served as a shield to protect electronic setup from chips generated during the machining process.

Results

Table 1. represents the calculated data of active force RMS & active vibration data for correlation analysis, where EX1, EX2 means Experiment number and S1:S5 indicates the surface number as shown in Fig. 2. In calculation of active forces during experiment 1 and experiment 2 it can be noticed that by keeping the same machining parameters the value of forces is near to each other, results of resultants of vibration also indicates same, it can be noticed that values of vibration RMS are also repeating, in some cases like, EX1S4 and EX2S4 the values are nearly same.

Materials Research Proceedings 46 (2024) 184-191 https://doi.org/10.21741/9781644903377-24

Table 1. Feed, Active Forces RMS and Active Vibrations RMS data.

Experiment		Active	
Number	Feed Rate	Active Force RMS	Active Vibration RMS
EX1S1	0.1	243.57	10.49
EX1S2	0.15	353.95	10.49
EX1S3	0.2	435.27	10.54
EX1S4	0.25	510.82	10.7
EX1S5	0.3	563.41	10.82
EX2S1	0.1	236.26	10.48
EX2S2	0.15	345.36	10.52
EX2S3	0.2	408.12	10.51
EX2S4	0.25	462.95	10.71
EX2S5	0.3	Tool Break	Tool Break

To understand the relationship between variables, Pearson correlation analysis was studied, where correlation coefficient lies between -1 and 1. In case of coefficient of relation $r > 0$ the two variables show a positive correlation, the value of r is near to +1 means the correlation is strong positive correlation, when the value is near to 0 it shows the correlation is weak. If $r < 0$ there a negative correlation between the variables which shows that by changing the value of the variable will not have the direct effect on the other. Table 2. represents the correlation between 3 main key variables of the study, feed, active force RMS and active vibration RMS, correlation coefficient between feed rate and active force RMS is 0.986, indicating very strong positive linear relationship which suggests that and increase in feed rate is strongly influences increase in active force, and there is less than 1% probability that this observed correlation is due to random variation. Correlation between feed and active vibration RMS also suggests strong positive linear relationship with correlation coefficient of 0.910 as feed rate increases the active vibration RMS also increases. Correlation coefficient between active force RMS and active vibration RMS is 0.864 with statically significant at 0.01 level, suggests that higher active force influences higher active vibration. According to experimental results in Table 2. first order regression model of active Force RMS as a function of feed was developed at a 95% of confidence level. The Eq. 3, is given by

$$y = 96.16 + 1.58E3\, x \quad (3)$$

In the Eq. (3) (4), y represents one variable and x represents the other variable, the Equation, $y = ax + b$ where a is the slope and b is the y intercept, which can be used to predict y based on the on the observed level of x, which highlights the impact of one variable on another. In Eq. 3. x is feed and y is active force RMS with $R^2 = 97.2\%$, y is active force RMS is in [N] and feed is in [mm/r], The analysis of variance (ANOVA) is represented in Table 3. , model is found statically significant with significance of probability *p*- value less than 0.05, In this case 0.000, since the exact significance is 0.000 so the effects will be statically significant, means high significance. [a] Dependent Variable – active force RMS, [b] Predictors - (Constant), feed rate, **"df" is** degrees of freedom which shows independent variables numbers in the model. The analysis of variance (ANOVA) of regression was analysed. using SPSSStatistics academic licence software.

Table 2. Correlation Analysis Feed, Active Force RMS & Active Vibration RMS.

Parameters	Correlation Method	Feed Rate	Active Force RMS	Active Vibration RMS
Feed Rate	Pearson Correlation	1	.986**	.910**
Active Force Magnitude	Pearson Correlation	.986**	1	.864**
Active Vibration Magnitude	Pearson Correlation	.910**	.864**	1

** The correlation is significant at the 0.01 level (two-tailed).

Table 3. ANOVA of Feed and Active Force RMS.

ANOVA[a]						
Model		Sum of Squares	df	Mean Square	F	Sig. (*p*)
1	Regression	97680.029	1	97680.029	245.658	.000b
	Residual	2783.381	7	397.626		
	Total	100463.41	8			

[a] Dependent Variable: active force RMS
[b] Predictors: (Constant), feed rate

The first order regression model for active vibration RMS as a function of feed rate is given by Eq. 4.

$$y = 10.8 + 1.63\,x \tag{4}$$

Table 4. ANOVA of Feed and Active Vibration RMS.

ANOVA[a]						
Model		Sum of Squares	df	Mean Square	F	Sig. (*p*)
1	Regression	0.104	1	0.104	33.672	.001b
	Residual	0.022	7	0.003		
	Total	0.125	8			

a Dependent Variable: active vibration RMS
b Predictors: (Constant), feed rate

With $R^2 = 82.8$, the value of R^2 is very satisfactory, the statical significance of model is studied by ANOVA as shown in Table 4. The significance of probability *p*- value is less than 0.05, the exact value 0.001 which is statically significant or high significance. [a] Dependent Variable – active vibration RMS, [b] Predictors - (Constant), feed rate. The value of R^2 is acceptable in both cases.

Summary

The introduced approach can be used to understand the machinability of material and relationship between cutting forces and vibration, and several findings can be drawn from this study.

The direction of active forces and active vibration differs, active force is the sum of tangential and axial, whereas active vibration considers the sum of tangential and radial direction.

Given the same machining parameters, the repetition of values of vibration is noticeable and values of forces are identical, this suggests that from machining parameters the vibration and force values may be predicted.

Pearson correlation suggests there are 99% of possibility of a real relationship between feed rate and active force with coefficient of correlation 0.986, and between feed rate and active vibration with the coefficient of correlation is 0.846.

ANOVA results for both model, feed rate and active force RMS as well as feed rate and active vibration RMS suggests that the model is highly significant. The results can improve manufacturing efficiency, predictive maintenance, and build more robust vibration resistance tools specific to materials and machining condition.

Acknowledgement

This research was supported by project No. 2020-1.2.3-EUREKA-2022-00025 "Deep Learning Based Process Monitoring for Complex Machining Tasks – DeepProMach" which was realized with financial help of the National Research Development and Innovation Fund of the Ministry of Culture and Innovation of Hungary.

References

[1] D. Eraslan et al., "Machinability evaluations of austempered ductile iron and cast steel with similar mechanical properties under eco-friendly milling conditions," Journal of Materials Research and Technology, vol. 11, pp. 1443–1456, Mar. 2021. https//doi.org/10.1016/J.JMRT.2021.01.123

[2] V. Sharma, P. Kumar, and J. P. Misra, "Cutting force predictive modelling of hard turning operation using fuzzy logic," in Materials Today: Proceedings, Elsevier Ltd, 2019, pp. 740–744. doi: 10.1016/j.matpr.2020.01.018

[3] D. I. Lalwani, N. K. Mehta, and P. K. Jain, "Experimental investigations of cutting parameters influence on cutting forces and surface roughness in finish hard turning of MDN250 steel," J Mater Process Technol, vol. 206, no. 1–3, pp. 167–179, Sep. 2008. https//doi.org/10.1016/j.jmatprotec.2007.12.018

[4] R. S. Pawade, S. S. Joshi, P. K. Brahmankar, and M. Rahman, "An investigation of cutting forces and surface damage in high-speed turning of Inconel 718," J Mater Process Technol, vol. 192–193, pp. 139–146, Oct. 2007. https//doi.org/10.1016/j.jmatprotec.2007.04.049

[5] Y. Wu, Q. Song, Z. Liu, and B. Wang, "Stability of turning process with a distributed cutting force model," International Journal of Advanced Manufacturing Technology, vol. 102, no. 5–8, pp. 1215–1225, Jun. 2019. https//doi.org/10.1007/s00170-018-2949-y

[6] H. S. Kim and K. F. Ehmann, 'A cutting force model for face milling operations', Int J Mach Tools Manuf, vol. 33, no. 5, pp. 651–673, 1993. https//doi.org/10.1016/0890-6955(93)90099-G

[7] M. Rizal, J. A. Ghani, and A. Z. Mubarak, "Design and Development of a Tri-Axial Turning Dynamometer Utilizing Cross-Beam Type Force Transducer for Fine-Turning Cutting Force Measurement," Sensors, vol. 22, no. 22, Nov. 2022. https//doi.org/10.3390/s22228751

[8] Q. Yu, S. Li, X. Zhang, and M. Shao, "Experimental study on correlation between turning temperature rise and turning vibration in dry turning on aluminum alloy," International Journal of Advanced Manufacturing Technology, vol. 103, no. 1–4, pp. 453–469, Jul. 2019. https//doi.org/10.1007/s00170-019-03506-7

[9] Nayak, P. K. Kankar, N. Jain, and P. K. Jain, "Force and vibration correlation analysis in the self-adjusting file during root canal shaping: An in-vitro study," J Dent Sci, vol. 13, no. 3, pp. 184–189, Sep. 2018. https//doi.org/10.1016/j.jds.2018.01.002

[10] S. Prasad and M. P. Babu, "Correlation between vibration amplitude and tool wear in turning: Numerical and experimental analysis," Engineering Science and Technology, an International Journal, vol. 20, no. 1, pp. 197–211, Feb. 2017. https//doi.org/10.1016/j.jestch.2016.06.011

[11] Ma, E. Shamoto, T. Moriwaki, and L. Wang, "Study of machining accuracy in ultrasonic elliptical vibration cutting," Int J Mach Tools Manuf, vol. 44, no. 12–13, pp. 1305–1310, Oct. 2004. https//doi.org/10.1016/j.ijmachtools.2004.04.014

[12] S. Xia, Z. Yin, C. Huang, Y. Guo, and C. Zhang, "Characteristics of Elliptical Vibration-Assisted Cutting with Variations in Tilt Angle of Elliptical Locus," Micromachines (Basel), vol. 14, no. 7, Jul. 2023. https//doi.org/10.3390/mi14071426

[13] W. Bai, R. Sun, Y. Gao, and J. Leopold, "Analysis and modeling of force in orthogonal elliptical vibration cutting," International Journal of Advanced Manufacturing Technology, vol. 83, no. 5–8, pp. 1025–1036, Mar. 2016. https//doi.org/10.1007/s00170-015-7645-6

[14] M. Hanief, M. F. Wani, and M. S. Charoo, "Modeling and prediction of cutting forces during the turning of red brass (C23000) using ANN and regression analysis," Engineering Science and Technology, an International Journal, vol. 20, no. 3, pp. 1220–1226, 2017. https//doi.org/10.1016/j.jestch.2016.10.019

[15] V. Guzeev, I. Deryabin, and D. Ardashev, 'Mathematical models of cutting forces in turning operations', Machine Science Journal, vol. 1, no. 1, pp. 36–45, May 2023. https//doi.org/10.61413/KZNC2821

[16] T. Dorlin, G. Fromentin, and J. P. Costes, "Analysis and modelling of the contact radius effect on the cutting forces in cylindrical and face turning of Ti6Al4V titanium alloy," in Procedia CIRP, Elsevier B.V., 2015, pp. 185–190. https//doi.org/10.1016/j.procir.2015.03.017

[17] EU Steel and Alloy Grades Number, 2024. Available: http://www.steelnumber.com/en/

[18] F. Ridwan and X. Xu, "Advanced CNC system with in-process feed-rate optimisation," Robot Comput Integr Manuf, vol. 29, no. 3, pp. 12–20, 2013. https//doi.org/10.1016/j.rcim.2012.04.008

Optimization of the CNC machining process using digital twins

Dimka Vasileva[1,a]*, Mariya Konsulova-Bakalova[1,b] and Iva Stoyanova[1,c]

[1]Technical University of Varna, Bulgaria

[a]d.vasileva@tu-varna.bg, [b]mbakalova@tu-varna.bg, [c]iva.stoyanova@tu-varna.bg

Keywords: Digital Twin, CNC Machine Tool, Industry 4.0, CNC Machine Processes, Big Data, Digital Models

Abstract. The optimization of machining processes, performed by CNC machine tools, using digital twin technology, is gaining a lot of popularity. Digital twin technology can be integrated into every stage of the workpiece life cycle, in order for defective parts prevention and high quality production realization. The human factor is the main source of failures at the production companies. For this reason, human activities like designing, operating, processing, monitoring, analyzing and controlling are being automated through the integration of digital twin technologies. With their help, CNC machine tools are becoming more and more intelligent, independent and autonomous, for the reason of reducing production costs associated with the human factor.

Introduction

The mechanical engineering industry is constantly improved, going through various industrial revolutions. At first it was started with manual labour, through mechanization, then mass production and automation, nevertheless the development does not end here – Fig.1. Solutions are being sought to further increased productivity, reduced production failures and to be able to store and process the amount of collected information.

As a result, an increasing number of manufacturing industries are being reorganized and optimized through the implementation of Internet Technologies, in already established automated systems for design, operation and control.

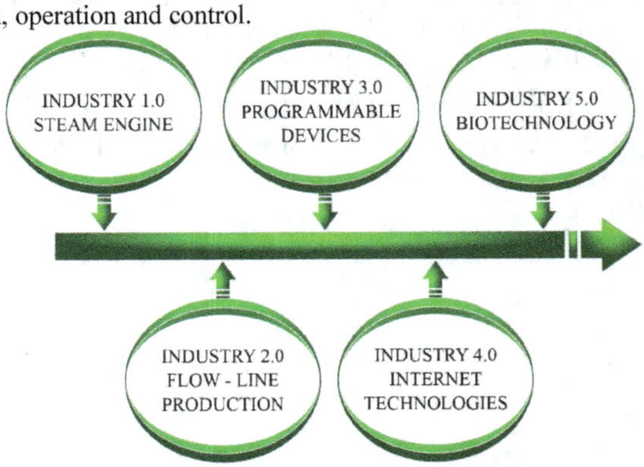

Fig. 1. Historical development of world industrial revolution

The real aim of Digital Twins implementing is to be created a digitalized smart industry, which is the base idea of Industry 4.0 This leads to "smart manufacturing", which can be achieved through the term "Digital Twin" (DT) – a digital copy of a potential or physical product and process. The main function is its ability to work in real – time, enabling it to help the improvement of timely optimization and anticipatory maintenance.

To accomplish this, information gathering devices must be available first, to submit the necessary information to the digital twin software. Next comes a stage of predictions and simulations in the safe digital environment. The goal is the final knowledge and results, obtained from the digital twin, to be applied to the potential or actual object or system. Therefore, a digital twin can be applied during all stages of the workpiece life cycle. Some of these stages also include supporting the creation of optimized computer numerical control (CNC) programs, controlled operation management of CNC machine tools (CNCMT) and the related cutting tools.

Optimizing CNC Machine Tools Using Digital Twins

The application of digital twins in mechanical engineering makes it possible in the first place to create a computer simulation of the operation of a machine tool. For the automation of production processes, the implementation of digital twins in the construction of robotic cells, for the study of the geometry of mechanical units of the machines using the finite element method, as well as for the avoidance and prediction of possible accidents, is also of significant interest.

Optimizing CNC Machine Tools Using Digital Twins needs information about the manufacturing processes, the parameters that the final product must meet and the overall digital visualization of these processes. All this and much more data is inserted, used and extracted from the digital twin in order to digitally transform and optimize the performance of the CNC machine that processes the relevant specific parts [1]. The availability of a wide database predisposes the possibility of creating an intelligent system. However, before it can become intelligent, a digital twin must be adaptable – Fig. 2

Fig. 2. Types of digital twins compared to the stage of development

The digital twin of CNC machine tools [2] have the ability to operate with newly added information and features in response to changes in the machining environment or the goals of producing quality products. This happens with the help of using Machine Learning algorithms and optimization techniques, without the need for human intervention [3], which in turn helps in increasing the efficiency and reliability of the physical system. On other hand, to make the digital twin more intelligent, "virtual sensors" are added to it, integrated into the digital visualizations of the CNC machine tool. This enables the collection of information without needing to be installed expensive sensors that need constant calibration or additional maintenance [4]. In order to carry out the optimization process using a digital twin, first of all, accurate digital models of the CNC machines must be built, consisting if their structure and behavior, at their all types of condition. CNCMT is a term used for machine tool equipped with a CNC system [5]. The reason for this, is the ability of CNC control to influence the entire manufacturing process of a product, since the various phases of CNCMT's life cycle affect its efficiency. These are the stages of supporting design, production, operation, maintenance and others [6]. It is also necessary to make a realistic virtual machine control panel (MCP), because it is used by operators to operate with virtual twin that can be used for simulation of the machine and to manipulate the execution of the CNC program [7]. The digital representation of the machining process presents important data, that are for a specific moment in time. For the optimization of this representation, the optimal online tracking of the process is done thanks to the digital twin of the process. The current data gathered via standard communication protocols that are carried out between the devices of the CNC machine by various types of information gathering devices. This communication network can be utilized for simulating the actions of CNC machines [8]. One of the significant advantages of digital twin technology is the ability to control the life of the tools in the machine tool throughout the production process. The application of this technology requires that in the machine tool ATC to have to have two of each kind cutting tools that is necessary for the production of a certain part. When the machine calculates that the life of a tool is over, it changes with the other tool and continues its work without interruption. The optimization of process data update is enabled by the capability of sensors, cameras and other devices having Internet of Things (IoT) technology. Due to the limited computational capacity of current CNC systems to evaluate and handle large amounts of data, many businesses have taken to uploading all of the data gathered from equipment to distant cloud datacenters [9]. This allows continuous, two – way, online communication between the machine tool and the operator and/or engineer responsible for the correct operation of the machine. The implementation of the control of the CNCMT is carried out by CNC – Fig. 3.

Fig. 3. Working principle of CNC system of a machine tools

In this case, the digital twin is used to ensure that the numerical control (NC) programs are error – free. This is done precisely through online simulations and visualizations of the real and already built DTs, of all the equipment of the CNCMT. Therefore, the CNC machine tool only executes the steps pre – planned by the NC code (Fig. 4).

In such a case, it lacks the ability to see and anticipate undesirable situations, which in turn leads to a waste of time and resources. Therefore, it is necessary to optimize the machining processes with CNCMT, by using digital twins.

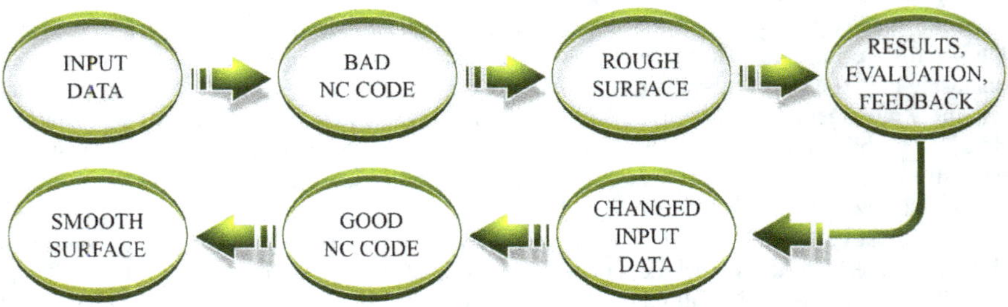

Fig.4. Optimization and feedback strategy of processing technology

Optimization of CNC Machining Processes Using Digital Twins

An essential component of the current machinery modernization development is CNC machine tool processing [10]. Recently, researchers have conducted data mining and analysis using the process and inspection data gathered during the manufacturing process. The analytical results are then utilized to anticipate part quality and optimize process parameters [11]. This is done through a CNCMT digital model, but the analyses that are done with it are set with static parameters, under ideal conditions.

This leads to a limitation of possibilities, which is why, for the purpose of optimization it is necessary to improve processes in real time, as this can be done thanks to digital twin technology. Specifically, the improvement process itself is carried out in the direction of manufacturing products that meet the requirements of customers, which requires optimal operating modes, reducing the presence of sudden collisions, various types of errors (geometric, thermal, vibration, etc.) and improving the overall stability of CNCMT.

Combined with the functionalities of the digital twin, the CNCMT gets the opportunity to be called the CNCMT Digital Twin (CNCMT DT) model – that is the complex's carrier, who have variable changes over time and data that is coupled [6]. The possibility of analysis and self – regulation assists in optimizing the required parameters and processes, helping engineers and operators to make faster and error – free decisions. General principle of diagnosis of malfunctions in machining processes by DT. The correlation between the digital twin model and the actual machine tool is realized by the fault identification of the CNCMT driven by digital twin, which gives the intelligent decision service access to real-time machine tool status data [12]. DT runs forward simulations and at some point a fault type appears to be present. For instance, tool wear will shorten a part's service life by decreasing both the surface quality and dimensional accuracy of the part [5]. In such a case, the DT receives present and past data on the parameters and status of the fault and related objects. Once a solution to the problem is identified, it is simulated in the digital environment, and if the result is unsatisfactory, then the analysis and calculation process is repeated until it is satisfied. The final corrective actions, from the forward calculations, are applied to the relevant process, shown on Fig. 5.

Fig. 5. Optimizing principle of digital twin

All of this is done in a minimum of time before the malfunction has occurred, thus preventing it from having negative consequences. Optimizing process parameters during the operation of CNCMT is highly essential for enhancing processing effectiveness, maintaining processing goodness, minimizing energy consumption, and reducing costs [13].

Optimizing thermal error prevention. Another prevention that a digital CNCMT DT can accomplish over a physical CNCNT is thermal error. Traditional prediction methods ignore the relationship between the data of the real world and the digital one, which leads to low prediction accuracy of thermal errors and affects the normal operation of the CNC machine tool. For this reason, in [14], in addition to using the digital twin technology, with its simulation capabilities to optimize traditional methods, a long short-term memory (LSTM) system is also integrated, with useful data processing capabilities that further optimizes the process for thermal error prevention. The reason is the constantly changing, time – varying values and location of these anomalies, which based on one modelling method, is difficult to achieve an accurate online reflection, that significantly impairs the successful performance of CNCNT. DT-LSTM enhances the diagnosis of the performance and reliability of thermal error by almost 11% when compared to the single approach. Specific measurement positions are defined for the parameters to be acquired through finite element simulation. In the presence of anomalies of the values at the required characteristics, they are corrected on the physical, through the digital system.

Another aspect of the quest to optimize the performance of the CNC machine tool is in relation to the trajectory of movement during operation. This is done by modeling the tool's digital twin and integrating a database containing not only the characteristics of the machine tool, but also everything related to its operation for the specific situation. This is done through information gathering devices and work signals from the CNCMT itself (Fig. 6). Specifically, for the machine tool's motion path provides information about its operation such as its positions, orientation relative to the axes of motion, feed and cutting speed and etc.

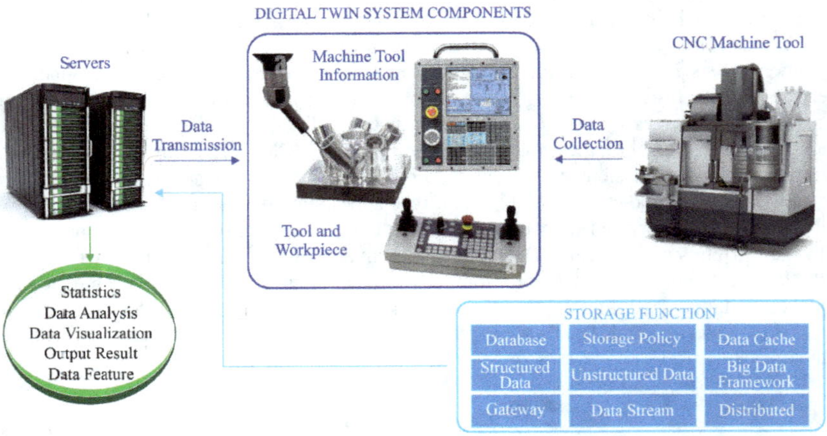

Fig. 6. Operation of Digital Twin System

Based on them, a processing optimization module is created, taking into account the mentioned characteristics and strategies that are proposed to improve specific parameters of the processing system in case of non – conforming instrument locations and to store the database's optimized knowledge, for later references. Furthermore, the machine-executable code created from the optimization data is utilized to manage the machining procedure [1]. Optimizing cutting parameters and toolpaths help reduce energy consumption and save costs.

Optimizing the required surface roughness. The roughness of the surfaces of the final product depends on the cutting specific requirements and the trajectory of the processing instrument or work table (depending of the type of processing). For this purpose, digital twin technology again helps to optimize the faster and qualitative achievement of the required surface roughness. This most often happens at the planning stage, by digitizing the working environment, parameters and participants, with the aim of predicting the final results of the processing, which can be heterogeneous. The reason for this is the excessive number of factors that influence this characteristic of the final product. Such are all cutting parameters, the type and condition of the tool, the clearances between the mechanical parts of the machine tool and etc. Finally, after the quality has been predicted and consequently established, if it is in negative aspect, it is brought back for and solution of the resulting problem [8].

Summary

The optimization process is always accompanied by initial difficulties, but eventually it proves to be essential for improving the effectiveness and performance level of work and the manufacturing process. As for CNC machine tool, it finds an increasingly common and vitally important application, and for its implementation, the digital twin technology is being integrated. It, with its varieties, supports manufacturing at every stage of the workpiece life cycle by supporting the creation of intelligent cyber – physical systems. Based on databases, from the physical CNC machine tool and simulations, on their digital replicas, they support the timely control, ensuring different types of errors. Problem situations will always be present in the manufacturing process, but at least the percentage of they occurrence is kept to a minimum. That is why new business models in the era of digitization and the Internet increasingly require the use of various software that are necessary to support the development and management of digital twins. The three types of digital twins are of significant interest for CNC machine tool enterprises, namely digital twins of the product, of the production process and of the equipment. Cloud technologies also cover machine servicing, which provides full access to the machine. To support the work of engineering enterprises, it is necessary to develop simulation models of the machine before it is manufactured, including with digital start-up of the machine.

References

[1] W. Zhao, R. Li, X. Liu, J Ni, C Wang, C Li, L Zhao, Construction Method of Digital Twin System for Thin-Walled Workpiece Machining Error Control Based on Analysis of Machine Tool Dynamic Characteristics. Machines 600, (2023), 11, https://doi.org/10.3390/machines11060600

[2] M. Grieves, Digital twin certified: employing virtual testing of digital twins in manufacturing to ensure quality products. Machines, 11(8), (2023). 808. EISSN 2075-1702, https://doi.org/10.3390/machines11080808.

[3] R. Ströbel, A. Bott, A Wortmann, J. Fleischer, Monitoring of Tool and Component Wear for Self-Adaptive Digital Twins: A Multi-Stage Approach through Anomaly Detection and Wear Cycle Analysis. Machines, 11, (2023), 1032. https://doi.org/10.3390/machines11111032.

[4] R. Ward, C Sun, J. Dominguez-Caballero. Machining Digital Twin using real-time model-based simulations and look ahead function for closed loop machining control. Int J Adv Manuf Technol 117, (2021), 3615–3629. https://doi.org/10.1007/s00170-021-07867-w.

[5] W. Luo, T. Hu, Y. Ye, C. Zhang, Y. Wei. A hybrid predictive maintenance approach for CNC machine tool driven by Digital Twin. Robotics and Computer-Integrated Manufacturing, 65, (2020), 101974. ISSN 0736-5845, https://doi.org/10.1016/j.rcim.2020.101974.

[6] Y. Wei, T. Hu, T. Zhou, Y. Ye, W Luo. Consistency retention method for CNC machine tool digital twin model. Journal of Manufacturing Systems, 58, (2021), 313-322. ISSN 0278-6125, https://doi.org/10.1016/j.jmsy.2020.06.002.

[7] Information on https://www.siemens.com/global/en/industries/machinebuilding/machine-tools/cnc4you/basics-articles/digit-twin-rarely-comes-alone.html

[8] V. S. Vishnu, K. G. Varghese, B. Gurumoorthy. A data-driven digital twin of CNC machining processes for predicting surface roughness. Procedia CIRP, 104, (2021), 1065-1070. ISSN 2212-8271, https://doi.org/10.1016/j.procir.2021.11.179.

[9] H. Yu, D. Yu, C. Wang, Y. Hu, Y. Li. Edge intelligence-driven digital twin of CNC system: Architecture and deployment. Robotics and Computer-Integrated Manufacturing, 79, (2023), 102418. ISSN 0736-5845, https://doi.org/10.1016/j.rcim.2022.102418.

[10] L. Pan, X. Guo, Y. Luan, H. Wang. Design and realization of cutting simulation function of digital twin system of CNC machine tool. Procedia Computer Science, 183, (2021), 261-266. ISSN 1877-0509, https://doi.org/10.1016/j.procs.2021.02.057.

[11] L. Liu, X. Zhang, X. Wan, S. Zhou, Z. Gao. Digital twin-driven surface roughness prediction and process parameter adaptive optimization. Advanced Engineering Informatics, 51, (2022), 101470. ISSN 1474-0346, https://doi.org/10.1016/j.aei.2021.101470.

[12] J. Liu, D. Yu, Y. Hu, H. Yu, W. He, L Zhang. CNC machine tool fault diagnosis integrated rescheduling approach supported by digital twin-driven interaction and cooperation framework. IEEE Access, 9, (2021), 118801-118814.

[13] L.C. Magalhães, L.C. Ramos, J.B. Moura, L.R. de Moraes, R.E.N. Gonçalves, J.B. Hisatugu, W.H. Souza, M.T. de Lacalle, L.N.L. Ferreira. Conceiving a Digital Twin for a Flexible Manufacturing System. Appl. Sci. 12, (2022), 9864. https://doi.org/10.3390/app12199864

[14] Q. Lu, D. Zhu, M. Wang, M. Li, Digital Twin-Driven Thermal Error Prediction for CNC Machine Tool Spindle. Lubricants 11, (2023), 219. https://doi.org/10.3390/lubricants11050219

Innovative Manufacturing Engineering and Energy - IManEE2024 Materials Research Forum LLC
Materials Research Proceedings 46 (2024) 199-203 https://doi.org/10.21741/9781644903377-26

Innovative, environmentally friendly decoating of metallised plastic products using laser technology

Alexandru SOVER[1,a *], Markus ZINK[1,b]

[1]Residenzstr. 8, 91522 Ansbach, Germany

[a]a.sover@hs-ansbach.de, [b]zink.markus@hs-ansbach.de

Keywords: Metallised Plastic Parts, Metallisation, Recycling, Paint Striping, Laser Technology, Laser Cleaning

Abstract Metallised plastic parts are used for a wide range of applications, for example, the automotive and packaging industry or household goods because of their specific functions, high-quality look, and hygienic properties. The use of these products has increased significantly in many sectors of industry which admittedly implicates certain materials and products drawbacks, especially with a view to sustainability issues. The recycling of waste metallised plastic products is currently very difficult and usually, the plastic material is incinerated for the recovery of the metallic components. Ansbach University of Applied Sciences has been working on the removal of the paint layers from various materials for several years. This article presents a study on how the metallised layer can be removed from the plastic parts using laser technology. The results show that the metallic layers can be separated from the plastic carrier (thermoplastic parts) without damaging the polymers which facilitates a high-quality recycling of each material (polymer and metal). Metallised plastic parts of made from different materials, featuring various thicknesses and treated with different metallisation techniques, were successfully tested. The laser processing of the metallised parts has proven to be an effective and environmentally friendly method for achieving a high-quality separation of the materials.

Introduction

In many branches of industry, the use of metallised plastic parts has increased significantly [1,2,3,4]. Metallised plastic parts are composite systems made from a metal layer and usually a thermoplastic polymer material [1, 4, 5]. They can be produced at low cost and in a wide range of geometries, colours, and surfaces, and they are used in many areas because of their high-quality look and feel, and for hygienic reasons. Galvanised parts have a very wide and comprehensive range of applications, especially in the automotive industry [1, 2].

Through metallisation, the specific properties of plastics - low weight, flexibility in design, and economical production – are combined with the properties usually seen in metals [1, 4, 5]. This includes high reflection, abrasion resistance, electrical and thermal conductivity, barrier properties, and decorative surface structures. [4, 5, 6]

After use, they are usually burned to recover just the metallic materials or used in crushed form as a low-value filler [5, 6]. The recycling routes for plastics urgently need to be improved for reasons of climate, resource, and environmental protection.

However, recycling of these hybrid parts is still a big issue for the recycling industry. The metallic layer and strong adhesion between layers make separation and recycling difficult. Without prior separation, the broken metal particles are integrated as an inactive filler in the polymer during recycling, potentially starting points for cracks. Components made from previously coated plastics have very low tensile strength, breaking strength, and breaking elongation compared to the original non-metallised plastics. In addition, these recycled plastics are not suitable for functional or visible components, because metal fragments appear on the component surface and cause unwanted

Materials Research Proceedings 46 (2024) 199-203 https://doi.org/10.21741/9781644903377-26

effects. Furthermore, during the processing of thermoplastics, the metallic particles can severely damage the processing equipment (injection moulding machines, extruders, etc.).
Therefore, new methods must be developed to separate the used materials by type, because the most frequently used thermoplastic and metal materials are highly recyclable. [3, 7, 8]

A new technology for removing metallisation from plastic components.
Current methods for removing coatings from plastic components, such as cleaning with chemical solutions or mechanical processes such as blasting, nitrogen, or high-pressure water jet paint stripping, have so far produced technically poor results, economic disadvantages and/or considerable environmental pollution. [7,8]

The Ansbach University of Applied Sciences has been working for many years on the removal of different coatings from various materials as thermoplastics, metals, wood, etc. and has researched and developed a new environmentally friendly, cost-effective, and material-compatible process for removing coatings, using laser technology. [7, 8, 9, 10]

A laser beam brings high localised energy to the coated or metallised surface; when correctly dosed, the short, strong laser pulses have only a minimal thermal effect on the underlaying base material. A big advantage of this technology is that the laser parameters can be exactly adjusted to suit the specific material, optimising the process and its speed while thermal stress on the thermoplastic surface. [7, 8]

A decor metallised plastic strip from automotives was used for the first tests to remove the metallised layer from thermoplastic polymer. This is made of a, acrylonitrile-butadiene-styrene copolymer (ABS) that is galvanised and chrome-plated.

For the characterisation of the layer structure, the metallised plastic parts were cut, embedded in resin and analysed under a digital microscope.

The layer structure can be clearly seen in the cross-section (see Figure 1). The part consists of 3 layers, the carrier material (ABS), the galvanisation (Cu) and a chrome layer (Cr).

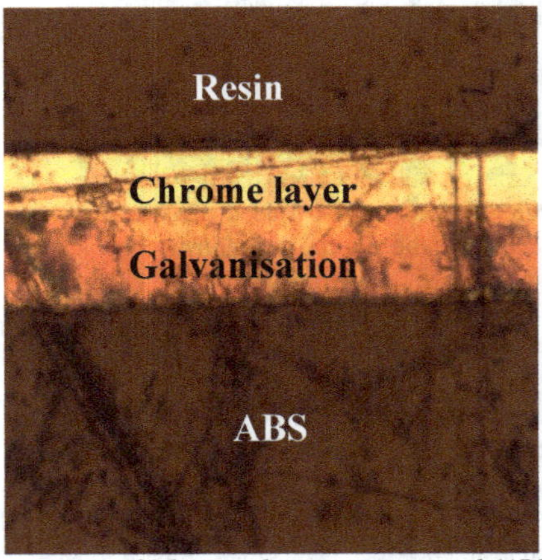

Figure 1 Cross-section consisting of 3 layers, the carrier material (ABS), the galvanisation (Cu) and a chrome layer (Cr).

Materials Research Proceedings 46 (2024) 199-203 https://doi.org/10.21741/9781644903377-26

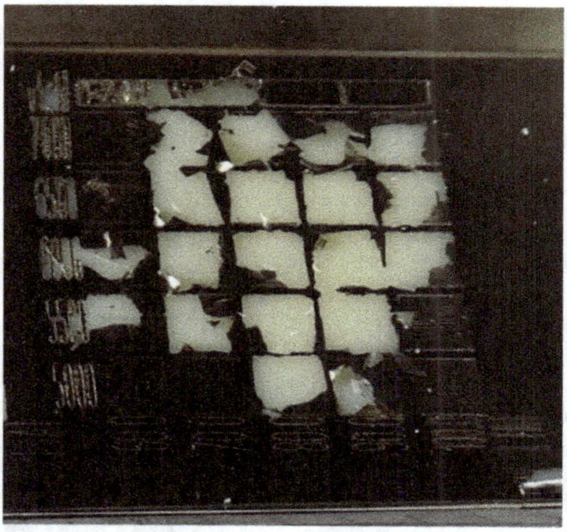

Figure 2 Metallised part after laser processing with different parameters.

After cutting and cleaning the parts, several square surfaces with different laser parameters were applied on the metallised surface.

The tests are performed with a 50 W fibre laser from Coherent-ROFIN. Five squares by 5 mm x 5 mm (see figure 2) are placed horizontally and vertically, changing the laser power (horizontal) and the speed of the laser (vertical). The additional parameters (e.g. laser frequency, pulse width, line width) are kept constant. On the horizontal axis, the power is varied from 60 % to 100 % in 10 % increments and on the vertical axis, the speed is changed from 5.000 mm/s to 7.000 mm/s in 500 mm/s increments. The pulse length is 50 ns with a line width of 0.1 mm at a frequency of 100 kHz. Figure 1 shows that at a power of 80 % and a speed of 6.000 mm/s, the metallisation can be easily removed from the copolymer carrier ABS.

With these parameters, the removal of the metallisation on a larger surface (20 mm x 20 mm) was tested. The metallisation can also be removed here without damaging the polymer substrate (see Figure 3).

For different materials and coating layers, individual laser parameters (speed, energy, pulse duration) are necessary to minimise thermal stress on the thermoplastic polymers, especially standard polymers, and the resulting surface damage.

The main objective of this research was to effectively separate the metal layer from commonly used standard thermoplastics, which can only be high quality recycled if the different material layers are separated.

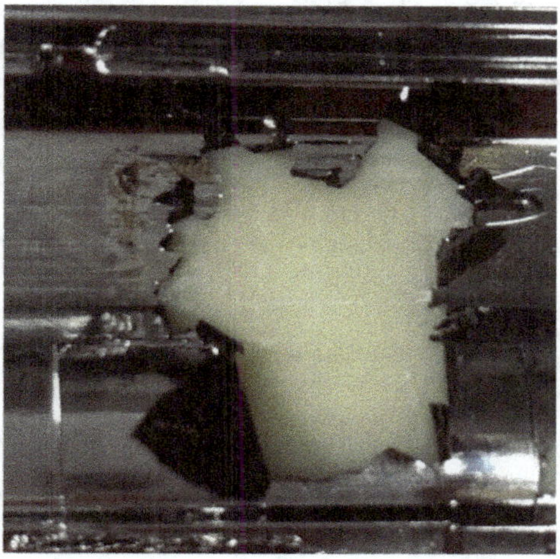

Figure 3 Laser processing on an area of 20 mm x 20 mm, with the metal layer separation.

Studies of metal stripping using the example of a showerhead

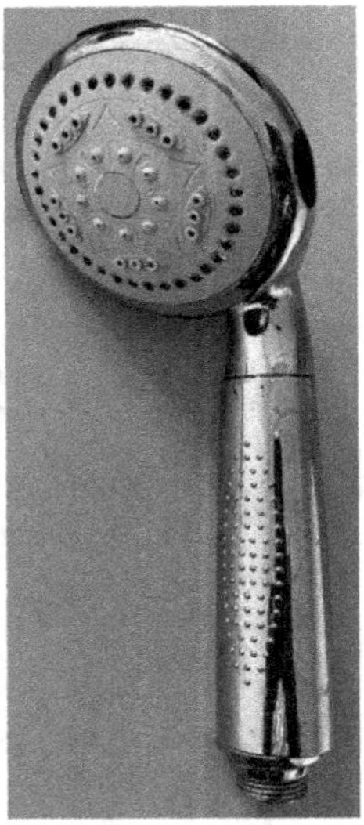

Figure 4 - Shower head part as a metallised

Research work analysed the removal of the metallic layer on the example of a showerhead, a common metallised ABS plastic product manufactured in large quantities (see Figure 4).

The first tests show that it is also possible to remove the metallised layer in a material-friendly way using the laser technology. The optimum laser processing parameters for removing the metallisation without thermal damage to the carrier material are also determined in various tests. The microscopic images (see Figure 5) clearly show the removal of the metallisation. In addition, the polymer carrier material is not burnt or damaged by the laser process and can be reused without any other quality problems.

After the laser process of the surface, it was observed that the adhesion between the metallic layer and the thermoplastic polymer was massively reduce, allowing the detachment of the metallic layer. The separating of the two materials offers new possibilities for the complete recycling of the hybrid metallised parts. Both separated materials can be recycled with high material efficiency.

Figure 5 - Microscopic image of the removed metal layer from a thermoplastic polymer

The study will be continued by using 3-dimensional laser processing techniques for a complete separation of the geometry in complex metallised parts.

Conclusions

To reuse the materials from the metallised thermoplastic parts such as door handles, trim strips, radiator grilles, hubcaps etc. with no burning/destroying of the polymer material, it is important to separate the metallised coating from the thermoplastic carrier material. The laser technology demonstrates excellent performance in removing the metal layer from sensitive thermoplastic polymers without damaging them.

Compared to other separation methods, this technology shows a big advantage: it is easy to use and quiet with low operating costs and requires no consumables or even environmentally hazardous media (no abrasives, chemicals, or solvents). No special safety and environmental measures are required, apart from radiation protection. Also, the productivity of the laser process can be improved by using powerful lasers and the automatically focus adjustment for 3 dimensional surfaces. An intelligent optical (camera) system can adjust the laser parameter according to the parts dimensions and other characteristics.

In future, the laser technology can perform an important role in the environmentally friendly removal of metallisation from polymers enableing the reuse of large volumes of waste metallised plastic components while offering a sustainable processing solution.

In summary, this research shows that it is possible to remove metallisation from thermoplastics using laser technology without damaging the material, which is essential for the reusing of the polymer parts or materials.

References

[1] Gorjizadeh N., Pahlevani F., Sahajwalla V. (2017), Metallised plastics waste as a new resource – Study of electronic structure and chemical bond, https://jmaterenvironsci.com/Document/vol8/vol8_N5/170-JMES-2676-Gorjizadeh.pdf

[2] Uglum P. (2020), The Future of Metallic Finishes on Plastic in the Automotive Industry, https://plasticsdecorating.com/articles/2020/the-future-of-metallic-finishes-on-plastic-in-the-automotive-industry/

[3] Knappich F., Schlummer M., Mäurer A., Harald Prestel H. (2018), A new approach to metal- and polymer-recovery from metallized plastic waste using mechanical treatment and subcritical solvents, https://link.springer.com/article/10.1007/s10163-018-0717-6

[4] Songyan Y., Sustainable Approach to Transform Metallized Plastic Waste and Automotive Plastic Waste into Value-added Resources, https://doi.org/10.26190/unsworks/21611

[5] Yin S., Rajarao R., Gong B., Wang Y., Kong C., Sahajwalla V., Thermo-delamination of metallised composite plastic: An innovative approach to generate Aluminium from packaging plastic waste, Journal of Cleaner Production, Volume 211, 2019, Pages 321-329, https://doi.org/10.1016/j.jclepro.2018.11.158.

[6] Mahmood A. A., Hossain R., Bhattacharyya S., Sahajwalla V., Recycling of polymer laminated aluminum packaging (PLAP) materials into carbonaceous metallic microparticles, Journal of Cleaner Production, Volume 269, 2020, https://doi.org/10.1016/j.jclepro.2020.122157

[7] Sover, A. (2019). Research on the Process of Paint Removal from Thermoplastic Materials by Laser. IOP Conference Series: Materials Science and Engineering. 564. 012028. 10.1088/1757-899X/564/1/012028. https://iopscience.iop.org/article/10.1088/1757-899X/564/1/012028

[8] Zink, M., Sover, A., Study on the Process of Paint Removal from Thermoplastic Materials by Laser Technology, Proceedings of the 2nd International Conference of the University of Applied Sciences Ansbach 3rd - 4th of July 2020 https://forschung.hs-ansbach.de/de/publikationen/113

[9] A. Sover, M. Zink, T. Vogel and R. C. Ciobanu, "The Use of Laser Technology for Recycling Hybrid Electronic Waste," 2022 International Conference and Exposition on Electrical and Power Engineering (EPE), 2022, pp. 360-363, doi: 10.1109/EPE56121.2022.9959803

[10] Zink M., Sover A. and Vogel T., Innovative process for material separation of printed circuit boards (PCB) with laser technology, The 25th edition of Innovative Manufacturing Engineering and Energy International Conference (IManEE 2021), Iasi, October 2021 https://iopscience.iop.org/article/10.1088/1757-899X/1235/1/012009

The impact of surface topography after milling process on tribological characteristics of 36CrNiMo4 steel

Andrzej DZIERWA[1,a *], Nataliia STELMAKH[2,b] and Nika TIKANASHVILI[3,c]

[1]Department of Manufacturing Technology and Production Engineering, Rzeszow University of Technology, Powstancow Warszawy 8, 35-959 Rzeszow, Poland

[2]Igor Sikorsky Kyiv Polytechnic Institute, National Technical University of Ukraine, 37 Prosp. Peremohy, 03056 Kyiv, Ukraine

[3]Faculty of Engineering, Georgian Aviation University, 16 Ketevan Dedofali Ave., 0144 Tbilisi, Georgia

[a]adzierwa@prz.edu.pl, [b]n.stelmakh@kpi.ua, [c]n.tikanashvili@ssu.edu.ge

Keywords: Surface Topography, Friction, Wear, Tribological Properties

Abstract. Tribological investigations were conducted under a dry contact conditions. Milled steel discs were put in contact with ceramic balls made of SiC. Disc surfaces were characterized by different values of arithmetical mean height of the surface (Sa) from 0.182 to 5.17 µm. Wear volume of the discs, distance required to achieve stable value of friction force as well as coefficient of friction were calculated. The results revealed that the increase in surface roughness led to a reduction in wear loss. Dependency between several surface topography and tribological parameters were also presented.

Introduction

Reducing friction and wear is crucial for enhancing operating efficiency and service life of engineering materials [1]. Effective ways of reducing friction and wear can not only improve performance but also contribute to safety and sustainability in various applications. Various authors have presented various methods that can be helpful in reducing them. These methods focus on surface engineering, material composition, and operational strategies. For instance, surface treatments as roller burnishing [2, 3], slide burnishing [4] shot peening [5] or vapour blasting [6] can delay wear processes and reduce friction. Also, the milling process significantly affects the friction and wear characteristics of various materials.

The authors of [7] investigated the effect of temperature on the wear of ground and milled samples made from EN 31 bearing steel. Despite similar surface roughness, milled samples were characterized by lower wear regardless of temperature. Li et.al. [8] found that the milling process affects the wear of GH4169 material, and the main factors influencing this are the spindle speed, feed per tooth and depth of cut. Authors of [9] reported that ultrasonic vibration-assisted milling reduces friction and wear on titanium alloy surfaces compared to common milling, enhancing tribological properties under oil lubrication. Bermudez et.al. [10] indicate that milling conditions, such as atmosphere and process control agent percentage, affect the wear resistance of mechanically alloyed aluminum materials. The authors of [11] also found a positive effect of the milling process in the study of a newly developed Ti-Nb-Zr alloy for biomedical applications.

Since surface topography has a significant effect on friction and wear, and the milling process allows obtaining surfaces with different roughness, the aim of the research was to determine the effect of surface topography created as a result of the milling process on the tribological properties of steel under dry friction conditions.

 Materials Research Forum LLC
https://doi.org/10.21741/9781644903377-27

Materials and methods

Tribological tests were carried out using a pin/ball-on-disc tribological tester T-11, manufactured by Institute for Sustainable Technologies – National Research Institute in Radom (Poland). In the experiments, a ball-on-disc configuration was utilized. The frictional couples consisted of a stationary ceramic ball made from silicon carbide (SiC) with a hardness of 72 ± 2 HRC and a rotating steel disc. The disc material was 42CrMo4 steel with a hardness of 40 ± 2 HRC. Milling process was finishing treatment of all disc surfaces. Surface of the discs was characterized by different values of the arithmetical mean height Sa in the range of 0.182 – 5.17 µm (Table 1). In addition to the Sa parameter, the milled surfaces were characterized by other surface topography parameters like skewness (Ssk), kurtosis (Sku), maximum height (Sz), auto-correlation length (Sal), peak density (Spd), core height (Sk), reduced peak height (Spk) and reduced valley depth (Svk) [12]. Fig. 1 presents isometric views of selected disc samples.

Table 1. Selected surface topography parameters of milled disc samples

Surface parameter	Designation of tested discs								
	M1	M2	M3	M4	M5	M6	M7	M8	M9
Sa [µm]	0,182	0.424	0.906	1.23	1.37	2.07	3.91	4.59	5.17
Ssk	-0.431	0.0585	0.544	0.0077	-0.457	-0.633	0.13	0.109	-0.083
Sku	3.56	2.41	2.2	1.93	3.12	2.3	3.28	2.82	2.44
Sz [µm]	2.11	2.61	5.28	7.2	11	11.6	37.7	40.3	44.7
Sal [mm]	0.0413	0.0293	0.0609	0.106	0.0941	0.0485	0.0943	0.119	0.0852
Spd [1/mm^2]	205	173	103	18.4	11.5	12.6	29.6	22.3	24
Sk [µm]	0.529	1.06	2.12	3.34	3.57	4.61	10.1	10.5	13.7
Spk [µm]	0.167	0.207	1.49	0.239	0.712	0.104	5.37	5.06	5.19
Svk [µm]	0.329	0.233	0.123	0.586	2.82	2.98	3.82	3.82	4.05

Tribological experiments were performed under dry sliding conditions at room temperature between 20 and 22 °C. The sliding distance was set to 200 m, the sliding speeds were 0.16 m/s, 0.28 m/s and 0.40 m/s while the normal load was set to 9.81 N. The number of test repetitions was three. During the friction processes, the friction force was recorded over the course of time. Following the tribological tests the wear level was assessed using a white light interferometer Talysurf CCI Lite. The measurement area was 3.3 mm × 3.3 mm, with wear track readings taken at four locations, each 90° apart. Following this, profiles were gathered perpendicular to the wear tracks, and the wear cross-sectional areas were determined using TalyMap Gold 6.0 software. At least seven profiles were taken from each measured area. The next step involved calculating the wear volume of the disc using Equation (1).

$$VD = \pi ds \tag{1}$$

in which *d* represents the wear track diameter [mm] and *s* indicates the cross-sectional area of the wear track [mm^2].

Results and discussion

Figs. 2-3 and Table 2 present the results of the tests. Fig. 2 presents volumetric wear levels of disc samples obtained at the different sliding speeds. Table 2 presents the average values of the friction coefficient and the distance after which the friction force reached a stabilized value (DSS). Fig. 3 shows cross-sectional areas of selected worn discs achieved at the sliding distance of 0.16 m/s.

Materials Research Proceedings 46 (2024) 204-210 https://doi.org/10.21741/9781644903377-27

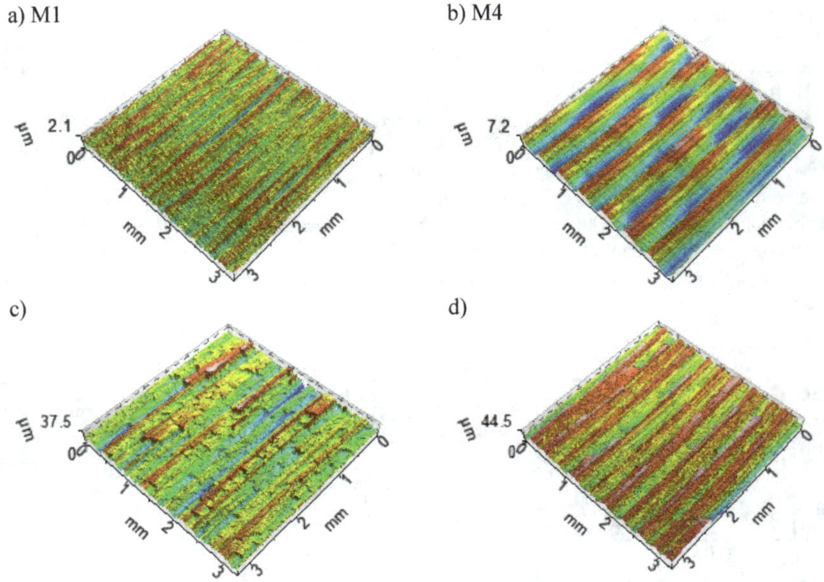

Fig. 1. Isometric views of selected disc surfaces: M1 (a), M4 (b), M7 (c), M9 (d)

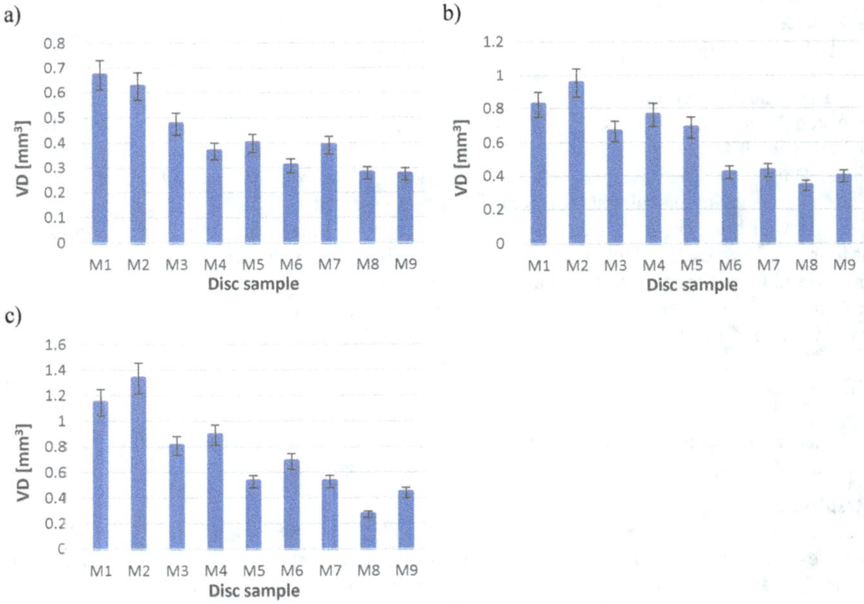

Fig. 2. Wear volume levels of disc samples

Table 2. Results of tribological tests

	Designation of tested discs								
v=0.16 [m/s]	**M1**	**M2**	**M3**	**M4**	**M5**	**M6**	**M7**	**M8**	**M9**
CoF av	0.811	0.806	0.779	0.801	0.790	0.788	0.794	0.785	0.768
DSS	18	24	35	32	32	34	58	61	77
v=0.28 [m/s]	**M1**	**M2**	**M3**	**M4**	**M5**	**M6**	**M7**	**M8**	**M9**
CoF av	0.824	0.808	0.799	0.787	0.802	0.782	0.790	0.766	0.762
DSS	16	19	28	28	26	31	49	62	70
v=0.40 [m/s]	**M1**	**M2**	**M3**	**M4**	**M5**	**M6**	**M7**	**M8**	**M9**
CoF av	0.801	0.774	0.802	0.791	0.795	0.768	0.771	0.754	0.756
DSS	15	20	25	24	25	28	48	66	62

The M1 sample exhibited the largest wear volume of the discs (VD = 0.671 mm^3) at a sliding speed of v = 0.16 m/s. This value was related to the maximum average friction coefficient and the least distance required to obtain a stable friction force. On the other hand, frictional pair with the disc M9 presented the smallest volumetric wear (VD = 0.273 mm^3) as well as average value of coefficient of friction and the highest DSS parameter.

At the sliding speed of 0.28 m/s we achieved a slightly different tendency. Although the coefficient of friction and DSS parameter reached the highest values for the same samples as at a lower speed, the volumetric wear was slightly different. The smallest value of wear volume was observed for sample M8 (VD = 0.342 mm^3) and the highest for frictional couple with the disc M2 (VD = 0.954 mm^3).

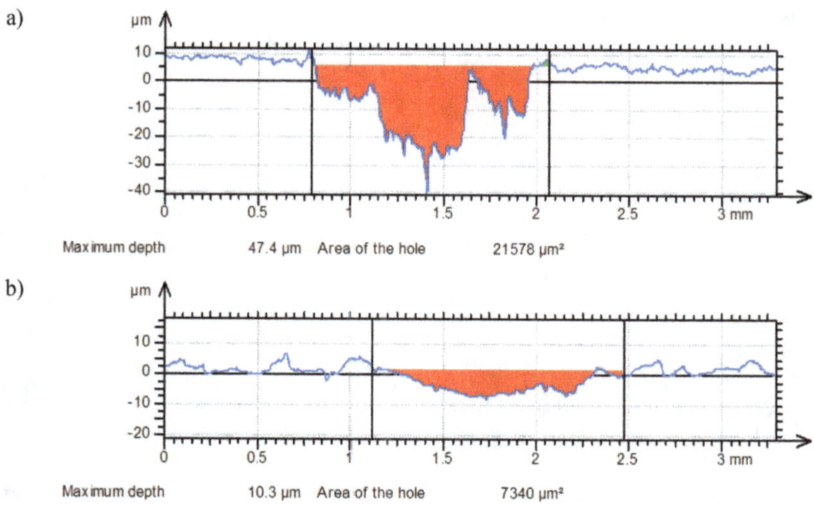

Fig. 3. Cross-sectional areas of selected worn discs: M2 (a), M8 (b) at the sliding speed of v = 0.16 m/s

An increase in the sliding speed to 0.40 m/s showed similar tendency when the sliding speed was 0.28 m/s. Similarly, sample M8 presented the highest wear volume (VD = 1.335 mm^3) and the disc M8 the lowest (VD = 0.272 mm^3). Frictional pair with the disc M8 was also characterized by the smallest value of friction coefficient (μ=0.754). Generally, higher sliding speeds resulted in greater wear volume of the discs. Moreover, the rise in sliding speed also led to a greater dispersion of wear results and rougher disc surfaces led to its lower wear volume. Similar behavior in case of the coefficient of friction was achieved by the authors [13, 14]. A significant correlation was also noted between wear volume and DSS parameter.

The linear correlation coefficient measures the strength of the linear relationship between two variables. In our study it was used to find the relationship between volumetric wear and attributes defining the surface topography (Fig. 3). The high coefficient of determination at all sliding speed was found between the Sa parameter and wear volume. At the sliding speed of 0.16 m/s the coefficient of linear correlation was -0.78, at the sliding speed of 0.28 m/s achieved -0.85 and at the 0.4 m/s reached -0.83. A visible dependency was also achieved between wear volume of the discs and the peak density (Spd). According to the authors of [15] an increase in the Spd parameter value results in a greater bearing surface. At the lowest sliding speed the coefficient of linear correlation exceeded 0.9 reaching a value of 0.92. Slightly lower values were recorded for the remaining sliding speeds, however the relationship between the parameters Spd and VD was still strong.

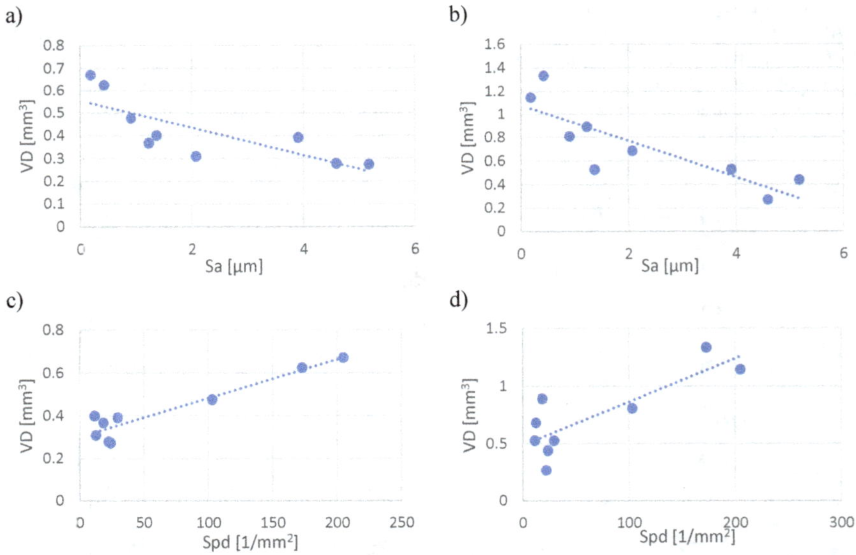

Fig. 4. Dependencies between volumetric wear VD and Sa (a, b) as well as between volumetric wear VD and Spd (c, d) at the sliding speed of: v=0.16 m/s (a, c) and = 0.40 m/s; (b, d)

Authors didn't find significant relationship between wear volume and skewness (Ssk) and kurtosis (Sku) which is often presented in scientific studies [13, 16, 17]. This may be due to its sensitivity to the occurrence of occasional deep valleys or high peaks on the surface [18].

The correlations obtained between surface topography parameters and tribological characteristics provided information that can be used in practical applications. The increase in roughness leads to a decrease in wear and friction coefficient. Also the distance necessary to achieve a stable friction force value increases with increasing surface roughness. Under the same load, the smooth surfaces have a larger contact area and more touchpoints than a rougher ones [19]. In that case, surfaces with a large maximum surface height can play an important role in capturing wear products thus preventing further severe wear. It seems that the valleys of surface topography can play a role in trapping certain wear debris and consequently contribute to a reduction in the wear intensity.

The wear of the balls used in the tests was negligible compared to the discs, which was the result of the significantly greater hardness of the balls compared to the hardness of the discs.

Conclusions

This research was intended to study the impact of surface topography resulting from the milling process on tribological behaviors of steel surfaces. The study was carried out on 9 different sets of surfaces. The following conclusions can be draw.

A rise in the Sa parameter is associated with a decline in the wear volume of the discs independently of the sliding speed. Significant linear correlations were also observed between the Spd parameter and the wear volume of the discs. The relationship between the skewness Ssk, kurtosis Sku and wear volume VD was not confirmed in the conducted studies. This could be attributed to the unpredictable nature of the Ssk and Sku parameters.

Friction forces stabilized very quickly when the disc surfaces were smooth. As the surface roughness increased, the distance after which the friction force stabilized increased.

A rise in sliding speed from 0.16 m/s to 0.40 m/s generally resulted in greater volumetric wear of the discs.

References

[1] L.O. Ajuka, T.S. Ogedengbe, T. Adeyi, O.M. Ikumapayi, E.T. Akinlabi, Wear characteristics, reduction techniques and its application in automotive parts – A review. Cogent Eng. 10 (2023) 2170741. https://doi.org/10.1080/23311916.2023.2170741

[2] Dzierwa, N. Stelmakh, N. Tikanashvili, Application of Taguchi technique to study tribological properties of roller-burnished 36CrNiMo4 steel, Lubricants, 11 (2023) 227. https://doi.org/10.3390/lubricants11050227

[3] A.M. Hassan, The effects of ball and roller burnishing on the surface roughness and hardness of some non-ferrous metals, J. Mater. Process. Technol. 72 (1997) 385-391. https://doi.org/10.1016/S0924-0136(97)00199-4

[4] Z.F. Kovács, Z.J. Viharos, J. Kodácsy, Determination of the working gap and optimal machining parameters for magnetic assisted ball burnishing, Measurement. 118 (2018) 172-180. https://doi.org/10.1016/j.measurement.2018.01.033

[5] B.K.C Ganesh, W. Sha, N. Ramanaiah, A. Krishnaiah, Effect of shot peening on sliding wear and tensile behavior of titanium implant alloys, Mater. Des. 56 (2014) 480-486. https://doi.org/10.1016/j.matdes.2013.11.052

[6] Lenart, P. Pawlus, A. Dzierwa, The effect of steel disc surface texture in contact with ceramic ball on friction and wear in dry fretting, Surf. Topogr. Metrol. Prop. 6 (2018) 034004. https://doi.org/10.1088/2051-672X/aac1a2

[7] T. Anto, P.M. Anil, An analysis on the influence of temperature on the sliding wear of components finished by grinding and milling processes. International Conference on Energy

Efficient Technologies for Sustainability. Nagercoil, India, (2013) 912-917. doi:10.1109/iceets.2013.6533509

[8] X. Li, W. Zhang, L. Miao, Z. Pang, Analysis of tool wear in GH4169 material milling process, Lubricants. 11 (2023) 245. https://doi.org/10.3390/lubricants11060245

[9] K. Zheng, W. Liao, Q. Dong, L. Sun, Friction and wear on titanium alloy surface machined by ultrasonic vibration-assisted milling, J. Braz. Soc. Mech. Sci. Eng. 40 (2018) 411. https://doi.org/10.1007/s40430-018-1336-9

[10] M.D. Bermúdez, F.J. Carrion, P. Iglesias, G. Martínez-Nicolás, E.J. Herrera, J.A. Rodríguez, Influence of milling conditions on the wear resistance of mechanically alloyed aluminium, Wear. 258 (2005) 906-914. https://doi.org/10.1016/j.wear.2004.09.071

[11] F. Hamadi, M. Fellah, N. Hezil, D. Bouras, S.E. Laouini, A. Montagne, H.A.E. Khalifa, A. Obrosov, G.A. El-Hiti, K.K. Yadav, Effect of milling time on structural, physical and tribological behavior of a newly developed Ti-Nb-Zr alloy for biomedical applications, Adv. Powder Technol. 35 (2024) 104306. https://doi.org/10.1016/j.apt.2023.104306

[12] PN-EN ISO 25178-2:2012 Geometrical Product Specifications (GPS) - Surface texture: Areal - Part 2: Terms, definitions and surface texture parameters

[13] M. Sedlacek, B. Podgornik, J. Vizintin, Correlation between standard roughness parameters skewness and kurtosis and tribological behaviour of contact surfaces, Tribol. Int. 48 (2012) 102–112. https://doi.org/10.1016/j.triboint.2011.11.008

[14] M. Sedlacek, B. Podgornik, J. Vizintin, Influence of surface preparation on roughness parameters, friction and wear, Wear. 266 (2009) 482–487. https://doi.org/10.1016/j.wear.2008.04.017

[15] Y. Pu, Y. Zhao, H. Zhang, G. Zhao, J. Meng, P. Song, Study on the three-dimensional topography of the machined surface in laser-assisted machining of Si3N4 ceramics under different material removal modes, Ceram. Int. 46 (2020) 5695-5705. https://doi.org/10.1016/j.ceramint.2019.11.017

[16] A. Dzierwa, P. Pawlus,W Zelasko, The influence of disc surface topography after vapor blasting on wear of sliding pairs under dry sliding conditions, Coatings. 10 (2020) 102. https://doi.org/10.3390/coatings10020102

[17] G. Liang, S. Schmauder, M. Lyu, Y. Schneider, C. Zhang, Y. Han, An investigation of the influence of initial roughness on the friction and wear behavior of ground surfaces, Materials. 11 (2018) 237. https://doi.org/10.3390/ma11020237

[18] M. Sedlacek, P. Gregorcic, B. Podgornik, Use of the roughness parameters Ssk and Sku to control friction—A method for designing surface texturing, Tribol. T. 60 (2016) 260–266. https://doi.org/10.1080/10402004.2016.1159358

[19] P. Guglielmi, F. Davoodi, G. Palumbo, Tribological behaviour of microindented 100Cr6 steel surfaces in dry contact conditions. Int. J. Adv. Manuf. Technol. 133 (2024) 2381–2400. https://doi.org/10.1007/s00170-024-13868-2

Innovative Manufacturing Engineering and Energy - IManEE2024 Materials Research Forum LLC
Materials Research Proceedings 46 (2024) 211-218 https://doi.org/10.21741/9781644903377-28

A digital twin framework for prediction, monitoring, and diagnosis in copper tube extrusion

Kyriakos SABATAKAKIS [1,a], Panagiotis STAVROPOULOS [1,b*], Apostolos KAIMENOPOULOS [2,c], Dimitrios KARATASIOS [2,d]

[1]Laboratory for Manufacturing Systems and Automation (LMS), Department of Mechanical Engineering and Aeronautics, University of Patras, Rio Patras 26504, Greece

[2]Halcor, Copper & Alloys Extrusion Division of ElvalHalcor S.A., 32011 Oinofyta, Greece

[a]sabatakakis@lms.mech.upatras.gr, [b]pstavr@lms.mech.upatras.gr, [c]akaimen@halcor.com, [d]dkaratasios@halcor.com

Keywords: Metal Forming, Copper Extrusion, Digital Twins, Industry 4.0

Abstract. The strict quality requirements, the increased production rates, variability, and volume constitute challenges in the quality of extruded profiles. State-of-practice methods cannot track down or mitigate all the product features and defects, thus increasing scrap generation and reducing production flexibility. Nevertheless, State-of-the-art approaches are few and dedicated to the product rather than the actual process with the last being researched in terms of process modeling. This study assesses the adoption level of I4.0 technologies in extrusion applications and it applies well-established practices across manufacturing to outline a Digital Twin framework for Prediction, Monitoring, and Diagnosis in the case of copper tube extrusion.

Introduction

Extrusion is one of the most commonly used deforming processes in the metalworking industry [1]. Aluminum and copper are commonly used in profile extrusion, with aluminum favored in construction, automotive, and aerospace for its strength, corrosion resistance, and formability. Copper is used where high thermal and electrical conductivity is needed, such as in heat exchangers, electrical components, plumbing, and HVAC systems [2]. To this end, extruding metals like aluminum and copper present several challenges as regards the variability of the final part quality. The preheated billet temperature distribution evolution directly affects the quality features of the final product as local high temperature alters billet properties and can even lead to local melting [2]. Taper heating [3] and isothermal extrusion strategies [4] have been developed to mitigate these factors and other ones such as tools' temperature, extrusion ratio & shape, and friction conditions that alter the temperature homogeneity of the billet. Temperature variations have a direct impact on the extruded part quality when combined with unoptimized process design & process control, and tool wear & misalignment. Internal defects such as center burst cracks, pipping, and back-end defects [5] reduce the structural integrity of the profile. Surface defects include pick-ups, die-lines, blisters, hot tears charged welds, and other types of surface artifacts that increase the roughness and affect the aesthetical appearance of the extruded part. Complying with dimensional requirements is another significant challenge, particularly in complex or thin-walled profiles and tube extrusion. Variations in exit temperature, ram speed, and tool misalignment can lead to uneven profile thickness and non-compliant cross-sectional geometry. Finally, as hot extrusion is performed above the recrystallization temperature to prevent work hardening and allow significant material deformation, any variation in the process parameters such as ram speed, die temperature, and billet preheating influence the microstructure, potentially leading to non-uniform grain size and affecting hardness and plasticity of the extruded part [2].

Extrusion tooling and equipment are subjected to deformation and wear, and they are exposed to high temperatures and process loads. While special alloys are used to mitigate the downsides of hot-work tool steels, load cycles can still affect the service life of a die [6] and lead to fatigue-based fracture, surface wear, and plastic deformation or deflection, depending also on other factors such as die shape complexity fillet radius and bearing length [7]. Mandrels design, particularly the cone angle and position of the sizing section, impacts the energy, force, and deformation conditions during tube extrusion. Additionally, while mandrel coating wear is typically minimal, failures have occurred due to slow ram speeds, which cause prolonged contact with the billet and overheating of the mandrel. Container's inner wall during extrusion builds up an oxide layer, which can contaminate the extruded part in terms of surface defects, dimensional inaccuracies, and reduced profile strength. This oxide buildup also acts as an abrasive, gradually wearing down the inner liner surface of the container [2].

State-of-the-Art

Research in metal extrusion mainly focuses on offline process optimization. This is broken down into several different research directions. The first one concerns tool and equipment optimization and focuses on the enhancement of the die performance and service life. Finite Element Models (FEM) have been used to predict the thermomechanical behavior of dies during extrusion aiding the design of more durable and efficient die [8]. Process analysis and optimization is another field of interest. Regarding the case of the numerical model, this has been done by developing a process model, which is used to calculate process state variables such as material flow, stress, strain, strain rate, and temperature of the billet and features of the extruded part such as the microstructure, temperature distribution, and dimensional features [9], [10]. FEM approaches have been utilized also to predict defects such as charge welds [11] and back-end defects [5]. Beyond the computationally costly FEMs, however, accurate analytical models have been developed for predicting the extrusion temperature taper and enabling the isothermal extrusion process [4]. Data-driven modeling has also been used for process optimization by utilizing Artificial Neural Networks (ANN). In [12] the authors developed 2 ANNs to enable the prediction of the billet's temperature at the end of the induction heater and of the extruded profile at the die exit using process and sensor data, while in [13] a similar study used an ANN-based method to link the extrusion pressure with the process parameters. Hybrid FEM-ANN approaches have also been used for developing surrogate models to predict the grain size [14] and temperature of the profile [15] using the FEM data for training. Finally, another research direction in metal extrusion concerns the inspection of the process, which however is still in its infancy. In [16] for example, the authors by using a multimodal camera system allocated surface defects and temperature patterns on the extruded part surface right, while in more recent studies [17] the use of ANNs enabled also the classification of these defects.

Profile extrusion research primarily focuses on prediction methods targeting specific state variables, Key Performance Indicators (KPIs), and applications. While these studies provide useful insights, their scope is narrow, overlooking factors like tool wear, equipment status, and part design. Additionally, real-time process monitoring is limited, likely due to the difficulty of measuring state variables such as material flow, stress, and strain in real time and although data-driven models exist, they lack real-time inputs and thus variability. The Digital Twin (DT) concept [18] could address these issues by integrating prediction, monitoring, and root cause analysis, but DTs have yet to be applied to profile extrusion in research. This is not however true for other metal-forming processes where the application of DT technology has emerged as a tool for real-time monitoring, predictive modeling, and process optimization. In sheet metal forming, a hybrid DT framework combines simulation, CAD, and measurement data, allowing real-time adjustments to process parameters by comparing virtual models with physical results, thus improving production accuracy and handling stochastic variations [19]. Similarly, an approach for stamping monitors mold wear and adjusts process conditions based on real-time punch force data to improve quality and reduce energy consumption [20] while in [21] soft sensors are created to monitor the elasto-mechanical behavior of forming presses, enabling better process transparency and more efficient tool transfers. Real-time quality prediction in sheet metal forming integrates FEM, Machine Learning (ML), and optimization algorithms to predict part quality and suggest adjustments before the next press stroke, enhancing speed and accuracy [22]. Additionally, press hardening processes benefited from DTs that were used to predict cooling curves using FEM simulations, enabling real-time adjustments for improved efficiency [23]. In tube bending, DTs incorporated real-time multi-source inputs to optimize bending outcomes. A Transfer Incremental Learning approach allows the model to adapt during production, reducing training time and improving predictive accuracy [24]. Another approach refines low-fidelity models into high-fidelity representations of plastic deformation using spatiotemporal reconstruction, enhancing real-time monitoring efficiency [25].

These studies demonstrate the broad applicability of DTs in improving real-time monitoring, predictive accuracy, and process optimization in metal-forming, driving the adoption of intelligent manufacturing systems in Industry 4.0. However, even metal forming lacks the level of DT adoption compared to other processes such as welding [26] and additive manufacturing [27]. The purpose of this study is to fill the gap regarding the absence of studies regarding DT frameworks in profile metal extrusion. Based on the current digital tools and the available DT approach in metal-forming the authors are rendering a DT framework for the case of direct hot extrusion of copper tubes.

A Digital Twin Framework for Tube Extrusion

In the context of this study, the initial definition of DT lists three major functions according to [28]. Prediction refers to the calculation of the KPIs of interest before the actual system runtime. Monitoring concerns the real-time monitoring of KPIs by using external sensor signals and triggers. This function concerns a narrow process window compared to the Prediction in favor of real-time performance. Finally, Diagnosis refers to the analysis of unpredicted failures and the update of the underlying models found in the previous functions.

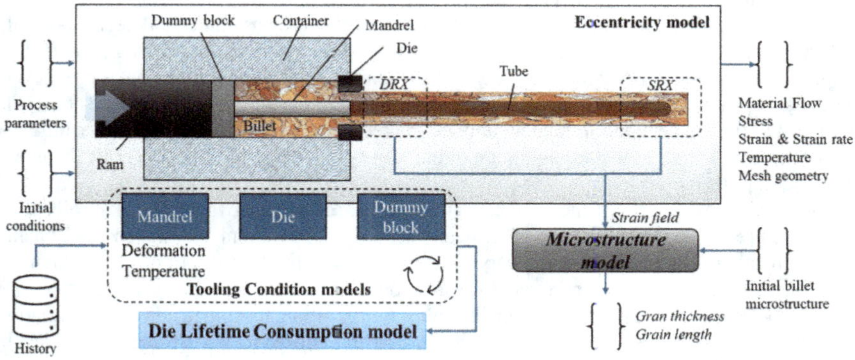

Fig. 1: DT Prediction function

Prediction. In Fig. 1 the top level of the DT's Prediction function is depicted. The tube eccentricity and microstructure are identified as the KPIs of interest. Based on that, the Prediction function is divided into two main parts: (1) the prediction of the eccentricity & (2) the prediction of the microstructure. In the literature, relevant FEMs have been investigated, as described in [9] for the case of direct hot extrusion of copper tubes to analyze how the tool misalignment and the upsetting percentage affect the eccentricity. The model enabled stress, strain, strain rate, and temperature conditions prediction within the billet across different processing conditions. Based on this approach a similar model is proposed herein with some additions and modifications, called the Eccentricity model. As such, the billet is modeled herein as a plastic body while the die, the piercer, and the dummy block as elastic bodies. Finally, the ram, the die holder, and the container are modeled as rigid bodies. This approach enables the decoupling of the elastic bodies to study their deformation under multiple load cycles as demonstrated in [6] The Eccentricity model aims to serve as a tool for analyzing different process designs and their effect on the eccentricity & deflection of the tools during the process. The model I/Os are rendered in Table 1.

The Eccentricity model is complemented by the Tooling condition models which are comprised of three (3) numerical models representing the die, piercer, and dummy block. The purpose of these models, unlike in [6] is to simulate only the residual deformation of the piercer, die, and dummy block during their service life. The models use as input data the temperature and stress data obtained by the Eccentricity model and provide back an output the resulting temperature fields and the plastic deformations after each extrusion stroke. In this way, the Eccentricity model can enhance its prediction by considering any geometrical and temperature variations that may build up over time on these three models.

Table 1: Eccentricity model I/Os

Inputs	Outputs
-Billet: Dimensions, temperature, material properties, mesh geometry	-Material flow
-Fitting: Billet-Liner gap, tube outside diameter, tube wall thickness	-Stress
-Parameters/variables: Ram speed, Container temperature, upsetting %	-Strain rate
-Die, Piercer, Dummy block status: temperature profile, mesh geometry	-Temperature
-Initial Equipment's status (Ram, Liner, Die holder): temperature profile	-Mesh geometry
-Heat & Shear friction coefficients: between the billet, tooling, & equipment	

Another model linked to the Eccentricity model is called the Die Lifetime Consumption model and concerns the prediction of die failure due to rising local overloads, material defects, tool-billet contact surface wear, and creep-fatigue damage. The contribution of creep and fatigue to the total lifetime consumption depends on the billet and die temperature, the duration of the press cycle, and the stress amplitudes resulting from the loading and unloading within each extrusion cycle. The Die Lifetime Consumption model sums the increment of a damage variable over time and defines material failure as the accumulation of the resulting damage variable to a critical value as described in [29].

Finer grain size improves the mechanical properties of extruded parts, so controlling the microstructure of extruded tubes is crucial. In [10] the authors developed a Dynamic Recrystallization (DRX) and Static Recrystallization (SRX) model to optimize aluminum rod extrusion. DRX results from strain and friction, producing fibrous grains. SRX can occur after extrusion based on factors like temperature and alloy composition, leading to new grain formation. The corresponding model, called herein the "Microstructure Model", predicts grain size by combining DRX and SRX processes, using experimental data and strain field predictions.

Monitoring. The DT Monitoring function is depicted in Fig 2. In literature and commercial systems, similar functions are based on data-driven models. In this case, that means somehow transforming the existing models or creating new ones by using sensor signals that monitor a part of the process. Reduced Order Modeling (ROM) is a method for reducing the computational complexity of a numerical model while preserving the expected fidelity within a controlled error and is also utilized to combine multiple component-level models into system-level simulations.

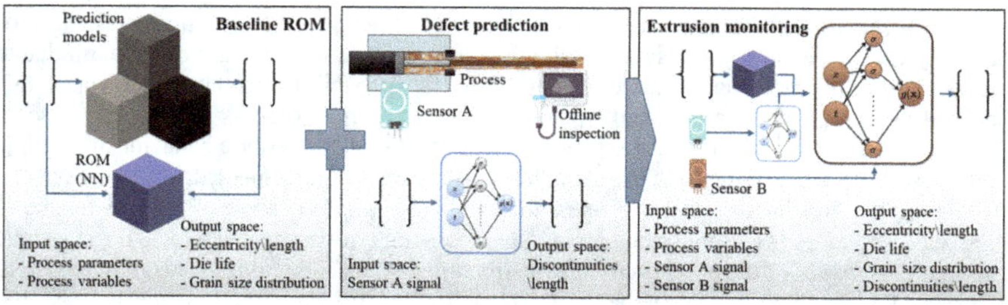

Fig. 2: DT Monitoring function

In this case, the DT Monitoring function is developed using the inputs/outputs of the DT Prediction models as has already been implemented using ROM in several studies [30]. The corresponding ROM defines the baseline for real-time monitoring of the extrusion where an ANN is trained to map an input to an output space and is called Baseline ROM (Fig. 2). The input space is a vector comprised of all the input parameters and variables found of the DT Prediction function. The Eccentricity\tube length, the Die life\extrusion cycle, and grain size distribution\tube length define the output space. The accuracy for a given process window length (distance of points in the input space) heavily depends on the amount of data collected from several simulation runs (number of data points). Operating the model in real-time requires providing the input vector representing the current extrusion cycle. The described Baseline ROM is a deterministic model; thus, it will not predict random events as in real production.

Defect prediction can be considered as a random event for a model like the previous one and requires tremendous effort just to implement even for a very narrow case study by using a numerical model [31]. To resolve that, supervised ML is used to correlate features extracted from sensor signals when monitoring the process or the manufactured part to the KPI of interest [32]. In this case, an ML approach is realized to identify discontinuities on the extruded tube. An ANN namely Defect prediction is trained based on the features extracted from sensor data - Sensor A (Fig. 2) and the identified discontinuities across the length of the tube. Sensor A could be anything that measures in real-time an input variable of the process (e.g. extrusion pressure/displacement), emission of the process (billet temperature before inserted into the container), or the actual value of a nominal parameter that is internally controlled.

The last part of the Monitoring function concerns the aggregation of the Baseline ROM and Defect prediction for the first one to be able to resolve random events and enhance the second one. To achieve this the feature extraction layers of these models are connected to a new ANN which is also fed by another sensor – Sensor B (Fig. 2). The resulting model, named Extrusion monitoring takes as an input the inputs of the Baseline ROM, the inputs of Defect detection, and the features of the Sensor B.

Diagnosis. The DT Diagnosis includes both the error identification as well as tracing back to the source of the error. These are rendering the 1st feature of the Diagnosis function. Empirical rules and statistical process control are realized for identifying deviations from a predefined pattern by constantly checking the DT Monitoring outputs. Additionally, this function allows running the entire DT Monitoring for simulating future abnormalities. Then as with the initial case, errors can be identified and traced back and thus help the operators create strategies for avoiding them in the first place. This renders the 2nd feature of the Diagnosis function. Finally, the DT Diagnosis function includes also a feature to compensate for small bias or non-linear deviations found between the DT Monitoring outputs and the actual KPIs of interest. This renders the 3rd feature of the Diagnosis function.

Starting with the 1st feature, the eccentricity, microstructure, and discontinuities must be constantly checked against predefined thresholds using control charts. In case of an alarm due to a single outlier, heuristic algorithms can point out using probabilities which model caused this abnormal event e.g. the excessive tool wear, use of inappropriate process parameters, or some variation that has been measured by the sensors. For identifying the erratic behavior of multiple points, several heuristic algorithms based on the Nelson rules are implemented.

Skipping the 2nd feature which is operation is obvious, the operation of the 3rd feature is described herein. The DT Monitoring output may present an error compared to actual test sample measurements made for the eccentricity, microstructure, and discontinuities. This may be due to the nature of the prediction problem that changes over time, or the assumptions made and captured to the model no longer hold. This is referred to as “concept drift” where the underlying probability distributions of variables and relationships between them change over time and can negatively impact the model built from the data [33]. To update the corresponding models of the DT Monitoring, methods like Transfer Learning [34], Ensemble Learning [35] are adopted. For the 1st case, the corresponding models are retrained when few data are available, and errors are random and small. For the 2nd case, an additional model is trained and combined with the existing one when the available data are more, and the error is systematic and prominent.

Conclusions & future work

The existing research in profile extrusion has focused on offline process optimization and has largely ignored the integration of real-time monitoring, addressing what-if and root-cause analysis scenarios. The proposed DT framework introduces 3 key functions for bridging this gap. *Prediction*: includes models to forecast KPIs like eccentricity and microstructure. They offer insights into stress, strain, temperature distribution, and material flow fields during the process considering the status of the involved tools and equipment. *Monitoring:* integrates real-time monitoring by fusing ROMs and ANNs. KPIs such as eccentricity and grain-size distribution are monitored in real-time enabling process & production control. *Diagnosis*: includes tools to diagnose unpredicted failures and to update the Prediction & Monitoring models based on real-time data, improving the overall system's robustness. Future work concerns the development of the DT’s physics-based and data-driven models. The first does not require experimental data, but only a few points for validation while the second one requires a vast amount of experimental data captured by sensors monitoring the process. As a next step, the development of the Eccentricity prediction model, along with the identification of the available SCADA/ERP that can provide real-time data are rendering the next milestones.

Acknowledgments

This research was partially supported by: the Industrial PhDs-UPatras IQ programme, through the project Implementation of a Hierarchical Digital Twin for extrusion process at industrial scale & the European Union's Horizon Europe research and innovation program under grant agreement No 101091869 (EU Project R3GROUP)

References

[1] G. Chryssolouris, Manufacturing systems: theory and practice, second ed., Springer New York, New York, 2006.

[2] M. Bauser, G. Sauer, K. Siegert, Extrusion, second ed., ASM International, Ohio, 2006. https://doi.org/10.31399/asm.tb.ex2.9781627083423

[3] V. N. Danilin et al., Simulation of taper heating and variable pressing rate to improve extrusion performance for high-strength aluminum alloys, Model. Simul. Mat. Sci. Eng. 32 (2024) 065006. https://doi.org/10.1088/1361-651X/ad56a6

[4] S. Barella et al., (2021). Predictive tools for in-line isothermal extrusion of 6xxx aluminum alloys, Mater. Proc. 3 (2021) 24. https://doi.org/10.3390/IEC2M-09239

[5] M. Negozio et al., FEM validation of front end and back end defects evolution in AA6063 and AA6082 aluminum alloys profiles, Procedia Manuf. 47(2020)202-208. https://doi.org/10.1016/j.promfg.2020.04.178

[6] S. Lechner, R. Nitschke, S. Müller, Numerical analysis of plastic die deformation during high temperature copper extrusion, Paper presented at 24th International Conference on Material Forming (2021). https://doi.org/10.25518/esaform21.4785

[7] A. F. M. Arif, A. K. Sheikh, S. Z. Qamar,(2003) A study of die failure mechanisms in aluminum extrusion, J. Mater. Process. Technol. 134 (2003) 318-328. https://doi.org/10.1016/S0924-0136(02)01116-0

[8] E. Giarmas, D. Tzetzis, Optimization of die design for extrusion of 6xxx series aluminum alloys through finite element analysis: a critical review, J. Adv. Manuf. Technol. 119 (2022) 5529-5551. https://doi.org/10.1007/s00170-022-08694-3

[9] L. De Pari Jr., W. Z. Misiolek, Numerical modeling of copper tube extrusion: Process and eccentricity analysis, J. Manuf. Sci. Eng. 134 (2012) 051005. https://doi.org/10.1115/1.4007283

[10] M. Schikorra et al., Microstructure analysis of aluminum extrusion: Prediction of microstructure on AA6060 alloy, J. Mater. Process. Technol. 201 (2008) 156-162. https://doi.org/10.1016/j.jmatprotec.2007.11.160

[11] E. C. Sariyarlioglu et al., Charge weld evolution in hollow aluminum extrusion: Experiments and modeling, CIRP-JMST 49 (2024) 14-27. https://doi.org/10.1016/j.cirpj.2023.12.007

[12] C. Lucignano et al., Artificial neural networks to optimize the extrusion of an aluminium alloy, J. Intell. Manuf. 21 (2010) 569-574. https://doi.org/10.1007/s10845-009-0239-0

[13] A. K. Abdul Jawwad, M. A. Barghash, Evaluating the effects of process parameters on maximum extrusion pressure using a new artificial neural network-based (ANN-based) partial-modeling technique, J. Adv. Manuf. Technol. 68 (2013) 2547-2564. https://doi.org/10.1007/s00170-013-4852-x

[14] M. Negozio et al., Predicting grain size in extruded AA6063 profiles: A unified approach based on finite element analysis and machine learning. J. Adv. Manuf. Technol. 133 (2024) 4543-4560. https://doi.org/10.1007/s00170-024-14021-9

[15] C. Guo et al., Optimization strategy for the indirect isothermal extrusion of high-strength aluminum alloy based on temporal fusion transformer, J. Adv. Manuf. Technol. 126 (2023), 5179-5189. https://doi.org/10.1007/s00170-023-11370-9

[16] T. Giesko, P. Garbacz, Application of hybrid vision method for the hot aluminium surface inspection, SSP 199 (2013) 267-272. https://doi.org/10.4028/www.scientific.net/SSP.199.267

[17] F. M. Neuhauser, G. Bachmann, P. Hora, Surface defect classification and detection on extruded aluminum profiles using convolutional neural networks, Int. J. Mater. Form. 13 (2020) 591-603. https://doi.org/10.1007/s12289-019-01496-1

[18] P. Stavropoulos et al., Metamodelling of manufacturing processes and automation workflows towards designing and operating digital twins, Appl. Sci. 13 (2023) 1945. https://doi.org/10.3390/app13031945

[19] L. Maier et al., Simulation-based data reduction and data processing for sheet metal forming in the hybrid twin framework, J. Adv. Manuf. Technol 134 (2024) 261-269. https://doi.org/10.1007/s00170-024-14135-0

[20] L. Gan, L. Li, H. Huang, Digital twin-driven sheet metal forming: modeling and application for stamping considering mold wear, J. Manuf. Sci. Eng. 144 (2023) 121003. https://doi.org/10.1115/1.4054902

[21] R. Kurth et al., Towards Cognitive Forming Machines: Utilization of Digital Twin-Based Virtual Sensors, Eng. Proc. 26 (2022) 10. https://doi.org/10.3390/engproc2022026010

[22] P. Link et al., Real-time quality prediction and local adjustment of friction with digital twin in sheet metal forming, Robot. Comput. Integr. Manuf. 91 (2025)102848. https://doi.org/10.1016/j.rcim.2024.102848

[23] F. Neubürger et al., Coupled finite-element-method-simulations for real-time-process monitoring in metal forming digital-twins, IEEE (ICCMA) 10 (2022) 260-265. https://doi.org/10.1109/ICCMA56665.2022.10011608

[24] J. Li et al., Task incremental learning-driven Digital-Twin predictive modeling for customized metal forming product manufacturing process, Robot. Comput.-Integr. Manuf. 85 (2024) 102647. https://doi.org/10.1016/j.rcim.2023.102647

[25] J. Li et al, Digital-Twin virtual model real-time construction via spatio-temporal cascade reconstruction for full-field plastic deformation monitoring in metal tube bending manufacturing, Robot. Comput.-Integr. Manuf. 91 (2025) 102860. https://doi.org/10.1016/j.rcim.2024.102860

[26] P. Stavropoulos K. Sabatakakis, Quality assurance in resistance spot welding: state of practice, state of the art, and prospects, Metals 14 (2024) 185. https://doi.org/10.3390/met14020185

[27] P. Stavropoulos et al., Robust additive manufacturing performance through a control oriented digital twin, Metals 11 (2021) 708. https://doi.org/10.3390/met11050708

[28] M. Glatt et al., Modeling and implementation of a digital twin of material flows based on physics simulation, J. Manuf. Syst. 58 (2021) 231-245. https://doi.org/10.1016/j.jmsy.2020.04.015

[29] C. Sommitsch et al., Modelling of creep-fatigue in containers during aluminium and copper extrusion, Comput. Mater. Sci. 39 (2007) 55-64. https://doi.org/10.1016/j.commatsci.2006.03.024

[30] C. Papaioannou et al., A Regression-Based Method for Reduced Order Modelling of Laser Welding Process. In Proceedings of the Changeable, Agile, Reconfigurable and Virtual Production Conference and the World Mass Customization & Personalization Conference (2023) 598-607. https://doi.org/10.1007/978-3-031-34821-1_65

[31] P. Esmati et al., Simulating chemical mixing and molten pool shape in dissimilar welds using thermal fluid dynamics, Int. J. Heat Mass Transf. 226 (2024) 125449. https://doi.org/10.1016/j.ijheatmasstransfer.2024.125449

[32] John D. Kechagias et al., A generalised approach on kerf geometry prediction during CO2 laser cut of PMMA thin plates using neural networks, Lasers Manuf. Mater. Process 8 (2021) 372-393. https://doi.org/10.1007/s40516-021-00152-4

[33] J. Gama et. al., A survey on concept drift adaptation, ACM Comput. Surv. 46 (2014) 1-37. https://doi.org/10.1145/2523813

[34] A. Papacharalampopoulos et al., Deep quality assessment of a solar reflector based on synthetic data: detecting surficial defects from manufacturing and use phase, Sensors 20 (2020), 5481. https://doi.org/10.3390/s20195481

[35] P. Stavropoulos, A. Papacharalampopoulos, K. Sabatakakis, Online Quality Inspection Approach for Submerged Arc Welding (SAW) by Utilizing IR-RGB Multimodal Monitoring and Deep Learning, In International Conference on Flexible Automation and Intelligent Manufacturing (2022) 160-16. https://doi.org/10.1007/978-3-031-18326-3_16

Materials Research Proceedings 46 (2024) 219-226 https://doi.org/10.21741/9781644903377-29

Experimental investigation of direct co-extrusion for bimetallic tubes

Konstantina D. KARANTZA[1,a*], Sotirios GEORGIOPOULOS[1,b], Protesilaos K. KOSTAZOS[1,c] and Dimitrios E. MANOLAKOS[1,d]

[1]Laboratory of Manufacturing Technology, School of Mechanical Engineering, National Technical University of Athens, Heroon Polytechniou 9, Athens 15780, Greece

[a] konstantinakarantza@mail.ntua.gr, [b] sotiris@mail.ntua.gr, [c] swing@central.ntua.gr, [d] manolako@central.ntua.gr

Keywords: Direct Extrusion, Bimetallic Tubes

Abstract. The current study aims to investigate the parallel direct extrusion of bimetallic tubes with a variable thickness ratio of the tubular layers. The bimetallic specimens consist of an aluminum tube as the inner layer and a copper one as the outer layer. Initially, quasi-static compressive tests are performed for the two utilized materials to estimate the fundamental material compressive properties. Following, the extrusion experiments are conducted and the load-displacement curves and the morphology of the final bitubes are examined. Also, the sliding of each tubular layer is evaluated by measuring their length. Finally, greater thickness in the outer pure-Cu tube reveals higher peak force and later material exit from the die, while the extrusion load depends proportionally on the total thickness. The aluminum inner tubes showed greater elongation than the outer copper ones, while reduced sliding was observed in the case of increased extrusion load due to the implementation of co-extruded bonding between the two layers.

Introduction

The current industrial and research interest is focused on developing and using materials that provide advanced properties. Bimetallic materials constitute a fundamental application of material combination, aiming to achieve the desirable properties and material characteristics of both metals. The bimetallics concentrate a wide range of benefits such as high specific strength, improved crashworthiness characteristics [1,2], reduced structural weight, high corrosion resistance, increased structural rigidity at high temperatures and capability of continuous production. For this reason, their industrial applications cover a wide range from crashworthy energy absorbers, heat exchangers in power plants and tubular heat devices like evaporators and condensers in high-corrosive environments to automotive cooling and exhaust systems, industrial pipelines, hydraulic pumps and nuclear applications. In fact, their continuous production is enhanced with the extrusion process, where the high pressure and friction levels provide stresses that create a strong joint in the bimetallic interface, producing, in that way, the desirable bimetallic structure.

Many studies have been conducted in the field of bimaterial tubes co-extrusion, intending to contribute both on process evolution and product optimization. Alcaraz and Sevillano [3] proceeded to an analytical study to connect the bimetallic extrusion parameters with failure conditions and determine the respective safety maps, utilizing the upper bound method and minimum energy principle. Also, Kazanowski et. al [4] focused on the bimaterial rod extrusion regarding the effect of the initial billet geometry on the final product geometry, aiming to optimize the process, using both experimental and numerical methods. Vamsi Krishna et al. [5] studied the fabrication of bimetallic tubes, with initial material properties determined by powder metallurgy and cold extrusion, with the purpose of determining a density ratio for safe extrusion on every material combination. In addition, an experimental and theoretical investigation into the influence of extrusion ratio and die geometry in the extrusion process was conducted by Berski et al. [6]. Besides, Knezevic et al. [7] studied different die designs for bimetallic tube extrusion, using finite

Materials Research Proceedings 46 (2024) 219-226 https://doi.org/10.21741/9781644903377-29

element analysis and crystal plasticity modeling in order to decrease the peak load during the process and optimize the performance. A review of cold extrusion on bulk materials regarding their mechanical behavior, the microstructure and some fundamental process parameters was conducted by Navaneetha and Lakshmi [8]. Moreover, Jang and Hwang [9] examined the contact effect of the layers interface on the bonding for bimetallic co-extrusion using the finite element method, especially regarding the dependence on material arrangements, extrusion ratios and bonding conditions. Sajuri et al. [10] performed a material characterization via metallography and tensile, bending and fatigue tests of bimetallic tubes focusing on their interlayer.

Furthermore, the influence of extrusion ratio and die angle on hot hydrostatic extrusion was analyzed experimentally in bimetallic tubes by Rhee et al. [11], with particular interest in the diffusion layer of metals interface, utilizing microscopic analyses. Tian et al. [12] conducted finite element analysis on the tube extrusion shearing process for fabricating bimetallic tubes at different temperatures, accompanied by microstructure characterization and microscopic observation of their bonding layers. Additionally, Kordeyazdi et al. [13] performed simulations on bimetallic tube extrusion process for different angles and internal profiles of the mold and also proceeded to validation with respective experimental data. The impact of tube shear extrusion and conventional extrusion processes on the mechanical properties, microstructure and bonding layer of bimetallic tubes were examined by Zhao et al. [14], applying several preheating temperatures on the metals and combining experimental tests and simulations. Also, Alaie et al. [15] investigated experimentally the parallel extrusion of bimetallic tubes through angular channels while also they conducted microscopic observations and microhardness measurements on the interlayer bonding. Finally, the co-extrusion process of bimetallic tubes and the interface of the layers were studied physically and numerically by Misiolek and Sikka [16], aiming to define a new billet design.

The current experimental study includes quasi-static compressive tests on Al-6063 F25 and pure Cu for estimating the fundamental material compressive properties and extrusion experiments for recording the load-displacement curves and observing the morphology of the extruded bitubes, as well as evaluating the sliding of each tubular layer. Besides, different thickness ratios of the inner and outer tubes are examined for concluding their effect on the extrusion load curve.

Materials and Methods

Experimental Setup. The extrusion experimental tests are conducted in an SMG pressing machine of 100tn nominal load capacity. Fig. 1 illustrates the testing configuration, which consists of a conical die, an outer die housing and a pressing punch powered by a hydraulic system.

(a) (b) (c)

Fig. 1. Experimental setup:(a) die configuration, (b) die housing, (c) pressing machine.

During the extrusion process, the two metallic tubes are pushed through the conical internal configuration of the die, providing a bimetallic extruded tube. In fact, the tool design allows for a parallel extrusion process, during which the two tubular layers simultaneously enter the conical die section, as Fig. 2 depicts. The conical section of the die is designed under 9.26 mm height, 20 mm internal diameter, 25 mm external diameter and 15° semi-angle, while the cylindrical punch is 25 mm and 80 mm in diameter and height, respectively. Finally, all extrusion tests are carried out at a velocity of 10 mm/min.

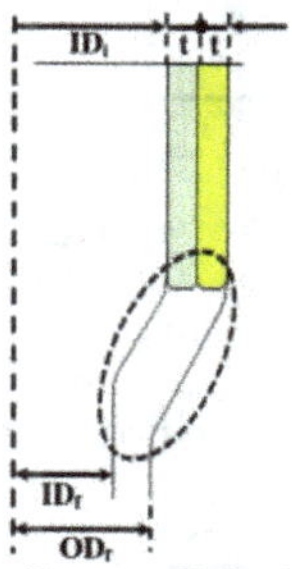

Fig. 2. Design for performing bimetallic parallel tube extrusion (Reprinted from [7] with permission from Elsevier).

Specimen Description. The initial specimen consists of two tubular layers, as shown in Fig. 3. Specifically, the external layer (sleeve metal) is pure Cu, while the inner one (core metal) is Al-6063 F25. The initial specimens were fabricated through pressing to achieve an interference fit between the two layers.

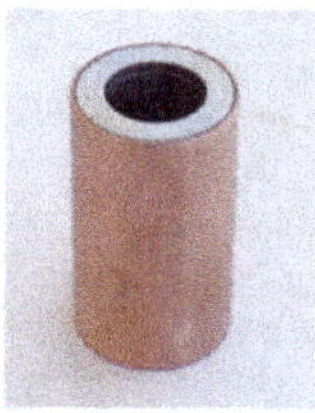

Fig. 3. Initial specimen of Al-Cu bimaterial tube.

Table 1 contains the dimensions of the initial specimens. All specimens are 40 mm long, with a 25 mm outer diameter, while the thickness ratio of the two tubes varies, providing four different internal diameter values of 15-17-19-21 mm.

Table 1. Initial dimensions of specimens.

Number of specimen	Inner tube (Al-6063 F25)			Outer tube (pure Cu)			Height [mm]
	Thickness [mm]	Inner diameter [mm]	Outer diameter [mm]	Thickness [mm]	Inner diameter [mm]	Outer diameter [mm]	
1	4	15	23	1	23	25	40
2	3	15	21	2	21	25	40
3	2.5	15	20	2.5	20	25	40
4	2	15	19	3	19	25	40
5	1	15	17	4	17	25	40
6	3	17	23	1	23	25	40
7	2	17	21	2	21	25	40

Materials Research Proceedings 46 (2024) 219-226 https://doi.org/10.21741/9781644903377-29

8	1	17	19	3	19	25	40
9	2	19	23	1	23	25	40
10	1.5	19	22	1.5	22	25	40
11	1	19	21	2	21	25	40
12	1	21	23	1	23	25	40

Material Characterization. To obtain the mechanical behavior of the two examined materials, two experimental compressive tests were conducted under quasi-static conditions for each material. In fact, different sample dimensions were examined in both cases. The two tests aim to minimize the uncertainty and deviations during predicting the mechanical material behavior, facing any experimental, measuring or recording errors. Fig. 4 depicts the compressive stress-strain diagram, while Table 2 lists the mechanical properties of each material.

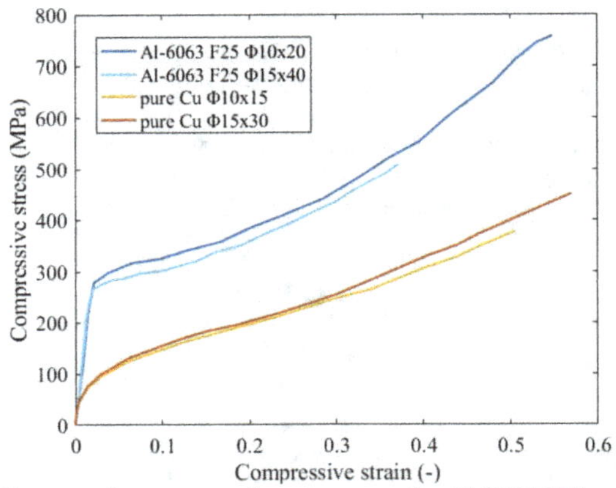

Fig. 4. Compressive stress and strain curves for Al-6063 F25 and pure Cu.

Table 2. Mechanical material properties for compression.

	Density [gr/cm^3]	Young modulus [GPa]	Poisson ratio [–]	Yield stress [MPa]
Al-6063 F25	2.7	70	0.34	275
pure Cu	8.913	119	0.35	70

Results and Discussion

Fig. 5 illustrates the experimental force-displacement curves during the extrusion tests for each specimen. In all cases, the force increases stably during the initial stage at early deformation until the material flow starts to exit from the die, reaching its local peak value at the end of the first stage, which coincides with the kink of the force-displacement curves. Following, the force fluctuation seems to be stabilized, or even slightly decrease, as the plastic flow of the material during exiting the die is facilitated. As Table 3 describes, the process load increases with the total thickness increase, while greater thickness in outer pure-Cu reacts to a higher load at the end of the first stage, which is exhibited later due to the higher necessary load.

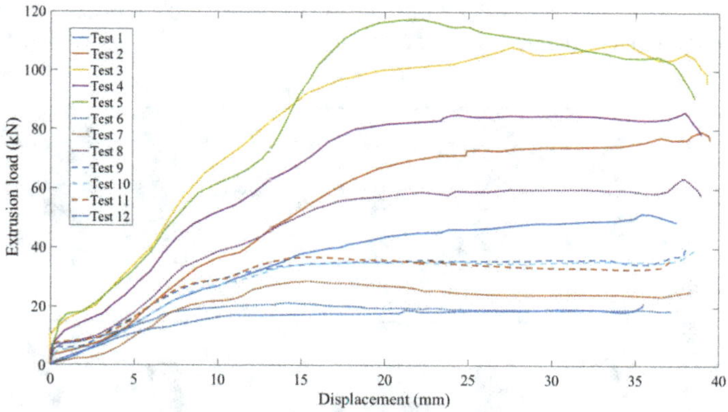

Fig. 5. Extrusion load and displacement for each test (different line-style for each total thickness of specimens, different color for each outer diameter of the inner tube/inner diameter of the outer tube).

Table 3. Load characteristics of extrusion tests.

Test	Load at the end of 1st stage [kN]	Punch stroke at the end of 1st stage [mm]	Maximum load [kN]	Total punch stroke [mm]
1	38.15	14.84	51.50	38.00
2	65.88	19.16	79.30	39.70
3	99.29	19.00	108.70	40.10
4	80.33	18.32	85.70	39.25
5	116.82	19.84	116.80	38.80
6	18.28	7.50	21.70	37.50
7	26.25	12.28	28.30	38.30
8	56.73	17.82	63.80	38.90
9	33.28	13.57	41.26	38.30
10	33.75	14.16	39.60	38.80
11	33.63	14.67	36.56	37.70
12	17.00	10.61	21.52	35.67

Fig. 6 depicts the final views of the extruded bimetallic tubes, showing that the main part of the specimens has been elongated due to the cross-sectional area decrease, while the upper part has received a conical shape from the die geometry. In more detail, the upper part has been simply compressed rather than been extruded, revealing a geometrical defect in the final product. Table 4 summarizes the height increase of the two tubular layers, showing that the Al-6063 F25 inner layer elongates more than the pure-Cu outer one, which is additionally restricted by the friction with the die except from the one in the interface. Fig. 7 illustrates the variance in the height increase, revealing that the elongation of the aluminum layer increases at a higher total thickness due to the increased load and the higher cross-sectional area decrease, while the outer pure-Cu layer reveals a lower length change. Finally, sliding between the inner and the outer layers is measured, which is attributed to the fact that the two layers manifest different longitudinal strains. In fact, lower

Materials Research Proceedings 46 (2024) 219-226 https://doi.org/10.21741/9781644903377-29

sliding is observed at higher extrusion loads due to the increased die pressure, which reacts to higher shear stress in the interface, creating a co-extruded bond resulting in a sliding decrease.

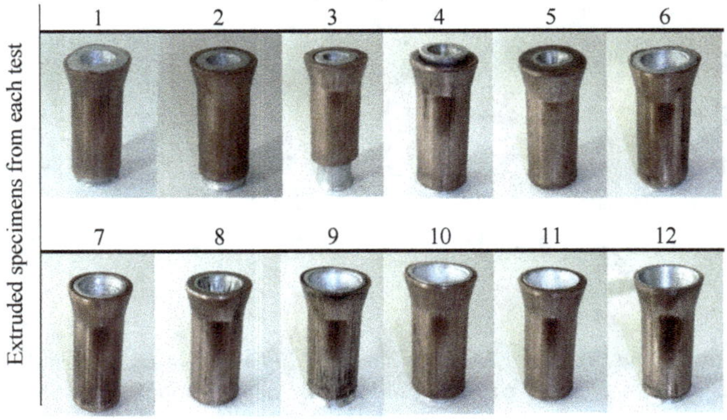

Fig. 6. Final views of extruded bitubes.

Table 4. Final dimensions of specimens

Test	Final height of Al [mm]	Height increase of Al [%]	Final height of Cu [mm]	Height increase of Cu [%]	Sliding [mm]
1	45.8	+14.5	41.6	+4.0	4.2
2	46.1	+15.3	40.9	+2.3	5.2
3	51.3	+28.3	40.5	+1.3	10.8
4	48.7	+21.8	43.9	+9.8	4.8
5	42.9	+7.3	42.3	+5.8	0.6
6	46.4	+16.0	42.5	+6.3	3.9
7	45.3	+13.3	43.5	+8.8	1.8
8	43.0	+7.5	42.1	+5.3	0.9
9	44.5	+11.3	40.4	+1.0	4.1
10	44.2	+10.5	41.8	+4.5	2.4
11	44.0	+10.0	42.5	+6.3	1.5
12	43.4	+8.5	41.0	+2.5	2.4

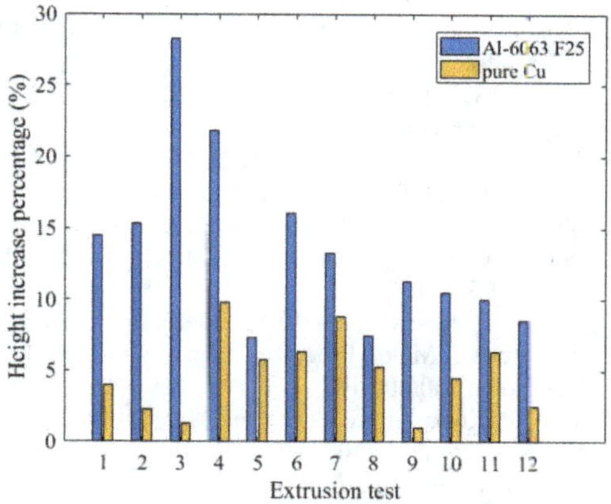

Fig. 7. Height increase percentage for each extrusion specimen.

Conclusions

This experimental study investigated the parallel tube direct extrusion of bimetallic tubes with different thickness ratios of the two tubular layers. The fundamental mechanical material properties for compression were estimated by conducting quasi-static axial compression tests for each material. Through the quantification of the extrusion force-displacement curves, the load and punch stroke were detected at the end of the first stage and the maximum load and total punch displacement of each experiment. Thereafter, the morphology of the extruded bitubes was investigated, measuring the final height of each layer and comparing the height increase percentage of them, as well as the siding between the inner and outer tube of each specimen.

The results revealed that the extrusion force was increasing stably during the initial stage until it reached its local peak value at the end of the first stage, where the material flow started to exit from the die. Subsequently, as the plastic flow of the material was exiting the die, the force fluctuation seemed to be stabilized or even slightly decreased. The process load showed a proportional behavior with the total thickness of specimens, while greater thickness in outer pure-Cu reacted to a higher load at the end of the first stage. Moreover, the main part of the specimens was elongated due to the cross-sectional area decrease. The Al-6063 F25 inner layer elongated more than the pure-Cu outer one, because the last was further restricted by the friction with the die, except from the friction in the interface. At higher total thickness, the elongation of the aluminum layer increased, while the copper layer revealed a lower length change. Finally, the sliding between the inner and the outer layers was ascribed to their different longitudinal strains caused by the different flow stresses, while higher extrusion loads in case of thicker outer Cu tube led to higher shear stress in the interface, creating a co-extruded bond and resulting in a sliding decrease. Therefore, the optimum thickness ratio between the two layers is detected when designing with minimum inner Al-6063 F25 thickness and maximum outer pure-Cu thickness, providing the minimum sliding between the two layers considering the test cases 5-8-11.

References

[1] K.D. Karantza, I.G. Papantoniou, P.K. Kostazos, D.E. Manolakos, Crashworthiness behavior of thin-walled bimaterial tubes under axial collapse, Mater. Today Proc. 93 (2023) 575–582. https://doi.org/10.1016/j.matpr.2023.02 178

[2] K.D. Karantza, D.E. Manolakos, Crashworthiness behavior of thin-walled bimaterial tubes under axial collapse, 38th Danubia-Adria Symposium on Advances in Experimental Mechanics, DAS 2022, ISBN 978-618862780-2.
[3] J.L. Alcaraz, J. Gil-Sevillano, Safety maps in bimetallic extrusions, J. Mater. Process. Technol. 60 (1996) 133–40. https://doi.org/10.1016/0924-0136(96)02318-7
[4] P. Kazanowski, M.E. Epler, W.Z. Misiolek, Bi-metal rod extrusion—process and product optimization, Mater. Sci. Eng. A 369 (2004) 170–80. https://doi.org/10.1016/j.msea.2003.11.002
[5] B. Vamsi Krishna, P. Venugopal, K. Prasad Rao, Co-extrusion of dissimilar sintered P/M preforms—An explored route to produce bimetallic tubes. Mater. Sci. Eng. A 407 (2005) 77–83. https://doi.org/10.1016/j.msea.2005.06.025
[6] S. Berski, H. Dyja, A. Maranda, J. Nowaczewski, G. Banaszek, Analysis of quality of bimetallic rod after extrusion process, Mater. Process. Technol. 177 (2006) 582–586. https://doi.org/10.1016/j.jmatprotec.2006.04.107
[7] M. Knezevic, M. Jahedi, Y.P. Korkolis, I.J. Beyerlein, Material-based design of the extrusion of bimetallic tubes, Comput. Mater. Sci. 95 (2014) 63–73. https://doi.org/10.1016/j.commatsci.2014.07.021
[8] E. Navaneetha, A.A. Lakshmi, Cold extrusion on bulk materials: A review, Mater. Today Proc. (2023). https://doi.org/10.1016/j.matpr.2023.09.168
[9] D.H. Jang, B.B. Hwang, Deformation Analysis of Co-Extrusion Process of Aluminum Alloy and Copper Alloy, KEM 340-341 (2007) 645–648. https://doi.org/10.4028/www.scientific.net/kem.340-341.645
[10] Z. Sajuri, A. Baghdadi, M.F. Mahmod, J. Syarif, Fatigue and Mechanical Properties of Aluminium-Copper Bi-Metal Tubes, Adv. Mater. Res. 896 (2014) 626–629. https://doi.org/10.4028/www.scientific.net/amr.896.626
[11] K.Y. Rhee, W.Y. Han, H.J. Park, S.S. Kim, Fabrication of aluminum/copper clad composite using hot hydrostatic extrusion process and its material characteristics, Mater. Sci. Eng. A 384 (2004) 70–76. https://doi.org/10.1016/j.msea.2004.05.051
[12] Y. Tian, H. Hu, D. Zhang, A novel severe plastic deformation method for manufacturing Al/Mg bimetallic tube, J. Adv. Manuf. Technol. 116 (2021) 2569–2575. https://doi.org/10.1007/s00170-021-07513-5
[13] M. Kordeyazdi, A.B. Kheradmand, Z. Lalegani, A. Haghani, Investigating Aluminum and Copper Role in Simulating the Extrusion Process of Bimetal Tubes by Finite Element Method, Trans. Indian. Inst. Met. 74 (2021) 1179–1192. https://doi.org/10.1007/s12666-021-02286-7
[14] J. Zhao, H. Zhang, W. Zhang, H. Zhao, Y. Li, X. Qin et al., Influence of Extrusion Method on Formation of Magnesium-Aluminum Bimetallic Composite Tube, J. Mater. Eng. Perform. 32 (2023) 7134–7148. https://doi.org/10.1007/s11665-022-07635-1
[15] M.A. Alaie, M. Kasaeian-Naeini, R. Hashemi, M. Rajabi, A. Hosseini, Fabrication of AA1050/CP-Cu Bimetallic Tubes using Parallel Tube-Shaped Channel Angular Pressing Technique and Assessing its Mechanical Properties and Microstructure, Trans. Indian. Inst. Met. 76 (2023) 1937–1948. https://doi.org/10.1007/s12666-023-02897-2
[16] W.Z. Misiolek, V.K. Sikka, Physical and Numerical Analysis of Extrusion Process for Production of Bimetallic Tubes (2006). https://www.osti.gov/servlets/purl/889030

Innovative Manufacturing Engineering and Energy - IManEE2024 Materials Research Forum LLC
Materials Research Proceedings 46 (2024) 227-234 https://doi.org/10.21741/9781644903377-30

A computational study on the adiabatic shear banding development during plane-strain uniaxial and angled-biaxial dynamic compression of AISI 1045 steel

Konstantina D. KARANTZA[1,a*] and Dimitrios E. MANOLAKOS[1,b]

[1]Laboratory of Manufacturing Technology, School of Mechanical Engineering, National Technical University of Athens, Heroon Polytechniou 9, Athens 15780, Greece

[a]konstantinakarantza@mail.ntua.gr, [b]manolako@central.ntua.gr

Keywords: Adiabatic Shear Bands, Shearing Instability, Plane-Strain Compression, Angled-Biaxial Compression, Structural-Thermal-Damage Coupling, LS-DYNA

Abstract. This work studies the adiabatic shear banding (ASB) development during dynamic plane strain compression of AISI 1045 steel under uniaxial and biaxial loading. In the case of biaxial loading, the examined steel sample is of 5° inclination in order to provide a shear-compression stress state due to its angled loading state. The numerical analysis is conducted through structural-thermal-damage doubly coupled finite element models in LS-DYNA by implementing the thermo-viscoplastic Modified Johnson-Cook plasticity flow rule and damage law. The simulations revealed that angled compression increases the propensity to ASB formation due to the introduction of a shear-compression stress state, which facilitates the shear strain localization, leading to lower critical strain for ASB initiation. Also, ASBs initially form under X-shape while following their trace forms S-shaped ASBs at higher deformation levels, while in case of biaxial compression the ASBs can be clearly distinguished to a stronger and a weaker one according to the magnitude of their interior strain and temperature. Finally, thermal softening is observed to occur prior to damage softening, determining in this way the ASB genesis, in contrast to the damage softening, which dominates its later evolution and its progression to fracture.

Introduction

Plastic strain evolution at low and medium strain rates is dominated by stable deformation mechanisms such as slipping and twinning, in contrast to high strain-rate plastic deformation, which is often driven by unstable mechanisms such as adiabatic shear banding (ASB). More specifically, ASB is initially manifested through intense shear strain localization along narrow bands, providing significant plastic work within the bands. That plastic work is converted to internal heat by the Taylor-Quinney coefficient, which often lies about 0.9 [1]. Thus, the internal heat is trapped in the ASB core owing to the high strain rate that represents adiabatic conditions without allowing enough time for heat diffusion. Therefore, the adiabatic heat leads to a remarkable temperature increase within ASB, enhancing the thermal softening effect. In fact, ASB macroscopically coincides with the strain instability point which comes when the thermal softening magnitude overcomes the one of strain hardening [2].

ASB development during dynamic plane strain compression has concentrated a significant amount of research interest. Anand and Spitzig [3] studied the plane strain compression of aged maraging steel, revealing an ASB orientation of 55°, while further, a theoretical expression for critical strain prediction was derived for the ASB genesis. Also, Prakash et al. [4] investigated the formation of macro-shear bands (MSBs) in dynamically compressed aluminum alloys at room temperature and at 573 K. MSBs were captured in case of Al-6% wt. Mg alloy at 573 K and 20% compressive strain, while on the contrary, lower aluminum series did not show any MSB formation. Moreover, Tang et al. [5] studied Inconel 718 nickel superalloy subjected to plane strain compression. The results showed that X-shaped ASBs formed at mild deformation conditions of lower strain rate and higher temperature, while in contrast S-shaped ASBs were developed at higher strain rates and lower temperature. In addition, Meyer et al. [6] conducted compression/shear tests in cylindrical samples with 6° inclination, showing that the presence of inclination promotes shear localization and ASB appearance by decreasing critical strain. Mateos et al. [7] studied AA2024-T351 against biaxial angled compression, concluding that the angled loading results in a multi-axial stress state, while further, the introduced shear stress increases the propensity to ASB genesis, which is promoted as the shear-to-compression ratio increases. In that way, a deep understanding of ASB mechanism has concentrated significant research interest regarding its appearance in manufacturing and metal forming processes, affecting the process progress and influencing the material deformation, process load, material failure and finally, product quality. Therefore, forging process [8], extrusion [9] and rolling [10] have been studied regarding bulk forming, while blanking [11,12] and deep drawing [13] have been investigated for sheet forming. Eventually, ASB phenomenon has also been analyzed during high-speed machining [14] and ballistic impact [15].

This work investigates numerically the ASB development during plane strain dynamic compression of AISI 1045 steel under uniaxial and biaxial loading which refers to angled compression under 5° inclination. Doubly structural-thermal-damage finite element (FE) models are developed for the numerical simulations in LS-DYNA, implementing the thermo-viscoplastic Modified Johnson-Cook (MJC) constitutive relation and damage criterion, while damage-equivalent stress/strain fields are implemented for the structural-damage coupling. This work focuses on assessing the ASB formation during plane strain compression, aiming to highlight ASB trajectories, while also main aim of the current study remains to confer individual roles to thermal and damage softening mechanisms. Finally, the influence of sample inclination on ASB initiation and evolution is examined through the simulation of biaxial angled compression.

Finite Element Modeling

This work studies numerically the ASB development during plane strain dynamic compression of AISI 1045 steel under uniaxial and biaxial loading. In the first case, a rectangular steel sample of 5 mm x 5mm is simulated under plane strain compression, while in the case of biaxial loading, a steel sample of 5 mm x 5mm with 5° inclination is studied under plane strain angled shear-compression as Fig. 1 depicts. The shear-compression comes from the introduced shear loading due to the sample inclination which transforms the original compressive loading to shear one due to the 5° angled sample geometry. Both simulations are conducted at a constant compressing velocity of 20 m/s, providing an initial strain rate of 4000 s^{-1}.

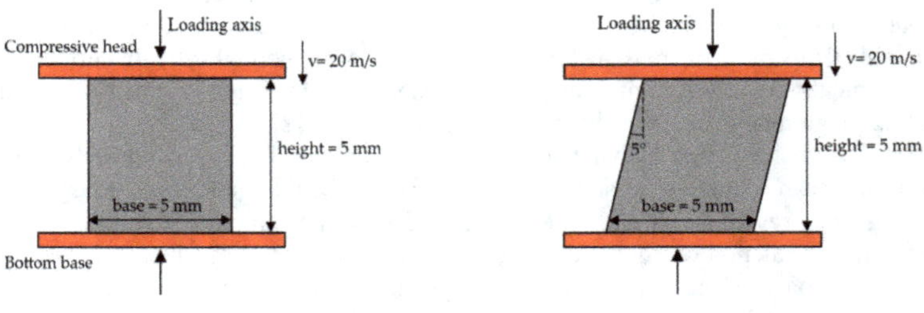

(a) (b)

Fig. 1. Plane strain compression under (a) uniaxial and (b) angled/biaxial loading.

This study implements a doubly coupled analysis by developing structural-thermal-damage coupled FE models in LS-DYNA. A Lagrangian FE mesh of 10 μm in size is implemented via 2D

Materials Research Proceedings 46 (2024) 227-234 https://doi.org/10.21741/9781644903377-30

4-node solid elements, while 2D R-adaptive remeshing technique is utilized in order to prevent computational instabilities due to extremely severe mesh distortion which is detected at ASB region. Regarding material modeling, the thermo-viscoplastic MJC constitutive relation is applied for modeling the material flow plasticity as Eq. 1 describes, while further, the thermo-viscoplastic MJC damage criterion is considered for evaluating the damage evolution according to Eq. 2. In more detail, σ is the plastic flow stress, T is the local material temperature, T_r is room temperature and T_m the melting point, while A, B, C, n and m are the MJC material parameters and $\dot{\varepsilon}_0$ the reference strain rate. Regarding MJC damage law, ε_f is the plastic strain failure, σ^* the stress triaxiality, $\dot{\bar{\varepsilon}}^p$ the equivalent plastic strain rate, and D_1, D_2, D_3, D_4 and D_5 are the MJC damage parameters. Moreover, the damage-equivalent strain r is introduced for the structural-damage coupling which is achieved by adjusting β-parameter equal to 1 for full damage coupling form as shown in Eq.3. Further, the damage parameter D is considered evaluating the damage evolution and providing the damage-equivalent strain field, while also it determines the fracture as Eq. 4 describes through the damage extent ΔD which reacts to failure and element erosion when it reaches the critical damage D_c which equals to 1. Also, another temperature-dependent damage criterion is applied by implementing the melting point as the critical temperature for failure, capturing so the local melting, which initiates failure via micro-voiding. Material failure is manifested via element erosion when one of the two damage criteria is satisfied.

$$\sigma = (A + B \cdot r^n) \cdot \left(1 + \frac{\dot{r}}{\dot{\varepsilon}_0}\right)^C \cdot \left(1 - \left\{\frac{T-T_r}{T_m-T_r}\right\}^m\right) \tag{1}$$

$$\varepsilon_f = (D_1 + D_2 \cdot \exp\,[D_3 \cdot \sigma^*]) \cdot \left(1 + \frac{\dot{\bar{\varepsilon}}^p}{\dot{\varepsilon}_0}\right)^{D_4} \cdot \left(1 + D_5 \frac{T-T_r}{T_m-T_r}\right) \tag{2}$$

$$r = \bar{\varepsilon}^p (1 - \beta \cdot D) \tag{3}$$

$$\Delta D = D_c \frac{\Delta\bar{\varepsilon}^p}{\varepsilon_f} \tag{4}$$

Therefore, the structural-damage coupling is achieved via the β-parameter, which provides a damage-equivalent field proportionally to the damage extent, which is, in turn, computed through the MJC damage law. Regarding the structural-thermal coupling, the computed plastic work is converted to internal adiabatic heat by the Taylor-Quinney coefficient χ reacting to a temperature rise, which affects the plastic flow stress through the thermal softening term. Thus, the structural-thermal coupling is achieved due to the implementation of a thermo-viscoplastic material model and the Taylor-Quinney coefficient χ, which calculates the temperature increase rate $\partial T/\partial t$ as Eq. 5 describes via the plastic work rate $\dot{W}_p$, the density ρ and the specific heat C_p.

$$\partial T/_{\partial t} = \chi \cdot \dot{W}_p/_{\rho \cdot C_p} \tag{5}$$

Eventually, the contact algorithms of 2D automatic-single-surface and 2D surface-to-surface are implemented for preventing penetration in the interfaces, with the first one applied for the self-penetration of the sample between the cracking edges which develop and come in contact, while the second one is applied for the interfaces in contact between the sample and the compressing plates. Finally, Table 1 lists the MJC material and damage parameters for 1045 steel according to literature data [16,17], which are utilized during modeling development.

Table 1. MJC material and damage parameters for 1045 steel.

Material Properties		MJC constitutive relation		MJC damage criterion	
ρ [kg/m^3]	7800	A [MPa]	553.1	D_1 [–]	0.06
E [GPa]	200	B [MPa]	600.8	D_2 [–]	3.31
v [–]	0.3	n [–]	0.234	D_3 [–]	-1.96
χ [–]	0.9	C [–]	0.0122	D_4 [–]	0.0018
C_p [J/kgK]	432.6	m [–]	1	D_5 [–]	0.58
α [μm/m°C]	11	$\dot{\varepsilon}_0$ [1/s]	1	D_c [–]	1
		T_m [K]	1733		
		T_r [K]	293		

Results

This work studies numerically the ASB development during dynamic plane strain compression of AISI 1045 steel under uniaxial and biaxial/angled loading in which the steel sample is of 5° inclination. Fig. 2 depicts the time variation of equivalent stress and effective plastic strain within the shear localization zone for the two simulations of the uniaxial and the biaxial compressions. Considering that the ASB onset coincides macroscopically with the instability point of peak stress, Fig. 2 demonstrates that biaxial compression reveals an earlier ASB initiation at 33 μs at compressive true strain of 12.4% in contrast to uniaxial compression, which shows a delayed ASB genesis at 40 μs and compressive true strain of 14.8%. Therefore, biaxial compression accelerates the ASB initiation by enhancing the shear stress state due to the sample inclination, facilitating so shear strain localization and promoting ASB formation, as experimental findings have also confirmed [6,7]. Further, the additional shear stress field due to the inclination reacts to lower peak stress. Fig. 2 also demonstrates the critical effective plastic strain values, which reveal the ASB genesis inside the shear localization zone, as the band extent with effective plastic strain greater than the critical strain reflects the ASB region. In that way, the uniaxial compression reveals a critical strain for ASB initiation of 1.265, while the biaxial compression shows a lower critical strain of 0.996, promoting more the ASB genesis.

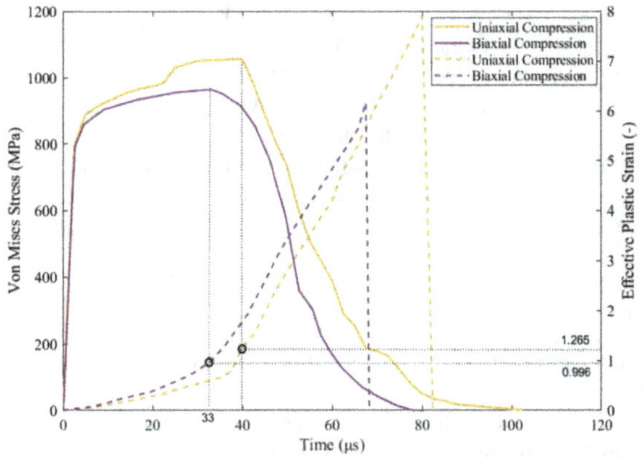

Fig. 2. Equivalent stress (solid lines) and effective plastic strain (dashed lines) time fluctuations for uniaxial and biaxial pane strain compressions.

Fig. 3 depicts the ASB trajectories at two different times, one after their genesis and one at their final evolution stage. The axes of the diagrams refer to the non-dimensional cartesian coordinates as the ratios with sample width B and height H, while the origin is located in the sample center. ASBs are initially developed as straight bands along the sample diagonals of maximum shear direction until the central kink which acts as their meeting point, forming X-shaped ASBs in both uniaxial and biaxial compressions. As deformation increases, the central kink tends to elongate horizontally, providing S-shaped ASBs, which is more intense in case of uniaxial compression. In fact, the above tendency has been further ensured by experimental results regarding the ASB developing trajectories during deformation progress [5].

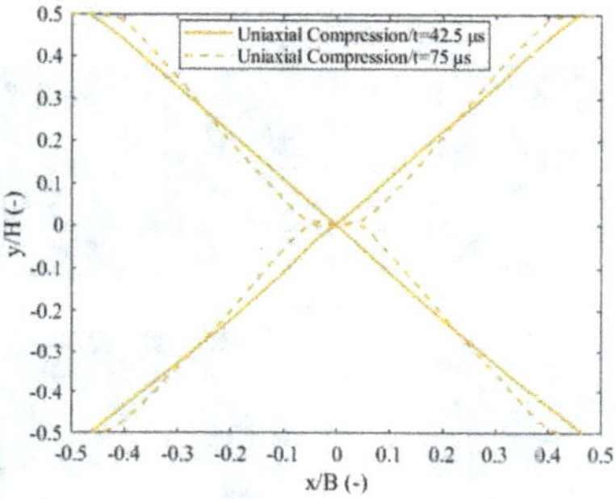

(a)

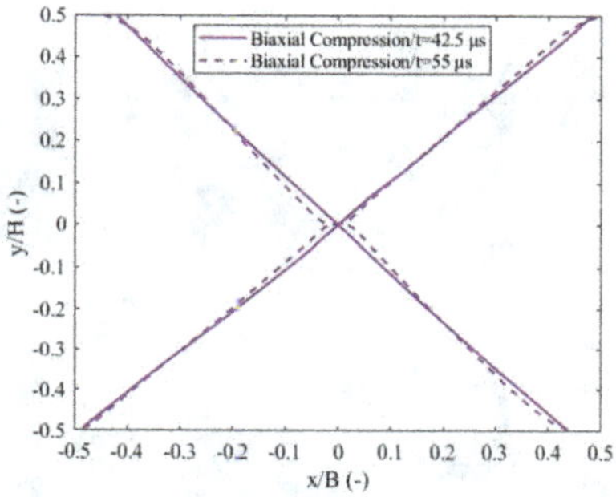

(b)

Fig. 3. ASB trajectories with time for (a) uniaxial and (b) angled/biaxial loading.

Fig. 4 visualizes the strain, temperature and damage fields during plane strain uniaxial and biaxial compressions at different times at the initial and at the final ASB evolution stages. In particular, Fig. 4a shows the two diagonal X-shaped ASBs, indicating their greater intensity around sample corners where higher localized strain, temperature and damage are obtained. On the contrary, ASBs tend to receive an S-shape at higher deformation due to the elongation of their central meeting point, as Fig. 4b shows, while also the strain and temperature magnitude have further increased and propagated, as well as the damage localization. Fig. 4c and 4d depict that ASBs initially form as X-shaped bands and then develop into S-shaped ones. However, the two ASBs can be clearly distinguished between a stronger and a weaker ASB from their stain magnitude and, in consequence, their interior temperature. Thus, the ASB propagating from the top left corner to the bottom right one seems stronger with higher strain and temperature magnitudes.

Materials Research Proceedings 46 (2024) 227-234 https://doi.org/10.21741/9781644903377-30

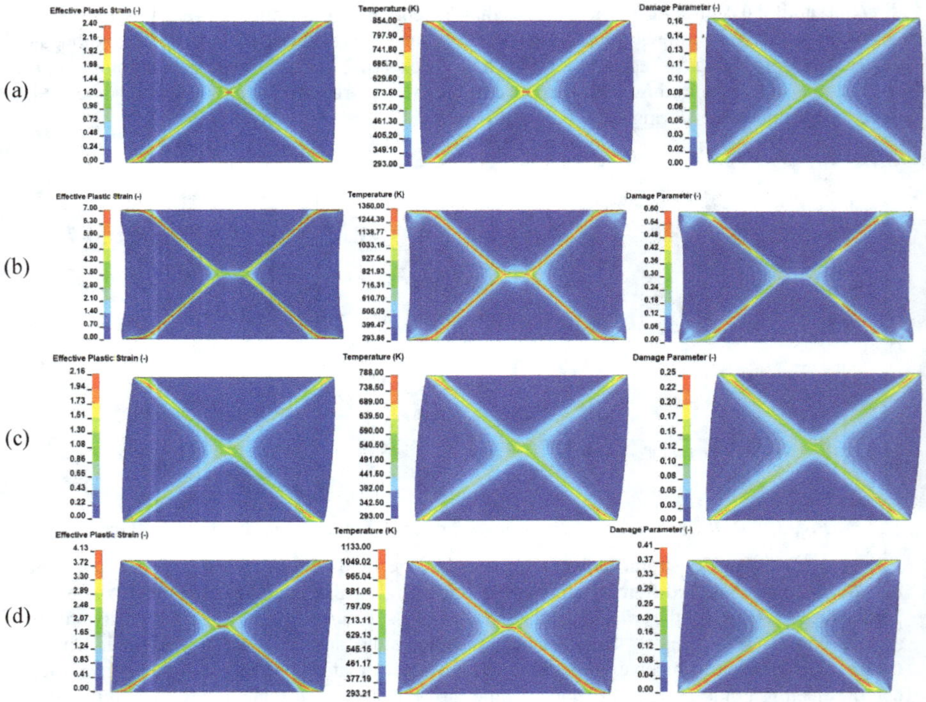

Fig. 4. Effective plastic strain, temperature and damage for: (a) uniaxial compression at 47.5 μs, (b) uniaxial compression at 75 μs, (c) biaxial compression at 42.5 μs and (d) biaxial compression at 55 μs.

Finally, Fig. 5 demonstrates the competition between thermal and damage softening mechanisms with time, as the damage softening magnitude depends proportionally on damage parameter D, while the homologous temperature affects the thermal softening magnitude proportionally due to that m-parameter equals 1 as Table 1 describes. In fact, the homologous temperature is specified as the non-dimensional ratio of the temperature differences in the thermal part of the MJC constitutive relation. Therefore, Fig. 5 shows that thermal softening precedes the damage softening mechanism by being superior at early stages. In that way, thermal softening is responsible for ASB genesis and its initial evolution stages, whereas damage softening influences the later evolution of ASB and its progression to fracture, which is manifested when the damage parameter reaches its critical value of 1. Eventually, angled biaxial compression seems to accelerate both thermal and damage softening mechanisms, with its effect on the earlier damage increment being even more intense.

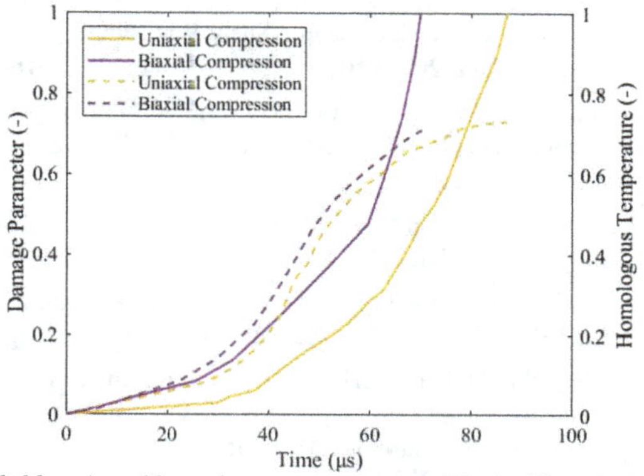

Fig. 5. Damage (solid lines) and homologous temperature (dashed lines) time fluctuations for uniaxial and biaxial pane strain compressions.

Conclusions

This work studied numerically the ASB development during dynamic plane strain compression of AISI 1045 steel under uniaxial and biaxial loading under 5° sample inclination. Structural-thermal-damage doubly coupled FE models were implemented in LS-DYNA, utilizing the thermo-viscoplastic MJC constitutive relation and damage criterion. The numerical simulations revealed that biaxial loading shows a greater propensity to ASB formation due to the shear-compression stress state in which the presence of additional shearing due to the sample inclination facilitates shear strain localization and, in consequence, the ASB genesis. Thus, the biaxial compression revealed earlier ASB onset and lower critical strain for ASB initiation. In addition, ASBs initially formed under an X-shape along the sample diagonals and the maximum shear direction, while at later deformation stages, they received an S-shape due to the elongation of their central kink. In fact, in case of biaxial compression, the two ASBs could be clearly distinguished into a stronger and a weaker ASB from the magnitude of their localized strain and temperature levels. Finally, thermal softening was detected to act as a precursor of damage softening, with the first one determining the ASB genesis and its earlier evolution, while the second one dominated the later propagation and its transition to fracture.

Acknowledgments

This research was funded by the Hellenic Foundation for Research and Innovation (HFRI) under the 4th Call for HFRI PhD Fellowships (Fellowship Number: 10838).

References

[1] D. Rittel, L. Zhang, S. Osovski, The Dependence of the Taylor–Quinney Coefficient on the Dynamic Loading Mode, J. Mech. Phys. Solids 107 (2017) 96–114. https://doi.org/10.1016/j.jmps.2017.06.016

[2] K.D. Karantza, D.E. Manolakos, A Review on the Adiabatic Shear Banding Mechanism in Metals and Alloys Considering Microstructural Characteristics, Morphology and Fracture, Metals 13 (2023) 1988. https://doi.org/10.3390/met13121988

[3] L. Anand, W.A. Spitzig, Initiation of Localized Shear Bands in Plane Strain, J. Mech. Phys. Solids 28 (1980) 113–128. https://doi.org/10.1016/0022-5096(80)90017-4

[4] A. Prakash, T.N. Tak, N.N. Pai, H. Seekala, S.V.S.N. Murty, P.S. Phani, S. Mahesh, P.J. Guruprasad, I. Samajdar, Inception of Macroscopic Shear Bands during Hot Working of Aluminum Alloys, Int. J. Plast. 166 (2023) 103632. https://doi.org/10.1016/j.ijplas.2023.103632

[5] B. Tang, L. Xiang, L. Cheng, D. Liu, H. Kou, J. Li, The Formation and Evolution of Shear Bands in Plane Strain Compressed Nickel-Base Superalloy, Metals 8 (2018) 141. https://doi.org/10.3390/met8020141

[6] L.W. Meyer, E. Staskewitsch, A. Burblies, Adiabatic shear failure under biaxial dynamic compression/shear loading, Mech. Mater. 17 (1994) 203–214.

[7] D. Mateos, S. Boakye-Yiadom, Effect of stress state on adiabatic shear banding in Al2024-T351, Int. J. Impact Eng. 178 (2023) 104602. https://doi.org/10.1016/j.ijimpeng.2023.104602

[8] M. Kang, J. Park, S.S. Sohn, H. Kim, K.H. Kim, S. Lee, Adiabatic shear banding and cracking phenomena occurring during cold-forging simulation tests of plain carbon steel wire rods by using a split Hopkinson's pressure bar, Met. Mater. Int. 21 (2015) 991-999. https://doi.org/10.1007/s12540-015-5252-6

[9] S. Liu, Y.Z. Guo, Z.L. Pan, X.Z. Liao, E.J. Lavernia, Y.T. Zhu, Q.M. Wei, Y. Zhao, Microstructural softening induced adiabatic shear banding in Ti-23Nb-0.7Ta-2Zr-O gum metal, J. Mater. Sci. Technol. 54 (2020) 31-39. https://doi.org/10.1016/j.jmst.2020.03.042

[10] Q. Wei, L.J. Kecskes, K.T. Ramesh, Effect of low-temperature rolling on the propensity to adiabatic shear banding of commercial purity tungsten, Mater. Sci. Eng. A 578 (2013) 394-401. https://doi.org/10.1016/j.msea.2013.04.109

[11] K.D. Karantza, S.A. Papaefthymiou, N.M. Vaxevanidis, D.E. Manolakos, Numerical investigation of the damage effect on the evolution of adiabatic shear banding and its transition to fracture during high-speed blanking of 304 stainless steel sheet, Materials 17 (2024) 1471. https://doi.org/10.3390/ma17071471

[12] K.D. Karantza, N.M. Vaxevanidis, D.E. Manolakos, A structural-thermal coupled modeling approach on the formation of adiabatic shear bands in steel sheet blanking process, J. Phys. Conf. Ser. 2692 (2024) 012038. https://doi.org/10.1088/1742-6596/2692/1/012038

[13] F.W. Bach, B.A. Behrens, M. Rodman, A. Rossberg, G. Kurz, Macroscopic damage by the formation of shear bands during the rolling and deep drawing of magnesium sheets, JOM 57 (2005) 57-61. https://doi.org/10.1007/s11837-005-0098-x

[14] G. Li, Y. Cai, H. Qi, M. Liu, Modelling of the relationship of adiabatic shear and cutting conditions for higher cutting speed ranges, Stroj. Vestn.- J.Mech.Eng. 63 (2017) 373-682. https://doi.org/10.5545/sv-jme.2017.4443

[15] J.J. DeMange, V. Prakash, J.M. Pereira, Effects of material microstructure on blunt projectile penetration of a nickel-based super alloy, Int. J. Impact Eng. 36 (2009) 102-1043. https://doi.org/10.1016/j.ijimpeng.2009.01.007

[16] M. Watremez, D. Méresse, L. Dubar, J. Brocail, Finite Element Modelling of Orthogonal Cutting: Sensitivity Analysis of Material and Contact Parameters, Int. J. Simul. Proc. Model. 7 (2012) 262. https://doi.org/10.1504/ijspm.2012.049820

[17] H.W. Hoffmeister, T. Grosse, A. Gerdes, Investigation of the Influence of Different Process Setting Parameters on the Surface Formation at Honing of Thermally Sprayed Layers, Procedia CIRP 1 (2012) 371–376. https://doi.org/10.1016/j.procir.2012.04.066

Innovative Manufacturing Engineering and Energy - IManEE2024 Materials Research Forum LLC
Materials Research Proceedings 46 (2024) 235-242 https://doi.org/10.21741/9781644903377-31

Laser machining with a high-power CW fiber laser

Alexandros GOSTA[1,a], Efstathios KAMPOSOS[1,b], Alexandros SKOULAKIS[1,c], Yannis ORPHANOS[1,d], Helen PAPADAKI[1,e], Nektarios A. PAPADOGIANNIS[1,f], Michael TATARAKIS[1,g], Evaggelos KASELOURIS[1,h], Vasilis DIMITRIOU[1,i,*]

[1]Institute of Plasma Physics and Lasers-IPPL, University Research and Innovation Centre, Hellenic Mediterranean University, Rethymno, Greece

[a]a.gosta@hotmail.com, [b]skamposos@gmail.com, [c]skoulakis@hmu.gr, [d]yorphanos@hmu.gr, [e]ddk134@edu.hmu.gr, [f]npapadogiannis@hmu.gr, [g]mictat@hmu.gr, [h]vagfem@hmu.gr, [i,*]dimvasi@hmu.gr

Keywords: Laser Machining & Manufacturing, 1050 Aluminum Alloy, High-Power Fiber Laser, Finite Element Method

Abstract. We investigate the CW fiber laser - aluminum 1050 alloy interactions. The manufacturing processes of laser marking, engraving, and cutting are studied by experiments and numerical simulations. The laser head Precitec LightCutter 2.0 motorized 3D is mounted on a 3-Axis CNC milling machine, assembled with the Pipa-Q connector and the powerful CW 2kW Trumpf fiber optic laser. The fiber laser beam interacts with the aluminum samples for varying laser flux and laser scanning speeds. The temperature distribution is monitored by a thermal camera, while thermocouples are used to measure the temperature at specific locations on the sample. A white light interferometric technique is applied for measuring the marking and engraving depths of the irradiated samples. A thermostructural finite element method model, accounting for elastoplastic effects and phase changes, is developed to simulate the laser machining processes, providing key insights into the thermomechanical behavior. A satisfactory agreement between experiments and simulations demonstrates the great potential of high-power fiber lasers in manufacturing.

Introduction

Early industrial laser applications primarily targeted welding, machining, and heat treatment processes. Over time, laser technology expanded to include applications such as forming, shock peening, additive manufacturing, micro- and nano- machining [1-4]. Laser power has increased while the energy consumption demanded from the new laser systems has decreased. High-power fiber lasers have become essential in industries such as aerospace and automotive for laser machining and material processing, due to their precision and efficiency [5]. The use of continuous wave (CW) fiber lasers, along with processing parameters like assist gas pressure and cutting speed, enhances the quality of the laser machining processes [6,7]. Additionally, computational methods like the finite element method (FEM), are employed to optimize laser processing techniques such as material ablation, engraving, drilling, and cutting [8-11].

We investigate the CW laser machining of Aluminum (Al) 1050 by experiments and FEM simulations, focusing on marking, engraving, etching and cutting processes, since this alloy is a very popular material in the industry of a very common commercial purity. It has 99.50% minimum purity and the remaining 0.5% is made up of mostly iron and silicon, providing good electrical, thermal conductivity and corrosion resistance at a low price. The fiber laser beam interacts with Al sheets under varying laser flux and scanning speeds. A thermal camera monitors the temperature distribution, while thermocouples are used to secure the temperature measurement precision at specific points of interest on the workpiece. A thermostructural FEM model is developed to simulate the laser machining processes of Al workpieces of thicknesses 0.6, 1, 2, and

3 mm. The machined parts are precisely measured by white light interferometry and stereo microscopy providing insights to the effects of each process parameters on the final Al products.

Experimental details

The TRUMPF TruFiber P compact fiber laser of 1.075 μm wavelength and maximum power of 2 kW along with the Precitec LightCutter 2.0 motorized 3D laser head, mounted on a 3-Axis CNC milling machine deliver the CW laser light on the workpiece. Mach3 [12] and the Fiberview laser control software, to control the machine scanning and the laser beam parameters respectively, by the help of a personal computer. To evaluate and validate the experimental and simulation results, precise thermal and diagnostics are employed. An Optris PI1M thermal camera (f=75 mm), is mounted on the CNC machine, focusing on the workpiece at 310 mm from the camera's sensor. The camera is equipped with two band-rejection filters Notch filters arranged in series to reduce the laser beam reflections with an optical density 12, and a long pass filter with a cut-on wavelength of 750 nm and set to capture thermal data at a framerate of 80 Hz. The emissivity of the material is calibrated in the camera's software, set to the value of 0.05 for Al 1050 [13]. An iris diaphragm is also used to mitigate reflections from the workpiece towards the thermal camera. Additionally, two thermocouples, a K- and an N-type, are mounted within a distance range from 0.5 to 2 mm from the irradiation point, to provide precisely the temperature values at predefined points, within the scanning path of the camera. The experimental setup is presented in Fig. 1.

The machined Al parts are furthermore optically evaluated and measured by the help of a Leica MZ6 stereomicroscope which provides image magnification ranging from 0.63x to 4.0x and the white light interferometry technique measures and evaluates the topographical and microstructural modifications of the workpiece [9].

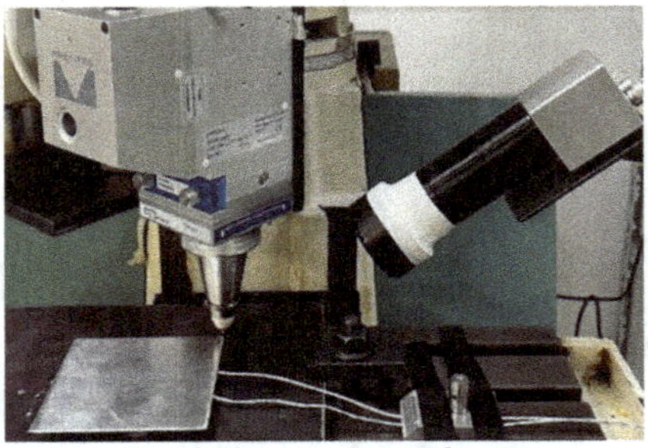

Fig. 1 The experimental setup

Numerical modeling and simulation

Mathematical equations. The FEM model is governed by the 3D transient heat conduction equation [9-11]:

$$\rho c\left(\frac{\partial T}{\partial t}+V_x\frac{\partial T}{\partial x}\right)=\frac{\partial}{\partial i}\left(\kappa(T)\frac{\partial T}{\partial i}\right)+\dot{Q}-a_T T_0(3\lambda+2\mu)\dot{\varepsilon}_{ii} \tag{1}$$

where, i=x,y,z coordinates, t time, ρ the density, c the specific heat, V_x the laser scanning speed along the X-axis direction, k thermal conductivity and $\dot{Q}$ the power generation per unit volume, T the temperature, T_0 the ambient temperature, α_T the thermal expansion coefficient, λ and μ lame coefficients and $\dot{\varepsilon}_{ii}$ the strain rate coupled with the 3D transient mechanical equation:

$$\rho\frac{\partial^2 U_i}{\partial^2 t}=\mu\frac{\partial^2 U_i}{\partial^2 k}+(\lambda+\mu)\frac{\partial}{\partial i}\left(\frac{\partial U_k}{\partial k}\right)-(3\lambda+2\mu)\alpha_T\frac{\partial T}{\partial i} \tag{2}$$

where, U is the displacement.

The CW laser scanning beam is introduced as a heat source described by the Gaussian distribution of the $q(r)$ absorbed laser heat flux or laser power intensity:

$$q(x,z) = \frac{2P_{tot}}{\pi r_b^2} e^{-\left(\frac{2((x-tV_x)^2+z^2)}{r_b^2}\right)} \tag{3}$$

where, P_{tot} is the total absorbed power and r_b is the laser beam radius. It holds that:

$$P_{tot} = \eta P_{inc} \tag{4}$$

where, P_{inc} is the incident laser power, and η is the average absorptivity of the Al workpiece. Since the laser beam irradiates the top surface of the Al sheet, heat flux is applied, and convection and radiation are also considered by:

$$-k\frac{\partial T}{\partial y} = q(x,z) - h(T - T_0) - \sigma\varepsilon(T^4 - T_0^4) \tag{5}$$

where h is the convective heat transfer coefficient, σ is the Stefan–Boltzmann constant being 5.67 $10^8 W/m^2K^4$ and ε is the emissivity. Heat flux is normal to the laser irradiated surface, while the motion of the laser beam is along the X-axis. The focused CW laser beam diameter is 250 μm with power ranging from 0.20 to 1.00 kW and the corresponding heat flux from 4 to 20 kW/mm^2.

Material Properties: Al 1050 belongs to the 1000 series of commercially pure aluminum alloys (99.5% pure), recognized for its high corrosion resistance, excellent electrical and thermal conductivity, and good formability. The mechanical properties of Al 1050 vary depending on its temper (heat treatment or strain-hardening). A common temper is H14, which is strain-hardened to a half-hard state. Table 1 presents the mechanical and physical properties of Al 1050 H14 sheet.

Table 1 Material properties of Al 1050 [14,15]

Tensile strength:	100 [Mpa]	n (strain hardening exponent):	0.42
Yield strength:	75 [Mpa]	m (thermal softening exponent):	1.34
Elongation:	~10%	Specific heat:	898.7 [J/kg*kelvin]
Density:	2.71 [kg/m^3]	Melting point:	650 [°C]
Young modulus:	68 [Gpa]	Thermal conductivity (20 °C):	222 [W/mK]
Bulk modulus:	75 [Gpa]	Latent heat:	400 [Joule/gr]
Shear modulus:	26 [Gpa]	D1 (fracture parameter):	-0.77
Poisson ratio:	0.33	D2 (fracture parameter):	1.45
λ (1st Lame coefficient):	25.46 [Pa]	D3 (fracture parameter):	-0.47
μ (2nd Lame coefficient):	26.10 [Pa]	D4 (fracture parameter):	0.0
A (initial yield of the material):	324 [Mpa]	D5 (fracture parameter):	1.6
B (strain hardening coefficient):	114 [Mpa]	Liquidus temperature:	657 [°C]
C (strain rate sensitivity):	0.002	Solidus temperature:	646 [°C]

Numerical Model & Simulation Parameters. A rectangular Al solid of size 10 x 6 mm^2 and thickness of 1 mm is modeled by 480.000 solid finite elements. Non-reflecting outflow boundary conditions are applied at the edges of the workpiece, and the ambient temperature is set at 20 °C. The dynamic elastoplastic behavior of the workpiece is described by the Johnson-Cook constitutive strength material model, that considers the effect of plastic strain, strain rate and temperature. The flow stress is expressed as:

Materials Research Proceedings 46 (2024) 235-242 https://doi.org/10.21741/9781644903377-31

$$\sigma_y = (A + B\varepsilon^n)(1 + C ln\frac{\dot{\varepsilon}}{\dot{\varepsilon}_0})(1 - \frac{T - T_r}{T_m - T_r})^m \quad (6)$$

where ε is the equivalent plastic strain, $\dot{\varepsilon}/\dot{\varepsilon}_0$ is the dimensionless plastic strain rate, $\dot{\varepsilon}_0$=1s^{-1} is a reference strain rate used to normalize the strain rate, T_m is the melting temperature of the workpiece and T_r the room temperature. The material constants A, B, C, n, m in Table 1 is determined from literature experimental results. The 3D transient coupled thermal-structural FEM model is simulated at the GRNET high-performance computer (HPC), ARIS [16]. It is noticed that for a laser machining simulation time of 0.5 sec, the runtime on 2 nodes of 40 cores ranges from 6 to 24 hours, with respect to the laser scanning speed parameter, ranging from 60 to 1200 mm/min.

Results & discussion

Laser marking is initially investigated by the laser machining of a 1 mm thick Al 1050 H14 sheet. In laser marking, thermal stresses induced by the laser cause plastic deformation of the metal without melting or removing material. Instead, it results in localized heating that leads to a color change due to oxidation. This color change is typically a surface effect caused by controlled oxidation [17]. The laser scanning speed is set to 1200 mm/min for a linear distance of 10 mm along the X-Axis, utilizing a constant power of 200 W to the CW laser, at a room temperature of 20 °C, under atmospheric conditions. The same experimental data are introduced to the FEM model, along with the absorptivity coefficient which is adopted from [18], since it is temperature, ambient atmospheric conditions, surface roughness and debris, dependent like the AA 1100 metal. The FEM simulation results reveal a maximum temperature of approximately 600 °C during the laser radiation process. During the experiment, the thermal camera recorded a maximum temperature of 600 °C in accordance with the computational results, securing that the marking process is under the melting point of the solid Al, as presented in Fig. 2. The two thermocouples measured a maximum temperature of 45 °C, 15 ms after the laser irradiation at 1 mm far from the center of the initial laser irradiation point, agreeing with the computed temperature distributions of the simulation.

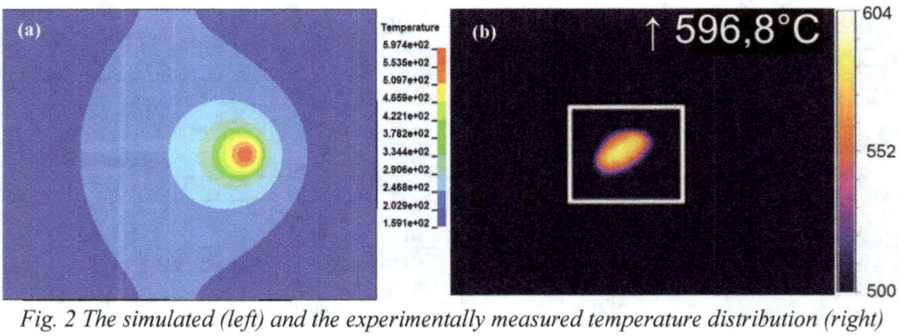

Fig. 2 The simulated (left) and the experimentally measured temperature distribution (right)

The computed plastic deformation approaches a maximum value of 0.023 for the effective plastic strain, on the Al samples surface, as shown in Fig. 3 (a), along with the stereomicroscopy image of the machined Al surface Fig. 3 (b), measured after the experiment. Additionally, the white light interferometry measurements depicted in Fig. 3 (c-d), reveal a normal displacement of 0.25 μm on the Al surface, in agreement with the simulation results.

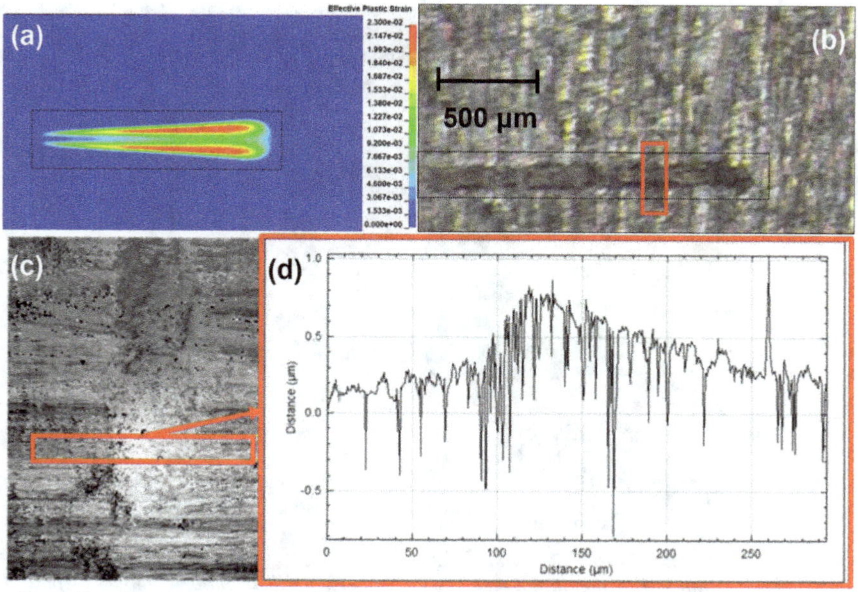

Fig. 3 Simulated plastic strain (a) and a stereomicroscopic image of the linear marking (b). White light interferometry measurement of the normal displacement (c,d). The red outlined rectangles indicate the area of measurement

The same laser power of 200 W but with a scanning speed of 75 mm/min, is used for the study of laser etching/engraving on the same experimental set-up. Under these irradiation conditions, the laser beam induces to the workpiece thermal stresses, which result in plastic deformations of the metal while additionally the processes of melting and material removal are involved. The laser etching occurs when material is removed by less than 25 μm. The laser cutting speed is updated in the FEM model to meet the new experimental parameters. The simulation results regarding the computed temperature field on the machined Al surface, 0.5 s after the start of laser irradiation, are demonstrated in Fig 4.a. The maximum temperature value of 660 °C is computed at the center of the laser beam while etching. The machined workpiece is shown in Fig. 4b. The stereomicroscopic imaging measurements determine that a slot of 250 μm is etched on the part. The same width is measured by the white light interferometry, which additionally reveals the maximum depth of the slot processed to be ~10 μm and a width of 250 μm. The experimentally measured results agree with the simulation results provided by the FEM thermostructural model. A representative experimental result is shown in Fig. 5. Corresponding simulation results shown in Fig. 4 deliver the same width and a depth of ~20 μm.

Materials Research Proceedings 46 (2024) 235-242 https://doi.org/10.21741/9781644903377-31

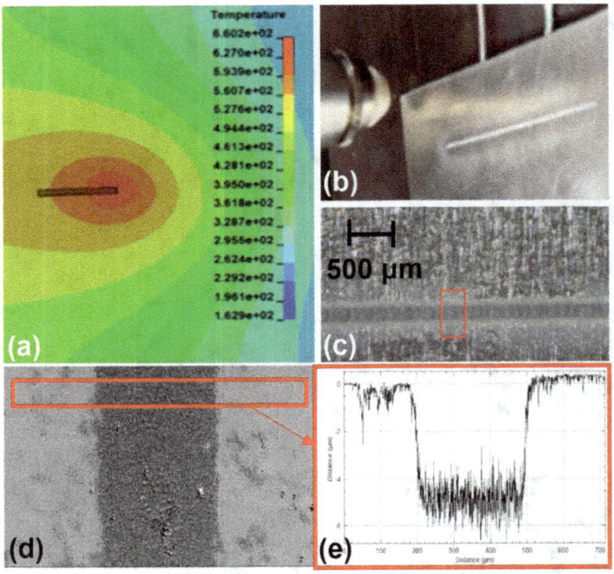

Fig. 4 Laser etching FEM simulation results with temperature distribution contours displayed 0.5 s after the start of laser irradiation (a), experimental results and stereoscopic imaging (b,c), and white light interferometry measurements (d,e)

Additionally, the K-type thermocouples are monitoring the temperature changes at 2 mm away from the laser spot center. The recorded Temperature (°C) vs Time (sec) graph is presented in Fig. 5 (right). The measured results denote a maximum temperature value of 220 °C. The nodal history of temperature evolution for a finite element at the same distance is demonstrated at the computed Temperature (°C) vs Time (msec) graph of Fig. 5 (left). The maximum temperature of 210 °C is computed by the FEM simulation at 0.5 sec.

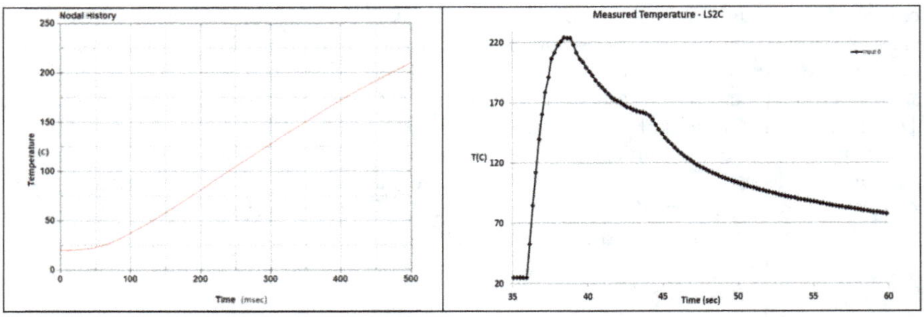

Fig. 5 Simulation (left) and experimental (right) results of temperature evolution vs time at 2 mm far from the laser spot center

Materials Research Proceedings 46 (2024) 235-242 https://doi.org/10.21741/9781644903377-31

Laser cutting is finally conducted with laser power in the range of 450 to 1000 W. Al 1050 alloy workpieces of 1-, 2- and 3-mm thickness are machined using the same experimental set-up and diagnostics. The agreement of the simulated vs the experimental results demonstrates the efficiency and accuracy of the high-power CW fiber laser. Representative results of the 1 mm thick Al sheet, laser machined by 450 W of the CW laser and a scanning speed of 60 mm/min on the X-Axis are demonstrated in Fig. 6. The FEM simulation results along with the stereomicroscopic image and the white light measurements result to a width of cut of 300 μm.

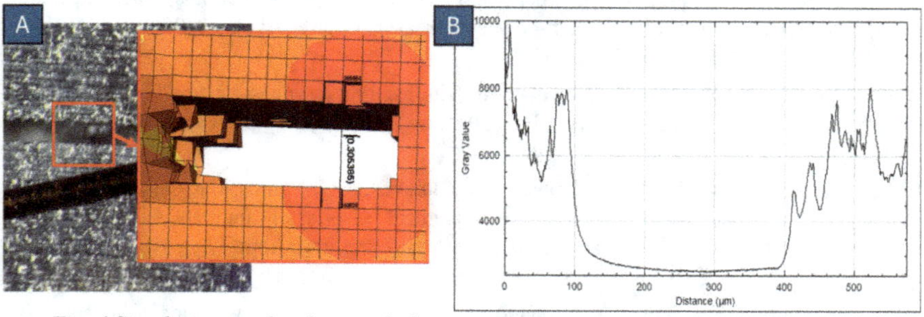

Fig. 6 Simulation results along with the stereomicroscopic image (A) and, white light measurement (B)

Conclusions

The laser machining capabilities and the accuracy of a high-power CW laser are experimentally and computationally investigated. The processes of laser marking, etching, engraving, and cutting of Al 1050 are studied. It is shown that laser machining can significantly modify the topography and material properties of metals, resulting to notable and controllable modifications and improvements in surface finish and material removal. The CW laser-matter interactions drive the thermo-elastic, plastic, melting, and cutting processes which are accurately simulated and experimentally validated. These verified FEM simulations allow for reliable predictions of the metal workpiece behavior and offer valuable insights for numerous applications, since high-power CW fiber lasers provide superior cutting speeds and quality, making them advantageous for machining and manufacturing.

Acknowledgements

This work was supported by computational time granted from the National Infrastructures for Research and Technology S.A. (GRNET) in the National HPC facility - ARIS - under project ID LaMPIOS III.

References

[1] J.R. Lawrence, ed. Advances in laser materials processing: technology, research and applications. Woodhead Publishing, 2018. https://doi.org/10.1016/C2015-0-05718-5

[2] Y.C. Shin, et al. Overview of laser applications in manufacturing and materials processing in recent years. J. Manuf. Sci. Eng.142 (2020) 110818. https://doi.org/10.1115/1.4048397

[3] S.P. Murzin, C. Stiglbrunner. Fabrication of Smart Materials Using Laser Processing: Analysis and Prospects. Appl. Sci. 14 (2023) 85. https://doi.org/10.3390/app14010085

[4] J. Dutta Majumdar, I. Manna. Laser material processing. Int. Mater. Rev. 56 (2011) 341-388. https://doi.org/10.1179/1743280411Y.0000000003

[5] A.K. Nath, High power lasers in material processing applications: an overview of recent developments. in: Laser-assisted fabrication of materials, Springer, Berlin, Heidelberg 2012, pp. 69-111. https://doi.org/10.1007/978-3-642-28359-8_2

[6] L. D. Scintilla, L. Tricarico. Fusion cutting of aluminum, magnesium, and titanium alloys using high-power fiber laser. Optical Engineering 52 (2013) 076115-076115. https://doi.org/10.1117/1.OE.52.7.076115

[7] C. Wandera, A. Salminen, V. Kujanpaa, Inert gas cutting of thick-section stainless steel and medium-section aluminum using a high-power fiber laser. J. Laser Appl. 21 (2009) 154-161. https://doi.org/10.2351/1.3184429

[8] Lauro, Carlos H., et al. Finite element method in machining processes: a review. In: Modern Manufacturing Engineering, Springer, Cham 2015, pp. 65-97. https://doi.org/10.1007/978-3-319-20152-8_3

[9] V. Dimitriou, E. Kaselouris, Y. Orphanos, M. Bakarezos, N. Vainos, I. K. Nikolos, M. Tatarakis, N. A. Papadogiannis, The thermo-mechanical behavior of thin metal films under nanosecond laser pulse excitation above the thermoelastic regime, Appl. Phys. A 118 (2015) 739-748. https://doi.org/10.1007/s00339-014-8792-6

[10] E. Kaselouris, I.K. Nikolos, Y. Orphanos, E Bakarezos, N. A. Papadogiannis, M. Tatarakis, V. Dimitriou, Elastoplastic study of nanosecond-pulsed laser interaction with metallic films using 3D multiphysics fem modeling, Int. J. Damage Mech. 25 (2016) 42-55. https://doi.org/10.1177/105678951557655

[11] E. Kaselouris, A. Skoulakis, Y. Orphanos et al. Analysis of the heat affected zone and surface roughness during laser micromachining of metals. Key Engineering Materials 827 (2020) 122-127. https://doi.org/10.4028/www.scientific.net/KEM.827.122

[12] Mach3 CNC Controller Software Installation and Configuration Version 3, ArtSoft USA, November 6, 2008, http://www.machsupport.com.

[13] Information on https://www.engineeringtoolbox.com/radiation-heat-emissivity-aluminum-d_433.html

[14] Information on https://www.jcmetalchina.com/blog/what-is-1050-aluminum-alloy/

[15] S. Akram, S. H. Imran Jaffery, M. Khan, M. Fahad, A. Mubashar and L. Ali, Numerical and experimental investigation of Johnson–Cook material models for aluminum (Al 6061-T6) alloy using orthogonal machining approach Adv. Mech. Eng. 10 (2018) 1-14. https://doi.org/10.1177/1687814018797794

[16] Aris Documentation. Available online: http://doc.aris.grnet.gr/system/hardware/ (accessed on 26 September 2023).

[17] H. Liu, L. Wenxiong, H. Minghui, Surface coloring by laser irradiation of solid substrates. Apl Photonics 4 (2019) 051101. https://doi.org/10.1063/1.5089778

[18] R. Indhu, et al. Overview of laser absorptivity measurement techniques for material processing. Lasers Manuf. Mater. Process. 5 (2018) 458-481. https://doi.org/10.1007/s40516-018-0075-1

Innovative Manufacturing Engineering and Energy - IManEE2024 Materials Research Forum LLC
Materials Research Proceedings 46 (2024) 243-251 https://doi.org/10.21741/9781644903377-32

Methodology for identifying the intersection of mutually intersecting surfaces to optimize sheet material cutting for the fabrication of parts and structures

Zoya TSONEVA [1,a*], Daniela SPASOVA [1,b]

[1] Department of Mechanics and Machine elements, Technical University of Varna, Bulgaria, "Studentska" street No. 1

[a] zoya_tsoneva@tu-varna.bg, [b] danielats@tu-varna.bg

Keywords: Engineering Graphics, Intersection of Surfaces, Descriptive Geometry, Cutting Sheet Material

Abstract. Sheet material is often used to create intricate details and complete structures in engineering, essential in both industry production and building construction. It is common to design and construct structures that involve various intersecting basic geometric surfaces. Introduced in the present paper is a validated methodology for identifying the intersection of two or more mutually intersecting surfaces Utilizing the developed methodology can greatly simplify the production of complex surface details and structures from sheet material, leading to improved accuracy in cutting maps, reduced waste generation, simplified work processes, and minimized errors in planning.

Introduction

The fabrication of structures from sheet material is sometimes the only possible solution for choosing the most viable technological option in engineering practice, particularly in the construction of tanks, pipelines, structures of metallurgical and hydromechanical equipment, and various types of industrial and technological pipe networks [1]. In certain circumstances, the production of details from sheet material not only reduces weight and cost significantly, but also simplifies the overall technology compared to alternative methods, such as casting or forging [2]. Sheet metal structures are becoming increasingly widespread due to the utilization of high-performance and strength-safe methods of arc seam welding. [2, 3]. Cost-benefit analyses of the production of parts by welding, casting and forging demonstrate that welding is a viable alternative, particularly for parts with intricate geometry [6, 7].

Sheet metal fabrications are versatile and easy to implement, whether for individual production of custom equipment or serial production. Proper structural and technological development streamlines the process of production, triggering substantial cost savings on materials, and enhanced safety and durability of the structure [1, 4, 5].

Sheet metal cutting is a crucial stage in numerous technological processes related to the construction of aircraft, ships, cars, furniture, and other items with complex geometries and strict design requirements [8, 9]. The primary goal of this process is the fabrication of quality blanks at minimal cost. When sizing the blanks on the sheet material, their optimal placement is necessary, with minimal waste and reduced cutting labour intensity [10, 11]. Created, to that effect, is a cutting map, which is a scaled drawing detailing the contours and dimensions of the blanks to ensure the most efficient cutting of the workpieces. The process can be done manually or with the assistance of software [12, 13]. Precise cutting of the sheet material is crucial as it reduces the amount of non-recoverable waste and has a direct impact on the final production costs [14, 15].

Outlined in the paper, through several stages, is a methodology for identifying the intersection of mutually intersecting surfaces, utilizing principles from engineering graphics, and applied geometry, designed to address common challenges encountered in engineering practice, with a focus on the optimization of the sheet material cutting for the fabrication of parts and structures with complex surfaces.

Exposition

When faced with engineering drawing problems involving the intersection of basic geometric surfaces, it is common to encounter challenges in determining where and how to begin solving such problems.

Prior to embarking on the process of solution, it is imperative to analyze and clarify the nature of the intersecting solids involved– be they edge, rotating, or cylindrical, and determine the most appropriate solution method. Awareness of the theorems that dictate the expected shape of the spatial intersection resulting from the intersection of solids is also relevant.

In practice, there are three primary methods for solving such problems:

- A method involving the intersection of solids with a bundle of planes that intersect at a finite line (Fig. 1).

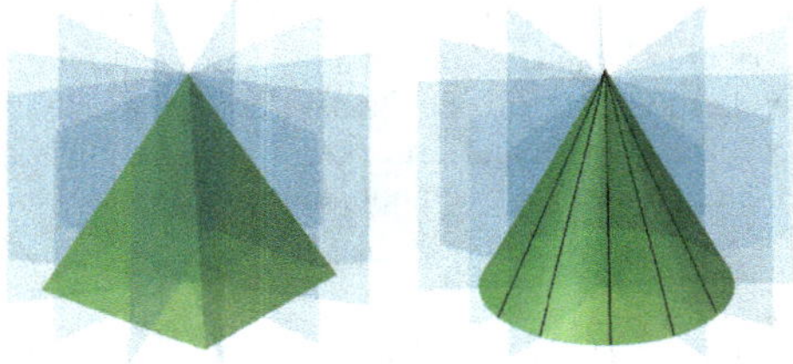

a b

Fig. 1 A bundle of planes intersecting in a finite line

- A method in which solid bodies intersect with a bundle of planes intersecting in an infinite line (intersect in a line at infinity). This method is commonly referred to as "The Method of Parallel Auxiliary Secant Planes" (Fig. 2);
- The method of auxiliary secant spheres (Fig.3).

a b

Fig. 2 A bundle of planes intersecting in an infinite line

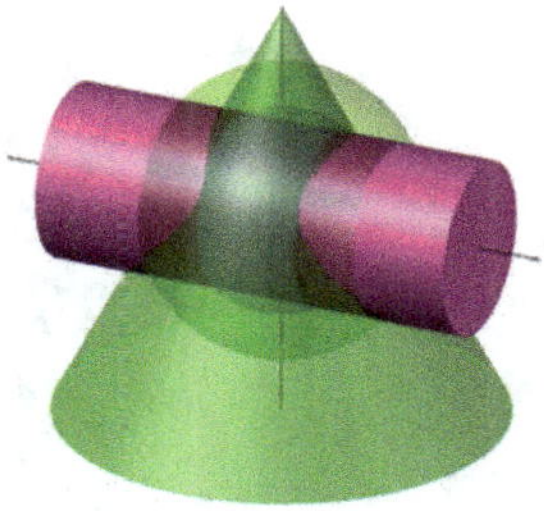

Fig. 3 Method of auxiliary secant spheres

While the first two methods are versatile and can be applied to finding the intersection between two or more bodies, regardless of whether they are intersecting rotating with rotating bodies, edge with edge, or a combination of both, the third method focuses solely on intersecting rotating bodies with intersecting axes.

METHODOLOGY:

1. The initial step in solving such types of problems is to *reconsider the intersecting bodies in a specific arrangement,* or adjustment, where the axes of the intersecting bodies are aligned parallel to a projection plane– either a principal or additionally introduced plane. Such a positioning simplifies the solution process significantly.

Introducing one or more additional projection planes strategically placed to ensure that the bodies are projected into a particular position during the final transformation. This alignment is, therefore, necessary only if the bodies are not yet in the desired position with respect to the principal projection planes. Should the solids be correctly positioned, this step can be left out.

2. The second step is *to determine the starting projection for the solution*. How should the secant planes be positioned so that they intersect both solids in the most effective manner? What will these intersections look like in varied projections?

This is where the most challenging aspect of the process arises.

It is, thus, recommended to select a projection in which the appearance of the image remains consistent, i.e. the projection will retain its original outward look after the problem has been resolved. This means choosing a projection where the prisms and cylinders are represented solely by their main edges, with their surrounding surfaces perpendicular (projecting) to both the observer and the projection plane (i.e. in a specific position). For example, in the drawing drafted in Fig. 4, it is the frontal projection that is the ideal starting point. Displayed in Fig. 5 is the profile projection of the intersecting bodies where the solution does not alter the resulting image. The solution in Fig. 8 is initiated from the frontal projection to maintain consistency in the image, with the cylindrical surface being perpendicular (projecting) to the frontal projection plane. In all these cases, the intersection between the solids occurs along the projecting surrounding surface.

Materials Research Proceedings 46 (2024) 243-251 https://doi.org/10.21741/9781644903377-32

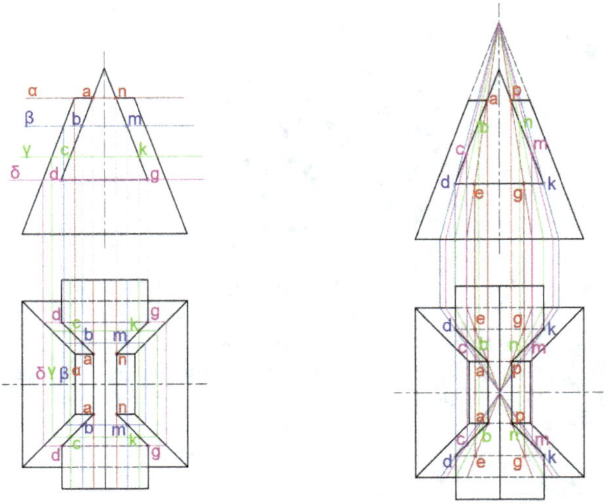

a b

Fig. 4 Intersection of edge bodies

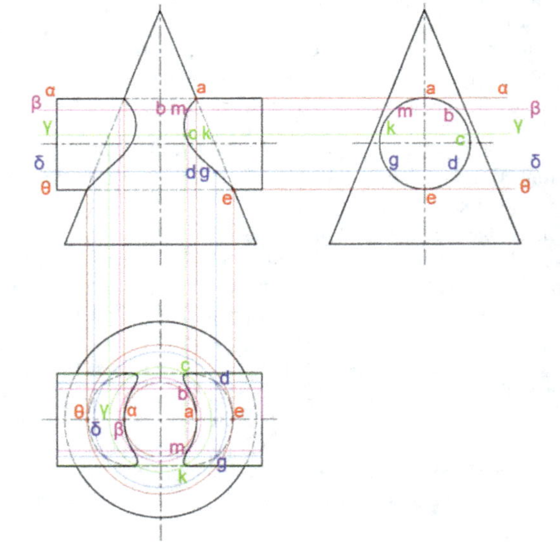

a

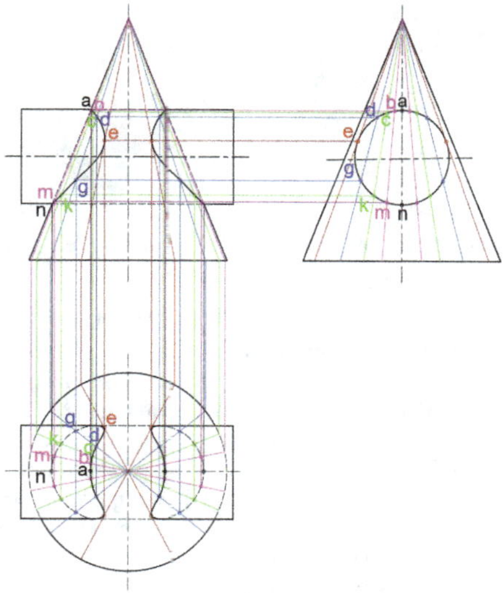

b

Fig. 5 Intersection of rotating bodies

3. In the third step, the bodies that intersect each other are *further cut across by a bundle of planes intersecting at a line that can be either finite or infinite*.

Let us now examine and compare the two methods.

3.1. Solution of the problem by the method "Bundle of planes intersecting in an infinite line"

Observed in Fig. 4a is a practical example where in the frontal projection of the drawing, the intersecting solids are cut through by four secant planes labelled as - α, β, γ, δ. When projected horizontally, these secant planes crosscut the pyramid in a shape that is homothetic to the base edge of the pyramid. The same planes intersect the triangular prism, resulting in the formation of rectangles. Where the two plane figures, formed by the intersection (of the two mutually intersecting bodies) intercross, obtained are the points of intersection referred to as a, b, c and d. These points are crucial in identifying the spatial intersection between the solids. The concepts that have been discussed thus far are visually represented in Fig. 6a.

The points of intersection of the two mutually intersecting solids with each of the secant planes α, β, γ, and δ, are marked with letters (Fig. 4a), namely:

- located on the plane α, are the points *a* and *n*;
- located on the plane β, are the points *b* and *m;*
- plane γ comprises *c* and *k;*
- plane δ holds *d* and *g*.

A similar problem-solving approach is illustrated in Fig. 5a. Here, as previously stated, the solution commences with the profile projection. The solids cut through multiple planes, including α, β, γ, δ and θ, requiring the focus of attention to be redirected exclusively to the region where the two bodies intersect. From the profile projection, the planes are then transferred to the adjacent projections. Formed, thus, is the way in which they will appear after cutting across the two solids. The intersection points of the intersecting plane figures created in the horizontal projection plane are labelled, and then connected in sequence. The resulting plane figures are depicted in Fig. 6b.

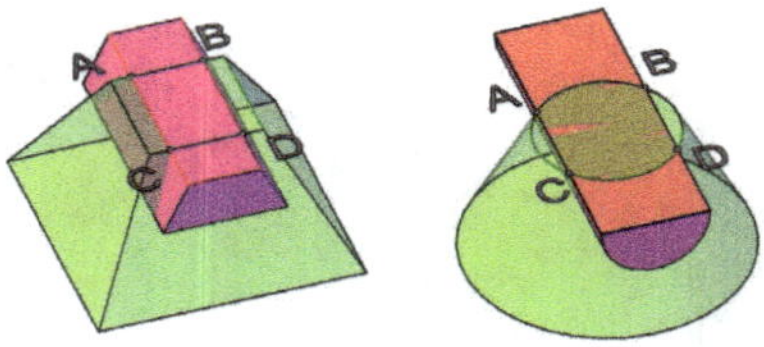

a b

Fig. 6 Visualization of the solid intersection method using a bundle of planes intersecting in an infinite line

3.2. Solution to the problem using the method of "Bundle of planes intersecting in a finite line"

The tasks in Fig. 4b and Fig. 5b have been resolved using this method. In the Monge drawing, the planes cutting across the mutually intersecting bodies are visualized as generatrices along the surrounding surface of the cone (Fig. 5b) or as lines connecting the apex to the base of the pyramid (Fig. 4 b). In such cases, the problem can be effectively resolved by constructing the generatrices along the surrounding surface of the cone in Fig. 5b, crossing the cylinder represented only by its base as a circle in the profile projection of the drawing. It is important to note that the creation of these generators along the cone's surface is only necessary at points where the cylinder intersects the cone. The presence of generatrices in other parts of the cone would only clutter the drawing as they will not intersect the cylindrical surface.

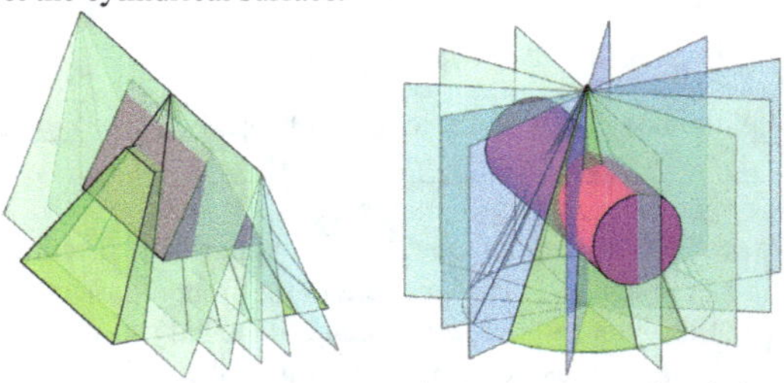

Fig. 7 Visualization of the solid intersection method using a bundle of planes intersecting in a finite line

What follows next is the shifting of the generators to adjacent projections in a projective correlation. In the drawing depicted in the paper, the generators, for clarity, are coded in colour. If this is not feasible, it is recommended that they be labeled appropriately.

The initial steps in the variant presented in Fig. 4b are the construction and location of the pyramid's apex, followed by descending lines (intersections) to the base, crossing only the region where the pyramid is intersected by the triangular prism. The lines are extended into the adjacent projection and in the drawing provided, they are colour-coded for distinction.

During this process, it is crucial to adequately indicate the specific positions where each plane identified by their intersections or generatrices along the surrounding surface of the solids, cut across the solids in a particular position. In the drawings, provided herein, this is done by a letter designation– points *a, b, c, d, e, g, ….*

4. The fourth step requires *shifting each of these points to their corresponding planes* (lines or generators) in the adjacent projections.

Examining the drawings, it becomes evident that these points mark the intersection of the two plane figures formed upon sectioning with any of the secant planes.

Materials Research Proceedings 46 (2024) 243-251 https://doi.org/10.21741/9781644903377-32

5. The fifth step involves *connecting all the letters carried over from the adjacent projection planes in order* so as to visualize what the intersection between the bodies looks like in each of the three projections.

Let us now apply this methodology to analyze and compare the solutions of the problem illustrated in Fig. 8a.

1. The first stage has already been completed in the assignment itself, as the cylinder in the frontal projection is represented as the one projecting towards the frontal projection plane.

2. It follows, thus, from the first stage that solving the problem should start from the frontal projection since this is where the cylinder is in its proper positioning.

3. In the third stage, the bodies are cut through with a bundle of planes intersecting in a finite or infinite line. The intersection of mutually intersecting bodies with planes is done only within the very area where the bodies cross. This process is exemplified in Fig. 8, emphasizing the validity of the two methods.

During this stage, the planes from the frontal projection are also transferred to the horizontal projection. In Fig. 8a, the planes cross across the pyramid, resulting in the homothetic to the base form of the intersections, i.e. that of hexagons. At this stage, the intersections of the plane with the cylinder should also be depicted. However, in order to save time and not to clutter the drawing, this step is omitted. The next stage will address how to proceed in this regard.

The third stage, furthermore, is also the stage where each of the planes α, β, γ, δ ..... in the frontal projection crosses the cylindrical surface at points *a, b, c, d, ...*

4. At the fourth stage of the process, each of the points *a, b, c, d, ...* is shifted to its corresponding plane within the horizontal projection plane.

To maintain clarity in the drawing, the labelled intersection points from the frontal projection are directly transferred onto their respective planes in the horizontal projection rather than plotting the intersections of the planes with the cylinder. This approach not only saves time and effort in graphic drawing but also prevents unnecessary clutter in the final representation.

Analogously, the solution to the problem depicted in Fig. 8 b follows a similar pattern. After constructing the secant planes whose intersection coincides with the axis of the pyramid, marked are the points where each plane (which in this method are presented as straight lines along the surrounding surface of the pyramid) intersects with the surrounding surface of the cylinder (arranged along the circle) in the frontal projections at points *a, b, c, d, ...* These points are then transferred from the frontal to the horizontal projection plane of their respective plane.

5. At the fifth stage, all the letters carried over from the frontal to the horizontal projection plane are connected in order, and determined, accordingly, is the intersection between the bodies in the projections using both methods.

Perspective and Application

The methodology has been tested in a learning environment with students from various mechanical engineering majors studying descriptive geometry. A mandatory stage in solving such problems is the drawing of unfolds of the surrounding surface of the intersecting bodies and the creation of a mock-up verifying the solution of the problem.

High-quality and professional construction documentation development optimizes the resources and time invested in each project, since even the smallest error or omission in solving such engineering tasks can lead to significant problems at the final stage of project implementation.

The methodology presented in the article precedes the creation of unfolds of the intersecting surrounding surfaces. After making the correctly constructed and drawn unfolds of the intersecting bodies, one can proceed to Application in software products such as "Orion" and "Astra" to make a map for cutting sheet material.

The methodology can also be used with software products such as AutoCAD, which is one of the most widely used in both mechanical engineering and construction.

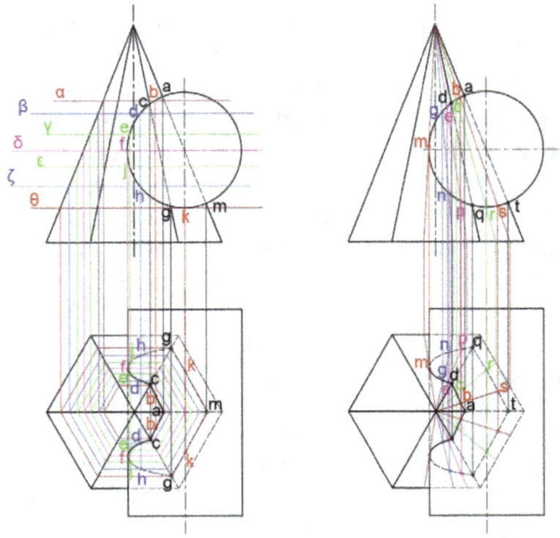

a b

Fig. 8 Comparison of the solutions for the intersection of an edge body with that of revolution using the two methods

Conclusions

It can be inferred from the analyses of the problem that the methodology developed for identifying the intersection of mutually intersecting surfaces and ensuring its consistency is applicable to both methods – "A bundle of planes intersecting in a finite line" and "A bundle of planes intersecting in an infinite line". The only exception is the method of "Auxiliary secant spheres", which falls outside the scope of the present study.

The developed methodology holds promise in aiding the creation of more accurate maps for sheet material cutting by optimizing the contour layout even for complex-relief blanks, leading to a reduction in final production costs.

Acknowledgments

The authors would like to kindly acknowledge the support of the scientific and research project NP 9/ 2024, financed on purpose from the state budget, within the framework of TU-Varna scientific research activity.

References

[1] V. Bundjulov, I. Dimovski, D. Petrov, Handbook on unfoldings of sheet structures, State publishing house"Technica", Sofia, 1966

[2] R. Migliari , Descriptive Geometry in the Era of the Computer, Roma: Kappa, 2008
https://www.academia.edu/1130842/Descriptive_geometry_in_the_Era_of_computers

[3] Radovan B. Štulić, Jelena T. Atanacković, Implementation Of Computer Technologiesin Descriptive Geometry Teaching:Surfaces Of Revolution, Facta universitatis - series:

Architecture and Civil Engineering 2003 Volume 2, Issue 5, pp. 379-385. https://doi.org/10.2298/FUACE0305379S

[4] L. Baranová, I. Katreničová, Role of Descriptive geometry course in development of students' spatial visualization skills, Annales Mathematicae et Informaticae 49 (2018) pp. 21–32

doi: 10.33039/ami.2018.04.001 http://ami.uni-eszterhazy.hu

[5] LAI, Chang Liang; LIU, Chuang; ZHANG, Xian Jie. A digital system implement for aircraft sheet metal cutting. Applied mechanics and materials, 2012, 110: 1637-1642. https://doi.org/10.4028/www.scientific.net/AMM.110-116.1637

[6] SCHMIDT, J.; DORNER, H.; TENCKHOFF, E. Manufacture of complex parts by shape welding. Journal of nuclear materials, 1990, 171.1: 120-127. https://doi.org/10.1016/0022-3115(90)90356-R

[7] HOANG, Nguyen-Hieu; MORIN, David; LANGSETH, Magnus. Testing and modelling of butt-welded connections in thin-walled aluminium structures. Thin-Walled Structures, 2022, 171: 108681. https://doi.org/10.1016/j.tws.2021.108681

[8] SHARMA, Amit; YADAVA, Vinod. Experimental analysis of Nd-YAG laser cutting of sheet materials–A review. Optics & Laser Technology, 2018, 98: 264-280. https://doi.org/10.1016/j.optlastec.2017.08.002

[9] DIMITROV, Diayn M.; MINCHEVA, Desislava; SLAVOV, Stoyan D. Influence of porosity to dynamic Young's modulus of sintered iron. Bayesian approach. Materials Today: Proceedings, 2022, 59: 1677-1682. https://doi.org/10.1016/j.matpr.2022.03.399

[10] Information on https://umecon.ru/press/view/320?ysclid=m0ofs2gout232210083

[11] Rao, Y., Huang, G., Li, P. et al. An integrated manufacturing information system for mass sheet metal cutting. Int J Adv Manuf Technol 33, 436–448 (2007). https://doi.org/10.1007/s00170-006-0484-8

[12] PAN, M.; RAO, Yunqing. An integrated knowledge based system for sheet metal cutting–punching combination processing. Knowledge-Based Systems, 2009, 22.5: 368-375. https://doi.org/10.1016/j.knosys.2009.02.008

[13] Wang, G. G., & Xie, S. Q. (2005). Optimal process planning for a combined punch-and-laser cutting machine using ant colony optimization. International Journal of Production Research, 43(11), 2195–2216 https://doi.org/10.1080/00207540500070376

[14] CHO, Yigil, et al. Engineering the shape and structure of materials by fractal cut. Proceedings of the National Academy of Sciences, 2014, 111.49: 17390-17395. https://doi.org/10.1073/pnas.1417276111

[15] ELSHEIKH, Ammar H.; DENG, Wu; SHOWAIB, Ezzat A. Improving laser cutting quality of polymethylmethacrylate sheet: experimental investigation and optimization. Journal of Materials Research and Technology, 2020, 9.2: 1325-1339. https://doi.org/10.1016/j.jmrt.2019.11.059

Wear resistance of as-cast ZnMg-Y biodegradable alloy

Alexandra-Tamara ȘUTIC[1,a], Cătălin PANAGHIE[1,b], Viorel PALEU[2,c], Ramona CIMPOEȘU[1,d], Ana-Maria ROMAN[1,e], Margareta COTEAȚĂ[3,f], Liviu ANDRUȘCĂ[2,g], Gheorghe BĂDĂRĂU[1,h], Petronela PARASCHIV[3,i], Nicanor CIMPOEȘU[1,j,*]

[1]Faculty of Materials Science and Engineering, "Gheorghe Asachi" Technical, University of Iasi, 41 Dimitrie Mangeron Blvd., 700050 Iasi, Romania

[2]Faculty of Mechanical Engineering, "Gheorghe Asachi" Technical University of Iasi, 43 Dimitrie Mangeron Blvd., 700050 Iasi, Romania

[3]Faculty of Machine Manufacturing and Industrial Management, "Gheorghe Asachi" Technical University of Iasi, 59A Dimitrie Mangeron Blvd., 700050 Iasi, Romania

[a]alexandra-tamara.sutic@student.tuiasi.ro, [b]eu_catalin2006@yahoo.com, [c]viorel.paleu@academic.tuiasi.ro, [d]ramona.cimpoesu@academic.tuiasi.ro, [e]ana-maria.roman@academic.tuiasi.ro, [f]margareta.coteata@academic.tuiasi.ro, [g]liviu.andrusca@academic.tuiasi.ro, [h]gheorghe.badarau@academic.tuiasi.ro, [i]petronela.paraschiv@academic.tuiasi.ro, [j]nicanor.cimpoesu@academic.tuiasi.ro

Keywords: Biodegradable Alloy, Biomedical, Friction, Wear, SEM, EDS

Abstract. Biodegradable metallic materials offer a viable option for numerous medical devices where mechanical support of the implant is not required indefinitely. In some practical applications, biocompatible/biodegradable metallic elements are mechanically stressed by contact friction, compression, shear, or tensile forces. Metal samples (50x10x3 mm) were fabricated from cast ZnMgY ingots. An Amsler machine was used to apply frictional stress to the samples by placing the test specimen at the upper point of the tribological contact of the machine. The rotating element of the mechanical equipment is made of bearing steel (ASTM A295). During the friction testing of the experimental ZnMgY samples, the frictional torque was recorded using tensiometric measurements in conjunction with a Vishay P3 strain gage (Vishay, Germany). The recorded values were further processed by Savitzky-Golay filtering. Wear tests (20 N and 30 N load applied) lasted 1800 s at a constant low disc speed of 100 rpm and were performed in dry mode. Scanning electron microscopy (SEM 2D and 3D) and energy-dispersive X-ray spectroscopy (EDS) were used to evaluate wear marks.

Introduction

Wear resistance is an essential property of materials used in various industrial applications, particularly in the automotive and aerospace sectors; it also plays an important role in medical applications [1]. The choice between zinc- and magnesium-based alloys is often based on their mechanical properties, including wear resistance [2]. Zinc alloys are widely used because of their excellent casting properties, corrosion resistance, and mechanical strength [3]. They are often alloyed with aluminum, copper, and magnesium to improve their properties [4]. Zinc alloys generally have a fine-grained structure that can be further refined via procedures such as heat treatment and alloying. The presence of alloying elements can lead to the formation of intermetallic compounds that improve wear resistance [5,6]. Zinc alloys typically have moderate hardness and tensile strength. For example, ZA-8 (a zinc-aluminum alloy) has a tensile strength of approximately 350 MPa and a hardness of 85 HB. The presence of alloying elements, particularly aluminum and magnesium, enhances these properties, leading to increased wear resistance [7].

Zinc alloys exhibit good wear resistance under certain conditions. Studies have shown that ZA alloys can exhibit wear rates as low as 0.05 mm^3/Nm when tested under specific conditions (e.g. dry sliding). Their performance is often improved by surface treatments such as zinc plating, which can further increase the hardness and reduce friction. Magnesium alloys are generally lighter than zinc alloys, but they may have a lower tensile strength [8]. For example, AZ31B (a magnesium-aluminum-zinc alloy) exhibits a tensile strength of approximately 260 MPa and a yield strength of 150 MPa. Certain magnesium alloys, particularly those alloyed with aluminum, exhibit excellent tensile properties and can compete with zinc alloys. Magnesium alloys can exhibit good wear resistance, particularly when alloyed with aluminum or treated by surface hardening techniques. For example, AZ91D (a magnesium-aluminum alloy) exhibits wear rates of about 0.1 mm^3/Nm under similar test conditions [9]. However, they are generally less wear resistant than high-performance zinc alloys [10].

Both zinc and magnesium alloys are subject to similar wear mechanisms, including adhesive wear, abrasive wear and fatigue wear. However, the severity of these mechanisms can vary depending on the alloy composition and microstructure [11]. Zinc alloys generally exhibit better performance in corrosive environments because of their inherent corrosion resistance (with a corrosion rate of 0.1 mm/year). Although they are light and strong, magnesium alloys can corrode more rapidly (with a corrosion rate of up to 1.5 mm/year in saline environments), which can affect their wear resistance over time. The choice between zinc- and magnesium-based alloys in terms of wear resistance depends on the specific application and environmental conditions. Zinc alloys tend to offer better wear resistance because of their hardness and corrosion resistance, making them suitable for various industrial applications. Magnesium alloys, on the other hand, offer significant advantages in terms of weight and strength but may require additional treatment or alloying to increase their wear resistance. Further research and development into alloy compositions and surface treatments may lead to improved wear resistance for both zinc and magnesium alloys, thus extending their applicability in advanced engineering.

In this study, the authors tested the wear resistance of three Zn-based materials to analyze the influence of alloying zinc with magnesium and yttrium. The behaviors of worn zones in the electrolyte solutions were also observed compared with those of unworn zones. Adding Mg and Y to Zn improves the wear resistance of pure zinc.

Materials and Experimental Details

The experimental materials (pure Zn, Zn3wt%Mg and Zn3wt%Mg0.6wt%Y) were used to investigate the behavior of the Zn-based material toward mechanical wear. The structure and chemical constituents of the alloy were investigated in a previous study [12,13].

The AMSLER tribometer allows different types of samples to be tested (disk on disk, shoe on disk and pad on disk). The parameter to be measured is the friction torque. Dead weights are used to apply the load. To measure the frictional torque, tensiometric measurements were performed in association with a Vishay P3 tension bridge (Vishay, Germany). The use of 6th order Savitzky-Golay filtering in the LabVIEW data post-processing interface provided the results shown in Fig. 1. The design and construction of the AMSLER tribometer are detailed in previous studies [14-17].

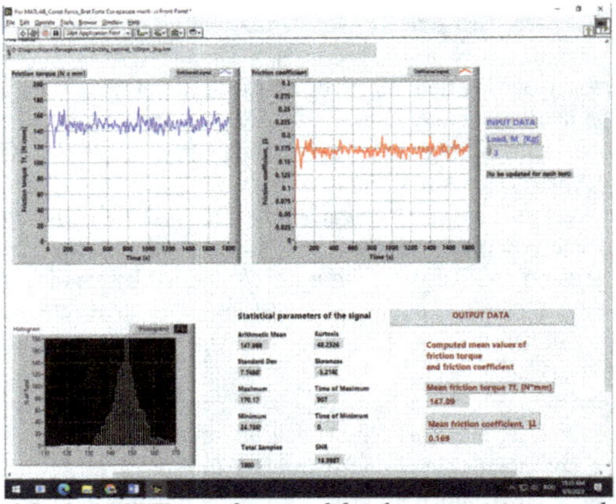

Fig. 1. Screenshot of the LabVIEW interface used for data acquisition on the AMSLER machine (dry lubrication, Zn3Mg sample, speed 100 rpm, load 30 N).

The input data consists of the applied load (which must be updated for each specific test) and the diameter of the lower rotating disk (which remains the same). The upper sample is the stationary coated pad being tested. The output data consists of the average friction torque, average friction coefficient and statistical parameters of the recorded signal. Each test lasted 30 minutes. A constant lower disk speed of N=100 rpm was used for all tests. The applied load for the dry tests was 20 N respectively 30 N. If the signal-to-noise ratio (SNR) is greater than 2, this indicates a high standard of friction torque signal measurement, with a high quality of signal acquisition around SNR=1. For Gaussian signals, the Kurtosis is 3; a K value > 3 indicates well weighted data with low noise.

The surface of the alloys after wear has been analyzed using a macro-optical instrument for macroscopic observations and scanning electron microscopy (SEM VegaTescan LMH II, 2D and 3D surface images) for micro-scale investigations. A Bruker EDS detector was used to determine the chemical composition after wear and immersion in a simulated body fluid (SBF).

Experimental Results

The cast Zn, ZnMg, and ZnMgY samples were subjected to wear tests on a steel roller, as described in the experimental setup section. After the wear tests, the behaviors of the samples were compared. Fig. 2 (Q=20N) and Fig. 3 (Q=30N) show a comparison of the frictional torque variations over time for S01 (pure Zn), S02 (Zn3Mg), and S05 (Alloy III Zn3Mg0.6Y) for both loads. The addition of Mg reduced the frictional torque, and a similar behavior was observed for the ZnMg and ZnMgY materials, with greater variation in the values for the Y-based alloy, probably due to the presence of YZn12 compounds, which are harder than the α-Zn solid solution or ZnMg-based compounds. Stabilization of the surface response was observed approximately 5-6 minutes after the start of the test (Fig. 2).

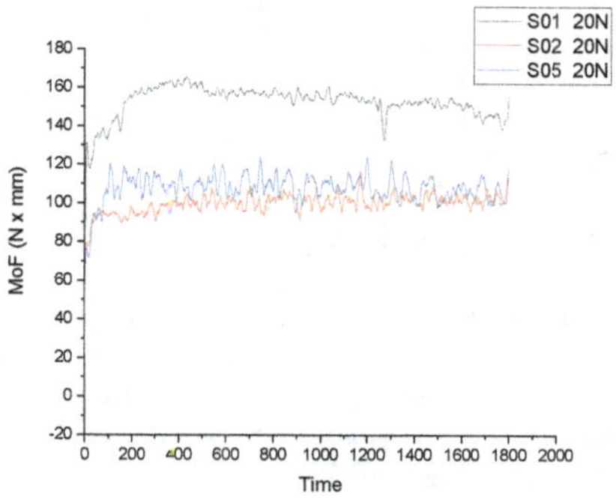

Fig. 2. Moment of friction versus time for S01 (pure Zn), S02 (Zn3Mg) and S05 (Zn3Mg0.6Y) at Q=20 N.

Fig. 2 and Fig. 3 show that for both 20N and 30N applied loads, sample S05 (Zn3Mg0.6Y) gives a stable and low coefficient of friction.

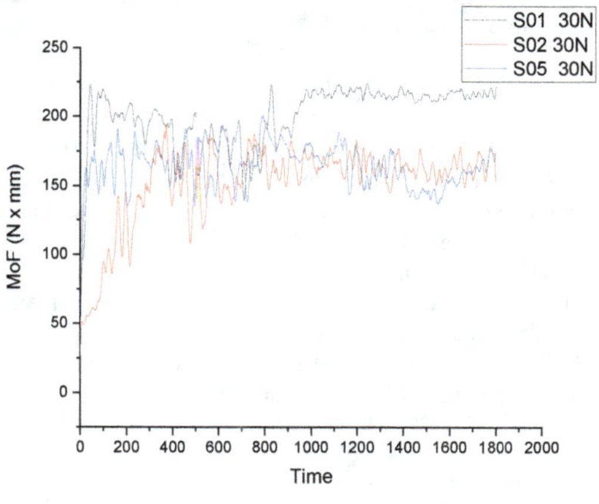

Fig. 3. Moment of friction versus time for S01 (Zn pure), S02 (Zn3Mg) and S05 (Zn3Mg0.6Y) at Q=30 N.

In order to determine the wear rate, each sample had to be weighed prior to and following the friction test. Under an uninterrupted rate of 100 rpm, each total sliding distance for all friction test was 556 m. Table 1 shows the data gathered for a constant load (Q = 30 N). Archad's widely recognised equation was used to assess the wear rate (Eq. 1) [16].

$$W = \frac{M}{Q \cdot D} \tag{1}$$

With W being the wear rate, M the mass loss [g], Q the applied normal load [N] and D the sliding distance (556 m).

Table 1. Mass loss and wear rate for different samples tested at Q=30 N.

Sample	Notation	Mass [g]		Mass loss, [g]	Relative mass loss, [wt. %]	Wear rate, $W = \frac{M}{Q \cdot D}$
		Initial	Final			
pure Zn	S01	18.4	18.353	0.020	0.1089	7.193E-04
Zn3Mg	S02	7.3	7.254	0.016	0.2201	5.755E-04
Zn3Mg0.6Y	S05	12.5	12.461	0.009	0.0722	3.237E-04

At Q = 30 N, the best wear resistance was obtained for sample S05 (Zn3Mg0.6Y), confirming the influence of Y addition. The lowest wear resistance and highest wear rate were obtained for sample S08 (pure Zn). The wear rate values were confirmed by surface staining (Fig. 4 - optical microscopy and Fig. 5 - scanning electron microscopy). The wear marks were larger for Zn and ZnMg and smaller for the ZnMgY alloy (Fig. 4 (c)).

(a) (b) (c)

Fig. 4. Optical images of the wear zone for pure Zn (a), ZnMg alloy (b) and ZnMgY (c).

All of the marks show material loss (Fig. 5 (b)), on the indented surface in the rolling direction of the steel roll. The affected surface also displays material fragments, oxides, and small pieces of material detached from the surface. Accumulation of overlapping materials can be observed at the edge of the wear mark, which is mainly caused by the difference in micro-hardness between the two contacting materials (Fig. 4 and 5). For Zn-based materials, no cracking was observed at both the macroscale (Fig. 4) and microscale (Fig. 5). This is favorable for maintaining the structural integrity of these materials and for the size of the wear debris or its subsequent degradation. Regarding biodegradable alloys, it is not desirable to break into large fragments, as this may cause severe discomfort or blockage in biological organisms.

(a)

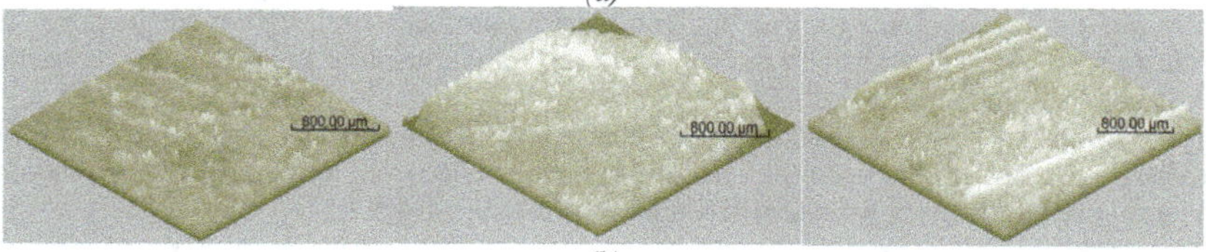

(b)

Fig. 5. SEM images of the worn surfaces: (a) 2D and (b) 3D from left to right: Zn, ZnMg and ZnMgY.

The samples were kept in SBF for 10 days following the mechanical wear testing. Figure 6 shows the emergence of chemical compounds, generally oxides and salts, but also locally phosphates, especially on the ZnMg sample (Fig. 6 (b)). On the pure zinc sample, Fig. 6 (a) shows a generalized oxidation of the surface exposed to the electrolytic solution, revealing the appearance of chloride- and sodium-based salts in the surface scratches.

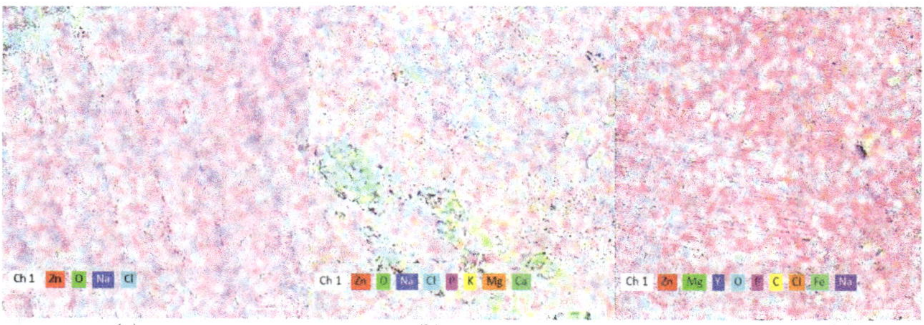

(a) *(b)* *(c)*

Fig. 6. Chemical element mapping on the worn zone: (a) pure Zn, (b) ZnMg alloy and (c) ZnMgY alloy.

The presence of salts leads to accelerated corrosion in this area due to the infiltration of Cl ions through the protective zinc oxide layer formed by surface passivation [18]. A more pronounced oxidation is observed in the worn area (Fig. 6 (c)) compared to the rest of the sample. On the worn surface, mainly due to the new surface profile created by the impregnation of the steel cylinder (Fig. 6 (b)), it can be observed that some compounds from the electrolyte solution were present on the surface and were retained, accelerating the corrosion process of the metallic material in this area compared to the unworn surface.

Conclusions

From the experimental results and discussion, the authors conclude as follows:

- Beside the improvement of hardness, based on the formation of different MgZn compounds the addition of Mg to pure Zn reduced the frictional moment and wear rate. This is particularly due to changes in mechanical properties.
- The addition of Y to ZnMg slightly increased the frictional moment and generally decreased the wear rate of pure Zn.
- The worn surface is more susceptible to corrosion due to the increased contact area with the electrolyte and the induced stresses on the surface;
- Maintaining a generalized corrosion, even on the worn zone, with no pitting observed, these alloys present the advantages of small corrosion products, useful for the degradation and elimination phase of these materials from the living organism after the mechanical objectives pursued by their implantation;
- In complex cases of bone regeneration, elements from different alloys may be used, for example, zinc-based and magnesium or iron-based, and the galvanic couples formed between these alloys can be studied by analyzing the electrochemical corrosion resistance.

Acknowledgement

This work was supported by a National Research Grants of the TUIASI, project number GNaC2023_285/2024.

References

[1] D. D. L. Lee, Wear Resistance of Zinc Alloys: A Review, J. Mater. Sci. 45 (2010) 4234-4242.

[2] T. Laurian, A. Tudor, I.V. Antoniac, F. Miculescu, A micro-scale abrasion test to study the influence of counterface roughness on the wear resistance of UHMWPE, J. Optoel. Adv. Mater. 9 (2007) 3383–3388.

[3] A.R. Khan, N.S. Grewal, C. Zhou, K. Yuan, H.J. Zhang, Z. Jun, Recent advances in biodegradable metals for implant ap-plications: Exploring in vivo and in vitro responses, Results Eng. 20 (2023) 101526. https://doi.org/10.1016/j.rineng.2023.101526

[4] D. Vojtech, J. Kubasek, J. Serak, P. Novak, Mechanical and corrosion properties of newly developed biodegradable Zn-based alloys for bone fixation, Acta Biomater. 7 (2011) 3515–3522. https://doi.org/10.1016/j.actbio.2011.05.008

[5] J. Venezuela, M. S. Dargusch, The influence of alloying and fabrication techniques on the mechanical properties, biodegradability and biocompatibility of zinc: A comprehensive review, Acta biomater. 87 (2019) 1-40. https://doi.org/10.1016/j.actbio.2019.01.035

[6] A.T. Şutic, R. Chelariu, R. Cimpoeşu, A.M. Roman, B. Istrate, V. Goanţă, M. Benchea, M. Moscu, A. Alexandru, N. Cimpoeşu, G. Zegan, Corrosion Behavior and Mechanical Properties of Zn–Ti Alloys as Biodegradable Materials, Metals 14 (2024) 764. https://doi.org/10.3390/met14070764

[7] N. Cimpoesu, V. Paleu, C. Panaghie, A.M. Roman, A.M. Cazac, L.I. Cioca, C. Bejinariu, S.C. Lupescu, M. Axinte, M. Popa, G. Zegan, Mechanical Properties and Wear Resistance of Biodegradable ZnMgY Alloy, Metals 14 (2024) 836. https://doi.org/10.3390/met14070836

[8] R. Vaira Vignesh, R. Padmanaban, M. Govindaraju, Study on the corrosion and wear characteristics of magnesium alloy AZ91D in simulated body fluids, Bull. Mater. Sci. 43 (2020) 8. https://doi.org/10.1007/s12034-019-1973-3

[9] B. Istrate, C. Munteanu, I.V. Antoniac, S.C. Lupescu, Current Research Studies of Mg–Ca–Zn Biodegradable Alloys Used as Orthopedic Implants—Review, Crystals 12 (2022) 1468. https://doi.org/10.3390/cryst12101468

[10] T. Wu, K. Zhang, Corrosion and Protection of Magnesium Alloys: Recent Advances and Future Perspectives, Coatings 13 (2023) 1533. https://doi.org/10.3390/coatings13091533

[11] B. Istrate, C. Munteanu, M.S. Bălțatu, R. Cimpoeșu, N. Ioanid, Microstructural and Electrochemical Influence of Zn in MgCaZn Biodegradable Alloys, Materials 16 (2023) 2487. https://doi.org/10.3390/ma16062487

[12] C. Panaghie, R. Cimpoesu, B. Istrate, N. Cimpoesu, M.A. Bernevig, G. Zegan, A.M. Roman, R. Chelariu, A. Sodor, New Zn3Mg-xY Alloys: Characteristics, Microstructural Evolution and Corrosion Behavior, Materials 14 (2021) 2505. https://doi.org/10.3390/ma14102505

[13] C. Panaghie, N. Cimpoeșu, M. Benchea, A.M. Roman, V. Manole, A. Alexandru, R. Cimpoesu, M.M. Cazacu, I. Wnuk, G. Zegan, "In-vitro" Tests on New Biodegradable Metallic Material Based on ZnMgY, Arch.Metall. Mater. 67 (2022) 587-594.

[14] C.C. Paleu, C. Munteanu, B. Istrate, S. Bhaumik, P. Vizureanu, M.S. Bălțatu, V. Paleu, Microstructural Analysis and Tribological Behavior of AMDRY 1371 (Mo–NiCrFeBSiC) Atmospheric Plasma Spray Deposited Thin Coatings, Coatings 10 (2020) 1186. https://doi.org/10.3390/coatings10121186

[15] C. Munteanu, V. Paleu, B. Istrate, A. Dascălu, C.C. Paleu, S. Bhaumik, A.D. Ancaș, Tribological Behavior and Microstructural Analysis of Atmospheric Plasma Spray Deposited Thin Coatings on Cardan Cross Spindles, Materials 14 (2021) 7322. https://doi.org/10.3390/ma14237322

[16] V. Paleu, S. Georgescu, C. Baciu, B. Istrate, E.R. Baciu, Preliminary experimental research on friction characteristics of a thick gravitational casted babbit layer on steel substrate, IOP Conf. Ser.: Mater. Sci. Eng. 147 (2016) 012028. https://iopscience.iop.org/article/10.1088/1757-899X/147/1/012028

[17] C. Paulin, D. Chicet, V. Paleu, M. Benchea, S. Lupescu, C. Munteanu, Dry friction aspects of Ni-based self-fluxing flame sprayed coatings, IOP Conf. Ser.: Mater. Sci. Eng. 227 (2017) 012091. https://iopscience.iop.org/article/10.1088/1757-899X/227/1/012091/meta

[18] T. Prosek, A. Nazarov, U. Bexell, D. Thierry, J. Serak, Corrosion mechanism of model zinc–magnesium alloys in atmospheric conditions, Corros. Sci. 50 (2008) 2216–2231. https://doi.org/10.1016/j.corsci.2008.06.008

Innovative Manufacturing Engineering and Energy - IManEE2024 Materials Research Forum LLC
Materials Research Proceedings 46 (2024) 260-268 https://doi.org/10.21741/9781644903377-34

Effect of deep cryogenic treatment on the mechanical properties of AlSi7Mg casting alloy

Daniela SPASOVA[1,a], Plamen STOYANOV[1,b], Plamen PETROV[1,c], Nikolay ATANASOV[1,d], Radostina YANKOVA[2,e]

[1] Materials Science and Technology Department, Technical University of Varna, Bulgaria

[2] Institute of Metal Science, Equipment and Technologies with Hydro- and Aerodynamics Centre "Acad. A. Balevski", Bulgarian Academy of Sciences, 1574 Sofia, Bulgaria

[a] danielats@abv.bg, [b]plamen_2002@abv.bg, [c]petpl@abv.bg, [d]nm_atanasov@abv.bg, [e]radyankova@ims.bas.bg

Keywords: Deep Cryogenic Treatment, AlSi7Mg Casting Alloy, Mg_2Si

Abstract. Alloys from the Al-Si system are not hardened by heat treatment and are mainly used for casting lightly stressed parts. If Mg (below 1%) is added to the Al-Si system, the hardening after quenching and aging is possible, as a strengthening secondary phase – Mg_2Si – is released. Therefore, the AlSi7Mg alloy is one of the most widely used alloys to produce cast automobile wheels, as well as for thin-walled castings with complex relief. In the present study, the possibilities of extra increasing of the mechanical properties of AlSi7Mg alloy by deep cryogenic treatment were investigated. The tests were carried out on samples of AlSi10Mg alloy previously hardened with T6 heat treatment mode (quenching at 530 °C temperature and artificial aging at 170 °C) and on specimens that have not been hardened by heat treatment.

Introduction

Aluminum alloys are one of the most widely used structural materials of nonferrous metals, possessing the following characteristics: low density, high corrosion resistance, high specific strength, good machinability, etc., and therefore are applied in various fields such as: aerospace and aviation industry, shipbuilding and boatbuilding, automotive industry, mechanical engineering, chemical industry and other fields [1,2]. With the rapid development of the industries mentioned above, the demands on the properties of aluminum alloys are increasing more and more. Therefore, it is particularly important to select appropriate processing methods to improve the mechanical and performance characteristics of aluminum alloys [3,4]. Deep cryogenic treatment (DCT) has a number of advantages in improving the properties of aluminum alloys compared to conventional heat treatment methods [5,6]. As an environmentally friendly, inexpensive and non-polluting new cold treatment process, DCT can greatly improve the microstructure and properties of metals and alloys with small deviations in their dimensions and shapes [7,8]. When the parameters of DCT are selected correctly, a temperature of -197°C can lead to a preferred grain orientation, and optimization of the secondary phase, in addition to increasing the dislocation density, distribution of aluminum alloys [9]. Furthermore, after DCT due to the microstructure improvements, the mechanical properties, corrosion resistance, wear resistance, electrical conductivity, and residual stresses of aluminum alloys can be significantly improved [10].

The alloys of the Al-Si system are not subjected to hardening heat treatment, but with the addition of Mg (less than 1%) to the Al-Si system, hardening by quenching and aging is possible, as a strengthening secondary phase - Mg_2Si - is released [11]. This makes the AlSi7Mg alloy one of the most used alloys for the production of automotive wheels, as well as for thin-walled castings and complex relief castings [12]. In low-pressure casting of AlSi7Mg alloy castings, a pressure of about 2 bar is used to obtain rapid mold filling and a dense structure, hence higher mechanical properties compared to other casting methods. This alloy possesses very good casting properties, especially in terms of mold filling and minimum shrinkage [13]. Another advantage is that this alloy is thermally ductile, which allows to further increase the strength of the workpiece. The most commonly used heat treatment is standardized as "T6" - quenching at 515-540°C and subsequent artificial aging at 170°C with a holding time of 8 hours [12, 14].

A number of scientists have investigated the mechanisms of the influence of DCT on the changes in the structure and mechanical characteristics of aluminum alloys. It is known that the microstructure of aluminum alloys consists mainly of aluminum matrix and secondary phases. In turn, the secondary phases determine the mechanical properties in aluminum alloys [6,15]. During DCT, due to the low temperatures (down to -197°C), the crystal lattice of aluminum alloys shrinks and it decreases in size, which leads to supersaturation of solute atoms and thus increases the driving force for their separation [10,16]. As going from high to low temperatures, the solubility of Si in aluminum alloys decreases, then Si may gradually separate from the Al matrix during DCT. Moreover, the super-low temperature hinders the diffusion of Si atom, which leads to the appearance of dispersed Si-rich region useful for the formation of Mg-Si clusters and β''-phase. After DCT, the separation of the secondary phases can not only be promoted but also their roughness can be reduced [11,17].

The present study analyzes the influence of DCT on the structure and mechanical properties of AlSi7Mg cast aluminum alloy in annealed and T6 mode heat treatment (HT) condition.

Materials and methodology of studies conducted

To investigate the influence of DCT on the structure and properties of AlSi7Mg cast alloy, microstructural and X-ray diffraction phase analysis, tensile testing and Vickers hardness measurement were carried out.

Investigated materials

AlSi7Mg alloy castings made by low-pressure die casting were used for the study. The chemical composition of the alloy is given in Table 1. To achieve the set tasks, specimens were made from the cast alloy for the various examinations and tests - reference specimens from a casting without thermal hardening and specimens from a thermally hardened casting with "T6" mode (quenching at 540°C and artificial aging at 170°C), some of which were subjected to DCT.

Table 1. Chemical composition of the AlSi7Mg alloy

Fe	Si	Mn	Ti	Cu	Be	Mg	Zn	Others	-
max 0.2	6.5 - 7.5	max 0.1	max 0.2	max 0.2	max 0.0003	0.3 - 0.45	max 0.1	each 0.05; total 0.15	Al - remainder

Deep cryogenic treatment methodology

AlSi7Mg alloy castings made by low-pressure die casting were used for the study. The chemical composition of the alloy is given in Table 1. In order to achieve the set tasks, specimens were made from the cast alloy for the various examinations and tests - reference specimens from a casting without thermal hardening and specimens from a thermally hardened casting with "T6" mode (quenching at 540°C and artificial aging at 170°C), some of which were subjected to DCT.

Materials Research Proceedings 46 (2024) 260-268 https://doi.org/10.21741/9781644903377-34

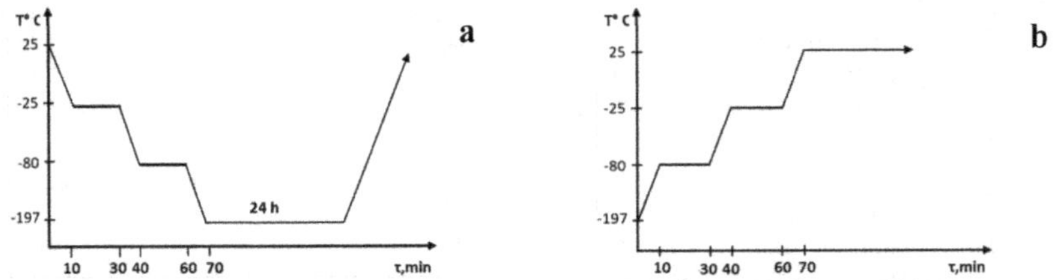

*Figure 1. Cryogenic treatment technology diagram- **a**- of cooling; **b**- on tempering to room temperature*

Tensile testing and hardness measurement

The tensile strength test was carried out according to BDS EN ISO 6892-1:2019 - Method for tensile testing of metallic materials at room temperature [18]. According to the standard guidelines, three specimens of each type were made from the cast AlSi7Mg alloy - reference specimens without heat treatment, specimens with heat treatment (T6 mode), specimens for cryogenic treatment without pre-hardening and hardened specimens for cryogenic treatment [19,20].

Hardness measurements were carried out by the Vickers method using an HVS-1000 digital hardness tester. For all specimens, the hardness was measured with a load of 1 kg in 3 different regions of the specimen and their value was averaged.

Structural analysis

The microstructure of the specimens was analyzed using Inverted Trinocular Microscope IM-3MET.

Qualitative (with phase identification) X-ray structure phase analysis was also carried out to identify phase changes. A standard digital-graphic radiograph recording methodology was used, using a TDM-10 Tongda X-ray diffractometer. X-ray emission with a copper tube with wavelength $\lambda = 1.79026$ Å at room temperature was used. XRD patterns were obtained in 2-Θ coupled mode in the range 30-90°, identifying and matching the peaks, using X'pert HighScore Plus software for XRD analysis.

Results and analyses

Tensile test results

Tables 2 show the obtained values of the mechanical properties after tensile testing and hardness measurements, and Figure 2 shows the graphical variations during tensile testing.

From the results obtained, it can be seen that the deep cryogenic treatment has a positive influence on the increase of the strength mechanical characteristics. After tensile strength testing, it was found that the influence of deep cryogenic treatment on the thermally treated specimens was more noticeable than for the specimens without heat treatment. In the non-hardened specimens, the tensile strength after DCT increased by only 5 units, while in the thermally hardened specimens the Rm value of 226 MPa increased to 245 MPa due to the additional dispersion of the chemical compound Mg_2Si due to the shrinkage of the crystal lattice of the α-solid solution during DCT. An increase in relative elongation of slightly more than 1% was also found, which is most likely due to a decrease in internal stresses.

Materials Research Proceedings 46 (2024) 260-268 https://doi.org/10.21741/9781644903377-34

Table 2. Values of mechanical properties after tensile strength test and hardness measurement

Samples	№	R_m [MPa]	Average R_m [MPa]	A [%]	Average A [%]	HV	Average HV
Reference sample (without HT)	1	150	149	6,64	6,82	65	67
	2	149		6,77		67	
	3	149		7,05		69	
Unhardened (without HAT) with DCT	1	154	154	5,52	4,65	68	68
	2	154		5,71		69	
	3	153		5,72		67	
with HT- T6 without DCT	1	222	224	5,35	5,73	91	93
	2	225		5,74		87	
	3	226		6,10		89	
with HT- T6 and with DCT	1	249	245	6,91	6,94	106	107
	2	243		6,57		107	
	3	244		7,34		108	

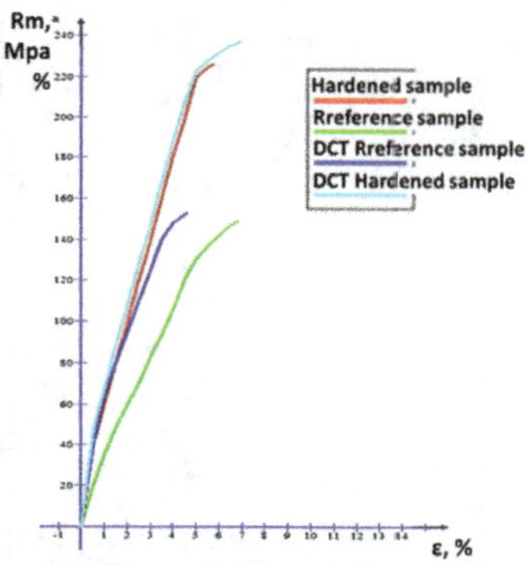

Figure 2. The tensile test diagram of the investigated samples

In the thermally hardened specimens subjected to DCT, the hardness increased significantly. On average, from 93HV measured hardness in the quenched specimens after DCT, the hardness increases to 107HV, i.e. 14 units. In contrast, in the thermally untreated AlSi7Mg specimens, the hardness after DCT remained unchanged.

Microstructural analysis results

From the microstructural analysis carried out, as expected, there is a thinning of the structure in the quenched specimens (Figure 3). In the specimens without heat treatment, a classical dendritic structure with a pronounced eutectic Eutectic(α+Si) around the α-solid solution is observed (Figure 3a, Figure 4a), with the structure being relatively uniform and homogeneous in the different zones. It is characteristic of the thermally hardened specimens that some of the Si from the eutectic dissolves in the α-solid solution, i.e., dispersed zones that are coherently bonded to the matrix are released upon reheating of the saturated α-solid solution. This classical mechanism leads to matrix hardening. The undissolved Si coagulated and led to the nucleation of the silicon crystals but was significantly reduced in quantity. The eutectic disintegrated and the coagulated Si crystals aligned along the α solid solution boundaries (Figure 4c).

Materials Research Proceedings 46 (2024) 260-268 https://doi.org/10.21741/9781644903377-34

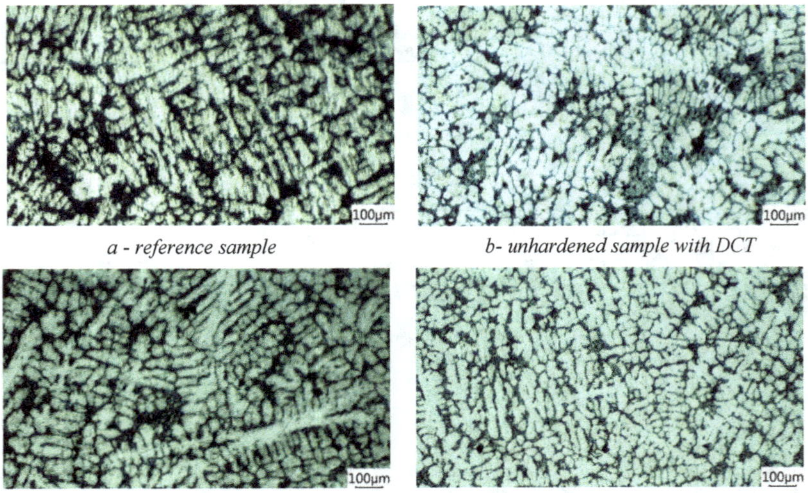

a - reference sample *b- unhardened sample with DCT*

c - sample with HT- T6 without DCT *d - sample with HT- T6 and with DCT*

Figure 3. Microstructures of cast aluminium alloy AlSi7Mg x40

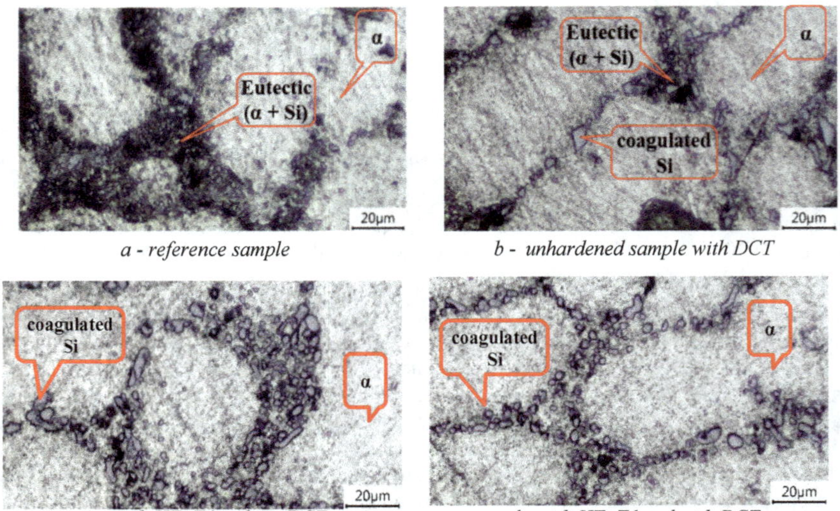

a - reference sample *b - unhardened sample with DCT*

c - with HT- T6 without DCT *d - with HT- T6 and with DCT*

Figure 4. Microstructures of cast aluminium alloy AlSi7Mg x400

In the specimens subjected to DCT without heat treatment (Figure 3b and Figure 4b), a partial decrease in eutectic was observed at the expense of the α-solid solution. The studies conducted by a number of world scientists proved that DCT leads to a shrinkage of the crystal lattice, which in turn increases the driving force for the precipitation of the atoms of the dissolved element at the grain boundary and leads to an increase in the strength mechanical characteristics [5]. It is characteristic of eutectic Eutectic(α+Si) that silicon forms a continuous spatial lattice in eutectic colonies. After DCT, due to the shrinkage of the crystal lattice, silicon fragments of the mechanical mixture are released and coagulate, resulting in a partial reduction in the quantity of mechanical mixture. Comparing the microstructures of the quenched specimens with and without DCT (Figures. 4c and 4d), it can be seen that the remaining small quantities of eutectic mixture after quenching decrease even more after conducting DCT. This fact is attributed to the grain shrinkage of the eutectic mechanical mixture, which in turn pushes the silicon out of it and leaves the α-solid solution from the eutectic and coagulated Si. The shrinkage of the crystal lattice of the α-solid solution also leads to the precipitation of the chemical compound Mg_2Si. Another observation is that some of the coagulated Si fragments suck up the smaller ones and thus become uniform in size and evenly distributed around the α-solid solution. This, in turn, accounts for the increase in strength and hardness in the DCT quenched specimens.

Results of the X-ray phase analysis
From the X-ray structural phase analysis carried out, no new phase states were detected in both the thermally unhardened and thermally hardened AlSi7Mg alloy specimens after the deep cryogenic treatment (Figure5, Figure 6). From Figure 5a,b it can be seen that there is not much difference between the phase diagrams of the thermally untreated specimens before and after DCT. From Figure 7a,b it can be observed that DCT leads to an additional precipitation of the β-phase Mg_2Si, the evidence for this process being a barely noticeable decrease in the broadening of the a-Al solid solution phase lines after DCT. Some of these peaks showing a splitting of the line into two distinct peaks (Figure 5b and 6b), which may be due to the formation of zones of solid solutions poorer in dissolved elements near the precipitations and others richer in such, with some difference in the lattice parameter, respectively.

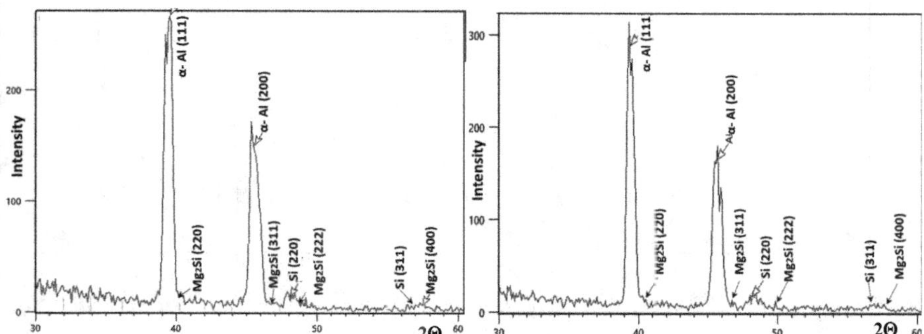

*Figure 5. X-ray diffraction patterns of: **a**- reference sample; **b**- unhardened sample with DCT*

There is also some shift of the peaks towards larger values of the 2Θ angle, i.e. towards smaller interfacial distances, suggesting a decrease in the lattice parameter of Al as the quantity of elements dissolved in it decreases. The increase in the quantity of precipitations is explained by the shrinkage of the crystal lattice of the α-solid solution due to DCT.Also, from Figure 7b, there is a broadening of the peak line in the DCT-conducted specimens compared to those in the reference thermally hardened specimens, which can be explained by the coherence relation of the precipitations with the crystal lattice of the alpha solid solution. A number of studies have shown that DCT leads to an increase in dislocation density, the area of interfacial boundaries blocking their movement and structure refinement, which explains the increased hardness and strength of the specimens after DCT.

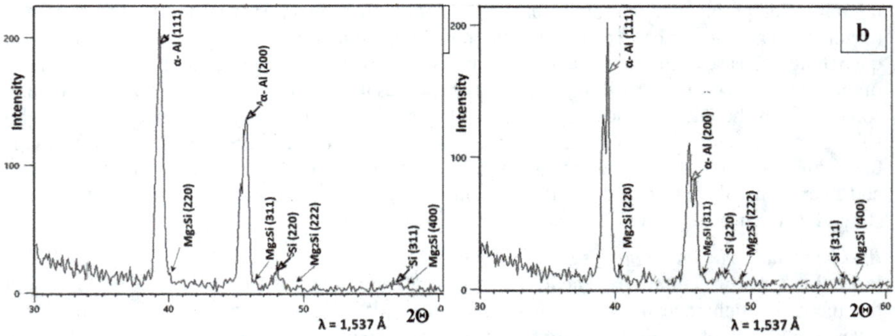

*Figure 6. X-ray diffraction patterns of: **a**- sample HT- T6 without DCT; **b**- sample with HT- T6 and with DCT*

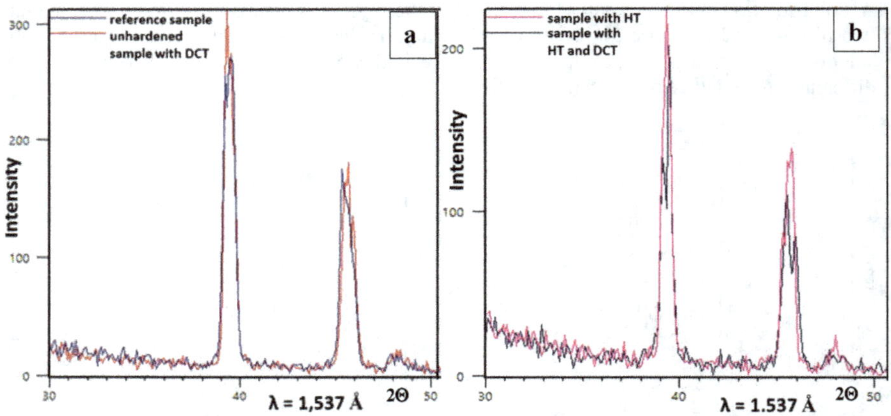

*Figure 7. X-ray diffraction patterns of: **a**- sample HT- T6 without DCT; **b**- sample with HT- T6 and with DCT*

Conclusions

On the basis of the experimental examinations carried out and on the basis of the results obtained, it can be summarized that the DCT treatment leads to better results in the T6 mode thermally treated specimens than in the specimens without preliminary heat treatment. This statement is supported by the mechanical tests, microstructural and X-ray phase analysis carried out. In the case of the non-hardened specimens, the tensile strength Rm after DCT increased by only 5 units, whereas in the case of the thermally hardened specimens the Rm value of 226 MPa increased to 245 MPa and the relative elongation increased by 1.2 %. Also, in the thermally hardened specimens subjected to DCT, the hardness increased significantly, on average from 93HV to 107HV, i.e. 14 units. In contrast, the hardness after DCT remained unchanged in the non-thermally treated AlSi7Mg specimens. The reasons for the increased mechanical properties were determined from the microstructural and X-ray phase analysis performed. From the microstructural analysis of the thermally hardened specimens, small amounts of eutectic mixture were found to remain, which after conducting DCT decomposed further, with the fragments of the precipitated coagulated silicon uniform in size and distributed uniformly quantity the α-solid solution. It was also found from the X-ray diffraction phase analysis that DCT leads to an increase in the quantity of β-phase Mg_2Si precipitations. This, in turn, was responsible for further increase in strength and hardness of the specimens already hardened by quenching after DCT was conducted.

Acknowledgments
The authors would like to kindly acknowledge the support of the scientific and research project NP 9/ 2024, financed on purpose from the state budget, within the framework of TU-Varna scientific research activity.

References

[1] M. Lech-Grega, W. Szymanski, B. Płonka, S. Boczkal, M. Gawlik, M. Bigaj, et al. The structure and properties of wrought aluminium alloys series 6xxx with vanadium for automotive industry. In: Light metals 2013. Hoboken, NJ, USA: John Wiley & Sons; 2013. pp. 527e32. https://doi.org/10.1007/978-3-319-65136-1_90

[2] A. Bankova, Z. Tsoneva, S.,Tenev, & M. Tachev,). Unfolds of non-expandable surfaces and some aspects of their practical application in transport engineering. In AIP Conference Proceedings (2023). https://doi.org/10.1063/5.0165446

[3] C. Song, B. Dong, S. Zhang, H. Yang, L. Liu, J. Kang,... & Q. Jiang. Recent progress of Al-Mg alloys: Forming and preparation process, microstructure manipulation and application. Journal of Materials Research and Technology, 2024., 25: pp. 411-418 https://doi.org/10.1016/j.jmrt.2024.07.051

[4] S. Zabunova, P. Petrov, D. Mincheva. Investigating the possibilities of nitrocarburizing of cast aluminium alloys in a gas environment. In: AIP Conference Proceedings. AIP Publishing, Vol. 3104, No. 1 (2024). https://doi.org/10.1063/5.0200338

[5] B. Pagidipalli, C. Nashine, & P. Bhansali. Influence of Deep Cryogenic Treatment on the Wear Behaviour of Different Al–Si Alloys. In Advances in Materials and Metallurgy: Select Proceedings of ICEMMM 2018 (2019), pp. 217-225. https://doi.org/10.1007/978-981-13-1780-4_22

[6] E. Yao, H. Zhang, K. Ma, C. Ai, Q. Gao, & X. Lin. Effect of deep cryogenic treatment on microstructures and performances of aluminum alloys: A review. Journal of Materials Research and Technology. (2023). https://doi.org/10.1016/j.jmrt.2023.08.140

[7] M. Jovičević-Klug, L. Tegg, P. Jovičević-Klug, G. Dražić, L. Almásy, B. Lim,... & B. Podgornik. (2022). Multiscale modification of aluminum alloys with deep cryogenic treatment for advanced properties. Journal of materials research and technology, 21, pp. 3062-3073. https://doi.org/10.1016/j.jmrt.2022.10.089

[8] S. Paraschiv, L. Paraschiv. Trends of carbon dioxide (CO2) emissions from fossil fuels combustion (coal, gas and oil) in the EU member states from 1960 to 2018. Energy Reports, 2020, 6: pp. 237-242. https://doi.org/10.1016/j.egyr.2020.11.116

[9] Z. Weng, X. Liu, K. Gu, J. Guo, C. Cui, J. Wang. Modification of residual stress and microstructure in aluminium alloy by cryogenic treatment. Mater Sci Technol 2020, 36:1547e55. https://doi.org/10.1080/02670836.2020.1800182.

[10] Z. Weng, X. Xu, B. Yang, K. Gu, L. Chen, J. Wang. Cryogenic thermal conductivity of 7050 aluminum alloy subjected to different heat treatments. Cryogenics (2021), 116: pp. 103305. https://doi.org/10.1016/j.cryogenics.2021.103305.

[11] D. Wang, S. Huang, Y. Yi, H. He, C. Li. Effects of cryogenic deformation on the microstructure and mechanical properties of high-strength aluminum alloys. Mater Char (2022), 187: pp. 111831. https://doi.org/10.1016/j.matchar.2022.111831.

[12] Information on https://www.castingarea.com/education/al/al-alloys-in-brief.htm

[13] B. Zhang, S. L. Cockcroft, D. M. Maijer, J. D. Zhu, & A. B. Phillion. Casting defects in low-pressure die-cast aluminum alloy wheels. Jom, 2005, 57: pp. 36-43. https://doi.org/10.1007/s11837-005-0025-1

[14] D. Spasova, Investigation of the Structural Improvements after Heat Treatment of the Aluminium AlSi7Mg Alloy Wheels, Volume I, Issue 4 pp. 526-531.

[15] W. GAO, X. Wang, J. Chen, C. Ban, J. Cui, & Z. Lu. Wenlin, et al. Influence of deep cryogenic treatment on microstructure and properties of 7A99 ultra-high strength aluminum alloy. Metals, 2019, 9.6: 631. https://doi.org/10.3390/met9060631

[16] Huang YC, Li Y, Ren XW, Xiao ZB. Effect of deep cryogenic treatment on aging processes of Al-Mg-Si alloy. Phys Met Metallogr 2019; 120:914e8. https://doi.org/10.1134/s0031918x19070111.

[17] D. Chen. Preferred grain orientation of aluminum and aluminum alloys after cryogenic treatment. Metall, (2008). 62(11), 744.

[18] Information on https://www.iso.org/standard/78322.html

[19] I. Mehmedov, S. Tenev, V. Manev, A. Bankova, & Z. Tsoneva. Static characteristic study of a modified Crankshaft pulley. In AIP Conference Proceedings Vol. 2868, No. 1, (2023). https://doi.org/10.1063/5.0165739

[20] A. Guzanová, E. Janoško, D. Draganovská, M. Vrabeľ, M. Tomáš, P. Horňak,... & N. Veligotskyi. Investigation of Applicability Flowdrill Technology for Joining Thin-Walled Metal Sheets. Metals, 2022, 12.4: 540. https://doi.org/10.3390/met12040540

Materials Research Proceedings 46 (2024) 269-276 https://doi.org/10.21741/9781644903377-35

Application of recycled carbon fibre to develop new products

R.C.M. SALES-CONTINI [1,3,a] *, H.H. BERNARDI[1,b], J.T. MATSUSHIMA[1,2,c], FRANCISCO J.G. SILVA[3,4,d], W.M.M. MENEZES[1,e]

[1]Technological College of São José dos Campos, Professor Jessen Vidal, Centro Paula Souza, São José dos Campos, São Paulo, Brazil

[2]National Institute for Space Research, INPE-SJC, São José dos Campos, São Paulo, Brazil

[3]CIDEM, ISEP, School of Engineering, Polytechnic of Porto, Porto, Portugal

[4]Associate Laboratory for Energy, Transports and Aerospace (LAETA-INEGI), Porto, Portugal

[a]rcms@isep.ipp.pt, [b]heide.bernardi@fatec.sp.gov.br, [c]jorge.tadao@fatec.sp.gov.br, [d]fgs@isep.ipp.pt, [e]william.menezes@cps.sp.gov.br

Keywords: Carbon Fibre, Recycling, Pyrolysis, Life Cycle Assessment, Waste Reduction, Product Development

Abstract. This work deals with the pyrolysis recycling process used to produce the recycled carbon fibres and their application in several areas: sustainable energy, aeronautical prototypes and even consumer goods developed by the DAMA Laboratory (Development and Advanced Materials Application Laboratory) at FATEC-SJC in Brazil. Composite materials made from recycled carbon fibre were produced using conventional manufacturing processes such as manual impregnation and vacuum bagging combined with oven heating. As the result shows, the recycled carbon fibre use is feasible, but the product's final validation requires testing to prove the performance of the recycled fibre.

Introduction

Structural composites based on epoxy and carbon fibre have been increasingly useful in the aeronautical industry and are a high-value-added product, whose use aims to reduce the weight and operating costs of aircraft. These have excellent mechanical properties such as low specific mass, high tensile strength, high specific resistance, excellent electrical properties and corrosion resistance [1]. These, in addition to having a long useful life, maintain structural quality throughout the period, however, their useful life is finite. However, its use generates waste, and extra costs for the company, as it must be discarded in landfills to be incinerated. The incorrect disposal of this waste is a worrying factor, as toxic compounds generated by these disposals can contaminate water and soil. A more feasible and sustainable approach would entail the recycling of the waste produced by the process in question, which would result in cost savings for the industry while simultaneously establishing a novel production methodology, thereby aligning with Goal 12 of the 2030 Agenda, which advocates that nations, corporations, and institutions "ensure sustainable consumption and production standards [2]. To reduce these costs, recycling is a more suitable process for reusing this material. So, the less noble way of reusing it has been through mechanical recycling, which is done by crushing the material and using it as reinforcement to process new composites. For sublime reuse, the thermal process by pyrolysis is used, but recycling does not occur directly and quickly, as the epoxy resin and carbon fibre are not recovered to reach their original characteristics, which leads to the use for other purposes [3-5].

The most diverse manufacturing processes generate various types of waste, which is a huge waste of material that ends up in the trash. In the aeronautical industry, it is no different. With the desire to make aircraft lighter to save fuel, composite material use has become widespread in this industry. Consequently, the amount of waste from these materials has increased in the same

Materials Research Proceedings 46 (2024) 269-276 https://doi.org/10.21741/9781644903377-35

proportion. Thus, aircraft manufacturers are looking for solutions to reuse this material. The pre-impregnated carbon fibre is one of the materials that companies sometimes lose, also known as prepreg, that is, a carbon fibre fabric that already comes with resin impregnated from the factory. If the company does not consume this composite within the expiration date, the resin can cure and the material becomes unusable. The same applies to scraps that are not used in the production process.

On an international scale, recycling took on a new dimension when the two biggest competitors in the aviation industry, Airbus and Boeing, joined forces. In 2008, they optimised the recycling process and added value to aircraft materials, reducing the amount of waste discarded at the end of the product cycle [6]. Recycled carbon fibre has the potential to be cheaper than virgin fibre, opening up a new market and new opportunities in different industries [6]. By 2030, it is expected that up to 8,000 aircraft will reach the end of their life cycle [7].

The use of carbon fibre reinforced carbon composite (CFRC) in the construction of an aircraft involves its use in interiors, as well as in structural components, and since the launch of the Boeing 787, it can exceed 50% in structural mass [8]. However, with the increased use of this material, an increase in waste is generated, which can occur through the expiration of prepregs, manufacturing scraps, test materials, production tools and the end of life of components [9]. The European aerospace industry, for example, produces around 8,000 tons of this type of waste per year [10].

According to AMS (Automotive Manufacturing Solutions), in 2015, around 18,000 tons of carbon fibre waste were produced from manufacturing operations worldwide, of which only 1,600 tons were recycled [11]. According to ABMACO [12], Brazil is responsible for producing 130 thousand tons of polymer composites, accounting for approximately 10% of losses in the production process and landfill disposal costs in the order of R$90 million. Its increasing use means that environmental legislation provides a strong incentive to seek new technologies for reusing this waste.

In general, the reuse of polymer composite fibres involves thermal or chemical processes, which are complicated for industries because they include undesirable solvents in recycling cycles [13-15].

Studies on the reuse of carbon fibres by thermal recycling show that it is possible to reuse carbon fibres while maintaining up to 80% of their original mechanical characteristics, as demonstrated by Park et al. [15], where researchers worked with a temperature range of 200 to 600°C, at ambient pressure with CRFC produced from epoxy resin, achieving favourable responses with monitoring of tensile, compression, impact and toughness properties in phenolic composites obtained from recycled CF, establishing 400°C as the best thermal condition for treatment. Intending to give a second use to prepreg waste, this work proposes to make prototypes using pre-impregnated carbon fibre to be used in several areas such as sustainable energy (supercapacitor electrodes and wind blades and spinners), aeronautical prototypes (speed brakes, drone parts (wing, wing spar and wheel)), and even consumer goods (sand ladder platform and skateboard).

Materials and Methods

Prepreg thermal treatment - Pyrolysis process

The pre-impregnated residues (Hexcel 8552/AS4) were thermally treated at 500 ± 10 °C for 4 h in an argon atmosphere using an Inforgel furnace. Figure 1 shows the pre-impregnated residue images as received and placed in the furnace before the pyrolysis process.

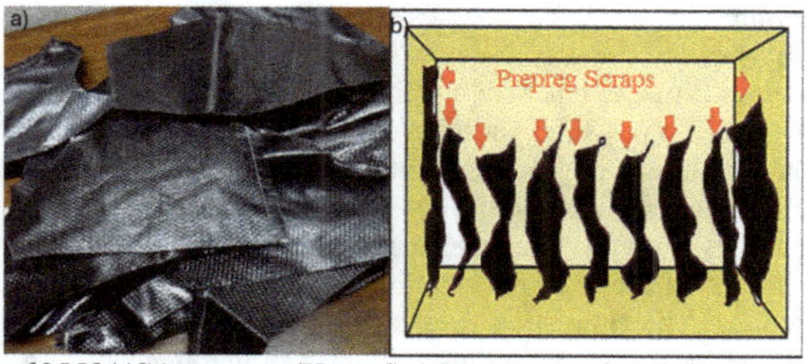

Figure 1. Scraps of 8552/AS4 prepreg (Hexcel) (a) as received and (b) placed in the furnance. Adapted from [1].

Procedures used for manufacturing prototypes

The common material used for prototype manufacturing was prepared plain wave fabric, AS4 3K 200 g/m^2 (HexPlyTM) [16] recycled via the pyrolysis method.

To manufacture the drone parts (wing, wing spar and wheel), wind spinner, and sand ladder platform, the recycled carbon fabric was combined with resin/hardener system Araldite LY 5052 (HuntsmanTM) and Aradur 5052 (HuntsmanTM) in a mix ratio of 100:38 parts by weight. To produce the skateboard prototype, recycled carbon fabric was combined with a resin/hardener system Araldite was added to an ivory wood lamina. All these laminates were produced by the hand lay-up process (HLUP) followed by vacuum bagging (VB). All laminates were cured at 60 °C for 2 h and post-cured at 80 °C for 2 h in a Makel double electric oven according to the resin data sheet [17]. Upon completion of the curing process, the cured laminates were demoulded and the finishing processes of cutting, drilling and milling were performed. After this, the prototypes were subjected to the tests. The prototype was modelled using the software *SolidWorks*. The deformation was calculated with a FEM (Finite Element Analysis). After the calculations, CFD (Computational Fluid Analysis) were made using the software SIEMENS Simcenter STAR-CCM+ and Hyperworks.

To produce the speed brake, one hundred (100) layers of FCR fabric were manually impregnated with a 60/40% by mass mixture of phenolic resin (Ph) and Si, stacked and cured in a vacuum bag at 180°C and 0.7 MPa pressure. The final FCR/Si polymer composite was heated at 2°C/min until reaching 1000°C in an inert Ar atmosphere. During this process, the phenolic resin was converted to glassy carbon, resulting in a vitreous carbon matrix FCR/Si composite. The FCR/Si composite was heat treated at 1600°C in an inert Ar atmosphere at a heating rate of 6°C/min. Once the temperature was reached, the furnace was turned off and the sample was cooled in the furnace, still under an inert atmosphere, to form the recycled FCR/SiC composite (CRFCR/SiC). The tribological test was carried out following the ASTM G99-05[18] standard, using pins obtained from the moulded composites and tracks made with material obtained from a stator disc of an AMX brake made in CRFC with a pyrolytic carbon matrix.

To produce supercapacitor electrodes, the recycled carbon fibres were subjected to a chemical oxidation process using 65% by volume nitric acid, an EVEN HJ-4 magnetic stirrer with heating and a Sterilifer SX 1.3 AS analogue drying oven. Thermal oxidation was carried out in a tube furnace. The electrochemical characterization was performed using the AUTOLAB potentiostat, model AUT84095 and type PGSTAT302N.

Materials Research Proceedings 46 (2024) 269-276 https://doi.org/10.21741/9781644903377-35

Results and Discussions

Aeronautical prototypes

In aviation, a new aircraft model has been gaining ground in recent times, the unmanned aerial vehicle, better known as UAV, which participates in various types of operations from rescue to monitoring large areas and has a high level of onboard and structural technology [19].

To reduce UAV production costs and have a sustainable production process, it is necessary to use sustainable materials. For this purpose, wing, spar, wheels and airbrake prototypes were developed to test the use of recycled carbon fibres through the pyrolysis process based on the AMT-600 GURI project. The simulation results using a prototype made of commercial carbon fibre as a database were compared with those obtained in the mechanical test for the prototype made of recycled carbon fibre.

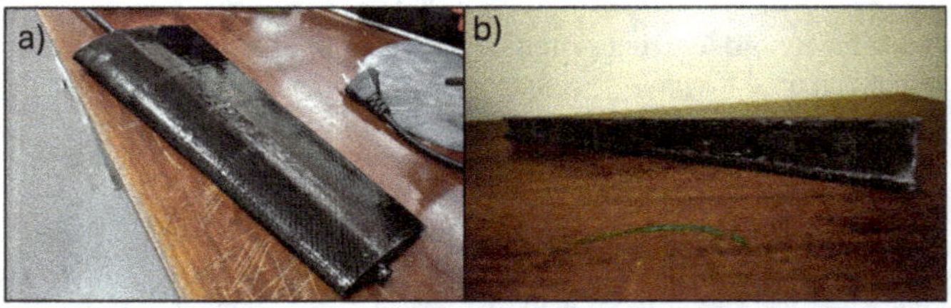

Figure 2: a) recycled carbon fibre wing and b) recycled carbon fibre spar.

Figure 2.a shows a wing made from recycled carbon fibre. The CFD and wind tunnel tests revealed that the wing deforms under aerodynamic load, with the spar's deformation differing by 22.86% from the expected theoretical value. When subjected to lift force, the recycled carbon fibre spar (Figure 2.b) shows a slight improvement compared to the original spar. However, when drag force is applied, the original spar outperforms the recycled carbon fibre version. In the end, the stress from the drag force seems to be why the recycled carbon fibre spar is less effective than the original. The simulation results from HyperWorks for both the recycled carbon fibre spar and the original spar are very similar. Therefore, it can be concluded that the spar made from recycled carbon fibres has nearly the same properties as the original AMT-600 GURI spar.

Several studies were carried out on some UAV models, such as the Hermes 450 used by the Brazilian Air Force [20], the vast majority of which use wheels with closed rims without the presence of through slots (Figure 3). This type of profile is more resistant to impact, does not have stress concentration points such as sharp corners and reduces drag in flight since the landing gear in the UAV is not retracted, in addition to being a less complex profile and easier to produce on a large scale.

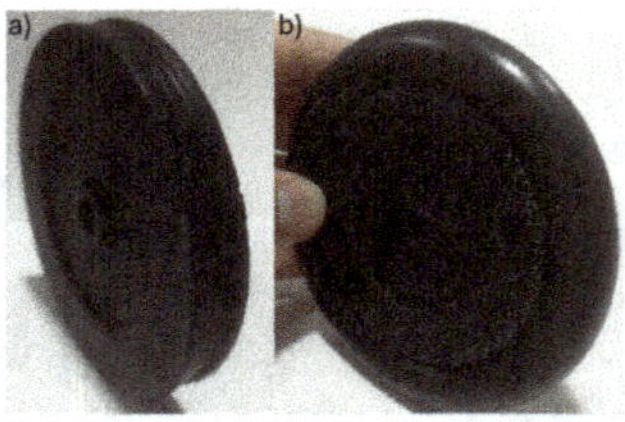

Figure 3: a) Wheel after machining and b) wheel with rubber tyre.

Tribological tests show that the recycled composite, despite similar friction coefficients, has a 40% higher wear rate when compared to the composite obtained with commercial carbon fibre (Figure 4).

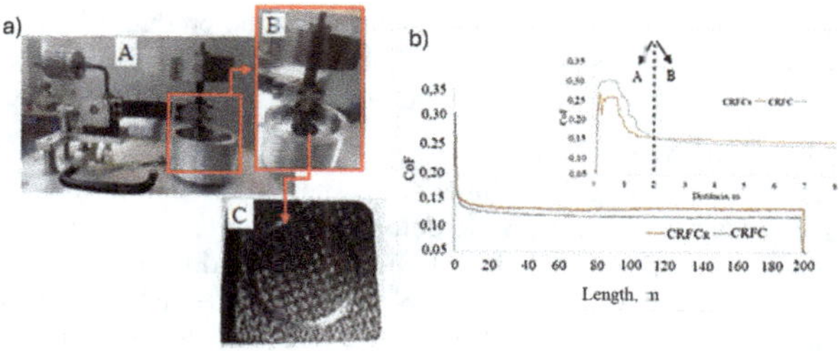

Figure 4: Pin-on-disc test: a) A) General view, B) Focus on pin-on-disc contact, C) Wear generated on the road surface and b) Comparative friction coefficient between CRFC$_R$ and CRFC composites.

Consumer goods prototypes

For the skateboard development, three laminate designs were examined. The first and third designs included two ivory wood laminates with varying resin types, one being aeronautical and the other commercial, along with recycled carbon fibre. The second design contained three ivory wood laminates with recycled carbon fibre and utilized commercial resin. The laminate samples underwent flexural testing [21]. Following the flexural tests, it was determined that the third design would be produced as it showed a mechanical strength improvement of over 18.65% compared to the first design, and it would be 87% cheaper to produce due to the use of commercial resin. The second design was not chosen even though it had a mechanical strength increase of 20.18% over the first design because it was oversized and added 30% more weight.

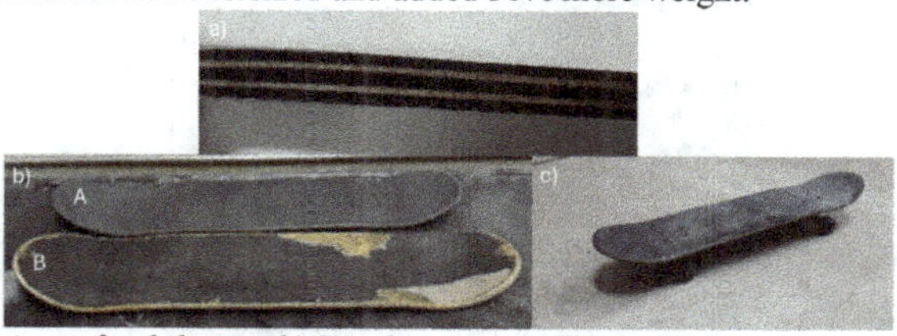

Figure 5: a) Laminate final design, b) Final prototype design (A) to market standard (B) and c) Skateboard made with recycled carbon fibre.

Figure 6. Field assessment: a) initiation phase; b,c) movement of the vehicle across the specified segment; d) termination of movement [1].

The engineered segment of the sand ladder platform (Figure 6) demonstrates the capacity to withstand the operational conditions for which it was originally designed; nevertheless, the parameters associated with the manufacturing process demand optimization to realise a product that embodies characteristics more congruent with the intended design and exhibits enhanced final quality. This undertaking necessitates further inquiries into the manufacturing process as well as the product design methodology. According to the sand ladder platform calculation, the product is anticipated to bear a load of 4110.39 N, equivalent to 420 kg. Considering that the vehicle employed has an estimated 433 kg, one may deduce that the product is suitably engineered. A meticulous visual assessment was performed on the prototype after the testing phase, disclosing no apparent damage to the product [1].

Sustainable energy prototypes

During the compressive test [22], the spinner was expected to split, however, this split did not occur. The spinner tested underwent plastic deformation approximately 60 mm from its initial height. Figure 7 shows the spinner before and after the test and the deformation suffered by the material. An elastic deformation on the prototype was also observed, confirmed by the machine displacement, which was approximately 74 mm. In addition to these deformations, it was possible to observe buckling in some areas near the base of the spinner. The maximum load supported by the prototype was 9630 N, which is 28% lower than that considered for the spinner manufactured with commercial carbon fibre fabric.

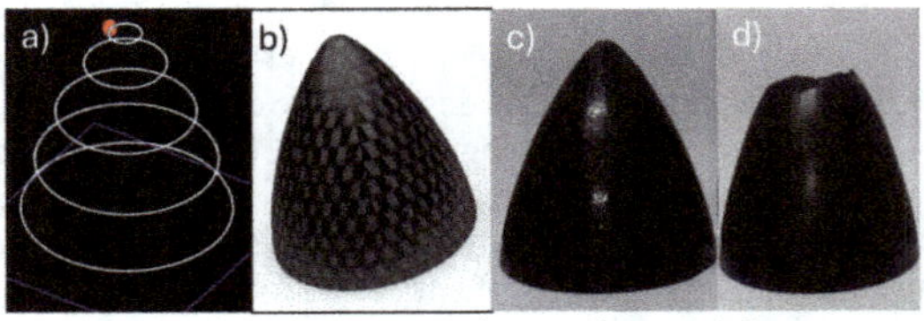

Figure 7: a) Reading, b) Modeling of the manufactured part, c) Spinner, and d) Spinner after compression test.

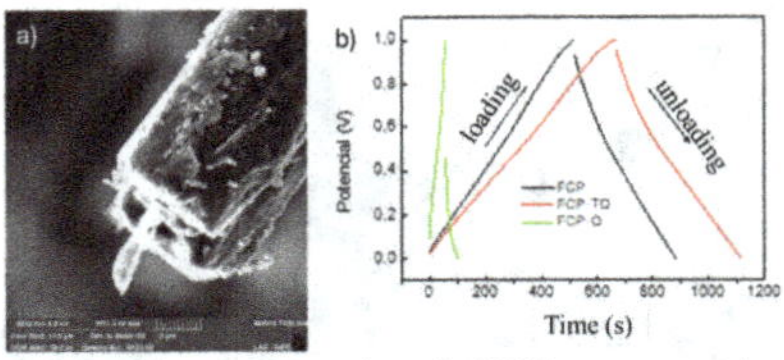

Figure 8: a) Pyrolysed carbon fibre oxidised with 65% nitric acid, detail at 20kx magnification, and b) Charge and discharge curves obtained in 2.0 mol L^{-1} H_2SO_4 solution established at a current of 1 mA [23].

Analysing the galvanostatic curves obtained during the third cycle, the charge and discharge profiles for the FCP (pyrolysed carbon fibre), FCP-TQ (pyrolysed carbon fibre subjected to thermal chemical treatment), and FCP-Q (pyrolysed carbon fibre exposed to chemical treatment) samples reveal that the charging and discharging processes transpire in an almost linear fashion for both FCP and FCP-TQ, as illustrated in Figure 8. The observed linearity of the charge/discharge curves is attributed to phenomena occurring at the purely capacitive electrode/electrolyte interface. Furthermore, an additional assessment can be made regarding the duration of the processes involved; extended timeframes for a specified current imply an enhanced capability for energy storage and delivery across a defined potential range, as evidenced in the FCP and FCP-TQ samples when contrasted with the FCP-Q.

Summary

The empirical findings derived from mechanical assessments indicate that recycled composites exhibit a greater degree of mechanical loss in comparison to those composed of commercial fibres. The degradation of sizing during the pyrolysis procedure significantly influences the outcomes of mechanical evaluations, thereby underscoring the criticality of a robust interface. The methodologies selected for the design, fabrication, and evaluation of the recycled carbon fibre prototypes were deemed suitable and yielded favourable results. The prototypes demonstrated exceptional final quality, characterised by notable mechanical strength. In the context of sustainable energy prototypes, the electrochemical assessments conducted on recycled fibres illustrated the feasibility of recycling, as these materials exhibited commendable capacitive responses, with characteristic values akin to those of electrochemical capacitors, thereby affirming their prospective application as electrodes for supercapacitors. Consequently, it can be inferred that the pyrolysis procedure implemented was effective in realising the anticipated reduction in mechanical strength for the recycled composite, suggesting that these materials may still possess utility in a range of applications beyond their initial intents.

Funding
This research was partially funded by scientific initiation scholarships (CPS #5498/2016, CNPq #800319/2016-8, #162840/2017-9, and #143500/2019-8).

Acknowledgements
Institutional support: Centro Paula Souza–FATEC/SJC-Brazil and IPP-CIDEM/ISEP-Portugal. Infrastructure and equipment: Center of Competence in Manufacturing (CCM), the Laboratory of Materials and Processes (LMP) at ITA, the Department of Aerospace Science and Technology – Materials Division (DCTA-AMR), and the Associated Laboratory of Sensors and Materials (LAS) at INPE, Brazil. Embraer for supplying prepreg scraps and Texiglass for providing commercial carbon fibres.

References

[1] R.C.M. Sales-Contini, H.M.S. Costa, H.H. Bernardi et al. Mechanical Strength and Surface Analysis of a Composite Made from Recycled Carbon Fibre Obtained via the Pyrolysis Process for reuse in the Manufacture of New Composites. Mater., 17(2), 2024, 423. https://doi.org/10.3390/ma17020423

[2] ONU, 2015 Transforming our world: the 2030 Agenda for Sustainable Development, 2015/2018

[3] A. Jacob, Composites can be recycled. Reinf. Plast., 55 (3) (2011) 45-60. https://doi.org/10.1016/S0034-3617(11)70079-0

[4] D. Åkesson, Z. Foltynowicz, J. Christéen et al. Microwave pyrolysis as a method of recycling glass fibre from used blades of wind turbines. *J. of Reinf. Plast. and Comp.* 31(17), (2012), 1136-1142. doi:10.1177/0731684412453512

[5] G. Oliveux, L.O. Dandy, G.A. Leeke, Current status recycling of fibre reinforced polymers: Review of Technologies, reuse and resulting properties, Prog. in Mater. Sci., 72 (2015) 61-99, https://doi.org/10.1016/j.pmatsci.2015.01.004

[6] W. Carberry, Airplane Recycling Efforts Benefit Boeing Operators. 2008. Information on: <http://www.boeing.com/commercial/aeromagazine/articles/qtr_4_08/article_02_1.html

[7] S.Y. Ye, A. Bounaceur, Y. Soudais *et al.* Parameter Optimization of the Steam Thermolysis: A Process to Recover Carbon Fibers from Polymer-Matrix Composites. Waste Biomass Valor, 4 (2013) 73–86. https://doi.org/10.1007/s12649-013-9220-4

[8] BNDS - Banco Nacional de Desenvolvimento Econômico. Potencial de diversificação da indústria química brasileira: Relatório 4: fibras de carbono. Rio de Janeiro: Bain & Company, nov. 2014.

[9] S. Pimenta, S.T. Pinho, The effect of recycling on the mechanical response of carbon fibres and their composites. Compos. Struct. 94 (12) (2012) 3669–3684. https://doi.org/10.1016/j.compstruct.2012.05.024

[10] F.A. Lopez, O. Rodriguez, F.J. Alguacil et al. Recovery of carbon fibres by the thermolysis and gasification of waste prepreg. J. Analyt. Appl. Pyrolysis, 104 (2013) 675–683. https://doi.org/10.1016/j.jaap.2013.04.012

[11] FLATOUT. Carbon fiber wheel manufacturing. Information on: https://www.flatout.com.br/koenigsegg-ensina-arte-de-fazer-rodas-de-fibra-de-carbono/

[12] ABMACO. Compósitos I – Materiais, Processos, Aplicações e Desempenho, 2008 – Ed. 1 – São Paulo, 2008.

[13] P. Xu, J. Li, J. Ding, Chemical recycling of carbon fibre/epoxy composites in a mixed solution of peroxide hydrogen and N, N-dimethylformamide. Composites Science and Technology. 82 (18) (2013) 54–59. https://doi.org/10.1016/j.compscitech.2013.04.002

[14] K. Yu, Q. Shi, M.L. Dunn, T. Wang et al. Carbon Fibre Reinforced Thermoset Composite with Near 100% Recyclability. Adv. Funct. Mater. 25 (33) (2016) 6098-6106. https://doi.org/10.1002/adfm.201602056

[15] J.-M. Park, D.-J. Kwon, Z.-J. Wang et al. Effect of Thermal Treatment Temperatures on the Reinforcing and Interfacial Properties of Recycled Carbon Fibre–phenolic Composites. Comp., 47 (2013) 156-164. https://doi.org/10.1016/j.compositesa.2012.12.002

[16] HEXCEL Hexply ® 8552—Product Data Sheet—EU Version. [s. 1.], 2016; pp. 1–6. Information on: https://www.hexcel.com/user_area/content_media/raw/HexPly_8552_eu_ DataSheet(1).pdf

[17] HUNTSMAN Araldite ® LY 5052 Aradur™5052 -Product Data Sheet—English Version. 2018; pp. 1–5. Information on: https://www.huntsman.com/docs/Documents/ARALDITE%2 0LY%205052%20ARADUR%205052%20Technical%20Datasheet%20%28US%29.pdf

[18] AMERICAN SOCIETY FOR TESTING AND MATERIAL. ASTM ASTM G99-05, Pin-on-Disk Apparatus Philadelphia, PA, 2017.

[19] J. M. de Moura Júnior, F. M. da Cruz, Y. V. Póvoas et al. Use of unmanned aerial vehicle for safety inspection at construction sites: a systematic literature review: Utilização de veículo aéreo não tripulado para inspeção de segurança em canteiros de obra: uma revisão sistemática da literatura. *Concilium*, *23*(2) (2023) 491–508. Information on: https://clium.org/index.php/edicoes/article/view/760

[20] FAB. Hermes 450, Information on: www.fab.mil.br/noticias/mostra/19817/

[21] AMERICAN SOCIETY FOR TESTING AND MATERIAL. ASTM D790. Standard Test Methods for Flexural Properties of Unreinforced and Reinforced Plastics and Electrical Insulating Materials, Philadelphia, PA 2010.

[22] AMERICAN SOCIETY FOR TESTING AND MATERIAL. ASTM D695-02 Standard Test Method for Compressive Properties of Rigid Plastics, Philadelphia, PA 2002.

[23] V.E. Rodrigues; H.M.S. Costa, M.R. Baldan, R. C. M. Sales, J.T. Matsushima· Reciclagem de fibras de carbono oriundas de compósitos estruturais de matriz epóxi por processo térmico de pirólise e aplicação como materiais de eletrodo em supercapacitores. Matéria (Rio J) [Internet]. 2021;26(2):e12970. Information on: https://doi.org/10.1590/S1517-707620210002.1270

Innovative Manufacturing Engineering and Energy - IManEE2024 Materials Research Forum LLC
Materials Research Proceedings 46 (2024) 277-282 https://doi.org/10.21741/9781644903377-36

Investigation of T6 heat treatment parameters for AA6061 matrix composite reinforced with 15% B_4C

Halil Karakoç[1,a*], Şener Karabulut[1,b]

Department of Mechanical Program, Hacettepe University, 06930 Ankara, Turkey

[a]halil.karakoc@hacettepe.edu.tr, [b]senerkarabulut@hacettepe.edu.tr

Keywords: AA6061 Matrix Composite, B_4C Reinforcement, T6 Heat Treatment, Microstructural Analysis, Mechanical Properties

Abstract. This study investigates the effects of solution temperature, aging temperature, and aging time on the hardness and transverse rupture strength (TRS) of Al6061/15 wt% B_4C composites. Detailed microstructural analysis revealed a uniform dispersion of B_4C particles within the AA6061 matrix, demonstrating a strong interface and good bonding. The hardness and TRS of the composites were evaluated under various heat treatment conditions. The results showed that the optimal mechanical properties were achieved at a solution temperature of 500°C, an aging temperature of 180°C, and an aging time of 4 hours, with a peak hardness of 89.00 HB and a TRS of 562.74 MPa. Lower aging temperatures resulted in increased hardness with extended aging times, while higher aging temperatures and prolonged aging led to significant reductions in both hardness and TRS due to over-aging and precipitate coarsening.

Introduction

AA6061 aluminum alloy is extensively utilized across various industries due to its favorable properties, such as high strength, formability, and weldability. However, the mechanical properties of AA6061 can be significantly enhanced by incorporating ceramic reinforcements, such as boron carbide (B_4C). B_4C is a ceramic material characterized by its exceptional hardness, high melting point, and excellent chemical resistance, making it ideal for applications in abrasive environments, cutting tools, and nuclear shielding [1]; [2]; [3]. To further improve the mechanical properties of AA6061 and its composites, the T6 heat treatment process is commonly employed [4]. Gatea et al. [5] xamined the effects of various heat treatments on the fracture mechanisms and microstructural changes in aluminum silicon carbide (SiC) metal matrix composites. Their study revealed that T6 treatment, while enhancing certain properties, reduced ductility due to precipitate formation, whereas O-condition annealing improved toughness and ductility by mitigating harmful intermetallic compounds and increasing sensitivity to strain rate. Rajasekaran et al. [6] investigated the aging process in AA6061 reinforced with 15% SiC, finding that peak hardness was achieved after 500 hours of aging under T4 conditions, with maximum hardness observed at 180 °C during T6 treatment. Similarly, Uyan et al. [7] evaluated the aging effects on Al7075 MMCs reinforced with 5% SiC, noting that aging duration played a critical role in achieving higher hardness at both high and low aging temperatures. Swamy et al. [8] explored the impact of heat treatment on Al6061 MMCs with varying SiC content, reporting a significant increase in tensile strength, particularly after ice quenching and artificial aging. Chen et al. [9] investigated the influence of heat treatment on the microstructure and corrosion behavior of multilayer AA6061 aluminum alloy, finding that while heat treatment induced abnormal grain growth, the re-precipitation of fine β″ phases enhanced corrosion resistance, even in the presence of corrosion-promoting Al5FeSi intermetallic compounds. Additionally, Wang et al.[10] studied the effect of repeated impact loading on AA6061-T6, concluding that impact fatigue life decreases exponentially with increasing energy, with experimental and numerical data supporting the applicability of the Gurson-Tvergaard-Needleman model.

Materials Research Proceedings 46 (2024) 277-282 https://doi.org/10.21741/9781644903377-36

This study aims to investigate the combined effects of T6 heat treatment parameters—specifically solution temperature, heating temperature, and aging time—on the hardness and TRS of resulting composite specimens. The findings contribute to a deeper understanding of how these parameters influence the hardness and overall performance of AA6061-B_4C composites, thereby providing insights into optimizing the processing conditions for enhanced material properties.

Materials and methods

Experimental procedure for fabrication of AA6061-B_4C composite

In this study, AA6061 aluminum alloy powder and B_4C ceramic powder, both with particle sizes under 10 µm, were used. The powders were precisely weighed using a Precisa microbalance to achieve a 15:1 weight ratio of AA6061 to B_4C. To ensure an even distribution of B_4C within the AA6061 matrix, the powders were mixed for 45 minutes in a Turbula three-dimensional mixer, based on preliminary studies to ensure effective particle interaction and minimize segregation. The mixed powders were cold compacted into cylindrical specimens using a hydraulic press at 300 MPa in a mold with a 100 mm internal diameter. This pressure was selected to achieve sufficient green density for successful sintering while preserving powder integrity. The green compacts were then sintered at 550°C for 60 minutes to promote interparticle bonding and enhance mechanical strength, ensuring adequate densification without excessive grain growth. Following sintering, the composite blocks underwent extrusion at 550°C using a die with a 1:4 reduction ratio, which helped achieve a uniform microstructure. The extruded specimens, now with a rectangular cross-section of 24 mm × 86 mm, were hot rolled at 550°C to a final thickness of 12.7 mm, improving densification and microstructure. Rectangular specimens measuring 12.7×6.35×31.7 mm were then cut from the hot-rolled samples using a Mitsubishi wire EDM machine.

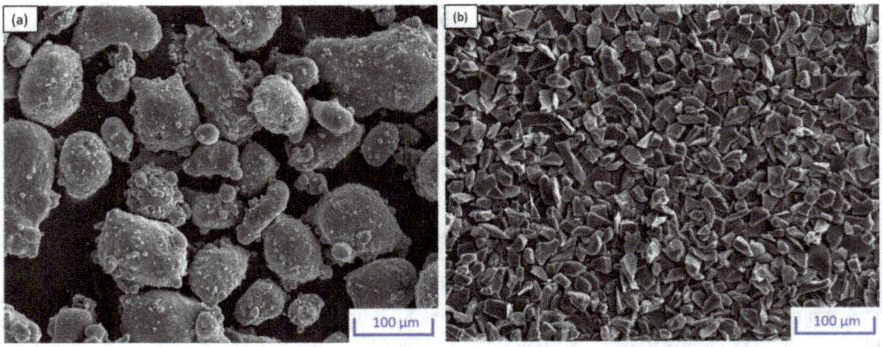

Figure 1. Particle size distributions of AA6061 and B_4C powders

T6 heat treatment procedure

The samples were solution-treated in a Protherm-type furnace at the temperatures listed in Table1 for 1 hour, then quenched in circulating tap water at room temperature. After quenching, the samples underwent aging at various temperatures and durations. Hardness was measured using an Emco-Test Duravision 200 hardness tester, following ASTM E10-23, with a 2.5 mm ball and a 31.25 kg load, averaging values from five distinct areas per sample. TRS was tested with an Instron 3369 universal testing machine, calculating the mean bending strength from three specimens per composite.

Samples were polished using SiC papers with grits of 600, 800, 1000, and 1200, followed by diamond suspensions (9, 6, and 3 microns) and a final polish with colloidal silica (0.001 µm). The samples were etched with Keller's reagent (1 mL HF + 200 mL H2O) for 15 seconds and analyzed

Materials Research Proceedings 46 (2024) 277-282 https://doi.org/10.21741/9781644903377-36

under a Leica DM4000 M optical microscope. Scanning electron microscopy (SEM) images and EDS analyses were obtained using a Tescan Maia3 XMU scanning electron microscope with secondary electron imaging.

Table 1. T6 heat treatment parameters

Test parameters	Remarks
Solution temperature. (°C)	500, 525,550,580
Aging temperature, (°C)	140,180,240,310
Aging time, (hour)	2,4,8,16

Experimental results and discussion

Figure 2 shows the distribution of B_4C particles within the AA6061 matrix, while Figure 3 provides a high-resolution SEM micrograph detailing the distribution and interfaces of these particles within the aluminum matrix. Both optical and SEM images confirm a uniform dispersion of B_4C particles, indicating a strong interface and effective bonding between the B_4C particles and the matrix.

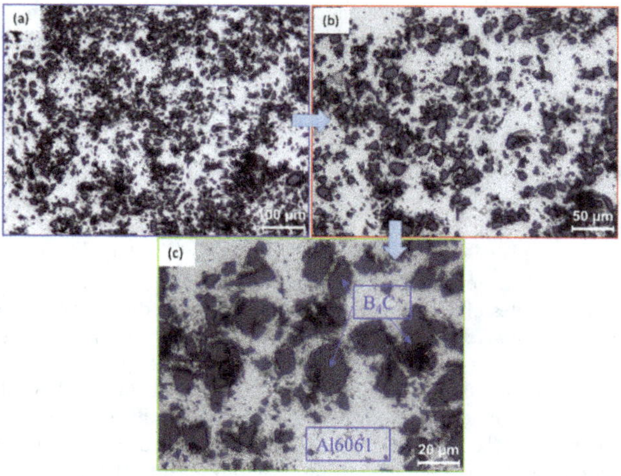

Figure 2. Optical image of resulting composite specimen

This uniform dispersion and robust bonding are essential for optimizing the composite's mechanical properties. A strong interface allows for efficient stress transfer, enabling the composite to bear higher loads without failure, while the even distribution reduces localized stress concentrations that could otherwise cause premature failure. The microstructural analysis confirms that the manufacturing process successfully produced a homogeneous composite with well-integrated B_4C particles, which likely plays a key role in enhancing the composite's hardness and strength.

 Materials Research Forum LLC
https://doi.org/10.21741/9781644903377-36

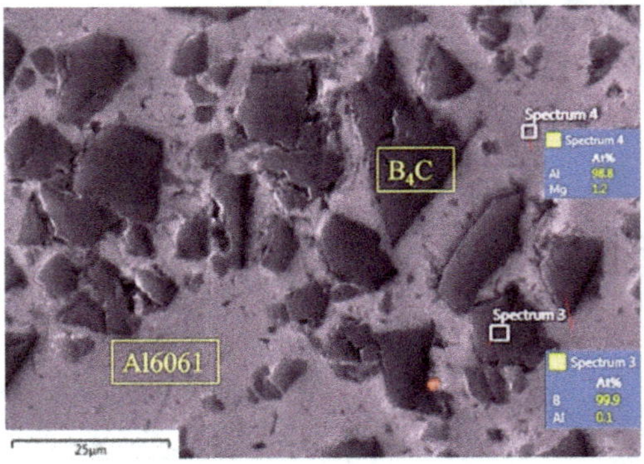

Figure 3. (a) Distributions and interfaces of B_4C particles in AA6061 matrix and (b), (c) EDS results

The study systematically examined the hardness and transverse rupture strength (TRS) of composite materials under varying solution and aging temperatures, as well as different aging durations. As depicted in Figure 4, these parameters significantly influence the material properties, underscoring the critical importance of optimizing heat treatment protocols for improved material performance. The analysis revealed distinct trends in both hardness and TRS across different solution and aging conditions. Solution temperature was found to have a relatively minor effect on hardness. The highest hardness value, approximately 72 HB, was observed at a solution temperature of 550°C. Specifically, the hardness at 500°C was about 70 HB, decreased slightly to 68 HB at 525°C, peaked at 72 HB at 550°C, and then declined again to 68 HB at 580°C. This variation indicates that within this temperature range, the solution temperature does not substantially alter the material's hardness, likely due to the stabilization of the microstructure. Notably, the maximum hardness of 89 HB was achieved at a solution temperature of 500°C, combined with an aging temperature of 180°C and an aging time of 4 hours. This optimal combination likely promotes effective precipitate formation and distribution, thereby enhancing hardness. In contrast, aging temperature had a pronounced impact on hardness. At an aging temperature of 140°C, hardness increased with aging time, peaking at 77.60 HB after 8 hours, before slightly decreasing to 70.90 HB at 16 hours. This suggests that prolonged aging at lower temperatures can result in some degree of over-aging, although this effect is less severe than at higher aging temperatures. The hardness reached its maximum at an aging temperature of 180°C, but generally declined at higher aging temperatures of 240°C and 310°C. For instance, at 310°C, the hardness significantly decreased to 53.38 HB after 16 hours, indicating substantial over-aging and precipitate coarsening, which reduce the material's ability to resist deformation.

Materials Research Proceedings 46 (2024) 277-282 https://doi.org/10.21741/9781644903377-36

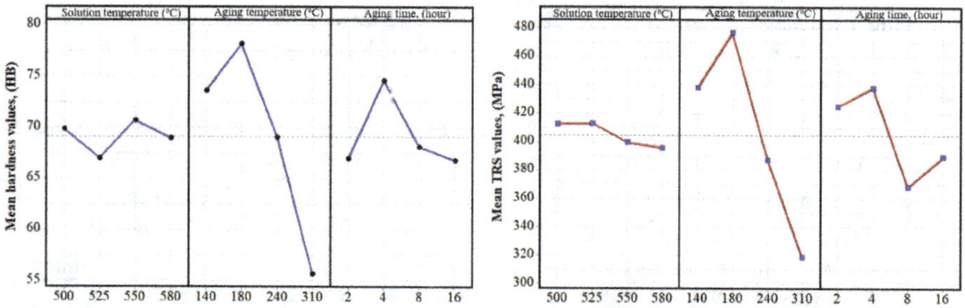

Figure 4. Effect of solution temperature, aging temperature, and time on average hardness and average TRS

The aging time also significantly affects the hardness values. At 2 hours, the hardness value is around 64 HB, increasing markedly to approximately 72 HB at 4 hours, then decreasing at 8 hours and further decreasing at 16 hours. The increase in hardness up to 4 hours of aging time suggests this duration is sufficient for forming and distributing strengthening precipitates. Beyond this period, the decline in hardness indicates over-aging, where prolonged exposure leads to coarsening and a reduction in the beneficial effects of the precipitates. The TRS values exhibit a similar pattern to the hardness results. The highest TRS of 562.74 MPa is observed at a solution temperature of 500°C, an aging temperature of 180°C, and an aging time of 4 hours. This condition aligns with the peak hardness, suggesting a strong correlation between hardness and TRS under optimal aging conditions. For an aging temperature of 140°C, TRS values vary but show the highest value of 484.89 MPa at 16 hours. This indicates that, unlike hardness, TRS can benefit from longer aging times at lower temperatures, possibly due to further stabilization of the microstructure. At higher aging temperatures of 240°C and 310°C, TRS values decrease significantly. For instance, at 240°C, the TRS drops to 335.35 MPa at 4 hours and continues to decline with longer aging times. The lowest TRS of 235.23 MPa is recorded at an aging temperature of 310°C and an aging time of 8 hours, demonstrating the detrimental effects of over-aging on the mechanical strength of the composites. The hardness and TRS are optimized at a solution temperature of 500°C and an aging temperature of 180°C, with intermediate aging times around 4 hours. Higher aging temperatures and prolonged aging times lead to over-aging, resulting in precipitate coarsening and reduced mechanical properties. These findings underscore the importance of precise control overheat treatment parameters to achieve the desired balance of hardness and TRS in composite materials.

Conclusions

This study evaluated the hardness and transverse rupture strength (TRS) of Al6061/15 wt% B_4C composite materials under various solution temperatures, aging temperatures, and aging times. The optimal mechanical properties were achieved at a solution temperature of 500°C, an aging temperature of 180°C, and an aging time of 4 hours, resulting in a peak hardness of 89.00 HB and a TRS of 562.74 MPa. These conditions facilitated optimal precipitate formation and distribution. Microstructural analysis showed a uniform dispersion of B_4C particles within the AA6061 matrix. The strong interface and good bonding between the B_4C particles and the matrix are critical for achieving optimal mechanical properties.

Aging temperature and time were crucial in determining mechanical properties. At 140°C, hardness increased with aging time, peaking at 77.60 HB at 8 hours, but decreased slightly with prolonged aging. Higher aging temperatures (240°C and 310°C) led to significant decreases in hardness due to over-aging and precipitate coarsening. TRS values exhibited similar trends, with the highest values achieved under optimal aging conditions and significant reductions observed at higher aging temperatures and prolonged times.

The optimal mechanical properties for the Al6061/15 wt% B_4C composite were achieved at a solution temperature of 500°C, an aging temperature of 180°C, and an aging time of 4 hours.

References

[1] Şener Karabulut, Halil Karakoç RÇ. Influence of B4C particle reinforcement on mechanical and machining properties of Al6061/B4C composites 2016;101:87–98.

[2] Canakci A. Microstructure and abrasive wear behaviour of B4C particle reinforced 2014 Al matrix composites. J Mater Sci 2011;46:2805–13. https://doi.org/10.1007/s10853-010-5156-2

[3] Ibrahim MF, Ammar HR, Samuel AM, Soliman MS, Samuel FH. On the impact toughness of Al-15 vol .% B 4 C metal matrix composites. Compos Part B 2015;79:83–94. https://doi.org/10.1016/j.compositesb.2015.04.018

[4] Venkatesh R, Srinivas S. Effect of heat treatment on hardness, tensile strength and microstructure of hot and cold forged Al6061 metal matrix composites reinforced with silicon carbide particles. Mater Res Express 2019;6. https://doi.org/10.1088/2053-1591/ab4297

[5] Gatea S, Ou H, McCartney G. Deformation and fracture characteristics of Al6092/SiC/17.5p metal matrix composite sheets due to heat treatments. Mater Charact 2018;142:365–76. https://doi.org/10.1016/j.matchar.2018.05.050

[6] Rajasekaran S, Udayashankar NK, Nayak J. T4 and T6 Treatment of 6061 Al-15 Vol. % SiC P Composite . ISRN Mater Sci 2012;2012:1–5. https://doi.org/10.5402/2012/374719

[7] Uyan P, Gürler R. Effect of Aging Heat Treatment on the Mechanical Properties of SiC Reinforced 7075 Al-Alloy Composites Manufactured by Vortex Casting Method. Univers J Mater Sci 2018;6:27–38. https://doi.org/10.13189/ujms.2018.060104

[8] Swamy NRP, Ramesh CS, Chandrashekar T. Effect of heat treatment on strength and abrasive wear behaviour of Al6061-SiCp composites. Bull Mater Sci 2010;33:49–54. https://doi.org/10.1007/s12034-010-0007-y

[9] Chen L, Zhu L, Lu L, Yang Z, Ren X, Zhang X. The effect of heat treatment on the microstructure and electrochemical corrosion behavior of multilayer AA6061 alloy fabricated by additive friction stir deposition. Appl Surf Sci 2024;650:159167. https://doi.org/10.1016/j.apsusc.2023.159167

[10] Wang B, Qiu Z, Gao P, Xu Y, Wang C, Han K, et al. The experimental and numerical study of AA6061-T6 aluminum alloy under repeated impacts. Int J Fatigue 2024;183:108213. https://doi.org/10.1016/j.ijfatigue.2024.108213

Materials Research Proceedings 46 (2024) 283-290 https://doi.org/10.21741/9781644903377-37

Experimental insights into PEG based new fluids thermal effusivity

Nicoleta COJOCARIU[1,a], Elena Ionela CHERECHEŞ[1,b], Alina Adriana MINEA[1,c*]

[1]Technical University "Gheorghe Asachi" of Iasi, ROMANIA

[a]nicoleta.cojocariu@academic.tuiasi.ro, [b]elena-ioanela.chereches@academic.tuiasi.ro, [c]alina-adriana.minea@academic.tuiasi.ro

Keywords: PEG, Nanocolloid, Thermal Effusivity, Heat Transfer

Abstract. Fluids thermal transfer is of utmost relevance in a large number of industrial or semi industrial applications. On the other hand, the need for environmental friendly new fluids plays a relevant role in the heat transfer agent selection. Poly(ethylene glycol) are synthetic, hydrophilic, biocompatible polymers with extensive use in a large number of applications. PEGs are synthesized by means of a ring-opening polymerization of ethylene oxide and a large range of molecular weights and polydispersity's are produced. From these products, only substances with lower molecular weights can be employed for thermal transfer., more exactly for molecular weight lower than 600. This paper presents the outcomes of an experimental investigation into the thermal effusivity of PEG 400, PEG 200 and water mixtures for convection heat transfer applications. Quite a lot of samples with different concentrations were manufactured, examined and the data were thoroughly discussed comparatively and in the context of cutting-edge advancements. The concept now known as thermal effusivity has been emplyed since the early stages of heat transfer research as a valuable feature for analyzing 1D heat transfer setups.

Introduction

Nowadays there is an imperious need to discover and study new heat transfer fluids, especially for applications beyond water evaporation point. There is a tremendous need for new fluids to be implemented at medium temperatures (i.e. larger than 80 – 100 °C). Several candidates for these temperatures are polyethylene glycols (PEG), that serves as a defoaming agent, lubricant, and viscosity modifier in a variety of products. PEGs are not a new material, but their mixtures and applications can be further studied (see [1] for further details).

PEGs, together with chemicals as paraffin and fatty acid, are organic phase change materials with a uniform phase transformation and reasonable nucleation frequency [2]. Basically, PEGs are polyether composite with immeasurable usages, from engineering to medical applications. Their structure is H–(O–CH2–CH2)n–OH, while their usages are:

- Various chemical applications (including use as a lubricant, in biochemistry or biomembrane experimental studies, as a surfactant, and as a calibration substance in mass spectrometry, among others);
- Medicine (utilized as an excipient, for instance);
- Biology (used as a crowding agent for protein crystallization);
- Commercial applications (for instance, in tattoos to monitor diabetes, as an anti-foaming agent in various food and beverage products, in skin creams, and as a dispersant in toothpaste)
- Industrial uses (employed as an anti-foaming agent, in technical ceramics, and as an insulator)
- Recreational applications.

The name PEG is followed by a numerical value that represents the average molecular weight. An in-depth discussion on these aspects is outlined in Minea [1], together with an outline of PEGs thermophysical properties at room temperature.

Summarizing, Poly(ethylene glycol) is a synthetic, hydrophilic, biocompatible polymer with extensive use. This paper will be focused on low mass PEGs, as for example POEG 200 and PEG 400 and several mixtures. These two polymers are mixtures of molecules with varying sizes that differ in the number of oxyethylene units, having average molecular weights of 200 and 400, respectively.

Research results for different PEG based fluids are ongoing and several authors (see [3-11]) studied low mass compounds for different thermophysical properties, outlining these fluids relevance for heat transfer applications.

PEG-200 offers a diverse array of potential applications and can be employed as defoaming agent, lubricant or viscosity converter in several applications. PEG-200 can additionally be utilized as a coating for fresh produce, a solvent in metalworking fluids, a binder and modifier in paint formulations, and as a humectant in both inks and abrasives [1, 2, 10]. Another important feature is that PEG 200 is environmentally friendly, which is a great quality nowadays, when special preoccupation on environmental issues is mandatory. Plus, PEG 200 is completely soluble in water and has an average molecular weight ranging from 190 to 210 (see Fig.1 for the chemical structure).

On the other hand, PEG 400 shares mostly the same features as PEG 200, having a larger molar mass (i.e. around 400) [3-9, 11].

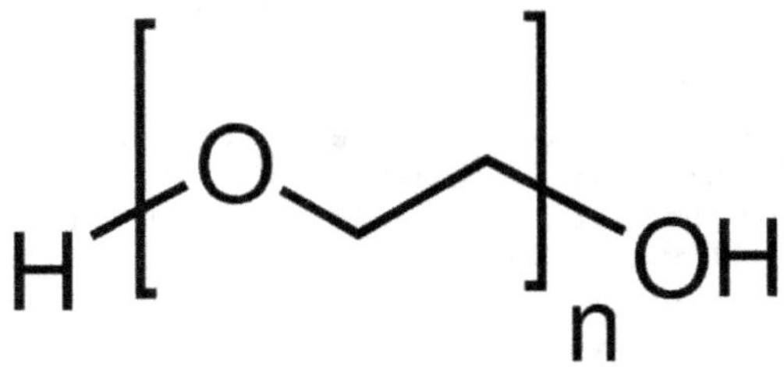

Figure 1. Chemical structure of PEG 200

This research is focused on PEG 200, PEG 400 and their several mixtures as the base fluids for heat transfer applications. Results were discussed in terms of thermal effusivity as a good indication for heat transfer behaviour, being one of the most relevant parameters when it discusses phase change materials applications and performances.

Experimental

Several fluids were manufactured using water, PEG 200 and PEG 400 as base. The substance were supplied by Sigma-Aldrich (St. Louis, USA) and the suspensions were manufactured as are identified in Table 1 and Table 2, using different mole fractions. For fluids manufacturing, the chemicals were measured on a KERN ADJ 100–4 balance (Kern, Germany), which provides an accuracy of 1 mg. Further, the new fluids, as well as the pure fluids (see Table 1) were tested using the C-Therm Trident Thermal Conductivity System (C-Therm, USA) is equipped with a Basic-k Module featuring a Modified Transient Plane Source (MTPS) sensor, in accordance with ASTM D7984. The equipment measures thermal effusivity with a precision of 0.999, as per manufacturer indications.Thermal effusivity testing has been standardized by ASTM International (American

Society for Testing and Materials). and C-Therm incorporated stringent procedures and experimental parameters to ensure the highest accuracy in measuring thermal effusivity.

The experimental data were collected at ambient temperature, since previous research demonstrated that there is minor variation when temperature rises (see Chereches et al. [12, 13])

Table 1. Manufactured fluids

Fluid	Code	PEG 200 mass in mixture [g]	PEG 400 mass in mixture [g]	Water mass [g]	Total mass [g]
distilled water	W	-	-	10	10
PEG 400	F0				
PEG 200	F1	7.5144	-	-	7.5144
0.75 PEG200+0.25 PEG400	F2	7.5214	5.0149	-	12.5363
0.5 PEG200+0.5 PEG400	F3	3.7619	7.52	-	11.2819
0.25 PEG200+0.75 PEG400	F4	1.495	9.0318	-	10.5268
0.25 PEG200+0.75 water	F5	7.502	-	2.0181	9.5201

Table 2. Fluids composition

Fluid	PEG 200		PEG 400		Water	
	mass fraction[-]	mole fraction[-]	mass fraction[-]	mole fraction[-]	mass fraction[-]	mole fraction[-]
distilled water	-	-	-	-	1	1
PEG 400	-	-	1	1	-	-
PEG 200	1	1	-	-	-	-
0.75 PEG200+ 0.25 PEG400	0.6	0.75	0.4	0.25	-	-
0.5 PEG200+ 0.5 PEG400	0.34	0.5	0.67	0.5	-	-
0.25 PEG200+ 0.75 PEG400	0.14	0.25	0.85	0.76	-	-
0.25 PEG200+ 0.75 water	0.79	0.25	-	-	0.21	0.75

The experimental setup assures the best collection of data for the investigated samples, as was recommended by the C-Therm training in using the equipment and C-Therm software for gathering data. Data was collected at ambient temperature, while C-Therm software register both the temperature, as well as thermal effusivity values.

Experimental accuracy is calculated as 0.1 % for mass weighting and maximum 0.01 % at thermal effusivity tests, as is calculated by the C-Therm software. The software calculates the R-squared value in comparison with a test sample, as is depicted in the C-Therm documentation.

Materials Research Proceedings 46 (2024) 283-290 https://doi.org/10.21741/9781644903377-37

Results and discussion

Firstly, the thermal effusivity of base fluids was measured, followed by the new fluids investigation.

The experimental results are described in Fig. 2, where the influence of the components can be observed. The thermal effusivity accuracy is $R^2 = 0.99$, as was measured by the equipment for each data set. Several measurements were achieved for each sample and the results are averaged. Basically, thermal effusivity reflects a material's capacity to exchange heat with its immediate environment at a surface, and it is expressed as:

$$e = \sqrt{k\rho c_p} \tag{1}$$

Here, e represents thermal effusivity, k denotes thermal conductivity, ρ indicates density, and cp refers to mass specific heat capacity.

Fig. 2 shows the experimental points for each fluid and is noticed that the higher value was measured for distilled water, while PEGs and their mixtures registered lower thermal effusivity and this is normal, since the thermal conductivity of PEGs is lower than the one of water. Plus, there is a correspondence between the thermal effusivity of each component and the one registered for the mixture.

On the other hand, the data were compared with the rule of mixture, which is mostly applied for all fluidic mixtures between two immiscible components. The rule of mixture for most of the properties, except viscosity, which follows Arrhenius law can be outlined as:

$$e_{mixture} = (N_1 \cdot e_1) + (N_2 \cdot e_2) \tag{2}$$

Where e is the thermal effusivity, mixture refers to the mixture property, while N is the molar fraction.

The results are comparatively plotted in Fig. 3 where it can notice that the rule of mixtures does not apply to thermal effusivity. The deviation was higher for the mixture PEG 200 + distilled water (62.39 %), while for the two PEGs mixture the variation lies into – 0.01 – 1.36 %. An observation arises that, when two similar fluids are mixed, the rule of mixtures can successfully estimate the thermal effusivity of the new compound.

Materials Research Proceedings 46 (2024) 283-290 https://doi.org/10.21741/9781644903377-37

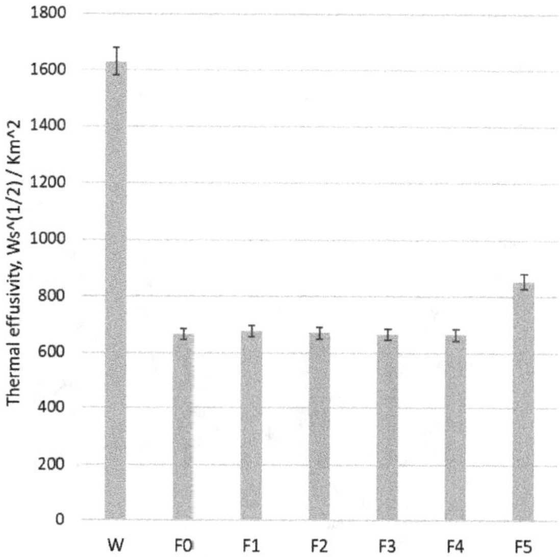

Fig. 2 Thermal effusivity experimental results

If it looks to Fig. 3 it can say that the rule of mixture can predict the thermal effusivity of the mixture between the two PEGs with a good precision. On the other hand, when it add water to PEG, the experimental thermal effusivity of the mixture is lesser than the one calculated. In this situation, a pattern needs to be observed and more experimental is necessary to be developed in order to set an experimental correlation.

In the next lines, also thermal diffusivity will be discussed for the developed fluidsThermal diffusivity, measured in m²/s, is calculated by dividing thermal conductivity (k) by the product of density (ρ) and specific heat capacity at constant pressure (cp). It offers important insights into the heat transfer processes occurring within the fluid. The equation writes as:

$$a = k/\rho c_p \tag{3}$$

The results on thermal diffusivity are depicted in Table 3, where the data were calculated based on thermal effusivity and density of all liquids and mixtures.

The density of a mixture from two liquids is the resulting density of the mixture formed when two liquids are mixed and is represented as:

$$\rho_{\text{mixture}} = (M_{\text{mixture}}) / \left(\frac{M_1}{\rho_1} + \frac{M_2}{\rho_2}\right) \tag{4}$$

Where ρ is the density, mixture refers to the mixture property, while M is the mass of each liquid or mixture, respectively.

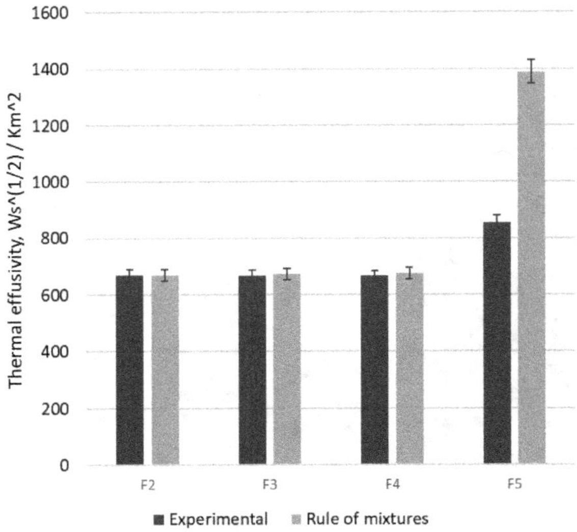

Fig. 3 Comparison with rule of mixture

Table 3 Data on thermal diffusivity

	PEG 200 mole fraction	PEG 400 (or water for F5) mole fraction	Thermal diffusivity [m^2/s]
water distilled	-	-	1.49E-07
PEG 400 – F0	-	-	9.49E-08
PEG 200 - F1	-	-	9.5E-08
F2	0.75	0.25	9.5E-08
F3	0.5	0.5	9.49E-08
F4	0.25	0.75	9.49E-08
F5	0.25	0.75	9.9E-08

From Table 3 data it can conclude that all the PEGs and their mixtures have similar behavior at heat transfer, while water and F5 possess better features in terms of thermal transport. However, it needs to outline here that the W and F5 fluids are suitable for temperatures up to 90 °C, while the PEGs and their mixture can support temperatures up to 200 °C.

As a prospectus, these authors can suggest several ways to increase ethe thermal effusivity, thus heat transfer behavior by using the concept of nanofluid. In these conditions, the addition of solid high conductive nanoparticles can improve the heat transfer comportment. Nevertheless, this method needs to be carefully studied and debated in order to control the drawbacks of an increased viscosity or sedimentation occurrence, that will affect the flow behavior and increase the pumping power.

Conclusion

Several mixtures based on polyethylene glycol were manufactured and the thermal effusivity was experimentally measured. Results indicated that the effusivity of the mixture depend on the compound, while the rule of mixture cannot predict the values of different chemicals mixture. The experiment outcomes can be summarized as:

- Water thermal effusivity is higher than that of PEGs
- PEG 200 and PEG 400 thermal effusivity depends on their molar mass
- PEG 200 has larger thermal effusivity than that of PEG 400, while all the mixtures have values between the two PEGs
- The mixture between PEG 200 and distilled water has an increased thermal effusivity, being strongly influenced by the water addition.
- The thermal diffusivity of all mixtures is lower than that of water

As a general conclusion, it may say that, even if the water addition increases the heat transfer capacity, the applications are different in terms of temperature, while PEGs area of application is beyond water boiling point. Finally, the addition of nanoparticles to PEG based fluids can be a solution to increase the thermal effusivity and further experimental insights are mandatory.

Acknowledgement

This research paper was supported by Boosting Ingenium for Excellence (BI4E) project, funded by the European Union's European Union's HORIZON-WIDERA-2021-ACCESS-05-01-European Excellence Initiative under the Grant Agreement No. 101071321

References

[1] AA. Minea State of the Art in PEG-Based Heat Transfer Fluids and Their Suspensions with Nanoparticles. *Nanomaterials*. 2021; 11(1):86. https://doi.org/10.3390/nano11010086

[2] L.F. Cabeza, Advances in Thermal Energy Storage Systems—Methods and Applications; Elsevier: Amsterdam, The Netherlands, 2014.

[3] M.A.; Marcos, D.; Cabaleiro, M.J.G.; Guimarey, M.J.P.; Comuñas, Fedele, L.; Fernández, J.; Lugo, L. PEG 400-Based Phase Change Materials Nano-Enhanced with Functionalized Graphene Nanoplatelets. Nanomaterials 2018, 8, 16.

[4] B.; Tang, C.; Wu, M.; Qiu, X.; Zhang, S.Zhang, PEG/SiO2—Al2O3 hybrid form-stable phase change materials with enhanced thermal conductivity. Mater. Chem. Phys. 2014, 144, 162–167.

[5] Y.; Kou, S.; Wang, J.; Luo, K.; Sun, J.; Zhang, Z.; Tan, Q. Shi, Thermal analysis and heat capacity study of polyethylene glycol (PEG) phase change materials for thermal energy storage applications. J. Chem. Thermodyn. 2019, 128, 259–274.

[6] M.A.; Marcos, Podolsky, N.E.; Cabaleiro, D.; Lugo, L.; Zakharov, A.O.; Postnov, V.N.; Charykov, N.A.; Ageev, S.V.; Semenov, K.N. MWCNT in PEG-400 nanofluids for thermal applications: A chemical, physical and thermal approach. J. Mol. Liq. 2019, 294, 111616.

[7] M.A.; Marcos, Cabaleiro, D.; Hamze, S.; Fedele, L.; Bobbo, S.; Estellé, P.; Lugo, L. NePCM Based on Silver Dispersions in Poly(Ethylene Glycol) as a Stable Solution for Thermal Storage. Nanomaterials 2020, 10, 19

[8] D.; Cabaleiro, Hamze, S.; Fal, J.; Marcos, M.A.; Estellé, P.; Zyła, G. Thermal and Physical Characterization of PEG Phase Change Materials Enhanced by Carbon-Based Nanoparticles. Nanomaterials 2020, 10, 1168.

[9] S.; Navidbakhsh, Majdan-Cegincara, R. Effect of c-Fe2O3 nanoparticles on rheological and volumetric properties of solutions containing polyethylene glycol. Int. J. Ind. Chem. 2017, 8, 433–445.

[10] K.; Yapici, Cakmak, N.K.; Ilhan, N.; Uludag, Y. Rheological characterization of polyethylene glycol based TiO2 nanofluids. Korea-Aust. Rheol. J. 2014, 26, 355–363.

[11] M.A.; Marcos, Lugo, L.; Ageev, S.V.; Podolsky, N.E.; Cabaleiro, D.; Postnov, V.N.; Semenov, K.N. Influence of molecular mass of PEG on rheological behaviour of MWCNT-based nanofluids for thermal energy storage. J. Mol. Liq. 2018, 318, 113965.

[12] M. Chereches, A. Vardaru, G. Huminic, E.I. Chereches, A.A. Minea, A. Huminic, Thermal conductivity of stabilized PEG 400 based nanofluids: An experimental approach, International Communications in Heat and Mass Transfer 130 (2022) 105798.

[13] M. Chereches, C. Ibanescu, M. Danu, E.I. Chereches, A.A. Minea, PEG 400-Based Phase Change Materials Nano-Enhanced with Alumina: an experimental approach, Alexandria Engineering Journal, 61 (9) (2022) 6819-6830.

Materials Research Proceedings 46 (2024) 291-299 https://doi.org/10.21741/9781644903377-38

Tools for the incremental deformation of sheet metal using a single point. presentation of an innovative new tool

Emanuel BĂDULESCU[1,a*], Eduard Laurențiu NIȚU[2,b], Daniela Monica IORDACHE[2,c]

[1]National University of Science and Technology Politehnica Bucharest, Bucharest University Center, Doctoral School of Industrial Engineering and Robotics, Splaiul Independenţei Street, no. 313, Bucharest, Romania

[2]National University of Science and Technology Politehnica Bucharest, Pitești University Center, Manufacturing and Industrial Management Department, Târgul din Vale Street, no. 1, Pitești, Romania

[a]emanuel.badulescu@stud.fiir.upb.ro, [b]eduard.l.nitu@upb.ro, [b]daniela.c.iordache@upb.ro

Keywords: Incremental Forming, Formability, SPIF Tools, Rigid Tools, Eccentric Tool with Variable Radius (ETVR)

Abstract. Incremental forming with a punch is a process of processing metal sheets in continuous development, due to the low cost of the tools and the possibility of its application on numerically controlled machines. Since many tool models have recently been developed for processing sheet metal by incremental forming, their choice is difficult. This article aims to present a brief synthesis of the types of tools proposed and tested by researchers for incremental forming, highlighting the usage fields and the advantages and disadvantages of their use. The presentation of the kinds of tools together with the specific performances associated with the different materials and geometric shapes of the processed parts, facilitates their optimal choice to obtain the desired part so that it meets the functional characteristics prescribed by the designer. Another purpose of this article is to present a new innovative tool with a specific geometry, which will enhance the single-point incremental forming process and the quality of the resulting parts.

Introduction

The industrial sector is facing immense pressure due to customer demands for customized products delivered in the shortest time possible. The need for diversification from final users compels the industrial environment to develop and optimize flexible manufacturing methods. This is necessary to ensure the ability to produce a wide range of products quickly and cost-effectively, without sacrificing quality. One of these flexible methods of manufacturing sheet metal parts is incremental sheet forming (ISF). The process of incremental plastic deformation can be implemented in various ways, with the two most important categories being Single Point Incremental Forming (SPIF) and Double Point Incremental Forming (DPIF). DPIF technological system utilizes two independent tools and offers various constructive variants, including those with a counterpunch, a partial mold, and a total mold [1]. The work published by Emmensa [2] presents methods for deforming metal sheets without using specific conventional tools (die and punches) for plastic deformation. These methods are considered to be the first incremental plastic deformation processes.

This flexible manufacturing process offers various advantages. For example, parts can be made from blank sheet metal without the need for specific tools, as required in conventional cold plastic deformation. The process's flexibility enables a quick transition from one part to another with a different shape and size by changing the trajectory. Additionally, the incremental deformation on small surfaces increases the deformability of materials. It is also possible to deform materials with high strength that are usually challenging to form using classical processes. Furthermore, the forces

Materials Research Proceedings 46 (2024) 291-299 https://doi.org/10.21741/9781644903377-38

that appear during the process are not dependent on the size of the piece, and the processing time generates minimal noise. ISF has advantages but also comes with some drawbacks. These include longer processing times compared to the classic process, suitability for prototypes and small batch parts only, the need for multiple steps or passes to obtain right angles of part walls, as well as lower surface and dimensional accuracy [1,3,4]. ISF has advantages but also comes with some drawbacks. These include longer processing times compared to the classic process, suitability for prototypes and small batch parts only, the need for multiple steps or passes to obtain right angles of part walls, as well as lower surface and dimensional accuracy [1,5,6].

The Single Point Incremental Forming (SPIF) process is generally utilized in various industrial sectors, including the automotive, aeronautical, marine, medical, and food industries [4,6-10]. The SPIF process requires a complete deformation system, which includes three basic elements: the metal sheet product to be deformed, a device for orienting and fixing the sheet blank, to prevent slipping during processing, and an incremental plastic deformation tool mounted in the tool holder of the machine CNC (Computer Numerical Control) that to follow the set trajectory and incrementally deform the metal foil.

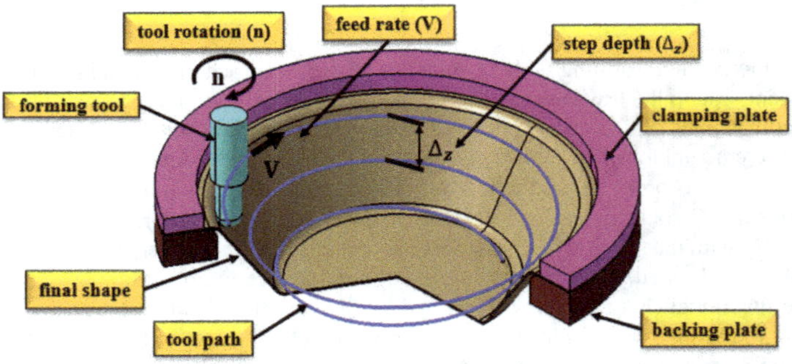

Fig. 1. The basic principle of the incremental sheet forming (ISF) process

Fig. 1 illustrates the functioning of the SPIF. The pre-cut sheet metal is placed on the backing plate and its movement is restricted using a clamping ring. Depending on the part's geometry, the specialized CNC software defines the trajectory for the deformation tool by setting the incremental step depth “Δ_z”, in the "Z" direction. Finally, the process parameters, feed speed “V” and spindle speed “n”, are set. After the input data has been properly entered, the CNC machine tool moves the deformation tool along the imposed path with the defined technological parameters. The blank undergoes incremental deformation in a localized area, specifically the contact surface between the tool and the blank sheet. Upon completing the entire trajectory, the final deformed part is obtained. The quality of the final part depends on several factors, including process parameters, the type of trajectory used, the dimensions of the metal sheet, the configuration of the final part, the rigidity of the technological system, and the type of tool used [11].

Tools used in SPIF

The tools used for incremental plastic deformation are not standardized, and the researchers are the ones who developed their punches, especially for incremental plastic deformation in a single point. Since these tools are not standardized, their variability makes classification according to a single criterion very difficult. Fig. 2 illustrates a categorization of the tools used SPIF based on various criteria such as number of components, type of friction between the tool and blank, head shape, type of contact, cooling type, and type of CNC machine tool. According to this proposed classification, a punch can be included in multiple categories, not just one. In the following, we will present the main categories of tools along with their advantages and disadvantages.

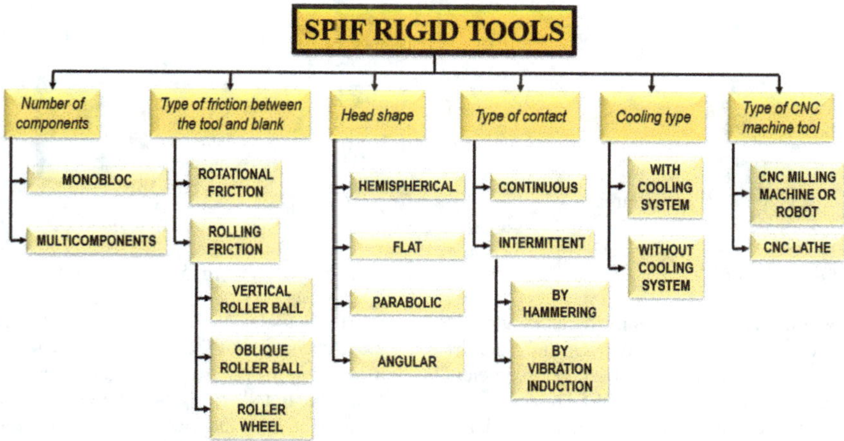

Fig. 2. Classification according to several criteria of tools used in SPIF.

Rigid hemispherical tools are the most used types of punches and the most diverse in terms of construction. Fig. 3 shows a tool with a hemispherical head in a 3D view. This is the most used punch at SPIF and the simplest as a constructive form. This punch is a cylindrical rod with a diameter "D". On this monobloc tool, two different areas are distinguished: *the body of the tool* that makes contact with the tool holder device follows the set trajectory of the CNC machine and *the tool head*, represented by the hemispherical surface of the tool, comes into direct contact with the surface of the sheet metal, and incrementally deforms it plastically. It represents the reference and comparison tool of most specific tools developed by researchers. The contact area of rigid tools with a hemispherical head is the same because they have a hemispherical surface. However, the body of the punch may vary and have different sections depending on the specific technological system in which it is used [12].

Depending on the shape of the tool body, we can identify in the specialized literature different constructive variants of tools with a hemispherical head, respectively: *hemispherical tool with constant section* (this is the simplest and most common type of punch, it has a hemispherical contact area, and the section of the punch body is constant), *hemispherical tool with variable conical section* and *hemispherical tool with multiple variable conical sections*. The sections have the role of stiffening the tool [6,13,14]. The diameter of punches with hemispherical heads ranges from 2 to 30 mm. For larger parts, the diameter can reach 100 mm, and the most commonly used diameter being 12 mm. The best formability was achieved using a 10 mm diameter punch. The radius of the active zone is the geometric parameter that differentiates the punches with hemispherical heads and that has the greatest influence on the respective process: the roughness of the parts obtained, the deformability of the material, deformation time, and deformation force [3,13-16]. This type of punch has the advantage that they are simple, easy to make, do not require special tools and consequently have a low cost. Their disadvantage is that they are limited in terms of performance, and requires strategies and additional equipment to overcome them.

Materials Research Proceedings 46 (2024) 291-299 https://doi.org/10.21741/9781644903377-38

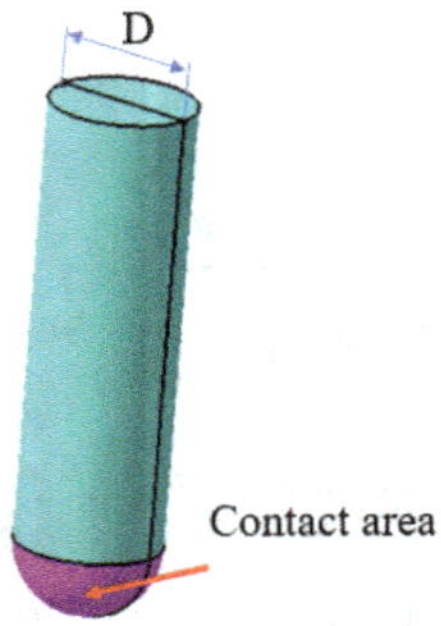

Fig. 3. Rigid hemispherical tool

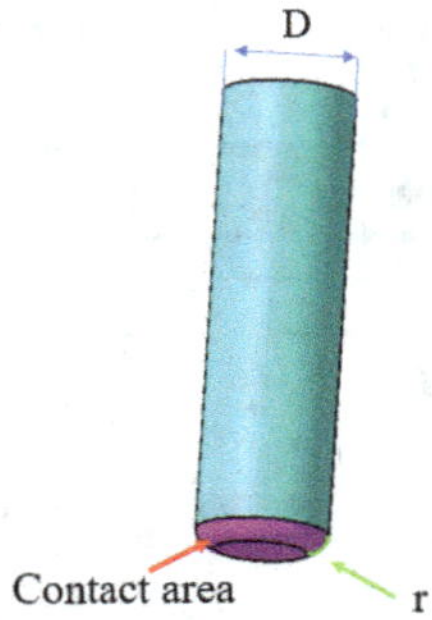

Fig. 4. Rigid flat tool

Fig. 5. Roller-ball tool

Rigid flat tools are simple punches and very similar to the hemispherical ones. They are formed by a circular rod with a diameter "D", and the flat surface at the end of the rod is connected with a connection radius "r". For a given diameter "D", various types of flat head tools can be created, depending on the value of the connection radius "r", as illustrated in Fig. 4. The contact area between the punch and the sheet metal is larger compared to the hemispherical head punch, which is a distinctive geometric feature of this type of tool. Such punches made of high-speed steel were tested by Ajay Kumara et al [11]. The advantages of such a tool are the following: the complexity of the tool is low, the precision of the parts obtained is higher, reduces the pillow effect, the deformation forces are lower, superior deformability if the combination between the diameter and the connection radius is chosen appropriately compared to tools with a hemispherical head [17]. This tool also has disadvantages, for example: parts with complex surfaces and small radii cannot be produced, and in some cases, dynamic recrystallization occurs due to high frictional forces, and the final parts require additional heat treatments [18].

Roller-ball tools are more complex tools compared to monobloc ones. This type of tool consists of a body and a mobile part, usually a ball but sometimes a roller wheel. The main distinction is that when deforming with a rigid tool, such as one with a hemispherical head, a sliding friction force occurs between the tool and the sheet blank. However, when using a roller ball tool, this force transitions into a rolling friction force. Fig. 5 shows a roller ball tool, in which the active part is a ball that rotates inside the tool body. The specialized literature presents various constructive variants of tools that convert sliding friction into rolling friction. We want to highlight the *roller ball tool*, similar in construction to the one in Fig. 5, which is the most commonly used tool. Additionally, there's the *oblique roller ball* (ORB), a *tool that efficiently transforms sliding friction into rolling friction* through the use of bearings and *rolling-wheel* technology. Among the main advantages of these types of punches are: the *reduction of the forces* that occur during the process, the *increase of the formability* of the material due to the achievement of larger angles of the part's walls, and better results both on the *roughness* and on the *thickness distribution.* As disadvantages, the complexity of the tool can be pointed out because it is made of several components and involves a higher cost [12,19-21].

Intermittent contact tools are tools with contact surfaces that vary based on the punch's angular position. This category includes tools that induce vibrations and come in various geometric shapes. A type of punch developed by Song Wu [22] and his collaborators is the *contact-induced vibration tool*. This punch is unique because it produces vibrations in the metal sheet when rotated. It is a simple monobloc flat end tool, with a connection radius of 2.5 mm, and an active area shaped to obtain an elliptical section. H. Long et al. [23] have developed a new type of vibration tool called *rosette tools* that have several grooves. These tools induce vibrations in the contact area between the tool and the blank. These induced vibration tools have the main advantages of *improving the formability* of materials and *reducing the deformation force*, which allows the processing of parts from materials with high hardness. disadvantage of these tools is that they require more specialized equipment to produce [22,23]. Another new tool model with intermittent contact used for incremental plastic deformation and that uses the hammering principle is the circumferential hammering tool, proposed by D. Nasulea and G. Oancea [24]. The punch has a monobloc hemispherical head with a variable section. In the active area, it has two parallel flat surfaces that are symmetrically positioned with respect to a plane containing the axis of symmetry of the tool. The distance between the two flat surfaces is 6 mm. The advantages of the circumferential hammering tool are the *increase of formability* of the material so that parts with deeper walls and slopes can be obtained and *improving the geometric precision* by reducing the effect of elastic return. The disadvantage of this type of tool is the fact that its production process is more complex, and it requires more efficient equipment for its production [24].

The tool that transforms sliding friction into rolling friction with a bent head for a lathe is a tool developed by Neagoe Ion and his collaborators. It presents all the advantages of the roller ball tools and has the particularity that it has a bent tail presenting the disadvantage that they can only be used on the lathe [25].

Tools with an internal cooling system are used for the deformation of parts on electricity-assisted incremental sheet forming (E-ISF) systems. Runze Liu and his collaborators developed and tested different versions of tools with an internal cooling system like: a conventional rigid tool with an inner water-cooling channel, a rolling-ball tool with an inner water-cooling channel, and rolling-wheel tool with an inner water-cooling channel. They have the advantage that reducing the temperature between the tool and the blank and consequently improving the quality of the final piece in the (E-ISF) system. One major disadvantage is that they are complex, expensive, difficult to make tools, and require additional installations [26].

As previously mentioned, the specialized literature presented the development of numerous tools for ISF, especially in SPIF. Each tool has its advantages and disadvantages which I mentioned briefly. Because the tools used in SPIF are numerous and varied, we have also presented a classification that includes them all.

Development of an innovative new tool

In the previous section, it was noted that there are various types of tools for each group, each with specific advantages and disadvantages. Some are easy to use and have fewer components, but they may not perform well in terms of formability, roughness, or pillow effect. Other tools have a beneficial impact on the formability of materials, reduce the deformation force but do not improve the pillow effect or are more complex, requiring more special tools for their manufacturing. For these reasons, it is necessary and important to develop new innovative tools that offer as many advantages and as few disadvantages as possible and to enhance the process of incremental plastic deformation.

In the following, we present a new innovative tool, called "*eccentric tool with variable radius (ETVR)*" that we developed. This tool is designed to have numerous advantages such as improving formability, surface roughness, simplicity of manufacturing, reducing deformation force, minimizing pillow effect, and being a monobloc punch.

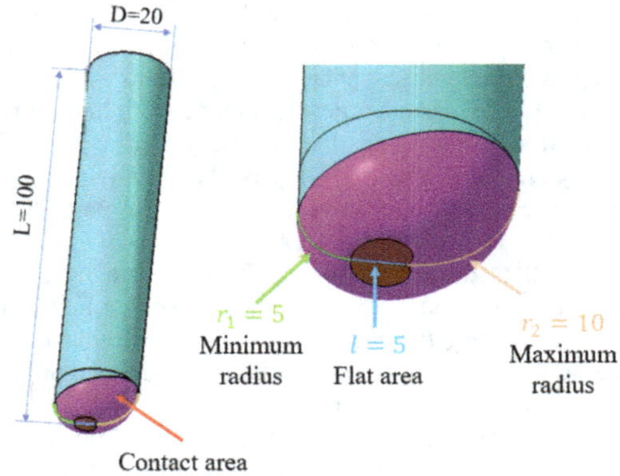

Fig. 6. The geometric characteristics of the ETVR

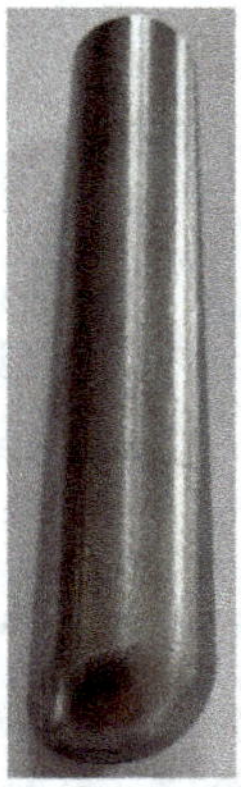

Fig. 7. Manufactured ETVR tool

Fig. 6 shows the innovative RCIVT, 3D tool with its dimensional values. This new innovative tool (ETVR), with a variable radius, is a simple cylindrical rod with a diameter of D=20 mm and a total length of L=100 mm. On the active area that comes into contact with the metal sheet, a variable connection radius was created between the outer cylindrical surface and the flat surface of the rod. It has a minimum value of $r_1 = 5$mm at one end, a maximum value of $r_2 = 10\ mm$ on the opposite side at 180° around the axis, and a flat area at the end of the rod that has a length of l=5mm. Also, in Fig. 7 the manufactured tool made on a CNC milling machine is presented.

This innovative tool aims to combine and possess the advantages of three different tools, respectively:

- *Rigid flat tool* which increases the formability of the material by increasing the temperature due to the friction between the flat area of the tool and the metal foil, and reduces the defect of the parts obtained by incremental plastic deformation called the pillow effect;
- *Rigid hemispherical tool with a big radius R=10mm* which influences the quality of the surface, obtaining a lower roughness and having a higher productivity;
- *Rigid hemispherical tool with a small radius R=5mm* which, due to the smaller contact area and implicitly smaller deformed surface, increases the formability of the sheet of metal that have to be processed and decreases the deformation force.

Another particularity of the ETVR tool is the geometric shape of the active zone that continuously changes the area of the contact surface between the tool and the blank sheet, depending on the angular position in relation to its own axis.

In Fig. 8, it is noticeable that the tool has the maximum contact surface with the metal sheet when the rotation angle (φ) of the tool in relation to its own axis is 0°. The contact surface is minimal at an angular position of 180°. Beyond this, the tool is only in contact on the front flat surface up to 360° if the tool rotation speed "n" is higher than the feed speed "V".

To more clearly understand the evolution of the contact surface between the ETVR and the metal sheet, a diagram was made for different angular positions of the tool for a part that has a wall angle α=45°, as can be seen in Fig. 8. The example was made by sectioning with a plane that passes through the axis of rotation of the tool and perpendicular to the plane tangent to the surface of the part in contact point between the punch and the deformed part whose wall angle α=45°, at a certain depth.

Materials Research Proceedings 46 (2024) 291-299 https://doi.org/10.21741/9781644903377-38

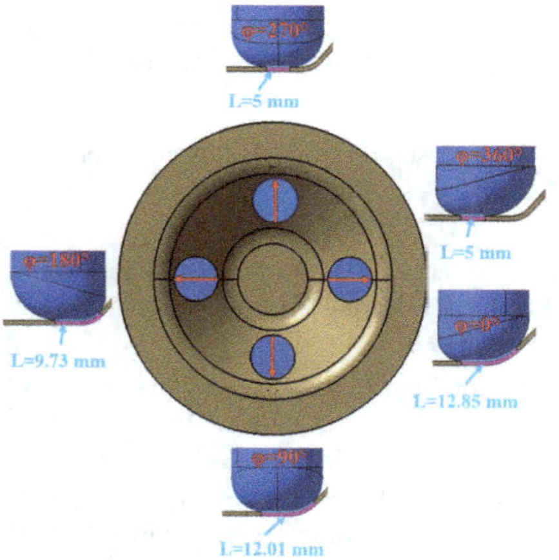

Fig. 8. The contact between the ETVR tool and the blank in different rotational positions for a part with the wall angle α=45°

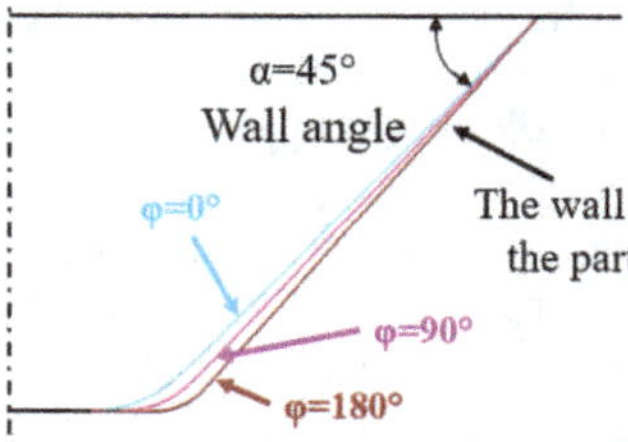

Fig. 9. The position of the generator of a part with the wall angle α=45°, and different angular positions of the ETVR *tool*

To highlight the variation of the contact surface between the tool and the punch, we took as an element of comparison the length of the contact line resulting from sectioning the punch and the part at a certain depth with the sectioning plane, and we have the following values:

- For φ=0°→ the length of the contact line is L=12.85 mm;
- For φ=90°→ the length of the contact line length is L=12.01 mm;
- For φ=180°→ the length of the contact line length L=9.73 mm;
- For φ=270°→ the length of the contact line length L=5 mm (if the tool rotation "n" is higher than the feed speed "V");
- For φ=360°→ the length of the contact line L=5 mm (if the tool rotation "n" is higher than the feed speed "V").

In Fig. 9 it was highlighted that the generator of the inclined wall of the part has different positions and inclinations, depending on the angular position φ of the punch. These variations of the position of the generator depending on the angular position "φ" of the punch were made on a part whose wall angle α=45°. It can be observed that at the same point of the trajectory, for the angular position of the tool φ=0°, the angle of the generator has the minimum value and for the angular position of the tool φ=180°, the angle of the generator has the maximum value.

Conclusions

In this paper, we briefly presented the basic principles of SPIF. The specialized literature presented the development of numerous tools for ISF, especially in SPIF. Each tool has its advantages and disadvantages which we mentioned in the paper. Because the tools used in SPIF are numerous and varied, we have also presented a classification that includes them all.

Starting from the advantages and disadvantages of the analysed tools, a new innovative tool, named "eccentric tool with a variable radius - ETVR", was developed. The particularity of the ETVR tool is the geometric shape of the active zone that continuously changes the area of the contact surface between the tool and the blank sheet, depending on the angular position of its axis. Therefore, this new tool combines the advantages of three rigid tools, respectively: flat tool, hemispherical tool with a radius R=10mm, and hemispherical tool with a radius R=5mm.

The research will continue by conducting experiments to compare the results obtained using the new tool with those obtained using the traditional tools and validate its advantages.

References

[1] T. Trzepieciński, B. Krasowski, A. Kubit, D. Wydrzyński, Possibilities of application of incremental sheet-forming technique in aircraft industry, Scientific Letters of Rzeszow University of Technology – Mechanics. (2018) 87-100. DOI: 10.7862/rm.2018.08

[2] W. C. Emmens, G. Sebastiani, A. H. van den Boogaard, The Technology of Incremental Sheet Forming - a Brief Review of the History, Journal of Materials Processing Technology. 210 (2010) 981–997. https://doi.org/10.1016/j.jmatprotec.2010.02.014

[3] Y.H. Kim, J.J. Park, Effect of process parameters on formability in incremental forming of sheet metal, Journal of Materials Processing Technology. 130–131 (2002) 42–46. https://doi.org/10.1016/S0924-0136(02)00788-4

[4] A. Mulay, B. S. Ben, S. Ismail, A. Kocanda, Experimental Investigation and Modeling of Single Point Incremental Forming for AA5052-H32 Aluminum Alloy, Arab J Sci Eng. 42 (2017) 4929–4940. https://doi.org/10.1007/s13369-017-2746-1

[5] D. Năsulea, G. Oancea, Incremental Deformation: A Literature Review, MATEC Web of Conferences, 121 (2017) 3017. https://doi.org/10.1051/matecconf/20171210

[6] M. Milutinović, R. Lenđel, M. Potran, D. Vilotić, P. Skakun, M. Plančak, Application of single point incremental forming for manufacturing of denture base, Adv Techn Mat. 39(2) (2014) 15–24.

[7] M. Amino, M. Mizoguchi, Y. Terauchi T. Maki, Current status of "Dieless" Amino's incremental forming, Procedia Engineering. 81 (2014) 54 – 62. https://doi.org/10.1016/j.proeng.2014.09.128

[8] P. Gupta, J. Jeswiet, Manufacture of an aerospace component by single point incremental forming, Procedia Manuf. 29 (2019) 112–119. https://doi.org/10.1016/j.promfg.2019.02.113

[9] P. De Waal Eksteen, Development of incrementally formed patient-specific titanium knee prosthesis. PhD Thesis. Stellenbosch: Stellenbosch University, 2013.

[10] J. Jeswiet, J. R. Duflou, A. Szekere, P. Lefebvre, Custom Manufacture of a Solar Cooker – A Case Study, Advanced Materials Research. 6-8 (2005) 487-492. doi:10.4028/www.scientific.net/AMR.6-8.487

[11] A. Kumar, V. Gulati, P. Kumar, V. Singh, B. Kumar, H. Singh, Parametric effects on formability of AA2024-Oaluminum alloy sheets in single point incremental forming, Journal of Materials Research and Technology. 8 (2019) 1461-1469. https://doi.org/10.1016/j.jmrt.2018.11.001

[12] M. Çalişkan, Ö. Çerlek, F. Özen, A comparative study: incremental sheet forming of AA1050 by roller-ball tool and standard tool with its optimization, Sādhanā. 49 (2024) 78. https://doi.org/10.1007/s12046-023-02420-9

[13] Y. Zeradam, A. Krishnaiah, Y. Zeradam, A. Krishnaiah, Numerical Simulation and Experimental Validation of Thickness Distribution in Single Point Incremental Forming for Drawing Quality Steel, International Journal of Applied Engineering Research. 15(1) (2020) 101-107

[14] R. Jagtap, V Sisodia, K. More, S. Kumar, Hybrid incremental forming: investigation on localized thinning and thickness distribution in formed parts, Lecture Notes in Mechanical Engineering. (2021) 151–61. https://doi.org/10.1007/978-981-15-6619-6_16.

[15] J. Jeswiet, F., Micari, G. Hirt, A. Bramley, J. Duflou, J. Allwood, Asymmetric Single Point Incremental Forming of Sheet Metal, CIRP Annals. 54(2) (2005) 88-114. https://doi.org/10.1016/S0007-8506(07)60021-3

[16] V. Gulati, A. Kumar, Investigation of some process parameters on forming force in single point incremental forming, Int Res J Eng Technol. 4(4) (2018) 784-791.

[17] X. Ziran, L. Gao, G. Hussain, Z. Cui, The performance of flat end and hemispherical end tools in single-point incremental forming, Int J Adv Manuf Technol. 46 (2010) 1113–1118. DOI 10.1007/s00170-009-2179-4

[18] P. Gupta, J. Jeswiet, Effect of temperatures during forming in single point incremental forming, The International Journal of Advanced Manufacturing Technology. 95 (2018) 3693–3706. https://doi.org/10.1007/s00170-017-1400-0

[19] B. Lu, Y. Fang, D.K. Xu, J. Chen, H. Ou, N.H. Moser, J. Cao, Mechanism investigation of friction-related effects in single point incremental forming using a developed oblique roller-ball tool, International Journal of Machine Tools and Manufacture. 85 (2014) 14–29. https://doi.org/10.1016/j.ijmachtools.2014.04.007

[20] A. Kumar, V. Gulati, P. Kumar, H. Singh, Forming force in incremental sheet forming: a comparative analysis of the state of the art, Journal of the Brazilian Society of Mechanical Sciences and Engineering. 41 (2019) 251. https://doi.org/10.1007/s40430-019-1755-2

[21] S. Hatori, S. Hatori, A. Sekiguvhi, A. Ozer, Conceptual design of multipurpose forming machine and experiments on force-controlled shear spinning of truncated cone, 17th International Conference on Metal Forming, Metal Forming 2018, 16-19 September 2018, Toyohashi, Japan. https://doi.org/10.1016/j.promfg.2018.07.361

[22] S. Wu, P. Geng, N. Ma, F. Lu Contact-induced vibration tool in incremental sheet forming for formability improvement of aluminum sheets, Journal of Materials Research and Technology. 17 (2022) 1363-1379. https://doi.org/10.1016/j.jmrt.2022.01.103

[23] H. Long, W.X. Peng, Z.D. Chang, H. Zhu, Y.J. Jiang and Z.H. Li, New rosette tools for developing rotational vibration-assisted incremental, Journal of Materials Processing Tech. 326 (2024) 118311. https://doi.org/10.1016/j.jmatprotec.2024.118311

[24] D. Nasulea, G. Oancea, Achieving Accuracy Improvements for Single-Point Incremental Forming Process Using a Circumferential Hammering Tool, Metals. 11 (2021) 482. https://doi.org/10.3390/met11030482

[25] I. Neagoe, A.C. Filip, M.A. Vasiloni, Experimental Researches on the Limit Angle of Conical Parts Manufactured By Single Stage Incremental Forming on CNC Lathes, Applied Mechanics and Materials. 371 (2013) 68-172

[26] R. Liu, B. Lu, D. Xu, J. Chen, F. Chen, H. Ou, H. Long, Development of novel tools for electricity-assisted incremental sheet forming of titanium alloy, Int J Adv Manuf Technol. 85 (2016) 1137–1144. https://doi.org/10.1007/s00170-015-8011-4

Materials Research Proceedings 46 (2024) 300-306 https://doi.org/10.21741/9781644903377-39

Investigation of the wear resistance of surface layered by MMA and carbonitriding layers on structural alloy steel

Apostol UCHERDZHIEV[1,a], Tatyana MECHKAROVA[1,b *], Desislava MINCHEVA[1,c], Nikolay VALCHEV[1,d]

[1]Technical university of Varna, Department of Materials Science and technology, Varna, Bulgaria

[a]Ucherdzhiev@tu-varna.bg, [b]t.mechkarova@tu-varna.bg, [c]dmincheva@tu-varna.bg, [d]armicronn@gmail.com

Keywords: Surface Layering by MMA, Carbonitriding, Wear Resistance

Abstract. This study investigates the wear resistance of surfaces layered by MMA welding followed by carbonitriding, deposited on a 40X structural alloy steel. The primary objective is to develop a technology for restoring worn surfaces of shafts operating in marine environments. Wear resistance was evaluated using a "block-on-ring" test. The microstructure and properties of the obtained layers were analyzed using standard metallographic techniques, including macro and microstructural analysis and microhardness testing. The results of this study will serve as a basis for developing an effective repair technology for shafts used in marine equipment.

Introduction

Vortex pumps are a common choice when transporting fluids at low flow rates and high hydraulic heads, as is often required in marine pumping systems [1, 2, 3]. The unique characteristic of these pumps is that due to the low fluid velocity, a high head is necessary to overcome significant hydraulic resistances and pressures [4, 5]. The extreme operating conditions experienced by the rotating components of vortex pumps lead to increased wear, particularly in the bearing seats (Fig. 1) [4, 5].

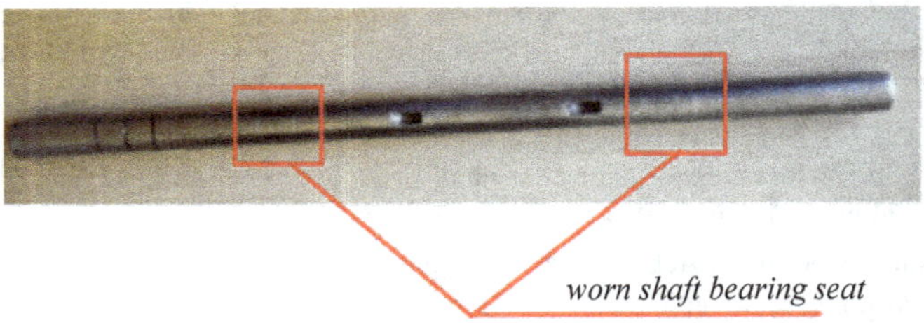

Fig.1. Location of the worn areas on the shaft

Methodology of the Research

Many known technologies exist for restoring worn-out bearing seats, but few are suitable for emergency repairs. The proposed technology of surface layering by MMA welding method [7] followed by carbonitriding offers a practical alternative for these urgent situations.

For the purposes of the study, two samples (Fig. 2) were prepared from the base material, structural alloy steel grade 40X (in accordance with DSTU 7806 and GOST 4543 standards). The chemical composition and mechanical properties of 40X steel are presented in Table 1[8].

Table 1. Chemical composition and mechanical properties of steel 40X [8]

Silicon [%]	Manganese [%]	Copper [%]	Nickel [%]	Sulfur [%]	Carbon	Phosphorus [%]	Chromium [%]
0,17-0,37	0,50-0,80	0,30	0,30	0,035	0,36-0,44	0,035	0,80-1,10
Heat treatment, delivery condition	Section [mm]	$\sigma_{0.2}$ [MPa]	σ_b [MPa]	δ_5 [%]	ψ [%]	KCU [J/m^2]	HB
Bar.Hardening 860°C, oil.Vacation 500°C, water or oil	25	780	980	10	45	59	-
Forging, normalization	<100	315	570	17	38	39	167-207

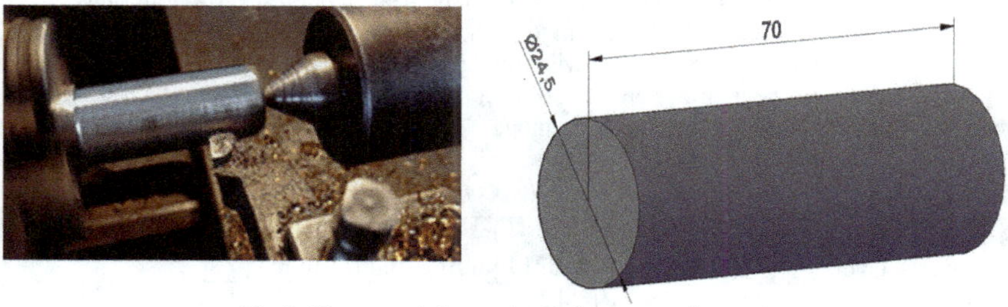

Fig.2. Shape and dimensions of the turned samples

Before conducting the main experiment, two cylindrical specimens were machined using a lathe (Figure 2) and subjected to preliminary heat treatment to transform the microstructure into granular sorbite. This heat treatment enhances the steel's toughness, ensuring high core toughness prior to surfacing by MMA and subsequent carbonitriding. Both processes increase surface hardness and wear resistance. Figure 3 illustrates the heat treatment cyclogram.

After reviewing the existing technological possibilities [1, 2, 3, 4, 7], the following electrodes for MMA were selected [9, 10], as shown in Table 2:

Table 2. Electrodes from the ESAB brand catalog and their chemical composition in % [9,10]

Electrodes	C [%]	Si [%]	Mn [%]	Cr [%]	Ni [%]	N [%]
OK 83.28	0.1	< 0.7	0.7	3.2	-	-
OK 61.20	0.02	0.4	1.9	19.8	9.8	0.05

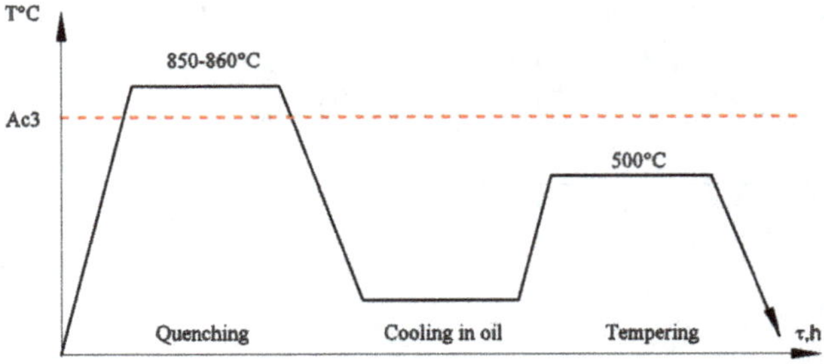

Fig.3. Cyclogram of the preliminary heat treatment of the samples

According to the experimental methodology shown in Table 3, one of the specimens was preheated in a resistance furnace to 200°C and then surfaced by MMA weld method with the OK 83.28 electrode. The preheating temperature was selected based on the calculation of the carbon equivalent of 40X steel. The second specimen was not preheated, but instead, a buffer layer of OK 61.20 was applied, followed by OK 83.28 on top. The buffer electrode is of an stainless austenitic steel grade and has lower hardness. Following the layering process, the specimen surfaces were machined to eliminate irregularities and reduce roughness. The specimens were then subjected to carbonitriding in a furnace for final strengthening.

Table 3. Methodology of the experiment

№	top layer	buffer layer	preheated	L [mm]	D_o [mm]	D_{min} [mm]	D_{max} [mm]	D_{equal} [mm]	
1	OK 83.28		200°C	72	24,5	27	28	27,5	layering+low temperature gas nitriding
2	OK 83.28	OK 61.20		72	24,5	29,3	30,4	29,85	layering+low temperature gas nitriding

Figure 4 shows the scheme of the carbonitriding process in a laboratory 2-liter furnace.

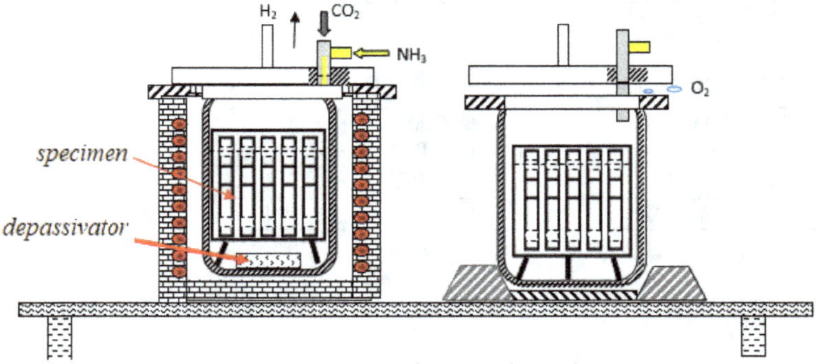

Fig.4. – Scheme of furnace for carbonitriding

The operating modes are according to the cyclogram in Figure 5:

Materials Research Proceedings 46 (2024) 300-306 https://doi.org/10.21741/9781644903377-39

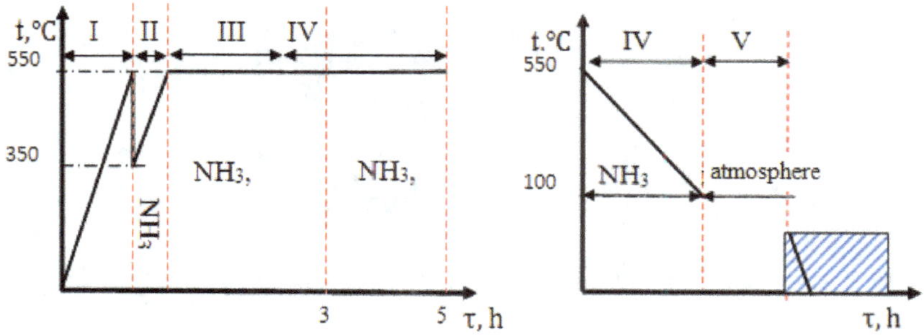

Fig.5. Carbonitriding cyclograms

To evaluate the wear resistance of the welded surfaces, enhanced by carbonitriding, a "block-on-ring" friction test was conducted. A rotating test specimen interacted with a stationary bronze counter body, simulating real work conditions of friction.

Temperature measurements were taken on the test body, counter body with type K thermocouple (Chromel / Alumel), and specimen surface using non-contact method.

The test specimen is secured in the collet chuck of the stand (Fig. 6) to ensure balanced rotation. The bronze counter body ($CuAl_8Fe_3$) is positioned on the specimen that its inner cylindrical surface aligns coaxially with the surface of the test specimen undergoing wear. The specimen is rotated at a speed of n = 1500 rpm. A load is applied to the specimen, calculated at the center of the contact area between the specimen and the counter body, and transmitted via a lever system. The investigation was conducted under identical conditions for both specimens, in a dry friction regime, with additional temperature measurements.

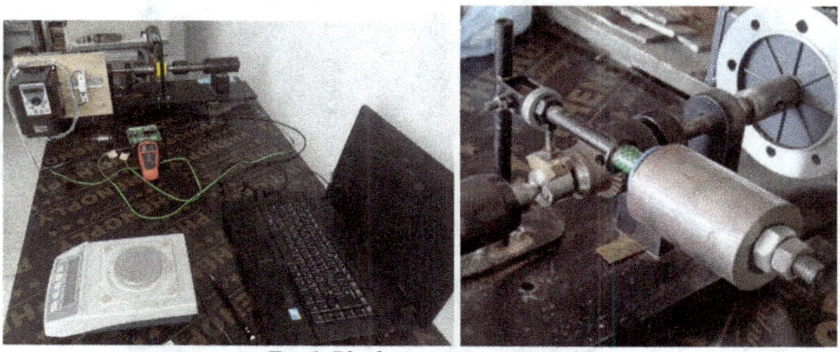

Fig.6. Block on ring wear stand

To evaluate the wear parameters, the weight loss of the specimens, temperature regime, corresponding to a certain traveled distance and friction time, were measured under equal conditions - load and sliding speed.

To perform the analysis of the structure and properties of the layered by MMA and subsequently strengthened by carbonitriding, metallographic analyses were conducted.

Materials Research Proceedings 46 (2024) 300-306 https://doi.org/10.21741/9781644903377-39

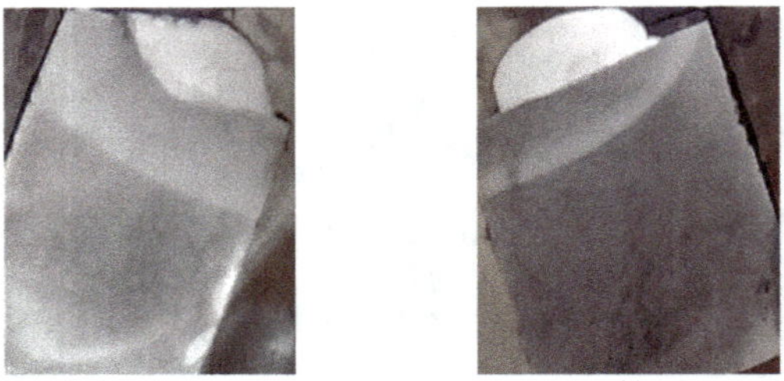

a) b)

Fig.7. Macrostructure of the surface layered by MMA method
a) specimen 1; b) specimen 2

As can be seen from the macrostructures in figure 7, specimen 2 (coated with a buffer layer) has a significantly smaller heat affected zone (HAZ) compared to specimen 1. This is related to the significantly lower thermal conductivity of the buffer layer (with an increased content of Cr).

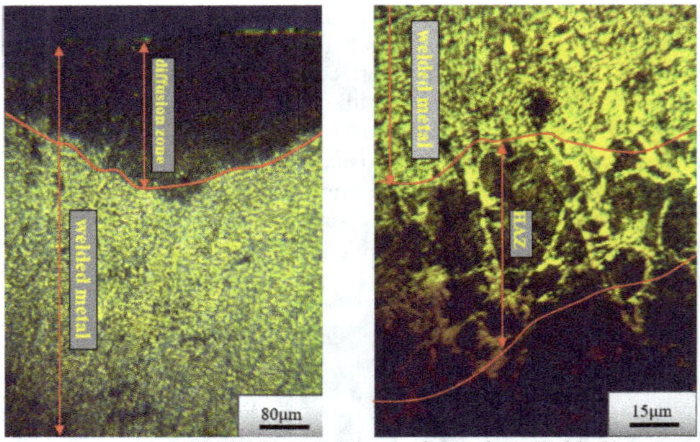

Fig.8 Microstructure of sample 1 (Table 3)

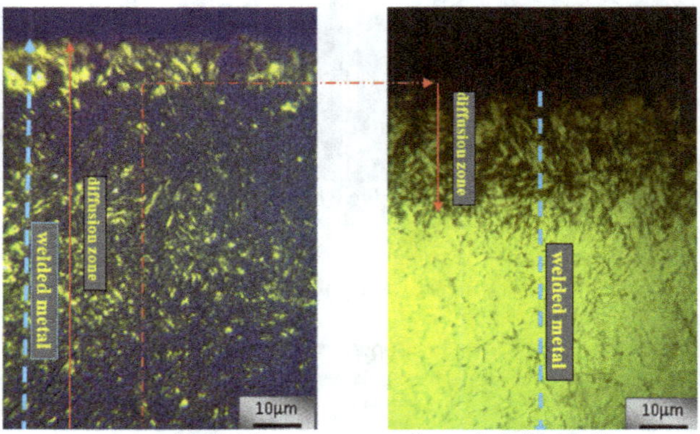

Fig.9 Microstructure of sample 2 (Table 3)

During the MMA layering process with a buffer layer of electrode OK 61.20 (specimen 2), a significant amount of chromium (18%) was formed on the surface. When the OK 83.28 electrode was applied on top, the liquid buffer layer causes the homogenization of chromium, increasing its concentration in the top layer. After nitriding specimen 2, the elevated chromium content formed an only diffusion zone with a significantly higher amount of chromium nitride, observable in the microstructure as light, needle-like grains.

To determine the extent of hardening of the samples, a Vickers microhardness test was conducted. The results of the test are shown graphically in Figure 10.

At carbonitriding, it can be observed that the surface layer with a buffer (specimen 2) develops a strengthened diffusion zone consisting of chromium nitride (a mechanical mixture of ferrite and chromium nitride) with significantly higher hardness, measured at 1020 HV. The thickness of this layer is relatively narrow, measuring 96 µm.

In difference, specimen 1 exhibits a considerably thicker HAZ layer, with a thickness of 150 µm, but with a lower hardness compared to specimen 2, measured at HV 750.

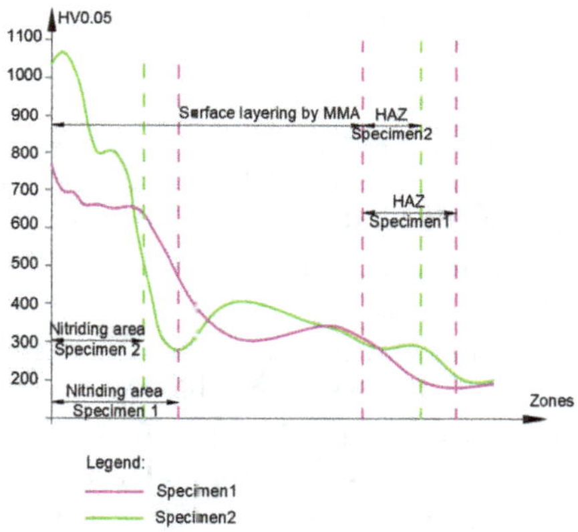

Fig. 10. Vickers Microhardness HV0.05 of the Specimens (as per Table 3)

Figures 11 show the results of the weight loss of the samples measured in the block-on-ring dry friction experiment.

The rapid metal loss observed after 15 minutes is attributed to the traces that have formed.

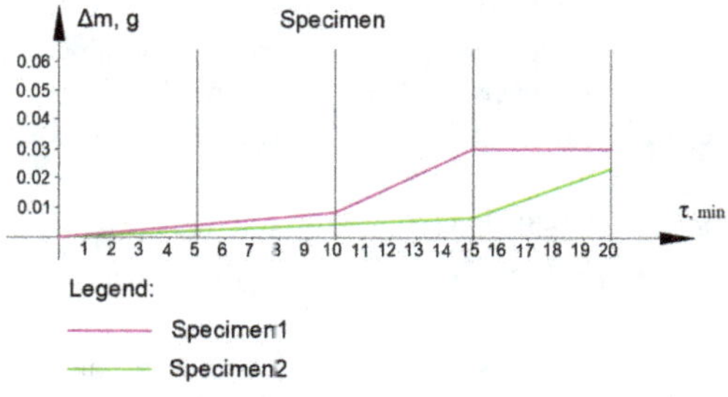

Fig. 11. Weight loss of the samples

Summary

It can be seen that in the sample with a buffer layer (sample 2), the thickness of the heat affected zone (HAZ) is significantly narrower and finer-grained (fig. 8) compared to sample 2.
From the measurements of the hardness after the surface layering using the MMA method, it was found that both of the proposed technologies for the repair and restoration of bearing seat in both samples was 450 HV.
After carbonitriding, sample 2 shows a significantly higher hardness (1020HV) compared to sample 1 (750HV), due to the higher chromium content in the welded layer.
The diffusion zone after carbonitriding in sample 2 (96μm) is smaller than in sample 1 (150μm) related to the increased content of chromium, which hinders nitrogen diffusion.
Both samples show good wear resistance under dry friction conditions. However, sample 2 demonstrated slightly better wear resistance due to its higher surface hardness.

The durability of the wear-resistant layer is ensured by the formed chromium nitride, the layer being relatively thick (about 100 microns), compared to layers obtained by diamond burnishing for example [6].

The proposed carbonitriding method is a finish after grinding and improves the resulting surface roughness

References

[1] K. Lyutskanov, (2018). Analysis and optimization of hydro abrasive wear of surfacing layers formed by using the method of gas-thermal welding with metal powders alloys. SCNVNA, 32, 62-65. https//doi.org/ 10.14748/scnvna.v32i0.4498

[2] Y. Denev, (2022). Application of hardfacing arc methods in Bulgarian ship repair SME. Trans Motauto World, 7(2), 50-52.

[3] K. Lyutskanov, Hristov, H., & Demirova, K. (2018). Experimental research of abrasive wear of surfacing layers. SCNVNA, 32, 52-56. https//doi.org/ 10.14748/scnvna.v32i0.4496

[4] Y. Argirov, Mechkarova, T., Luckanov, G., & Mincheva, D. (2024). Investigations of Wear Resistance of Hardfaced Layers on Steel 40X. TEM Journal, 13(1).

[5] J. Maximov, G. Duncheva, A. Anchev, V. Dunchev, Y. Argirov, V. Todorov, T. Mechkarova. (2022). Effects of Heat Treatment and Severe Surface Plastic Deformation on Mechanical Characteristics, Fatigue, and Wear of Cu-10Al-5Fe Bronze. Materials. 15(24). https//doi.org/ 10.3390/ma15248905

[6] G. Duncheva, J. Maximov, A. Anchev, V. Dunchev, Y. Argirov, S. Velkov. (2023). Modeling and Optimization of Surface Integrity and Sliding Wear Resistance of Diamond-Burnished Holes in Austenitic Stainless Steel Cylinder Lines. Machines.11(9), 872, ISSN 20754701. https//doi.org/ 10.3390/met13121930

[7] Y. Argirov, T. Mechkarova, A. Stoyanova, N. Atanasov, M. Konsulova-Bakalova. (2020). Study on structure and mechanical properties of 20X13 steel welding joints, IOP Conference Series: Materials Science and Engineering, 843(1). https//doi.org/ 10.1088/1757-899X/843/1/012012

[8] https://auremo.biz/materials/stal-40h.html

[9] https://esab.com/mn/apc_en/products-solutions/product/filler-metals/hardfacing/stick-electrodes-smaw/ok-83-28/

[10] https://esab.com/mn/apc_en/products-solutions/product/filler-metals/stainless-steel/stick-electrodes-smaw/ok-61-30/

Innovative Manufacturing Engineering and Energy - IManEE2024 Materials Research Forum LLC
Materials Research Proceedings 46 (2024) 307-313 https://doi.org/10.21741/9781644903377-40

Wear resistance of surface layers with subsequent diamond burnishing applied to 40X steel

Desislava MINCHEVA[1,a] *

[1]Technical University of Varna, Department of Materials Science and Technology, Varna, Bulgaria

[a]dmincheva@tu-varna.bg

Keywords: Surfaced Layers, OK83.28, Diamond Burnishing, Wear Resistance, Microhardness, Microstructures

Abstract. The subject of this study is weld overlays applied with the OK83.28 electrode on 40X steel, which underwent additional hardening through diamond burnishing with a specially designed diamond-tipped tool. The wear resistance of the welded and hardened layers was investigated under dry friction conditions. The wear characteristics of the samples were determined by measuring weight loss. A metallographic analysis of the structure formed in the welded layers after diamond burnishing was conducted. A residual stress analysis in the hardened zone was performed using X-ray diffraction. The microhardness of the hardened layers was measured using the Vickers method with a 50g load.

Introduction

Grinding is a conventional technology commonly employed as a finishing process when high surface quality and precise dimensional accuracy are essential. Microstructural flaws or inadequate surface quality can often result in significant damage to a component, and in severe cases, may lead to the failure of the entire assembly. Diamond burnishing (DB) presents a modern alternative to traditional grinding, offering enhancements in productivity, cost-effectiveness, and product quality. While the process is similar to roller pressure burnishing, it utilizes sliding contact instead of rolling contact. This technique is primarily applied to cylindrical components. The cold plastic deformation produced in the surface layer of the component by the tool achieves the desired surface characteristics: it reduces surface roughness, increases hardness, and induces compressive stresses on the surface. This method is especially suited for rotating components that adhere to high-quality standards, such as automotive crankshafts, bearings, and axles [1,2,3].

The publication [4] investigates the influence of technological parameters on the improvement ratio of surface roughness. The greatest improvement (~40%) was achieved at a lower burnishing speed (180 m/min) and force (45 N), with a high feed rate (0.1 mm/rev).

Authors [5] investigates the effect of slide diamond burnishing on the mechanical properties of GCR15 steel, comparing the performance of burnished and unburnished specimens. Slide diamond burnishing significantly enhances the fatigue strength of GCR15 steel, with a notable 36% improvement over unburnished specimens. The slight reduction in fatigue performance compared to literature values is likely due to the specimen geometry (small shoulder fillet). These findings highlight the potential benefits of burnishing for extending the fatigue life of steel components.

Diamond burnishing is a surface treatment technique that enhances the surface integrity and performance of components through intense plastic deformation. Although often seen as a simple smoothing process, DB can be applied in various forms: smoothing, deep, or mixed burnishing—depending on the process parameters. The authors [6] are focuses on optimizing DB for 41Cr4 steel under different conditions to identify the best parameters for different types of burnishing. Through planned experiments and regression analyses, key factors and objective functions, such as surface roughness and fatigue limits, are determined and justified. The authors investigate the roughness, microhardness, and fatigue strength of chromium-nickel steel AISI 316Ti, after DB. It has been established that: roughness of Ra = 0.055μm; microhardness increased by more than 32%. [7]. In numerous publications, Maximov and his team continue their research by examining the influence of applied diamond burnishing on surface quality. They found that diamond burnishing leads to an increase in the corrosion resistance and fatigue strength of the studied materials. [8,9]

Surfaced welding is the most widely used technology in the field of repair and restoration processes. It finds application in mechanical engineering, the oil and gas industry, as well as in shipbuilding. Repair and restoration of defective components are preferred due to their time and cost efficiency. Components that are most likely to fail are rotating parts subjected to wear from friction or overloading, such as shafts, axles, and working wheels, which are made from steel with excellent mechanical performance characteristics. [10-14]

In a previous research study [15,16], the wear resistance of weld overlays with the OK83.28 electrode on structural steel 40X was investigated. The present study is a continuation of that work, focusing on the wear resistance of the layers under dry friction conditions after diamond burnishing. The results obtained would be valuable for developing technologies related to the repair and restoration of machine components used in various industrial sectors.

Experimental part and Methodology

As the main material, structural steel 40X has been selected, and its chemical composition is presented in Table 1.

Table 1. Chemical composition of steel 40X [17]

C	Si	Mn	Ni	Cr	Cu	P	S
0.36 -0.41	0.17 – 0.37	0.5 – 0.8	≤ 0.3	0.8 – 1.1	≤ 0.3	≤ 0.035	≤ 0.035

The experimental part was carried out in few stages: Preparation of a cylindrical sample with dimensions D=24.5 mm and L=72 mm from 40X steel; Heat treatment of the sample to achieve a granular sorbite structure, ensuring high toughness of the steel core before the welding processes. Figure 1 illustrates the schematic of the heat treatment process, which involves quenching the samples at T = 850°C, followed by cooling in oil and high-temperature tempering at T = 500°C.

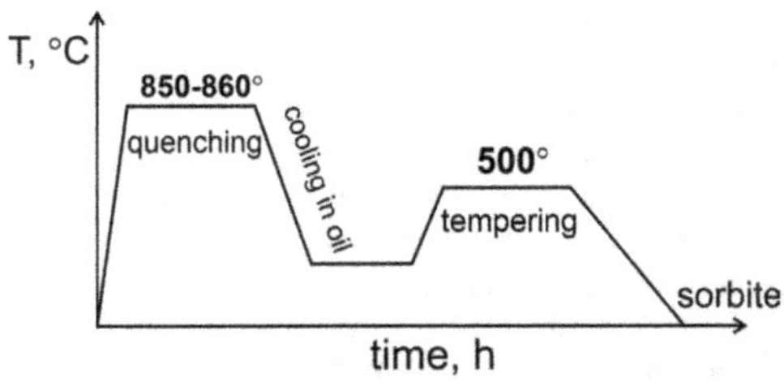

Figure 1. Scheme of the heat treatment process

The welding process of the sample with the OK 83.28 electrode involves the deposition of longitudinal layers. To avoid the formation of cold cracks, the sample is preheated to T = 200°C.

The chosen heating temperature is aligned with the carbon equivalent of the steel. The selected electrode OK 83.28, according to the technical specifications [18], provides good wear resistance to the welded layers. Table 2 presents the chemical composition of the weld layer produced with the selected electrode.

Table 2. Chemical composition of welded layer (OK83.28) [18]

Electrode	*C*	*Si*	*Mn*	*Cr*
OK 83.28	***0.1***	***< 0.7***	***0.7***	***3.2***

Diamond burnishing technology

The creation of a hardened surface layer through cold plastic deformation is done with the aim of altering the microgeometry of the original surface. This reduces the height of surface roughness, creates regularity in the microrelief, and introduces compressive residual stresses, which have a favorable effect on fatigue cyclic loading. The surface undergoes work hardening, increasing its wear resistance.

The welded test body was divided into two samples: Sample 1 - control sample; Sample 2 - which underwent diamond burnishing using a specially prepared tool (Fig. 2). The hardening of the surface layers through cold plastic deformation was performed with a deforming tool featuring a diamond spherical indenter at the tip. The diamond tool is attached to a standard setup on a CNC lathe, allowing for the adjustment of the tool's feed motion and the pressure force applied to the test body, which is secured between the lathe centers and rotated. The rotation speed of the test body is also controlled by the lathe's computer. The burnishing was conducted with a pressure force of Fb = 500 N. The tool also performed a feed motion of f = 0.044 mm/rev. The peripheral rotation speed of the test body was V = 40 mm/min.

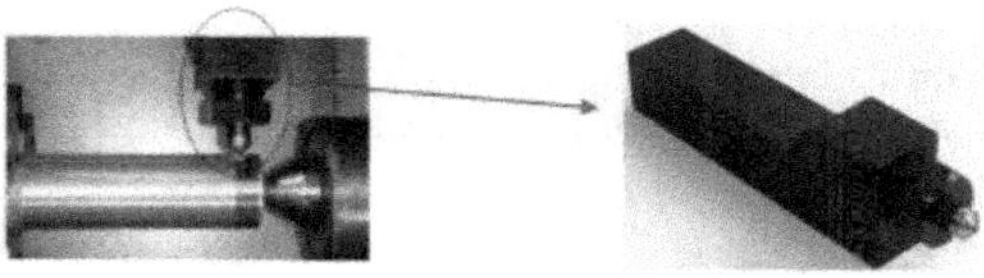

Figure 2. Diamond burnishing tool [9]

Wear resistance

To investigate the wear resistance of welded and subsequently hardened surface layers subjected to diamond burnishing, a methodology was employed to evaluate their wear through dry friction. This was achieved by using a rotating test specimen, mounted on a testing stand, in contact with a stationary counter body made of bronze The inner surface of the counter body contacts the surface of the rotating specimen, which rotates at a fixed speed of 1400 rpm. To evaluate the wear characteristics, the mass wear of the samples, the temperature regime, corresponding to a certain distance traveled and friction time, are measured under constant conditions - load and sliding speed. The relevant wear characteristics and wear resistance are determined by calculating the mass loss. Mass wear Δm (Eq.1) is the difference between the mass of the sample before friction m_0 and after a certain distance/time of friction m_1:

$$\Delta m = m_0 - m_1 \tag{1}$$

Results and Discussion

The microstructure analysis is performed using a NEOPHOT-2 metallographic microscope. For measuring the microhardness in the welded and hardened layers, a Hanemann attachment is used. A load of 50 grams was selected.

The structure of the layers welded with OK 83.28 consists of needle-like martensite, as expected according to the Schaeffler diagram (Fig.3).

Materials Research Proceedings 46 (2024) 307-313 https://doi.org/10.21741/9781644903377-40

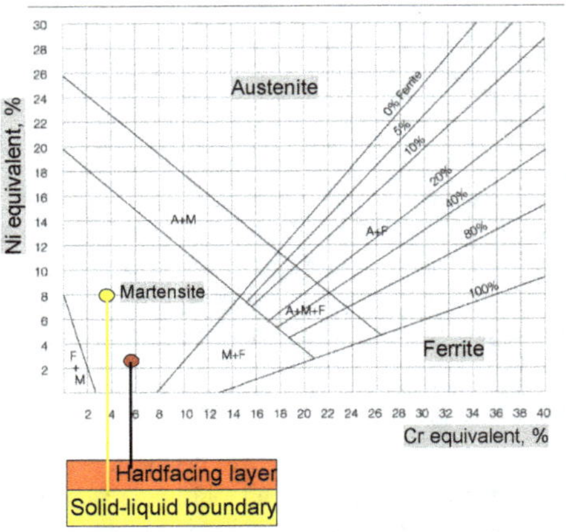

Figure 3. Microstructure of surfaced layer, welded with OK83.28 according to the Schaeffler diagram [19]

An X-ray diffractometer (Siemens) was used to assess the residual macro stresses resulting from the welding and subsequent hardening processes. The findings are illustrated in the histogram shown in Fig. 4. It is evident that substantial compressive stresses of -800 MPa were induced during the diamond burnishing (DB) process. In contrast, when evaluating residual stresses through X-ray diffraction, tensile stresses of +60 MPa were detected in the layers that were welded without undergoing diamond burnishing.

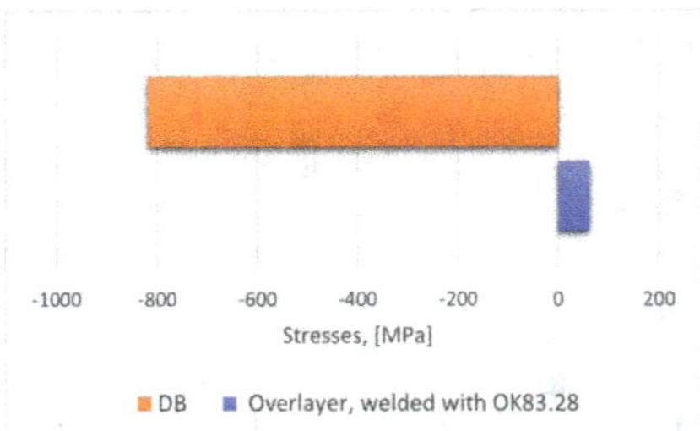

Figure 4. Histogram of residual stresses

The microhardness profile of Sample 1 (surfaced welded with OK83.28) is presented in Fig. 5. The graph shows that in the weld metal zone, the hardness increases to around 450 HV0.05. In the heat-affected zone, the hardness decreases to 250 HV0.05, which is close to that of the base metal (improved 40X steel).

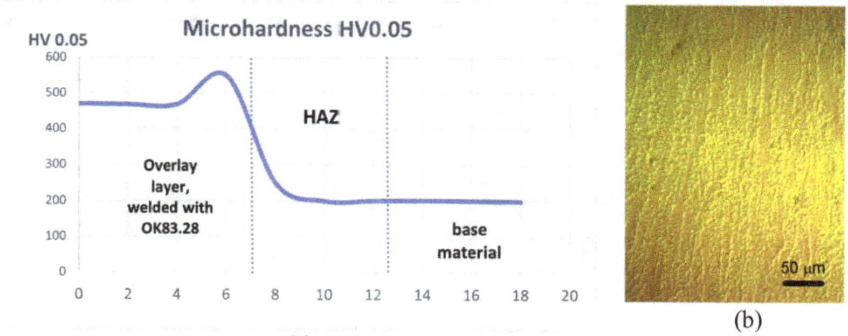

(a) (b)

Figure 5. Microhardness profile of surfaced layer, welded with OK83.28 (a); Microstructure on the surface (b)

The microhardness profile of Sample 2 (with DB process) is shown in the Fig. 6. The graph shows that the hardness of the diamond burnished layers is relatively high (450HV0.05), maintaining the high hardness obtained after hardfacing. The metallographic examination revealed that the dendritic crystals on the surface are deformed and oriented in the direction of the movement of the diamond tool. This is clearly illustrated in Figure 6.

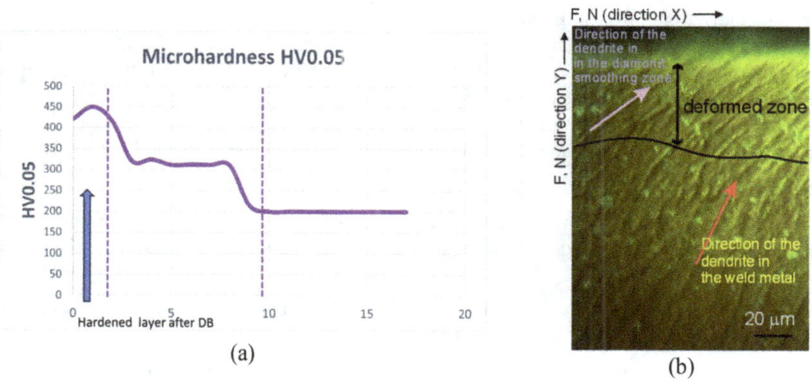

(a) (b)

Figure 6. Microhardness profile (a) of hardened layers after DB; Microstructure in the deformed zone after DB (b)

The dry friction test established the actual wear resistance of both samples. It was found that during the initial stage of testing, both samples wore evenly. However, after reaching 15 minutes of testing, sample 1 (hardfaced with OK83.28) showed higher values of wear resistance loss, with a relative wear of Δm= 0.8g (Fig.7a). Sample 2 (with diamond burnished layers) showed a smaller mass loss Δm=0.025g, which is likely due to the hardening resulting from the applied burnishing. (Fig.5a). Regarding the control of mass loss of the counter body during the dry friction process, it can be said that it wears slightly more when in contact with Sample 2 (DB). This is likely due to the work hardening created in the diamond-burnished layers, which affects the wear of the counter body. (Fig.7b).

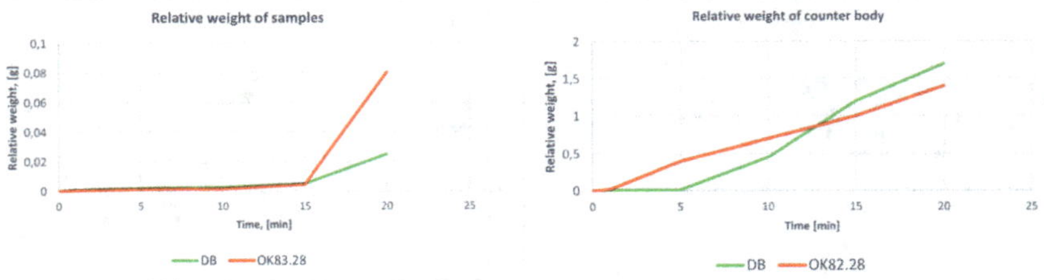

(a) (b)

Figure 7. Weight loss of the tested samples (a) and counter body (b)

Conclusions

The technological process for hard facing structural steel 40X with electrode OK83.28 has been correctly selected, with an appropriate preheating temperature of T=200°C established. The surface hardness of the welded layers increases twofold compared to that of the base material. The subsequent treatment of the welded layers using the diamond burnishing method significantly enhances the wear resistance of the surface layers. The wear resistance test showed a negligible weight loss of the diamond burnished samples (Δm=0.025g) compared to the weight loss in samples that underwent only hardfacing.

The samples subjected to diamond burnishing exhibit relatively high hardness in the contact zone of the welded specimens. A lower roughness was achieved, along with the potential for generating compressive stresses as a result of the applied diamond burnishing.

The technology can be used to improve the wear resistance of components where it is essential for the core to remain tough, while the surface layers exhibit low roughness and high hardness.

References

[1] M. Korzynski, Lubas, J., Swirad, S. and Dudek, K., Surface layer characteristics due to slide diamond burnishing with a cylindrical ended tool, J. Materials Processing Technology, Vol. 211, No. 1, (2010) pp.84–94.

[2] S. Swirad, The surface texture analysis after sliding burnishing with cylindrical elements, Wear, Vol. 271, Nos. 3–4, (2011), pp.576–581.

[3] G. Varga, Effects of technological parameters on the surface texture of burnished surfaces, Key Engineering Materials, Vol. 581 Precision Machining VII, pp.403–408, (2014) ISSN 1013-9826.

[4] J. Huuki, & Laakso, S. V., Surface improvement of shafts by the diamond burnishing and ultrasonic burnishing techniques, International Journal of Machining and Machinability of Materials, 19(3) (2017), 246-259.

[5] O. Taamallah, Hamadache, H., Mokas, N. et al., Investigation of the Effects of Slide Diamond Burnishing Process on the Mechanical Performance of GCr15 Steel. J Fail. Anal. and Preven. 23, 1101–1113 (2023). https://doi.org/10.1007/s11668-023-01652-5

[6] J.T. Maximov, Duncheva, G.V., Anchev, A.P. et al., Smoothing, deep, or mixed diamond burnishing of low-alloy steel components – optimization procedures. Int J Adv Manuf Technol 106, 1917–1929 (2020). https://doi.org/10.1007/s00170-019-04747-2

[7] J.T. Maximov, Duncheva, G. V., Anchev, A. P., Ganev, N., Amudjev, I. M., & Dunchev, V. P. Effect of slide burnishing method on the surface integrity of AISI 316Ti chromium–nickel steel. Journal of the Brazilian Society of Mechanical Sciences and Engineering, 40, (2018). 1-14.

[8] J. Maximov, Duncheva, G., Anchev, A., Dunchev, V., Argirov, Y., Nikolova, M., Effects of Heat Treatment and Diamond Burnishing on Fatigue Behaviour and Corrosion Resistance of AISI 304 Austenitic Stainless Steel, Applied Sciences (Switzerland), 2023, 13(4), 2570, ISSN 20763417. https://doi.org/10.3390/app13042570

[9] J. Maximov, Duncheva, G., Anchev, A., Dunchev, V., Argirov, Y., Effect of Diamond Burnishing on Fatigue Behaviour of AISI 304 Chromium-Nickel Austenitic Stainless Steel Materials, 2022, 15(14), 4768, ISSN 19961944. https://doi.org/10.3390/ma15144768

[10] K. Lyutskanov, Analysis and optimization of hydro abrasive wear of surfacing layers formed by using the method of gas-thermal welding with metal powders alloys. SCNVNA, 32, 62-65 (2018). https://doi.org/10.14748/scnvna.v32i0.4498

[11] Y. Denev, Application of hardfacing arc methods in Bulgarian ship repair SME. Trans Motauto World, 7(2), (2022). 50-52.

[12] U. Ozdemir, Sozeri, M., Findik, T., & Kilicli, V., Effect of buttering on the wear behavior of the SMA welded hardfacing layer in a low-carbon steel. Materials Testing, 65(4), 494-504. (2023). https://doi.org/10.1515/mt-2022-0438

[13] A. M. Stoyanova, Mechkarova, T.M., Argirov, Y.B., Konsulova-Bakalova, M.I., Atanasov, N.M., Study of structure and physico-mechanical properties of welding joints on vessel tank of austenite steel SS316, IOP Conference Series: Materials Science and Engineering, 2020, 843(1), 012013, ISSN 17578981. https://doi.org/10.1088/1757-899X/843/1/012013

[14] K. Lyutskanov, Hristov, H., & Demirova, K. Experimental research of abrasive wear of surfacing layers. SCNVNA, 32, 52-56. (2018). https://doi.org/10.14748/scnvna.v32i0.4496

[15] Y. Argirov, Mechkarova, T., Luckanov, G., & Mincheva, D. Investigations of Wear Resistance of Hardfaced Layers on Steel 40X. TEM Journal, 13(1) (2024)

[16] T. M. Mechkarova, A. Ucherdzhiev, N. Valchev,, & D. Mincheva,. Investigationof tribological characteristics of 40Xgradesteel after hardfaced MMA welding method. ANNUAL JOURNAL OF TECHNICAL UNIVERSITY OF VARNA, BULGARIA, 8(2), (2024). pp1-9.

[17] Retrieved from: https://auremo.biz/materials/stal-40h.html

[18] ESAB (n.d). Catalog of ESAB electrodes. Information on: https://esab.com/gb/eur_en/

[19] Science Direct. (n.d.). Schaeffler diagram. Science Direct. Retrieved from: Information on https://www.sciencedirect.com/topics/engineering/schaeffler-diagram [accessed: 02 August 2023]

Innovative Manufacturing Engineering and Energy - IManEE2024 Materials Research Forum LLC
Materials Research Proceedings 46 (2024) 314-321 https://doi.org/10.21741/9781644903377-41

Sustainability considerations in project management of new product development: Case study

Mihaela NICOLAU

Gheorghe Asachi Technical University of Iași, Faculty of Machine Manufacturing and Industrial Management, Bdul. Mangeron, 59A, 700050, Iași, România

mihaela.nicolau@student.tuiasi.ro

Keywords: Sustainability, Decision-Making, Supplier Selection, Project Management

Abstract. Due to an increased concern about global environmental changes, a product made of reusable environmentally friendly materials and processes can be seen valuable to the end user. Many companies use suppliers to design and manufacture sustainable components with certain technical, social, environmental, and economic constraints. The paper proposes a decision framework considering these constraints for ranking sustainable supplier alternatives. A multi-criteria decision-making (MCDM) technique such as Analytic Hierarchy Process (AHP) will be used to determine the weights of the evaluation criteria, and the Technique for Order Performance by Similarity to Ideal Solution (TOPSIS) will be used to rank the alternatives. The results confirm the usefulness of the proposed model in evaluating sustainable supplier selection strategies.

Introduction

In product development, the concept of sustainability is important to engineers, project managers, as well as business leaders. According to the World Commission on Environment and Development (the Brundtland Commission) in 1987 sustainability is focused on meeting present and future needs [1]. In the manufacturing industry sustainability issues are more relevant than in other industries. Therefore, sustainability can be obtained through sustainable product design, optimized manufacturing process, and purchasing [2].

There are many methodologies that help in optimization of products and processes. One popular method is Axiomatic Design. Through this method, customer requirements are decomposed along functional, design, and process parameters [3]. According to [4], Axiomatic Design has been utilized in the following applications: (1) product design, (2) system design, (3) manufacturing system design, (4) software design and (5) decision-making.

According to [5], the most important strategies in promoting sustainability in manufacturing include several different strategies such as material and resource efficiency [6], waste minimization, total quality and environmental management (TQEM), and green manufacturing.

According to [7], purchasing is a much more powerful agent of change than any other corporate function, while purchasing is sustaining environmental initiatives in the supply chain.

With all these variables to be considered, a structured decision-making approach will increase the efficiency of project management of the new product development.

Literature Review

Sustainable design practices: A literature review found that in the product development design for the environment or optimization the end of a product's life includes [8, 9]:

- *Design for environmental processing/extraction* include processing in a sustainable way.
- *Design for environmental manufacturing* includes use of non-hazardous materials and processes that minimize waste, toxic by-products, energy consumption and other factors.
- *Design for environmental packaging* includes environmentally friendly or recycled materials, reusable products, minimization of material and space.

Materials Research Proceedings 46 (2024) 314-321 https://doi.org/10.21741/9781644903377-41

- *Design for disposal or reuse* includes design for disassembly, reuse, refurbish, recycled.
- *Design for energy efficiency* focuses on designing for the minimization of energy consumption throughout the product's life.

Sustainable manufacturing practices: A literature study found that the most relevant ones are material and resource efficiency, total quality and environmental management, waste minimization, and green manufacturing.

Material efficiency is the consumption of materials without affecting the functions of a product, material efficiency is achieved [6].

Resource efficiency is the efficient use of fewer resources, reducing waste generation, reducing toxic production and extracting processes that have a negative impact on the environment [10].

Eco-efficiency combines the monetary value of product and services with lower ecological impacts and resource intensity throughout the life cycle [11].

Waste minimization focuses on reducing waste from various sources such as raw material, ingredient use, product loss, business consumables, water, energy and packaging consumption, as well as all solid, liquid and gaseous wastes and wasted effort [12].

Just-in-time (JIT) manufacturing focuses on a process that leads to elimination of waste created by overproduction, over inventory, over processing, repeating and unnecessary motion, over-transport, waiting, and defects [13, 14].

Total quality environmental management (TQEM) combines total quality management (TQM), which is a long-term, customer centered management approach focused on quality and environmental management systems (EMS), which is the organization's environmental programs management [15, 16].

Green manufacturing is defined by [17] as a manufacturing strategy focused on minimizing environmental impact and resource consumption.

Sustainable purchasing practices: A literature review revealed that sustainable purchasing practices can be grouped according to their quantitative or qualitative environmental criteria [18].

In another study, [19] grouped them by internal integration and supplier-related activities. In which internal integration refered to integration into specification development, design and process development, reverse logistics processes/investment recovery, environmental systems, while supplier-related activities include supplier evaluation and product restrictions, supplier assessment, supplier management and cooperation with suppliers.

In another study, [20] grouped sustainable practices according to their economic, environemental and social factors.

Methodology

Project managers regularly make decisions during any project involving the selection of the best option among a number of alternatives.

Multi-criteria decision-making (MCDM): As the decision-making process relies on a complex and non-numerical number of variables, MCDM approach is utilized to facilitate comparison of dissimilar elements.

According to [21] MCDM methods can be divided into multi-attribute decision making (MADM) and multi-objective decision making (MODM) methods.

MADM methods focus on decision making problems of limited number of alternatives.

Pairwise comparison methods are based on factors such as criteria and alternatives and include the most popular method Analytic Hierarchy Process (AHP), which was developed by [22] for independent factors applications. For dependency between factors, Analytic Network Process (ANP) [23] can be utilized.

Distance-based methods find the best alternative by considering the shortest distance to the ideal solution. One such method is The Technique for Order of Preference by Similarity to Ideal Solution (TOPSIS).

MODM applies to the decision making problems of unlimited number of alternatives and can be also further divided into no-preference methods, a priori methods, a posteriori methods and interactive methods. Each one of these methods can be further sub divided into other methods.

As the case study discussed here utilizes a limited number of alternatives and other factors, the methods AHP and TOPSIS will be used.

Selection process: The process of selection includes the following steps: 1) identification of alternatives; 2) selection criteria definition; 3) criteria weights definition; 4) alternatives ranking and best alternative is determined.

The alternatives are the suppliers for the components of the final product being developed.

Criteria: The selection process uses criteria that can be either quantitative or qualitative, as seen in Fig. 1. Quantitative criteria include costs or other units while qualitative criteria include linguistic factors, which can be converted to quantifiable scores. Quantitative analysis is done before performing a qualitative analysis. The results will show a ranking of alternative suppliers according to their score from which a selection can be made.

Environmental Criteria [9]:

- Quantitative Environmental Criteria: Cost of clean up, energy consumption, rework, sustainable products aqusition, recycling
- Qualitative Environmental Criteria: Competencies of environmental design, production, supply chain, marketing, management, quality system

Analytic hierarchy process (AHP): In order to identify the best category AHP method is used to estimate priorities of the elements of the decision being considered.

The AHP is a structured technique for analyzing complex decisions based on subjective and objective elements. It has been used as an MCDM approach to rank the alternatives by generating the relative weights of the criteria as well as the priority weights of alternatives with respect to each criterion [24].

The steps of the AHP method are:

1. Criteria importance definition

Once the criteria have been identified, their relative importance in achieving the objective is to be determined using Saaty's 1-9 scales [22] as shown in table 1.

Table1. Criteria scale, adapted from [22]

Criteria Importance	Equally importance	Moderate importance	Essential importance	Very strong importance	Extreme importance	Intermediate values, if needed
Scale	1	3	5	7	9	2, 4, 6, 8

2. Make pairwise comparison of criteria

Alternatives are to be compared with respect to one criterion at a time to find their relative weight using the above scale. According to [22], the scale needs to be used so that when one alternative is found more important than the one it is compared to, it receives one of the numbers shown in the table 1, then its corresponing element has the reciprocal value, as in Eq.1.
The xi represents the relative scale of importance used in

$$C_i = \begin{cases} x_{ij}, 1 < x_{ij} > 9 \\ 1,\ x_{ij} = x_{ji} \\ 1/x_{ij},\ x_{ji} \neq x_{ij} \end{cases}, \qquad i, j = 1, 2, 3, \ldots, n \tag{1}$$

3. Matrix normalization and Weight (Wi) calculation.

Normalize the data by dividing each value in the matrix of pairwise comparison with the total value of the column, following the Eq. (2). Weight is the ration of Ci, which is the value of the i-th row j-th column, divided by the total value of all pairwise comparisons of column j.

$$W_i = \frac{C_i}{\sum_{i=1}^{n} C_i} \quad (2)$$

4. Calculate consistency ration

This calculation determines if the estimated pairwise comparison is properly set for future analysis. According to [22, 23], a pairwise comparison matrix is consistent if its elements follow Eq. 3. In a consistent matrix, the largest principal eigenvalue (λ_{max}) is equal to the size of the matrix (n), Eq.4, and the Consistency index (CI) is zero, Eq. 5. Eigenvector is the ratio of the weight of each factor while eigenvalue represents the value of the division between matrix multiplication and eigenvector with the eigenvector value, Eq. 6.

$$x_{ij} * x_{jk} = x_{ik} \quad (3)$$

$$\lambda_{max} = n \quad (4)$$

$$CI = (\lambda_{max} - n)/(n-1) \quad (5)$$

$$\lambda_i = \frac{C_i * W_i}{W_i} \quad (6)$$

[22] defines departure from inconsistency as Consistency Ratio (CR), Eq. 7 andprovides the Mean Random CI as seen in table 2.

$$CR = CI/\ Mean\ random\ CI \quad (7)$$

Table 2. Mean Random CI, adopted from [22]

n	1	2	3	4	5	6	7	8	9	10	...	20
Mean	0	0	0.58	0.9	1.12	1.24	1.32	1.41	1.45	1.49	...	1.625

In case Eq. 7 is not satisfied, the process of evaluation of criteria is iterated until Consistency Ratio is less than 20%.

The Technique for Order of Preference by Similarity to Ideal Solution (TOPSIS): TOPSIS is used to rank the alternatives using Euclidean distance, for identifying the best alternative when dealing with a finite number of criteria [24]. The best decision is considered to be closest to the ideal solution and farthest from the non-ideal solution. The positive-ideal solution is a solution that maximizes the benefit criteria which is one of large value such as quality or experience.

The steps in TOPSIS are:

1. Construct the decision matrix

Create a matrix in which the rows represent the alternatives, and the columns represent the criteria. Evaluate their relative importance using a scale similar to the one shown in table 3.

Table3. Alternative Criteria Scale

Importance	Most Important	Second Choice	...	nth Choice
Score	1	2	...	n

2. Form a normalization decision matrix, using Eq. 8:

For I = 1,2,3 ,…,m

$$R_{ij} = \frac{x_{ij}}{\sqrt{\sum_{i=1}^{m} x_{ij}^2}} \tag{8}$$

3. Form a weighted normalized decision matrix, using Eq. 9:

$$y_{ij} = w_j * r_{ij} \tag{9}$$

In which w_j is the criteria weight.

4. Identify positive ideal solutions and ideal negative solutions using Eq.(10) and Eq.(11).

$$S^+ = (y_1^+, y_2^+, \dots, y_n^+) \tag{10}$$

$$S^- = (y_1^-, y_2^-, \dots, y_n^-) \tag{11}$$

5. Calculate the Euclidian distance between each alternative with a positive ideal solution and a negative ideal solution using Eq. (12) and Eq. (13)

$$D_i^+ = \sqrt{\sum_{j=1}^{n} (y_{ij} - y_i^+)^2} \tag{12}$$

$$D_i^- = \sqrt{\sum_{j=1}^{n} (y_{ij} - y_i^-)^2} \tag{13}$$

6. Calculate the closeness value of each alternative using Eq. (14).

$$V_i = \frac{D_i^-}{D_i^- + D_i^+} \tag{14}$$

7. Rank the results

The alternative with the largest closeness coefficient is selected.

Case Study

3D printed prototypes are often used in new product development for various applications. Due to the sustainable consideration imposed on the project, it becomes a necessity to use eco-friendly materials for green 3D printing. The most frequently used materials in sustainable 3D printing are thermoplastics such as Polylactic Acid (PLA) and Polyethylene terephthalate glycol (PETG) [25]. Plastics made from petroleum are demonstrated to pollute the environment. Therefore, the suppliers need to be aligned with the green initiative in terms of cost, time, quality and people. The selection process of these suppliers is discussed below.

Identification of suppliers: The project manager identifies at least three suppliers from available sources. These suppliers are identified as S1, S2, S3 and S4.

Selection criteria definition: In accordance with the "triple constraint" guiding project management, the criteria used in selection of a supplier should include time, cost, and quality. In this study, the selection criteria (C_i) used include policies (ISO certification) (C1), technical (Design) (C2), economic (environmental costs) (C3) and workforce diversity (C4). Design, costs, and workforce diversity are considered important characteristics offered by the suppliers.

The evaluation of the suppliers was done according to Table1 and a pairwise comparison of the criteria led to the development of the criteria matrix as seen in Table 4.

Table 4. Criteria matrix

Criteria	ISO certification	Design	Environmental costs	Workforce diversity
ISO certification	1	1/7	1/5	1/5
Design	7	1	2	3
Environmental costs	5	1/2	1	1
Workforce diversity	5	1/3	1	1

A normalized matrix is built using Eq. 2 and using Eq. 3 – 7, the criteria weights are found, such as 0.053481 for ISO certification; 0.487007 for Design; 0.240298189 for Environmental costs; and 0.219213852 for Workforce diversity.

Other important indicators are : The max eigenvalue, lambda = 4.055042; Consistency index, CI = 0.018347; Mean Random Index of 0.9, and the Consistency Ratio, CR = 0.020386038.

Applying Table 3 data to the specifics of this case, a scale of importance of criteria to suppliers is developed and then applied, as seen in Table 5.

Table 5 Criteria scale of importance

Alternative importance	Bad	Unsatisfactory	Satisfactory	Good	Very good
Scale	1	2	3	4	5

Upon surveying the suppliers, it was found that their biggest difference was in the workforce direction, as seen in Table 6.

Table 6 Decision matrix

Criteria/ Suppliers	ISO certification	Design	Environmental costs	Workforce diversity
S1	4	5	4	2
S2	5	4	4	3
S3	5	4	4	3
S4	5	5	4	5

A normalized decision matrix is built using Eq. 8 after which by applying Eq. 9 the weighted normalized decision matrix is obtained, as seen in Table 7.

Table 7. Weighted normalized decision matrix

Criteria/ Suppliers	ISO certification	Design	Environmental costs	Workforce diversity
S1	0.019862	0.235404	0.101886	0.055236699
S2	0.024828	0.188323	0.101886	0.082855048
S3	0.024828	0.188323	0.101886	0.082855048
S4	0.024828	0.235404	0.101886	0.138091747

Using Eq. 10 – 11, the positive and negative ideal solution are obtained as shown in Table 8.

Table 8. The positive and negative ideal solution

Criteria/ Suppliers	ISO certification	Design	Environmental costs	Workforce diversity
Positive	0.024828	0.235404	0.127358	0.138091747
Negative	0.019862	0.188323	0.101886	0.055236699

Using Eq. 12 and 13, the distance from positive ideal solution and the distance from negative ideal solution, are calculated, after which Eq. 14 is used to obtain the relative closeness to the ideal solution such as 0.350832for S1, 0.266463 for S2, 0.264002 for S3, 0.789314 for S4. These results are then ranked in order to obtain the best solution, such as 1 S4, 2 S1, 3 S2, and 4 S3.Therefore, the best alternative for this set of criteria is supplier S4. This result could have been estimated from Table 6, but this only proves that the method used in this article is generating good, objective results. The calculations were performed using Microsoft Excell spreadsheet. The method allows for combining criteria that use numerical values with non-numerical values. In this case study, there were used a small number of criteria and suppliers, but there are real life applications in which more criteria are used, and maybe more suppliers are evaluated.

Conclusions

Procurement management, an area of project management, includes supplier selection activities. Integration of sustainability considerations into procurement helps manufacturing organizations improve the value offered to their customers. In the selection process, by combining AHP, used for obtaining the best option of each decision category, with TOPSIS, used for identifying the best supplier, objective results can be generated regardless of the number or the type of criteria used. The method can be applied to other decision-making applications used in project management.

References

[1] G. H. Brundtland, Report of the world commission on environment and development: Our common future, United Nations, 1987. https://digitallibrary.un.org/record/139811?v=pdf#files

[2] K. Ramani, D. Ramanujan, W.Z. Bernstein, F. Zhao, J. Sutherland, C. Handwerker, J.K. Choi, H. Kim, D. Thurston, Integrated Sustainable Life Cycle Design: A Review, J. Mech. Des., 132(9) (2010) 091004. https://doi.org/10.1115/1.4002308

[3] N.P. Suh, Axiomatic design: Advances and Applications, Oxford University Press, New York, 2001.

[4] O. Kulak, S. Cebi, C. Kahraman, Applications of Axiomatic Design Principles: A Literature Review, Expert Syst. Appl. 37(9) (2010) 6705-6717. https://doi.org/10.1016/j.eswa.2010.03.061

[5] S. H. Abdul Rashid, S. Evans, P. Longhurst, A comparison of four sustainable manufacturing strategies, Int. J. Sustain. Eng. 1(3) (2008) 214–229. https://doi.org/10.1080/19397030802513836

[6] E. Worrell, L. Bernstein, J. Roy, L. Price, J. Harnisch, Industrial energy efficiency and climate change mitigation, Energ. Effic. 2 (2009) 109–123. https://doi.org/10.1007/s12053-008-9032-8

[7] K. Green, B. Morton, S. New, Green Purchasing and Supply Policies: Do They Improve Companies' Environmental Performance?, Supply Chain Manag. 3(2) (1998) 89-95. https://doi.org/10.1108/13598549810215405

[8] R. Sroufe, S. Curkovic, F. Montabon, S.A.Melnyk, The new product design process and design for environment: "Crossing the chasm". Int. J. Oper. Prod. Manag. 20(2) (2000) 267–291. https://doi.org/10.1108/01443570010304297

[9] Information on https://www.gdrc.org/uem/lca/guidelines.html

[10] F. Schmidt-Bleek, Factor 10: manifesto for a sustainable planet, in E.U. von Weizsacker, B. Stigson, J.D. Seiler Hausmann (Eds.) From Eco-Eficiency to Overall Sustainable Development in Enterprises, Wuppertal, 2001, pp.197-207.

[11] S. Schmidheiny, Changing course: A global business perspective on development and the environment, MIT Press, Cambridge, 1992.

[12] Environmental Technology Best Practice Program , Saving money through waste minimization: Raw material use, Crown, 1996. Information on https://tangram.co.uk/wp-content/uploads/GG-025-Saving-money-through-waste-minimisation-Raw-Materials.pdf

[13] R.B. Chase, N.J. Aquilano, Production and operations management: A Life Cycle Approach, Irwin, Chicago, 1989.

[14] S. Shingo, Study of Toyota Production System: From an Industrial Engineering Viewpoint, CRC Press, 1989.https://doi.org/10.4324/9781315136509

[15] Global Environmental Management Initiative (GEMI), Environmental reporting in a total quality management framework, 1994 Information on http://gemi.org/resources/REP_105.pdf

[16] R. Stroufe, Effects of environmental management systems on environmental management practices and operations, Prod. Oper. Manag. 12(3) (2009) 416–431.

[17] X. Tan, D. Liu, C. Li, A decision-making framework model of cutting tool selection for green manufacturing and its application, Adv. Manuf. Syst., 7(02) (2008) 257–260. https://doi.org/10.1142/S0219686708001516

[18] P. K. Humphreys, Y. K. Wong, F. T. S. Chan, Integrating Environmental Criteria into the Supplier Selection Process, J. Mater. Process Tech., 138(1) (2003) 349-356. https://doi.org/10.1016/S0924-0136(03)00097-9

[19] G. Vorosmarty, I. Dobos, The role of personal motivation in sustainable purchasing practices, Amfiteatru Econ. 21(50) (2019) 121-137.

[20] D. Kannan, Role of multiple stakeholders and the critical success factor theory for the sustainable supplier selection process, Int. J. Prod. Econ. 195 (2018) 391-418. https://doi.org/10.1016/j.ijpe.2017.02.020

[21] A. S. Yalcin, H. S. Kilic, D. Delen, The use of multi-criteria decision-making methods in business analytics: A comprehensive literature review, Technol. Forecast. Soc., 174 (2022) 121193. https://doi.org/10.1016/j.techfore.2021.121193

[22] R. W. Saaty, The analytic hierarchy process—what it is and how it is used. Math modelling, 9(3-5) (1987) 161-176. https://doi.org/10.1016/0270-0255(87)90473-8

[23] T. L. Saaty, Decision making—the analytic hierarchy and network processes (AHP/ANP), J. Syst. Sci. Syst. Eng. 13 (2004) 1-35.

[24] C. L. Hwang, A.S.M. Masud, Multiple attribute decision making: methods and applications - a state-of-the-art survey in Lecture notes in Economics and mathematical systems, M. Beckmann, H.P. Kunzi (Eds), Springer , New York, 1979.

[25] S.V. Kumar, U. Reddy, A. Nagpal, A. Kumar, S. Jayronia, R.A. Hussien, Towards Sustainable Additive Manufacturing: Exploring Ecofriendly Materials for Green 3D Printing, E3S Web Conf. 505 (2024) 01009. https://doi.org/10.1051/e3sconf/202450501009

Innovative Manufacturing Engineering and Energy - IManEE2024 Materials Research Forum LLC
Materials Research Proceedings 46 (2024) 322-329 https://doi.org/10.21741/9781644903377-42

Development of an algorithm for application for automated selection of a rational technological process

Teodora Peneva[1,a] *

[1]Department of Manufacturing Technologies and Machine Tools, 1Studentska Str., 9010 Varna, Bulgaria

[a] tdrpeneva@gmail.com

Keywords: Technological Process, Algorithm, Application, MCDM, FUCOM

Abstract. Within the framework of modern industrial requirements, the selection of rational technological processes is becoming more and more critical to maintain competitiveness and innovativeness of different types of production. This paper describes the development of an algorithm and software application designed for rational technological process selection. The algorithm and the application are using developed methodology with a mathematical model for comparative analysis and selection of a technological process. The software application facilitates the prioritization of different production aspects and supports strategic decision making through automated comparison and evaluation of alternative technological processes. In addition to providing accurate criteria evaluation, the algorithm will demonstrate high flexibility in adapting to changing production conditions and requirements. The study highlights the practical applicability of the algorithm in various manufacturing environments and discusses potential challenges and directions for future development of the system.

Introduction

Rapidly evolving technologies in recent years, the drive towards automation and the Industry 4.0 characteristic of integrating digital technologies into manufacturing environments are creating a prerequisite for the adaptation and construction of more and more software programs and applications. The integration of Manufacturing Execution Systems (MES), control systems and virtual tools for measuring design and geometric parameters [1,2], Enterprise Resource Planning and Management (ERP) and many others, are leading to increased accuracy and efficiency in manufacturing processes. This industrial era allows for better productivity, reduce costs, optimize resource management, reduce communication time and real-time decision making. This is why the quality of production, productivity and cost-effectiveness are essential for competitiveness.

The application of new automation technologies significantly speeds up the process and reduces the risk of human error. One of the most important challenges for systems supporting the evolution of Industry 4.0 is to provide the ability to adapt quickly to changing production conditions, to provide sustainability in the implementation of new technologies and to be compatible with the various software products on the market.

Loss function in manufacturing processes also play a key role in process optimization and control [3]. For this reason, managing the quality of production and monitoring the stability of the technological processes is not of minor importance. The need for fast and accurate decisions in complex manufacturing systems stimulates the search for new approaches, technologies, applications and algorithms [4,5] to support strategic decision making.

Although there have been significant advances in technological processes optimization methods, the selection of a rational technological process is often based on manual calculations or intuitive ones, leading to uncertainty and inefficiency. Developing an application to automate this selection through an algorithmic approach will allow for more accurate and rapid decision making tailored to a set of different criteria.

Materials Research Proceedings 46 (2024) 322-329 https://doi.org/10.21741/9781644903377-42

For the base of the algorithm is used developed methodology and mathematical model for the selection of rational technological process and implemented in them multi-criteria decision-making method (MCDM) - FUCOM for determining the weight of criteria [6]. These allow flexible adaptation of the process to the specific requirements of each production. Through them, the application will be able to perform precise evaluations and propose the most rational technological process according to the specific production and customer conditions and requirements.

The application for rational process selection would be an innovative and extremely useful tool for the benefit of not so experienced technologists, as well as for a quick reaction by the technology department when certain criteria in the process are changed. The purpose of the application is to reduce decision-making time, minimize errors, and ensure objectivity and accuracy in evaluating different alternatives. In this way, a basis for improving production efficiency and achieving higher profitability is created.

The main objective of this paper is to present an algorithm that will be the basis of developing an application for selecting a rational technological process and to help analyse and monitor the stability of the selected rational technological process.

Difficulties in creating an algorithm for production environments

Creating algorithms to control and optimise manufacturing processes is an important part of modern industry. Although they offer significant advantages, the development of such algorithms is accompanied by numerous challenges.

Manufacturing processes are typically highly dynamic - frequent order changes, varying customer demands, unforeseen machine downtime and failures, and material supply variations are just some of the challenges that algorithms must deal with. One of the important requirements in the design of algorithms is rapid adaptation to changing production cycles and unexpected events [7,8]. This requires the need for dynamic scheduling and flexible algorithms that can respond in real time and provide optimal resource management.

Another major problem of modern production environments is the enormous amount of data generated in the process. To be effective, algorithms need quality data that is not only reliable but also well-structured. The data collection and management process can be complicated by inconsistencies between the different systems that process the information. Ensuring data security and their integration across platforms represents an additional challenge for algorithm developers [9,10]. The implementation of standardised data management processes and tools is essential for the successful implementation of algorithms in industrial environments.

Another major challenge in automating manufacturing processes faced by software applications and programs algorithms is the requirement for a large number of data or database, as well as ensuring their security and integration into various manufacturing systems [11,12].

In manufacturing processes, often hybrid algorithms that combine different methods such as mathematical models are applied, however, this approach provides greater accuracy and flexibility, it also leads to high computing complexity and the need for significant computing resources [13,14]. Developing hybrid algorithms that can optimise production processes often requires a balance between decision accuracy and execution speed, especially in live applications.

Creating effective algorithms for production environments requires overcoming multiple challenges related to the dynamics of the environment, the complexity of the data, and the need for fast and accurate solutions. While the development of various types of algorithms satisfying different aspects of manufacturing needs, different kinds of difficulties arise.

Algorithm for automated selection of rational technological process

With the development of new digital technologies and the drive to automate all possible stages of the technological process, the integration of a variety of software applications supporting different stages of production does not remain in the background. This promotes a higher degree of connectivity and autonomy in processes, leading to more flexible and efficient production systems.

In this regard, an algorithm was developed based on an established methodology for selecting a rational technological process [6] in order to implement it in a software application for automatic selection of a rational technological process. The algorithm provides a detailed description of the steps of the created methodology and will serve as a framework for the project of an application that will analyse and select a rational technological process from "n" number of technological processes inputs.

Creating an algorithm is a process that requires systematic planning and structuring of the solution to a given problem. The main steps in developing the structure and logic of the algorithm are: analysing the problem, defining input and output data, logical sequence, using an appropriate data structure, defining iterations (loops).

Structure and logic of the algorithm

The algorithm is organized in consistent stages that include data entry, calculation of criteria weights, normalization of values, and final process evaluation. Its logic is based on developed mathematical relationships that ensure objectivity and precision in the selection of a rational technological process.

A flowchart of algorithm for automated selection of rational technological process is presented in Fig. 1. Its structure is formed by the following stages:

(1) Start. The start of the algorithm begins with the pre-preparation of the input data. At this stage, specialists in the field design various technological processes of the workpiece under examination, according to its design-technological requirements and those of the customer;

(2) Input data;

Input of technological processes. The user enters the technological processes for the part under examination that will be evaluated and compared;

Defining the evaluation criteria for the technological processes. The criteria by which the technological processes will be evaluated (accuracy, production time, cost, etc.) are defined. These criteria play a key role in the algorithm as they are the basis for the evaluation and ranking of the technological process alternatives;

Determine the weights of the criteria. Each criterion is assigned a weight according to its importance for the technological process evaluation;

(3) Determine the weighting coefficients of the criteria. The multi-criteria method - FUCOM is used as:

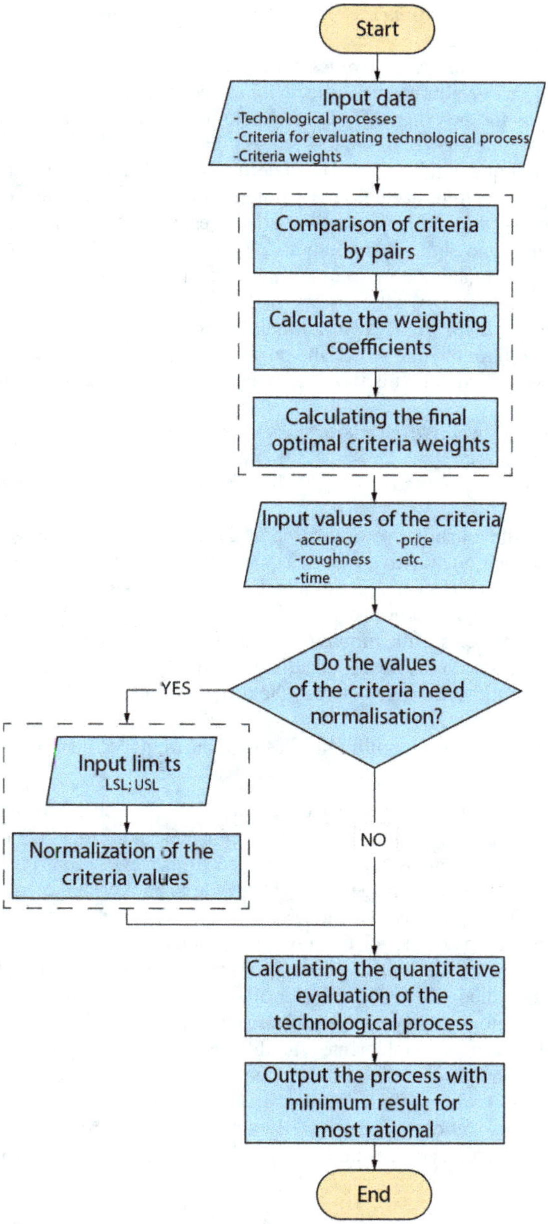

Fig.1. Flowchart of algorithm for automated selection of rational technological process.

Comparison of criteria by pairs. Comparative analysis of criteria weights by pairs is made;

Calculation of the weight coefficients of the criteria;

Calculation of the final optimal values of the criteria weights. This is a key stage as the correct determination of the weights ensures objectivity in decision making.

(4) Input values for the criteria of each technological process. The quantitative values for each criterion are input for every technological process that is considered;

(5) Do the input quantitative values need normalisation? At this stage, the algorithm checks if the criteria values need to be normalized. Normalisation allows the measured values to be brought to a common basis, thus achieving a generalized result for the different process criteria. This ensures consistency and high quality output;

(6) If "YES" - The quantitative values of the criteria need normalisation, the following loop is executed:

Input of criteria limits (LSL, USL). Limits of the considered criterion are entered. These boundaries set the limits within which the quantitative values of the criteria indicate how stable the process under consideration is (how many percent of the measured values fall within the set limits);

Normalisation of quantitative criteria values;

Continue to next stage.

(7) If "NO" - If the quantitative values of the criteria do not need to be normalised continue to the next stage;

(8) Calculating the quantitative evaluation of the technological process. Based on the weights of the criteria and their quantitative values, a quantitative score is calculated for each technological process. This involves the application of a mathematical model to obtain a summary quantitative score for each technological process;

(9) Output the process with minimum result for most rational. The algorithm selects the technological process with the lowest quantitative score as the most rational. This means that this process offers an optimal balance between the different criteria and best satisfies the production and customer requirements;

(10) End. The end of the algorithm finishes with the output of the results.

Advantages and disadvantages of the algorithm

The developed algorithm for automated selection of a rational technological process is an innovative tool that significantly eases strategic decision making in the manufacturing environment. Despite its many advantages related to process optimisation and automation, its application also has its disadvantages. In Table 1. the main advantages and disadvantages of the algorithm that define its effectiveness and potential for improvement are presented.

The algorithm automates the selection of a technological process, which significantly reduces decision-making time and eliminates the need for manual calculations. Thanks to the multi-criteria decision-making method - FUCOM, the algorithm provides an accurate and objective evaluation of alternative technological processes based on mathematical models.

The algorithm can be implemented to different production conditions and technological process evaluation requirements, allowing the use of a variety of evaluation criteria (cost, production time, etc.). This enables decisions to be made that are reflective of specific production process requirements and customer needs. By automating this process, the probability of human mistakes in the calculation and evaluation of the different technological processes is reduced.

Materials Research Proceedings 46 (2024) 322-329 https://doi.org/10.21741/9781644903377-42

Table 1. Main advantages and disadvantages of the algorithm for selection rational technological process.

Advantages	*Disadvantages*
✓ Automation and speed of the process; ✓ Precision and objectivity; ✓ Multiple criteria selection; ✓ Flexibility and adaptability; ✓ Reduction of the risk of errors; ✓ Integration with Industry 4.0.	× Implementation complexity; × Dependency on the quality of input data;

The easy adaptation and flexibility of the algorithm to different production conditions is achieved by normalising the quantitative criteria values. This allows the process evaluation to be relevant even when the criteria values vary significantly.

The algorithm is compatible with modern industrial requirements for digitalisation and automation of production processes, which improves the competitiveness of companies.

In smaller companies and those that have not implemented new digital technologies to automate technological processes, it can be a barrier because there is a lack of automated data transfer. They may require additional time and resources to adapt the existing processes to new technologies. The complexity of algorithm implementation can also vary depending on the level of digitisation and automation of the production process. For enterprises that use more traditional data management methods, the algorithm will work, but errors will occur from manual input data. This can be eliminated by making the transition to more modern platforms, which presents additional financial and time challenges.

If the values for the criteria are not correct or insufficient, this may lead to incorrect results in the evaluation of the technological processes. Inputting inaccurate data at this stage leads to an overall bias of the final result.

Another challenge in implementing the algorithm is to ensure efficient operation even in large production systems, which is why it is recommended to use modern computer architectures with parallel data processing capabilities as well as optimized resource management algorithms. This will allow the algorithm to run efficiently and provide real-time results, even with high computational complexity and large volumes of input data.

Summary

This paper presents an algorithm for a software application that is designed for the automated selection of a rational technological process. The algorithm is based on methodology and mathematical relations with an embedded multicriteria decision method, which provides an accurate and objective comparative analysis of alternative process decisions. This is what allows flexible adaptation of the process to the specific requirements of each production.

The developed algorithm offers significant advantages over traditional approaches by providing the ability to quickly and efficiently adapt to the dynamics of the modern manufacturing environment. Its main advantages are the reduction of the time required for the selection of a rational technological process and its integration into the production environment, flexibility and the possibility of implementation in different productions and the use of many criteria.

The algorithm serves as a tool for businesses and companies that want to improve their efficiency and competitiveness. By systematically and objectively evaluating alternative technological processes, it supports the strategic planning and optimisation of production activities.

The developed algorithm uses methods with great flexibility, which will make the application suitable for different production environments. By integrating the algorithm, enterprises can not only improve their production efficiency, but also remain competitive in a rapidly evolving industrial environment.

The author has started work on the development of the algorithm by implementing it in a software application that will be tested in real production conditions. At the time of writing, the algorithm has not yet been practically implemented due to finishing work on the under development application. The software application will be aimed at easing the decision making process by automated evaluation and comparison of different technological processes. It will also assist users to set their priorities, taking into account various criteria important for production and customers such as: cost, quality, production time, etc.

Acknowledgements
The algorithm presented in this paper was developed on computer equipment provided by the company "CERATIZIT Bulgaria" AG.

References

[1] G. Ivanova, A. Ivanov, V. Kozov, 3D learning tool for measuring of constructive and geometrical parameters of turning tools, ACM International Conference Proceeding Series, 431-436 (2016). https://doi.org/10.1145/2983468.2983472

[2] G. Ivanova, A. Ivanov, K. Kolarov, 3D virtual learning and measuring drill tools, ACM International Conference Proceeding Series, 337-343 (2013). https://doi.org/10.1145/2516775.2516779

[3] K. Kirov, T. Avramova, Optimal management of production processes using the function of losses in the production processes, IOP Conference Series: Materials Science and Engineering, 916 (2020) 012052. https://doi.org/10.1088/1757-899X/916/1/012052

[4] D. Ishenin, A. Govorkov, I. Teslenko, M. Klykov, O. Kabanov, A. Lyalin, A. Shaposhnikov, Z. Mukhamedova, An algorithm for computer-aided design of a technological process with preset manufacturability parameters, Engineering Systems Design and Analysis, 8(4) (2021) 733-738. ISSN: 2392-9537

[5] J. Song, B. Wang, X. Hao, Optimization Algorithms and Their Applications and Prospects in Manufacturing Engineering, Materials, 17 (2024) 4093. https://doi.org/10.3390/ma17164093

[6] T. Peneva, T. Avramova, Methodology for selecting a rational technological process, International Conference Modern Technologies in Industrial Engineering, International Journal of Modern Manufacturing Technologies, (2024), ISSN 2067-3604, Preprint.

[7] M. Ghaleb, H. Zolfagharinia, S. Taghipour, Real-time production scheduling in the Industry-4.0 context: addressing uncertainties in job arrivals and machine breakdowns, Computers & Operations Research, 123 (2020) 105031. https://doi.org/10.1016/j.cor.2020.105031

[8] Y. Fu, J. Ding, H. Wang, J. Wang, Two-objective stochastic flow-shop scheduling with deteriorating and learning effect in Industry 4.0-based manufacturing system, Applied Soft Computing, 68 (2018) 847-855. https://doi.org/10.1016/j.asoc.2017.12.009

[9] J. Xu, M. Kovatsch, D. Mattern, F. Mazza, M. Harasic, A. Paschke, S. Lucia, A Review on AI for Smart Manufacturing: Deep Learning Challenges and Solutions, Applied Sciences, 12 (2022) 8239. https://doi.org/10.3390/app12168239

[10] Y. Wilhelm, U. Schreier, P. Reimann, B. Mitschang, H. Ziekow, Data Science Approaches to Quality Control in Manufacturing: A Review of Problems, Challenges and Architecture, Engineering Systems and Analysis. https://doi.org/10.1007/978-3-030-64846-6_4

[11] H. Mimouni, A. Jalid, Solving Industrial Production Scheduling Challenges in the Era of Industry 4.0 and Green Manufacturing, Proceedings of the 7th International Conference on Logistics Operations Management, GOL (2024). Lecture Notes in Networks and Systems, Vol. 1105, Springer, Cham. https://doi.org/10.1007/978-3-031-68634-4_31

[12] E. Guzman, B. Andres, R. Poler, Matheuristic Algorithms for Production Planning in Manufacturing Enterprises, in Advances in Intelligent Systems and Computing, Vol. 1287 (2021) 121-134. https://doi.org/10.1007/978-3-030-78288-7_11

[13] F.M. Defersha, D. Rooyani, An efficient two-stage genetic algorithm for a flexible job-shop scheduling problem with sequence dependent setup times, Computers & Industrial Engineering, 147 (2020) 106605. https://doi.org/10.1016/j.cie.2020.106605

[14] B. Kocsi, M.M. Matonya, L.P. Pusztai, I. Budai, Real-time decision-support system for high-mix low-volume production scheduling in Industry 4.0, Processes, 8(8) (2020) 912. https://doi.org/10.3390/pr8080912

Studying the ergonomics of office furniture with specific art elements and office equipment

Ralitsa ZLATEVA [1,a]*, Penka ZLATEVA [1,b], Momchil TACHEV [1,c], Galina STANEVA [1,d], Kremena CANKOVA [1,e], Ginka ZHECHEVA [1,f]

[1]Technical University of Varna, 1 Studentska *Str.*, Bulgaria

[a]ralitsa.zlateva@tu-varna.bg, [b]pzlateva@tu-varna.bg, [c]m_tachev@tu-varna.bg, [d]c.staneva@tu-varna.bg, [e]kremenacankova@tu-varna.bg, [f]ginkaa@tu-varna.bg

Keywords: Ergonomic Furniture, Art Elements, Visual Ergonomics, Office Equipment, Artificial Intelligence (AI), 3D Printer Robots, Industry 4.0, Industry 5.0

Abstract. The quality of office furniture in today's computerized work environment significantly impacts workers' health and productivity. This study focuses on the production of ergonomic office furniture that incorporates artistic elements, balancing functionality with aesthetics. The aim is to analyze and evaluate these furnishings, explore their effects, and offer recommendations for enhancing design and functionality. The tasks include a literature review, identifying key characteristics and criteria for ergonomics and aesthetics of art-integrated furniture, analyzing current office furniture and equipment, and investigating the influence of ergonomic designs. Additionally, successful projects and manufacturing practices in Bulgaria and abroad will be examined to optimize the design and production of ergonomic furniture with artistic elements.

Introduction

The continuous use of office equipment and work furniture in the modern work environment significantly affects the health and productivity of workers. Therefore, ergonomic furniture and office equipment are becoming an increasingly important aspect in the work environment to reduce the risk of visual and musculoskeletal disorders. And despite the growing attention from an ergonomic point of view, there are also many challenges related to the integration of specific art elements on furniture design and the surrounding work environment.

Application of furniture with art elements in a computer environment

When creating work furniture with art elements, it is necessary to comply with basic principles. In addition to ergonomics and functionality, they must also be aesthetically well designed, in order for the workplace to be attractive, motivating and creative for employees. This, respectively, leads to an increase in the productivity of the users' work.

When selecting art elements for office furniture, it's essential to consider the type of work being performed, as well as collaboration with professionals like designers, artists, and ergonomics specialists. The choice of materials should prioritize sustainability, recyclability, and environmental impact, while also considering shape, texture, and color. Additionally, cultural influences, fashion trends, and consumer preferences play key roles in the decision-making process.[1,2,4]

The good design of work furniture is necessary to maintain the normal state of human life. Furniture plays an important role in everyday life and directly affects people's health and working capacity. They must be adapted to the physical structure and limitations of the human body. Therefore, they must correspond to human anthropometry, anatomical properties and biomechanical structure. [1,2,3,4]

With the advancement of Industry 4.0 and progress in 3D design and 3D printing of artistic elements for office furniture, traditional manufacturing limitations are being overcome through smart and robotic additive manufacturing technologies. These include 3D printing robots, which create new possibilities in furniture production. Today, the industry combines both subtractive and additive methods. Subtractive methods involve precision multi-axis milling of materials like wood, while additive manufacturing uses 3D printing, where recycled wood fibers are added to plastic to achieve a wood-like appearance.[4,5,6,7]

Research focuses on the surface roughness of 3D-printed furniture, as it affects users' tactile perception. By testing various materials, the structural properties of thermoplastics and composite materials are evaluated for printing, facilitating the integration of new technologies into office furniture production while enhancing both aesthetic and functional qualities. [7,8]

Artificial Intelligence (AI) also helps in selecting different user preferences, enhances creativity and personalization. Powered by artificial intelligence, the product life cycle, through assessment tools, analyzes environmental impact, helps designers make informed decisions, about materials, production and delivery. [4,9]

Modern design increasingly focuses on the sustainability of materials and reducing environmental impact, while preserving aesthetic and cultural traditions. Advanced technologies such as artificial intelligence (AIGC) are used to optimize processes, but inspiration comes from classic examples such as Ming-style furniture, characterized by simplicity and harmony, reflecting the connection with nature and crafted through masterful carving. (Fig.1)[10]

Another type of industrial furniture, with art elements, are computer tables made of wrought iron, "Rustic" style (Fig.2)[11]; a solid Chippendale-inspired Madison computer table (18th century) with elegant molded legs, wood veneers and solid beveled corners, with a dovetail drawer (Fig.3)[12]; a solid large Queen Anne style desk, crafted with many details including cabriole legs and palmette and leaf designs, with striking walnut veneer(Fig.4)[13,14]

Fig. 1. Desk "Ming style"

Fig. 2. Computer desk in "Rustic style"

Fig. 3. Madison Massive Computer Desk

Fig. 4. Massive large desk, Queen Anne

The Sayl chair by Herman Miller combines efficiency and sustainability by utilizing minimal materials while maintaining high ergonomic standards and comfort. Its elastomer strands, integrated into the intelligent 3D back, offer flexible support to the spine, promoting both freedom of movement and stability. The chair's guiding principle is: "provides the most with the least.". [15,16] (Fig. 5)

In another example of 3D design and 3D printing, is designed in dendritic form, a desk chair, using the "Grasshopper" graphics algorithm editor in "Rhino computer graphic" software and printed on a "Flashforge" 3D printer (Fig. 6)[17,18,47]

The Sihoo Doro Chair C300 is designed for ergonomic comfort during long sitting sessions, featuring adjustable height, backrest angle, and armrests. Made from breathable materials for enhanced airflow, it also includes self-adaptive lumbar support, a flexible backrest, and 3D coordinated armrests, as well as a wide adjustable headrest to enhance user comfort. (Fig.7)[19,20]

Inspired by nature, designs that incorporate curves, waves, and organic or biomorphic forms are integrated into work furniture, walls, shelves, and partition walls, reflecting a cyclical approach to design.(Fig.8)[21]

Fig. 5. Herman Miller - Sayl Chair

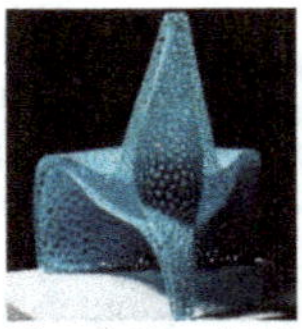

Fig. 6. Dendritic 3D-printed chair

Fig. 7. Chair Sihoo Doro-C300

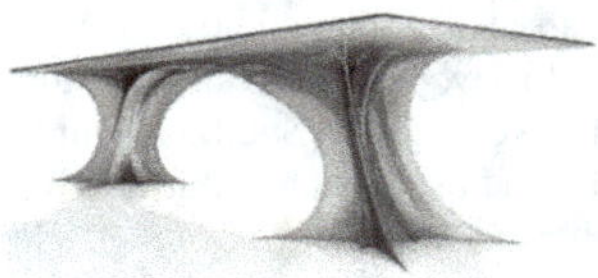

Fig. 8. Biomorphic, cyclical design

Furniture with art elements in Bulgaria and around the world. The use of innovative materials and design techniques in furniture construction is evident in various examples, such as computer desks featuring decorative 3D texture panels (Fig.9)[22,23] and art elements with cast iron legs.(Fig.10)[24] Laser-cut components(Fig.11)[25] and artistic pieces made from epoxy resin(Fig.12)[26] further enhance the aesthetic appeal. Wrought iron decorations [27] and carved wooden elements(Fig.13)[28], including teak mosaics(Fig.14)[29], showcase traditional craftsmanship alongside modern techniques. Additionally, advancements like 3D-printed furniture[30] and 3D veneered MDF[31] highlight the integration of technology in design. Easy-to-assemble workbenches suitable for both sitting and standing positions are often cut using CNC plotters(Fig.15)[32][33][34], utilizing software like VCarve Pro and Aspire for precise design execution.(Fig. 16)[35,36,37,38]

Fig.9 3D texture panels

Fig. 10. Cast iron table leg

Fig. 11. Laser cut elements and screens

Fig. 12. Art furniture made of epoxy resin

Fig. 13. Art elements of wood, carving

Fig. 14. Teak mosaic on computer table

Fig. 15. Assembled computer desk

Fig. 16. A table with cut art elements on a CNC plotter

Research methods

Anthropometric analysis. In the study of the ergonomics of work furniture with art elements, an anthropometric analysis was carried out, which evaluates the physical dimensions of the users and adapts the furniture to them. Field studies were conducted in an office building of a private company to collect data on the interaction of workers with furniture and digital devices (DDU) in a real work environment.

The study, analyzing the anthropometry and ergonomics of office furniture with art elements, included field research conducted on four floors of an office building. On the ground floor, the workplaces include 60x90 cm and 70x150 cm tables used by 6 and 5 workers respectively. These tables had no lumbar support or headrests and the legroom was 60cm. On the first floor, all 14 workers used a 70x150cm table, again without a support but with 70cm legroom. On the second floor, there were 12 workers using the same size table, also without lumbar support and with the same legroom. On the third floor, 3 workers were found at a table measuring 70x150 cm, equipped with a lumbar support. The work environment and furniture lacked art elements; however, employees indicated that their involvement could increase engagement and productivity.

In terms of digital devices, all floors included workers using PWM monitors at a distance of 50 cm. On the ground and first floors there were 5 PWM laptops at a distance of 40 cm. The study examined desk sizes and the availability of lumbar support, along with the impact of digital devices on workers. Only monitors with flashing backlight are observed on all floors, which can negatively affect the comfort and health of employees. It emphasizes the importance of integrating art elements in the workplace for the overall well-being of the workers, as well as their participation in decorating the offices, to express their identity.

According to a study by the University of Exeter, incorporating art and plants into work environments significantly enhances both productivity and employee well-being. Specifically, workers in such enriched settings showed up to 32% higher productivity, 45% improved well-being, and 60% greater engagement. Dr. Craig Knight, a researcher at the University, emphasized that "if you enrich a space, people feel much happier and work better," underscoring art as an effective means to achieve this. [45,46,47]

Analysis of work posture and work movements. Ergonomic software "ErgoFellow 3.0" was used to improve the visual ergonomics of computerized workplaces by calculating body dimensions based on the height of the worker, the average height was determined to be 170 cm.(Fig.17)[39,40]. In addition, the software used the REBA method, which evaluates risk factors, related to musculoskeletal disorders, by assessing different parts of the body, including hands, wrists and feet.(Fig.18)[39,41].

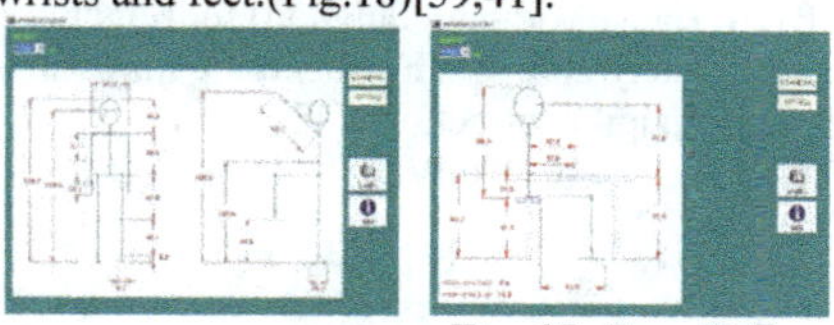

Fig. 17. Ergo Fellow 3.0 - Anthropometry

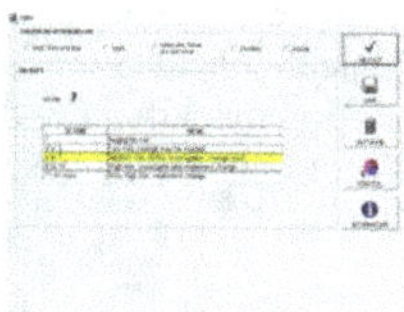

Fig. 18 Ergo Fellow 3.0 - REBA

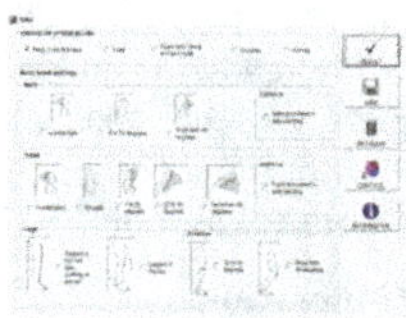

Fig. 19. Ergo Fellow 3.0 - REBA

The analysis reveals a risk of visual and musculoskeletal strain at work, which provides recommendations for monitoring and adjusting body and limb positions to mitigate these injuries.(Fig.19)[40]

Applying a specific element on working furniture

From the study, it was found that the material, texture, color and decoration of office furniture have a significant impact on the mental health of employees. The introduction of natural or artificial light, with adjustable color temperature, furniture and decorative elements made of natural wood improve concentration, comfort and increase the sense of security.

It is necessary to introduce an adjustable, mobile computer desk, with specific elements cut on a CNC machine. To determine its dimensions, Ergo Fellow 3.0 Ergonomic software - Anthropometry is used.

The computer desk can be used both in an office environment for a permanent workplace and in establishments where presentations can be made. Its convenience is that it can be used in a sitting and standing position, as well as being moved by means of wheels. For its stability, the wheels are equipped with stoppers. The computer desk can be raised in three ways: with a handle for manual height adjustment, with a pedal, by foot pressure or electric drive.

An art ornament is used, which is considered one of the most ancient and mystical signs of protection, dating back to the time of the proto-Bulgarians. It represents three interlocking eights (the signs for infinity) that depict the idea of perfection, balance and interconnectedness of everything in the universe. (Fig.20) [46,47,48,49]. The sizing of the computer desk was done according to anthropometric measurements taken with ergonomic software "ErgoFellow 3.0"(Fig.21). Rendering done using Blender and Adobe Photoshop. (Fig.22)

Fig. 20. Art element for a mobile computer desk

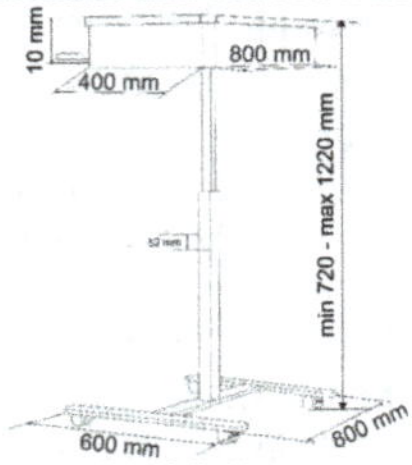

Fig.21. Dimensions of an adjustable computer desk

Fig.22. Adjustable mobile computer desk, with art ornament

Visual ergonomics and office equipment

Effective ergonomic design in the workplace requires proper monitor positioning to enhance comfort and mitigate health issues like poor posture and eye strain. While older VGA displays offer a pixel density of about 75 dpi, modern LCD monitors provide up to 115 dpi, improving reading speed by 17.4%. New LCDs are lighter, more energy-efficient, and produce less electromagnetic radiation. However, they often use PWM (pulse width modulation) for brightness control, which can lead to backlight flickering at approximately 200Hz. This flickering may cause migraines and headaches during extended use due to rapid pupil contractions.[42]

Monitors without PWM: BenQ BL2785TC 27" и 31,5"; BenQ GW2480; Dell 32 Curved Gaming Monitor – S3222DGM.

Conclusion

The ergonomic design of the work environment is essential for the health and well-being of employees. Through careful selection of furniture, proper positioning of equipment, and aesthetic enhancements, organizations can reduce the risks of visual and musculoskeletal strain while simultaneously increasing comfort and productivity. The integration of new technologies, such as 3D printing and artificial intelligence, allows for the personalization of the workspace, further enhancing employee satisfaction. In the context of Industry 4.0 and 5.0, the combination of art, ergonomics, and technology represents an innovative approach that meets the contemporary needs of the workforce.

Acknowledgement

The present article has been developed with the help of Science project 11 *Ergonomic and design research of interior furniture with art design elements, 2024.*

References

[1] R. Sang, F. Yang, Z. Fan, The Effect of Water-Based Primer Pretreatment on the Performance of Water-Based Inkjet Coatings on WoodSurfaces. Coatings 2023, 13, 1649. https://www.mdpi.com/2079-6412/13/9/1649

[2] Y. Liu; Wengang Hu; A. Kasal; Y. Erdil. The State of the Art of Biomechanics Applied in Ergonomic Furniture Design, 2023 https://doi.org/10.3390/app132212120

[3] L. Yu, S.; Q. Zheng, Y. Chen, T.; H. L. Zhang, R. Chen, X. Consumer personality traits vs. their preferences for the characteristics of wood furniture products. BioResources 2023,

https://bioresources.cnr.ncsu.edu/resources/consumer-personality-traits-vs-their-preferences-for-the-characteristics-of-wood-furniture-products/ DOI:10.15376/biores.18.4.7443-7459

[4] K. Chanchal, R. Deepti, nternational Journal of Housing and Human Settlement Planning, 2023. The Shaping of Spaces: A Journey Through Furniture Design History.

https://www.researchgate.net/publication/380854583

[5] A. Haghighi; A. Mohammed; A. Alhusin. at al. Energy Efficient Multi-Robotic 3D Printing for Large-Scale Construction – Framework, Challenges, and a Systematic Approach. June 2021; https://doi.org/10.1115/MSEC2021-63787

[6] M. Jovanović at al., September 2021. 3D printing large scale curved forms using heat formed coral reef tiles. https://www.researchgate.net/publication/355478353

[7] D. Reinhardt. May 2018. Structural Optimisation for 3D Printing Bespoke Geometries.https://www.researchgate.net/publication/325817399

[8] G. Struzikiewicz., M. Misiniec at al., 3 May 2024. Application of the TOPSIS decision-making method for selecting a manufacturing technique for children's furniture elements with therapeutic. Volume 82, pages 1251–1262, (2024). European Journal of Wood and Wood Products (2024) 82:1251–1262.https://link.springer.com/article/10.1007/s00107-024-02082-2

[9] S. Yu, Z. Zhen Citation: S. Yu, Z. Zhong, Y. Zhu, J. Sun, Green Emotion: Incorporating Emotional Perception in Green Marketing to Increase Green Furniture Purchase Intentions. Sustainability 2024, 16, 4935.DOI: 10.3390/su16124935

[10] Y. Wang, Xi. Yuchen, X. Liu. and Y. Gan. Exploring the Dual Potential of Artificial Intelligence-Generated Content in the Esthetic Reproduction and Sustainable Innovative Design of Ming-Style Furniture,2024. DOI: 10.3390/su16125173

[11] Industrial Writing Table Desk - Adjustable Height Iron Base - Marble Top https://www.rusticdeco.com/products/industrial-writing-table-cast-iron-base-adjustable-height-white-marble-top

[12] Madison Solid Computer Desk https://www.homelivingfurniture.com/i169140/4455-917-samuel-lawrence-furniture-madison---cherry-home-office-library-desk---cherry

[13] P. Wiggins, 2019. Queen Anne Furniture.https://www.thesprucecrafts.com/queen-anne-style-furniture-148838

[14] Restored Queen Anne Style Ornate Writing Desk. https://www.chairish.com/product/17522625/queen-anne-style-ornate-writing-desk

[15] Ceschin F.; Gaziulusoy G. August 2019. Design for Sustainability. ISBN: 9780429851964.DOI:10.4324/9780429456510

[16] T. Dovramadjiev, D. Dobreva, R. Zlateva, (2024). Improving the Visual Ergonomics of Computerised Workplaces Through the Use of Specialised Eye-Rest Software. In: Mirzazadeh, A., Molamohamadi, Z., Erdebilli, B., Babaee Tirkolaee, E., Weber, GW. (eds) Science, Engineering Management and Information Technology. SEMIT 2023. Communications in Computer and Information Science, vol 2198. Springer, Cham. https://doi.org/10.1007/978-3-031-72284-4_12

[17] Rhino 3D graphic software https://www.rhino3d.com/

[18] Doro C300 Ergonomic Office Chair by Sihoo. Dezeen staff , 21 June 2023. https://www.dezeen.com/2023/06/21/ergonomic-office-chair-sihoo-dezeen-showroom/

[19] Biomorphic, cyclical design. https://cyclic.design/biomorphic/

[20] 3D Embossed Glossy Faces. https://orhideamebel.com/

[21] Oxford – Curved Executive Desk with Optional Return. https://www.auraadesign.co.uk/oxford-curved-executive-desk-with-optional-return/

[22] Art elements from Metal fittings. https://elementi.hefest.bg/index.php?p=46&l=1; https://metaltablelegs.eu/

[23] Laser cut elements. https://arthdecor.com/product-category/lazerno-izryazani-elementi/; https://jelezar.com/bg/

[24] Art furniture made of epoxy resin https://milanovwoodart.com/masi/holni-masi/milanov-wood-art;

[25] Decorative ornaments for wrought iron furniture. https://hefest-plovdiv.com/index.php?p=8&l=1;

[26] Art elements of wood, carving.https://www.djesur.com/Darveni-Dekoracii-Darvorezbi-i-Ornamenti;

[27] Teak mosaic. https://styrodesign.eu/darvena-mozayka-ot-tikovo-darvo/;

[28] 3D printed furniture. https://3dbgprint.com/

[29] 3D фурнирован MDF. https://www.jaf-bulgaria.com/

[30] WFH Desk - Factory Second. https://workfromhomedesks.com/products/wfh-desk-factory-second

[31] Adjustable Wooden Standing Desk https://www.etsy.com/no-en/listing/869663884/adjustable-wooden-standing-desk-work

[32] Sit to Stand Up Desk https://www.amazon.sa/-/en/ecotribe-Modern-Height-Adjustable-Spaces/dp/B08MPCJV46

[33] Adjustable Standing Desk. https://www.avidcnc.com/adjustable-standing-desk-p-497.html

[34] Luxurious Style New Design CNC L Shape Functional CNC Executive Desk. https://vicbismilla1983.en.made-in-china.com/product/RJVUhMQbCwWj/China-Luxurious-Style-New-Design-CNC-L-Shape-Functional-CNC-Executive-Desk.html

[35] CNC router. https://cncrouter.bg/cnc-router-darvorezbi-mebeli

[36] 3D stl model for cnc office chair carved wooden 3d render. https://stlmax.com/catalog/3d_47587.html

[37] T. Dovramadjiev, R. Zlateva, Ts. Murzova, D. Dobreva, K. Markova, M. Murzova et.al. Scientific-Research Project TUV 2023 PD19 "Researching the possibilities of increasing the visual ergonomics of computerized workplaces".

[38] T. Dovramadjiev, et al. (2024). Interaction Between Artificial Intelligence, 2D and 3D Open Source Software, and Additive technologies for the Needs of Design Practice.
In: Iglesias, A., Shin, J., Patel, B., Joshi, A. (eds) Proceedings of World Conference on Information Systems for Business Management. ISBM 2023. Lecture Notes in Networks and Systems, vol 834. Springer, Singapore. https://doi.org/10.1007/978-981-99-8349-0_26; https://link.springer.com/chapter/10.1007/978-981-99-8349-0_26?fbclid=IwZXh0bgNhZW0CMTAAAR39C_V9fTWp_TdWx_flzf7UHmp0jgn2XNwnxAfPNB7VTiEbmox9QWIuCy4_aem_LBpMl0xeBreXipHQgyT6Mw; Print ISBN 978-981-99-8348-3; Online ISBN 978-981-99-8349-0

[39] D. Dobreva, T. Dovramadjiev, Ts. Murzova, M. Tachev, I. Iliev, K. Cankova, G. Jecheva, M. Murzova, G. Staneva, (2022). Ergonomic and Design Research of the Auxiliary Furnishings in High School of Education in the City of Varna. In: Tareq Ahram and Redha Taiar (eds)

Human Interaction & Emerging Technologies (IHIET 2022): Artificial Intelligence & Future Applications. AHFE (2022) International Conference. AHFE Open Access, vol 68. AHFE International, USA. DOI: 10.54941/ahfe1002813

[40] ErgoFellow 3.0 https://www.fbfsistemas.com/ergonomics.html

[41] C. Knight, & A. Haslam, S. (2010). The relative merits of lean, enriched, and empowered offices: An experimental examination of the impact of workspace management strategies on well-being and productivity. Journal of Experimental Psychology: Applied, 16(2), 158–172. https://doi.org/10.1037/a0019292

[42] Beyond Aesthetics: The Relationship Between Artwork & Productivity at Workspaces. https://maxestates.in/beyond-aesthetics-the-relationship-between-artwork-productivity-at-workspaces

A. Stoyanov V., A. Terziev, M. Uzunova, Numerical study of distribution of smoke and hazards in underground parking areas considering the operation of ventilation, 3rd International symposium on Environment Friendly Energies and Applications (EFEA 2014), 19 – 21 November, 2014, Paris, France, Inspect. Accession Number: 14985538; DOI: 10.1109/EFEA 2014 7059991;

Terziev, A., Stoyak, V., Ushakov, V., Brazhanova, D., Suleimanov, S., Analysis of the opportunities for improving energy efficiency in public buildings, (2021) 1019 (1), art. no. 012021, DOI: 10.1088/1757-899X/1019/1/012021, PUBLISHER: IOP Publishing Ltd, ISSN: 17578981

Materials Research Proceedings 46 (2024) 338-345 https://doi.org/10.21741/9781644903377-44

Analysis of the efficiency of pneumatic actuator regulation in selected production processes using artificial intelligence, graph algorithms, and probabilistic methods, taking into account failure rates and return on investment costs

Piotr SKUDLIK[1,a], Andrzej WRÓBEL[1]

[1] Department of Automation of Technological Processes and Integrated Manufacturing Systems, Faculty of Mechanical Engineering, Silesian University of Technology, Poland

[a] piotrek.11201@gmail.com

Keywords: Artificial Intelligence, Manufacturing, Mechanical Engineering, Automation, Robotic, Statistic, Neural Networks, Mathematical Algorithms, Management

Abstract. Modern production systems are characterized by a high degree of automation, where pneumatic actuators play a key role in assembly and transport processes. This paper presents an analysis of the performance of pneumatic actuator regulation using artificial intelligence (AI) algorithms and probabilistic methods in simulated production conditions. The study utilized LSTM neural networks for failure prediction, graph algorithms like Bayesian networks, and Monte Carlo simulations to assess the risk of downtime. The results show that AI models enable a significant reduction in unplanned downtimes and optimization of operational costs through dynamic regulation of actuator parameters. The simulation, based on real operational data, demonstrated that AI models outperform traditional control methods, such as PID controllers, in terms of efficiency and accuracy in failure prediction. However, the applied models have limitations, including high computational requirements and dependence on the quality of input data. The findings suggest that integrating AI algorithms into industrial automation systems can substantially improve production efficiency by reducing operational costs and failure risks. This study highlights the potential benefits of widespread AI adoption in industrial automation, particularly in ensuring long-term system reliability.

Introduction

The automation of production processes is a key component of modern industrial systems, where pneumatic actuators play a significant role. These mechanical elements are used in various processes, such as assembling small components, transferring different sizes of loads, and assisting in machining, where high force, precision, and quick response are required. Efficient regulation of pneumatic actuators directly impacts the performance, reliability, and operational costs of production.

Traditional control methods, such as proportional-integral-derivative (PID) controllers, while effective in many cases, may not be sufficiently flexible and adaptive under changing operational conditions. Even programmable logic controllers (PLCs), despite their many built-in functions and ease of programming custom instructions, are not inherently reliable devices and do not allow full real-time control. Therefore, there is growing interest in more advanced control methods that utilize artificial intelligence (AI), graph algorithms, and probabilistic methods. These modern approaches enable dynamic optimization of machine operating parameters, thereby increasing efficiency, reducing the number of unplanned downtimes, and minimizing risks associated with device operation.

The aim of this article is to present a comprehensive analysis of the effectiveness of pneumatic actuator regulation in simulated production conditions, taking into account failure rates and return on investment (ROI). The study is based on randomly generated data, but with appropriate criteria that reflect the actual working conditions of industrial plants.

Object

The study focuses on a double-acting linear pneumatic actuator, which serves as a critical component in many industrial automation systems, particularly in automotive assembly lines. This actuator was modeled in a simulated environment designed to closely mimic real-world production conditions. The actual actuator required adjusting the flow to get the right pressure on the piston, so it is possible to make a precise movement at the expected time. This machine unit is used to place the NoX sensor element to perform the welding operation on the end of the element. Double-acting actuators are characterized by their ability to move in both directions under the influence of compressed air, providing enhanced control, efficiency, and adaptability in automated processes. The simulation aimed to replicate the dynamic response of the actuator under various operational conditions to assess its performance, reliability, and economic impact in a controlled, cost-effective manner. The use of simulation data in this study stems from the ability to precisely control operational conditions, which is difficult to achieve in a real-world production environment. Simulations allow for rapid and cost-effective testing of various operational scenarios, minimizing both costs and risks associated with real-world experiments. As noted in the literature, simulations are a key research tool in industrial automation, particularly when well-defined mathematical models can reflect real working conditions. [1, 2]

Model assumptions:

1. **Operating Pressure:** The system operates at a stable pressure of 5 bar, which aligns with standard industrial settings for pneumatic systems, ensuring optimal performance and reliability of the actuator.
2. **Actuator Cycle Time:** Each complete cycle of the actuator, encompassing full extension and retraction of the piston, is set to 3 seconds. This cycle time is crucial for evaluating the responsiveness and efficiency of the control algorithms under realistic operational speeds.
3. **Downtime Cost:** The economic impact of system failures is quantified by assuming a downtime cost of 50,000 PLN per hour. This value is derived from industry data and serves as a key parameter for assessing the financial implications of actuator performance and maintenance strategies.
4. **Valve Configuration:** The actuator model incorporates a remote-controlled valve for precise air flow management and a non-return valve to prevent backflow, simulating a realistic pneumatic control setup. These components are critical for maintaining stable operation and preventing potential system failures due to pressure imbalances.
5. **Actuator Design Features:** The double-acting actuator is modeled with a piston capable of bidirectional movement, driven by alternating air pressure inputs. This configuration allows for detailed simulation of dynamic load conditions and provides insights into the actuator's operational limits, wear characteristics, and energy efficiency.

By leveraging these assumptions, the simulation aims to provide a comprehensive evaluation of the actuator's performance, offering valuable data for optimizing control systems and predicting maintenance needs in automotive assembly line applications.

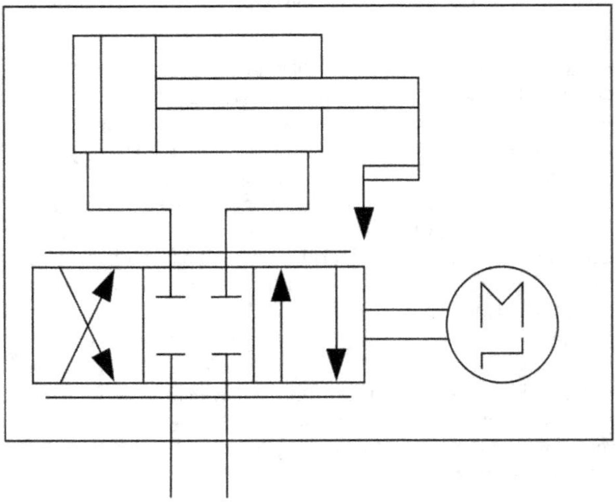

Fig. 1 Diagram of the simulated pneumatic valve. (custom graphic)

Metodology

The research was conducted in a simulated environment modeled after real automotive assembly lines, employing MATLAB and Python for the simulations. The simulations utilized advanced libraries dedicated to artificial intelligence, mathematical computations, and graph analysis. Models of pneumatic actuators were represented as data structures within the simulation, controlled by algorithms tailored to simulate real-world operations. The following tools and techniques were employed:

- **Programming Languages:** MATLAB and Python for simulation development
- **AI and Mathematical Libraries:** Utilized for optimizing the control algorithms and simulating complex system behaviors
- **Graph Analysis Tools:** Employed to visualize data and analyze actuator performance
- **Simulation of Pneumatic Actuators:** Detailed modeling of actuators to study their response and performance under various conditions
- **Algorithm Development:** Custom algorithms were coded to control the actuators, simulating real-time production line adjustments and fault management

This methodological approach ensured a comprehensive understanding of the system dynamics and allowed for the testing of various operational scenarios without the need for physical prototypes.

PID model

The PID controller (Proportional-Integral-Derivative) is a fundamental tool for regulating the parameters of pneumatic actuators. Its operation is based on calculating the control error e(t), which is the difference between the setpoint and the actual output value, and then appropriately controlling the parameters:

Materials Research Proceedings 46 (2024) 338-345 https://doi.org/10.21741/9781644903377-44

$$u(t) = K_p e(t) + K_i \int e(t)\,dt + k_d \frac{de(t)}{dt} \text{ (Eq. 1)}$$

where:

- K_p – proportional gain,
- K_i – integral gain,
- K_d – derivative gain.

The PID controller was used as a reference point to compare older and commonly used solutions with modern control methods. Despite its simplicity, PID has a limited ability to adapt to dynamically changing operational conditions, which can lead to downtime.

```
# PID function
class PIDController:
    def __init__(self, Kp, Ki, Kd):
        self.Kp = Kp
        self.Ki = Ki
        self.Kd = Kd
        self.integral = 0
        self.previous_error = 0

    def update(self, setpoint, measured_value, dt):
        error = setpoint - measured_value
        self.integral += error * dt
        derivative = (error - self.previous_error) / dt
        output = self.Kp * error + self.Ki * self.integral + self.Kd * derivative
        self.previous_error = error
        return output
```

Fig. 2 Part of the main programme for PID.

The output value u(t) was dynamically adjusted based on the current control error e(t). The algorithm was tested in various scenarios, including stable conditions (imposed variables) and dynamically changing conditions. The results of the algorithm were as follows (for 10000 cycles):

- Failure rate: 3.8%
- The estimated monthly downtime cost: 4000 PLN
- Average Delay Time: 1.2 s (in stable conditions), 1.5 s (under dynamic changes)
- Effectiveness: 65%

LSTM - neural network model

LSTM neural networks are specifically designed to work with sequential data, making them ideal for analyzing time series data and other physical quantities (pressure, flow) from pneumatic actuators. This model learns operational patterns based on historical data, allowing for the accurate prediction of future failures. The loss function used in the neural network training process is based on the mean squared error (MSE):

$$L(\theta) = \frac{1}{N}\sum_{i=1}^{N} (y_i - \hat{y}_i)^2 \text{ (Eq 2)}$$

where:

- y_i – actual output values,
- $\hat{y}_i$ – predicted output values,
- N – number of samples.

Materials Research Proceedings 46 (2024) 338-345 https://doi.org/10.21741/9781644903377-44

The LSTM model was trained on data containing information about pressure, air flow, and piston positioning, which enabled effective prediction of critical states. LSTMs are complex models that require significant computational resources, both during training and real-time predictions. Long training times and the need to adjust hyperparameters can pose significant challenges in industrial environments, where time is crucial.

```
import numpy as np
import tensorflow as tf
from tensorflow.keras.models import Sequential
from tensorflow.keras.layers import LSTM, Dense

# Operate data
time_steps = df[1]
features =df_feat[1]
X_train = np.random.rand(1000, time_steps, features)  # Train data
y_train = np.random.randint(2, size=(1000, 1))          #0 = no error, 1 = error

# LSTM
model = Sequential()
model.add(LSTM(50, activation='relu', input_shape=(time_steps, features)))
model.add(Dense(1, activation='sigmoid'))
model.compile(optimizer='adam', loss='binary_crossentropy', metrics=['accuracy'])

# Model's trainning
model.fit(X_train, y_train, epochs=20, batch_size=32, validation_split=0.2)
```

Fig. 3 Part of the main programme for LSTM.

The LSTM model significantly improved the accuracy of failure predictions, allowing for earlier intervention. As a result, the number of downtimes was reduced, and operational costs were lowered. Early detection of anomalies enabled quicker preventive actions, as confirmed by the following data (for 10000 cycles):

- Failure rate: 2.0%
- The estimated monthly downtime cost: 2500 PLN
- Average Delay Time: 0.3 s (in stable conditions), 0.9 s (under dynamic changes)
- Effectiveness: 92%

Graph Algorithms – Bayesian Networks

Bayesian networks were used to model complex dependencies between all actuator parameters. These algorithms enable dynamic prediction of the probability of critical states that could lead to failures. Bayes' theorem forms the basis of these calculations:

$$P(A|B) = \frac{P(B|A) \cdot P(A)}{P(B)} \text{ (Eq. 3)}$$

where:

- P(A|B) – probability of event A given event B,
- P(B|A) – probability of event B given A,
- P(A) – probability of event A,
- P(B) – probability of event B.

Based on the above formula, a multiplication of combinations of several programming codes was implemented, referring to various parameters and their dependencies.

```
import pyBN
from pyBN.learning.parameter import mle_estimateparams
from pyBN.learning.structure import chow_liu
from pyBN.inference.exact import belief_propagation

# Structure of net
bn = chow_liu(data)

# Learning
params = mle_estimateparams(bn, data)

# Prediction of Failure Probability Based on Input Data
evidence = {'Pressure': 7.0, 'Airflow': 100}
belief = belief_propagation(bn, evidence)
print(f"Prediction of Failur Probability: {belief['Failure']:.2f}")
```

Fig. 4 Part of the main programme for Bayesian Networks.

The obtained data indicated that the program's execution time was very long, causing the omission of some dependencies crucial for the correct functioning of the process. The average delay decreased, along with the costs, as confirmed by the following results (for 10000 cycles):

- Failure rate: 1.0%
- The estimated monthly downtime cost: 3500 PLN
- Average Delay Time: 1.3 s (in stable conditions), 1.0 s (under dynamic changes)
- Effectiveness: 72.5%

Monte Carlo simulation

The Monte Carlo simulation is one of the most versatile probabilistic methods used for risk assessment, uncertainty analysis, and decision-making in dynamic systems. In the context of regulating pneumatic actuators, this method allows for the evaluation of the impact of operational variables on the system's reliability and efficiency. The key premise of the Monte Carlo method is modeling random events based on the probability distributions of input variables and conducting multiple simulations, which enables the statistical assessment of outcomes.

$X = \frac{1}{N}\sum_{i=1}^{N} \quad f(x_i)$ (Eq. 4)

where:

- X – expected output value,
- $f(x_i)$– objective function,
- N – number of simulations.

The Monte Carlo simulation in the analysis of pneumatic actuators involves modeling random variations of key operational parameters, such as pressure, air flow, temperature, and environmental variables. This method allows for the simulation of hundreds of actuator operation scenarios, enabling the assessment of how frequently and under what conditions failures occur. Despite these advantages, the Monte Carlo method is not without its drawbacks. One of the main limitations of this method is its strong dependence on the quality of input data and probability distributions. The results of a Monte Carlo simulation are only as good as the data on which the

Materials Research Proceedings 46 (2024) 338-345 https://doi.org/10.21741/9781644903377-44

model is based. If the probability distributions of the input variables are imprecise or based on incomplete historical data, the simulation results can be misleading.

```
import numpy as np

def monte_carlo_simulation(n_simulations):
    failures = 0
    for _ in range(n_simulations):
        pressure = np.random.normal(5, 3.0)  # Average pressure: 7 bar, deviation: 0.5 bar.
        if pressure < 2.0 or pressure > 7.0:
            failures += 1
    return failures / n_simulations
```

Fig. 5 Part of the main programme for Monte Carlo simulation.

The results of the Monte Carlo simulation turned out to be surprisingly poor compared to the other methods. The results of the algorithm were as follows (for 10000 cycles):

- Failure rate: 4.2%
- The estimated monthly downtime cost: 5000 PLN
- Average Delay Time: 1.5 s (in stable conditions), 1.0 s (under dynamic changes)
- Effectiveness: 80%

Summary

The application of artificial intelligence, graph algorithms, and probabilistic methods for regulating pneumatic actuators in a production environment brings significantly increased the operational efficiency of this machine elements, though it did not always lead to a reduction in failure rates. However, failure prediction and adaptive control of parameters allowed for a reduction in downtime and a financial loss decrease of up to 1,500 PLN. Additionally, pressure and flow control enabled the reduction of average delay time by nearly one second. The reduction in the number of failures, increased efficiency, and resulting quick return on investment make these technologies a key element of modern production systems. Integrating these advanced methods with existing industrial automation systems can significantly improve long-term production performance and reliability. Notably, the system is easily configurable, and its replication is simply a matter of adding additional Ethernet modules connected to PLC controllers.

In addition to applying the advantages of artificial intelligence models, excluding limitations from their implementation. LSTM models, as pointed out in the work on time series analysis, require computational resources and runtime [6]. Probabilistic algorithms such as the Monte Carlo method are enhanced by the quality of the output data, as evidenced by research on probabilistic methods [7]. In industrial environments, imprecise data can cause incorrect predictions, which undermines the performance of the results. Key information is delivered to AI models is crucial. [8]

References

[1] A. Gupta. “Modeling and Simulation in Manufacturing Systems”, International Journal of Computer Integrated Manufacturing, 2003

[2] R. Merwe, "Simulation as a Tool in Manufacturing," Journal of Manufacturing Science and Engineering, 2001

[3] T. Oliphant, “A guide to NumPy“, Trelgol Publishing, 2006

[4] J. Hunter, “Matplotlib: A 2D graphics environment”, Computing in Science & Engineering, 9(2007) 90-95. https://doi.org/10.1109/MCSE.2007.55

[5] J. Muellbauer, "Economics and Consumer Behavior", Cambridge University Press, 1980

[6] F. Chollet, "Deep Learning with Python", Manning Publications, 2017

[7] M. Abadi, et al., "TensorFlow: A System for Large-Scale Machine Learning", OSDI, 2016

[8] A. Geron, „Hands-On Machine Learning with Scikit-Learn, Keras, and Tensorflow: Concepts, Tools, and Techniques to Build Intelligent", O'Reilly, 2019

[9] W. McKinney, "Data Structures for Statistical Computing in Python", Proceedings of the 9th Python in Science Conference, 2010. https://doi.org/10.25080/Majora-92bf1922-00a

[10] Information on https://numpy.org/

[11] Information on https://pandas.org/

[12] Information on https://www.python.org/

[13] Information on http://www.weld.labs.gov.cn

Simulation to adjust the draft of RO-RO pontoon

HALUS Andrei[1,a *], SEVERIN Irina [1,b] and ACOMI Nicoleta [2,c]

[1]National University of Science and Technology Politehnica Bucharest, Splaiul Independenței, No. 313, 060042, Bucharest, Romania

[2]Constanta Maritime University, St. Mircea cel Batran, No. 104, 900663, Constanta, Romania

[a]andreihalus0106@gmail.com, [b]irina.severin@upb.ro, [c]nicoleta.acomi@cmu-edu.eu

Keywords: Codesys, Programmable Logic Controller, Ballast System, Computer Simulation

Abstract. Floating pontoon berthing facilities are used by RO-RO ships and ferries to provide temporary or semi-permanent docking and loading/unloading capabilities. In order to accommodate ships having various freeboards, especially during operating heavy weights, it may be necessary to adjust the draft of the pontoon. In this article, the authors propose a software for calculating the draft of a RO-RO pontoon that retrieve, process and transmit data for the operations of adjusting the draft using of a programmable logic controller. The RO-RO pontoon we are referring to has actually a manual operation of the ballast installation. The software is implemented and simulated using the Codesys application. By introducing the command system, the work performed by the human operator is automatically facilitated, reducing the number of workers, decreasing the risks following the wrong maneuvers performed by the operator. This study establishes the foundations for future experimental procedures.

Introduction

Despite being under staff supervision, floating pontoons serve as public access points and maintain the connection between the land and the sea [1].

The pontoon types covered in this workpaper are RO-RO pontoons (Fig. 1). For RO-RO ships, they are utilized for vehicle loading and unloading. The stability of floating structures can be affected by various factors, including the object's weight, draft (It is represented by the vertical distance between the lower face of the bottom plates, without considering the keel or other fixed points and the plane of the water level), shape, and the high of the submerged portion. This contrasts with ships, which can only be operated efficiently at small heel angles [2,3].

The regulation of the pontoon's draught is done by a seawater ballasting/de-ballasting installation in/from the dedicated tanks in its construction. The ballast installation present on board the pontoon is a manually operated installation that consists of two electric pumps, a piping system for each tank and a system of electrically operated valves controlled from the control panel dedicated to this system. This actuation system is outdated compared to maritime vessels, including RO-RO ships that have an automated ballast/de-ballast system integrated on board. Additionally, the actuation of the ballast systems on board these vessels can also be managed through software that calculates cargo loads, which adds even greater value to the ballast system onboard the ships.

Despite the additional costs in the case of the automated system, but the reduction of human and crew errors, this paper will illustrate both types of installations applied in the case of the pontoon for mooring RO-RO ships. Draft stabilization and regulation installations that have an automation system consist of a PLC (Programmable Logic Controller) and a set of sensors: level sensors, pressure sensors, flow sensors, etc.

Fig. 1 Current pontoon structure in use for ro-ro ship mooring operations

Draft stabilization and regulation systems that have an automation system consist of a PLC (Programmable Logic Controller) and a set of sensors: pressure sensors, flow sensors, level sensors, etc. [4].

Manual operation of the RO-RO pontoon draft adjustment system

The draught control system, also known as the ballast system, with manual operation consists of one or more pumps that are operated locally or remotely and a set of manually operated valves. By combining closed and open valves, ballast tanks can be filled or emptied.

Fig. 2 Onboard ballast manually operated control valve system

Instead of the manual valves shown above (Fig. 2) electrically operated, electro-pneumatic, electro-hydraulic valves, etc. can be used. Although they are valves that bring a contribution to their ease of operation, the closing or opening operation falls into the category of manual actuation, since a human operator is needed to carry out these commands [5].

Ballast installation present on board the RO-RO pontoon

The ballast installation installed on board the RO-RO pontoon consists of two ballast pumps, remotely operated electric valves and the piping related to the installation.

The power supply and start/stop control are done by means of a soft starter, the electric motors related to the pumps being connected in a delta connection. The MSB [Main switchboard] is presented in Fig. 3.

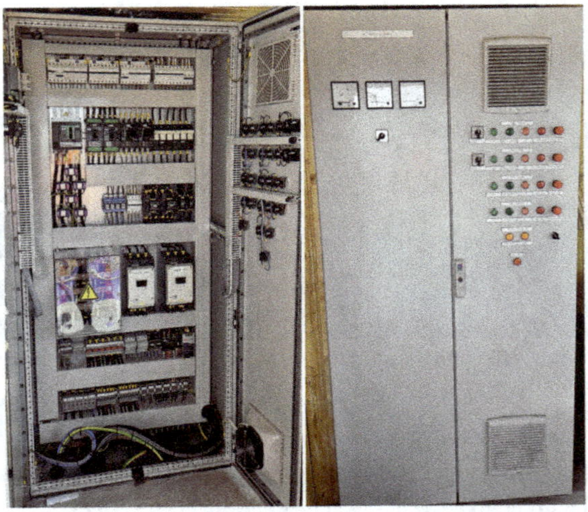

Fig. 3 Main electrical switchboard of the pontoon: open and closed states

Automatic operation of the RO-RO pontoon draft adjustment system

The ballast plant can be automated by introducing a PLC (Programmable Logic Controller) that can take data, process data and transmit data. The role of the PLC is to facilitate the work of the personnel, to display more information than that provided by the manual system and to perform more precise maneuvers than that of the human operator, eliminating human error. However, it must be taken into account that errors may occur within the system including hardware malfunctions, which would require the intervention of authorized personnel.

For the writing of the automatic ballasting software, the Codesys [6] program was used. The software is written in two programming languages:

1. Ladder Logic (LD): This programming language is primarily used for developing software for programmable logic controllers (PLCs). It employs graphical representation similar to hardware logic circuit diagrams, resembling a ladder with vertical rails and horizontal rungs [7].
2. Structured Text (ST): A high-level programming language with syntactical similarities to Pascal, ST is part of the IEC 61131-3 standard aimed at standardizing PLC programming languages. Its flexibility and intuitive design make it suitable for control algorithms and complex mathematical tasks, offering efficiency comparable to Ladder Logic [8].

For the PLC to be able to calculate the draft, it needs more input data: the capacity of the ballast storage tanks, input data from the pontoon stability calculation, information from sensors etc.

To enable the software to function in absence of physical level sensors, a sensor simulator was incorporated within the system. Given that multiple types of sensors can be utilized, the next section will introduce the ToughSonic 30 Level & Distance Sensor as a representative example. (Fig. 4)

ToughSonic 30 Level & Distance Sensor employs ultrasonic sound waves for measuring distances in the air. This innovative technology offers various benefits over optical, mechanical, and radar systems. A significant advantage is that ultrasonic waves remain unaffected by conditions such as ambient lighting, target hue, transparency, or reflectivity. Additionally, the sensor features a detection area that is considerably wider than that of optical sensors, facilitating the effective identification of both small and large objects [9]. The major disadvantage of using this type of sensor is that its optimal operating range is between 25.4 cm and 6.1 m, which requires

Materials Research Proceedings 46 (2024) 346-353 https //doi.org/10.21741/9781644903377-45

additional care in its positioning. Additionally, it is essential to consider the proximity of the signal cables to power lines, as the electromagnetic field generated by these cables can cause inaccurate sensor readings. Proper shielding and cable routing are critical to avoid such errors.

Fig. 4 ToughSonic 30 Sensor: Profile view and illustrative diagram of operation [10]

Following the presentation of the sensor example, the authors specify that it can be implemented in software but also other sensors that send information in current, voltage or on a communication protocol.

The code structure and diagram in the LD for each tank is identical, with differences in the name of the variables (Fig. 5).

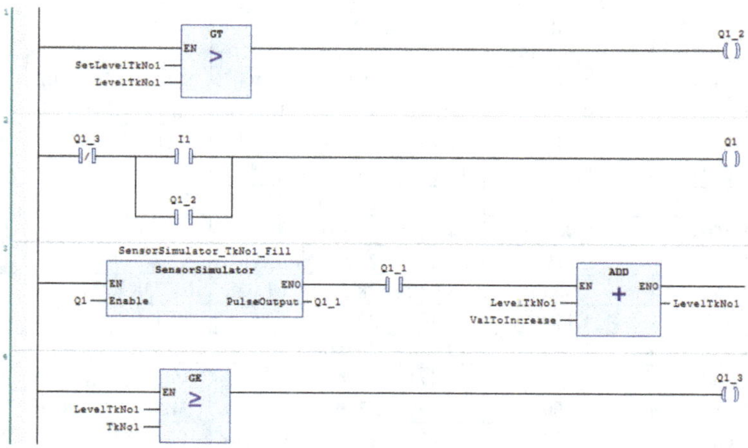

Fig. 5 Ladder logic diagram for tank level sensor simulation no.1 (Tank filling)

Filling ballast tanks according to the desired draft

The calculation of the filling of the tanks according to the draft set in the software is made based on the "SetMidDraft". This block calculates the filling of each tank separately by means of the code written in the Structured Text (Fig. 6). The calculations present in this block were made following several stability simulations with the Napa Software. This software has the possibility to simulate the filling of each tank, the calculation of intact stability and stability in case of failure. The output data from the simulations show us that the minimum draft of the pontoon is 0.719 [m], the maximum draft is 1.419 [m]. At these drafts the pontoon can bear an additional weight when unloading ships of 162.4 [t]. This unloading weight is divided as follows [11]:

- 50 [t] — Ramp of the RO-RO type vessel (divided into 25 [t] on the port side and 25 [t] on the starboard side);

- 48 [t] – Goods towing vehicle;
- 64.4 [t] – Trailer of the towing vehicle.

Loading the pontoon with the specified weights will change the minimum draft which will be 1.2 [m] and the maximum draft to the value of 1.7 [m].

```
(* Calculate BL2BB using the given formula *)
ValBL2BB:=((SetDraft-MinDraft)*(MaxBL2BB-MinBL2BB)/(MaxDraft-MinDraft))+MinBL2BB;

(* Calculate BL2TB using the given formula *)
ValBL2TB:=((SetDraft-MinDraft)*(MaxBL2TB-MinBL2TB)/(MaxDraft-MinDraft))+MinBL2TB;
```

Fig. 6 Code sequence in Structured Text of block "SetMidDraft"

For the safety of tank filling calculations, using extrapolation, it was found that the draught level and the ballast levels in the tanks are directly proportional, and the values increase linearly. The values are presented in Table 1.

Table 1 Calculation of ballast tank load at interval of 0.05 [m]

Pontoon draft with moored vessel and cargo [m]	Pontoon draught without moored vessel and without load [m]	BL2BB [t]	BL2TB [t]	BL3BB [t]	BL3TB [t]	BL4BB [t]	BL4TB [t]
1.000	0.719	2.500	2.500	2.390	2.390	23.440	1.630
1.050	0.769	6.997	6.228	10.771	10.771	25.115	3.856
1.100	0.819	11.494	9.955	19.152	19.152	26.790	6.081
1.150	0.869	15.991	13.683	27.533	27.533	28.466	8.307
1.200	0.919	20.489	17.410	35.915	35.915	30.141	10.532
1.250	0.969	24.986	21.138	44.296	44.296	31.816	12.758
1.300	1.019	29.483	24.865	52.677	52.677	33.491	14.983
1.350	1.069	33.980	28.593	61.058	61.058	35.167	17.209
1.400	1.119	38.477	32.320	69.439	69.439	36.842	19.434
1.450	1.169	42.974	36.048	77.820	77.820	38.517	21.660
1.500	1.219	47.471	39.775	86.201	86.201	40.192	23.885
1.550	1.269	51.969	43.503	94.583	94.583	41.867	26.111
1.600	1.319	56.466	47.230	102.964	102.964	43.543	28.336
1.650	1.369	60.963	50.958	111.345	111.345	45.218	30.562
1.700	1.419	65.460	54.685	119.726	119.726	46.893	32.787

According to data from Table 1, it is noted that the tanks with the number one on the port board and the starboard board are not used for adjusting the draft. This is verified with the help of NAPA software. According to Fig. 7 and Fig. 8 of the simulations carried out in the case of the pontoon without the ship moored and without cargo, the list of tanks and their loads can be seen in the upper left, the draft of the pontoon in the lower left and implicitly the graphic illustration of the tank loads on the right side. In the graphic illustration, the tank loads hatched or filled with magenta are seen (depending on the amount of liquid in the tanks), together with views from both sides, aft and top view. In the side views, the medium loading line of the pontoon, the purple line, is also present.

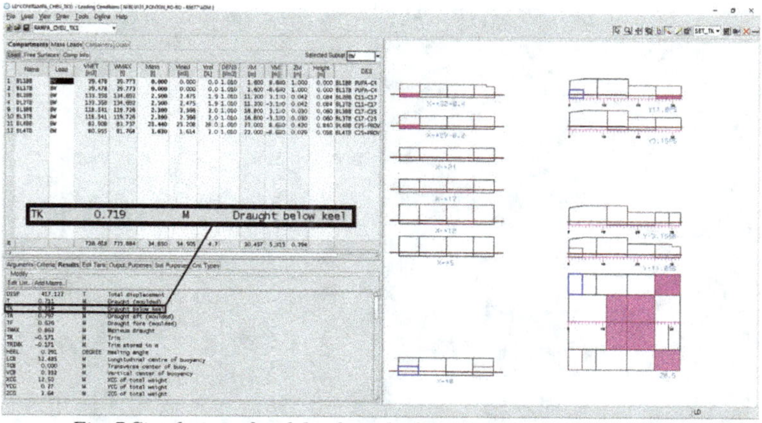

Fig. 7 Simulation of tank loads at draft 0.719 [m] using NAPA software

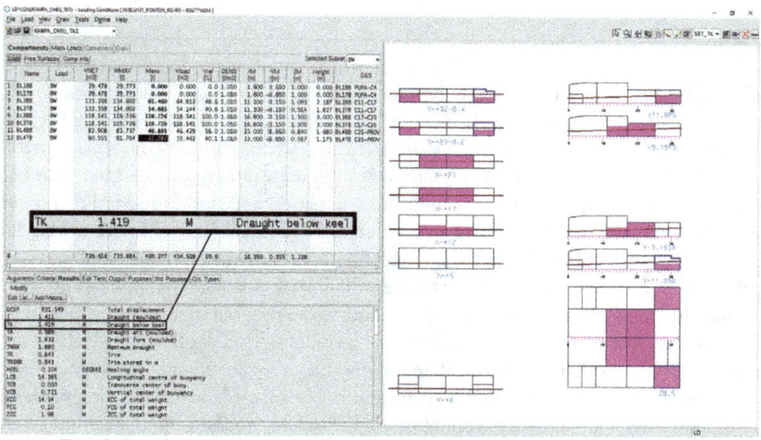

Fig. 8 Simulation of tank loads at draught 1,419 [m] using NAPA software

Graphical interface of the pontoon software

The graphical interface on the HMI (Human Machine Interface) (Fig. 9) is composed of the following elements:

- The diagram is seen from above of the pontoon with the efferent piping. The diagram highlights the ballast tanks and as the load level increases, a mime (Digital images of the assets and processes that have been created in the HMI/SCADA system) of alabaster color is observed that expresses the approximate load in percentages [12];
- The current filling level of each tank expressed in m^3;
- The set level resulting from the calculations performed in the block "SetMidDraft" expressed in m^3;
- The draught set by the operator expressed in m.

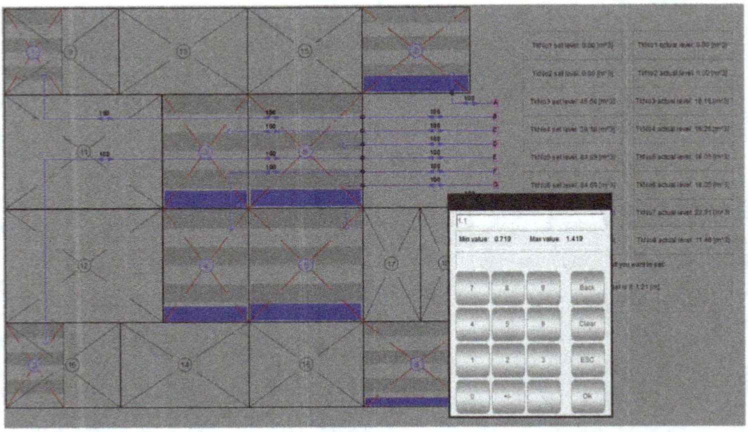

Fig. 9 HMI interface for pontoon control: top view and draft adjustment control

Conclusions

The automation of the pontoon ballast installation offers significant operational benefits, but also presents certain challenges that should be considered. The key advantages include increased efficiency in ballast operations, simplified control through the replacement of manual systems with an automated interface, and a reduction in human error. Additionally, the need for separate draft monitoring systems is eliminated, as the automation system integrates real-time data collection and processing. These improvements contribute to safer and more streamlined operations, ultimately enhancing the overall performance and reliability of RO-RO pontoon berthing facilities.

However, the implementation of such an automated system incurs considerable costs. These include expenses related to the acquisition and installation of sensors, communication cabling, and modifications to the main switchboard to accommodate the programmable logic controller (PLC). Additional costs stem from labor required for installation, as well as the integration of a touchscreen interface, which may necessitate modifications to the existing infrastructure.

While the benefits of automation are clear, the associated costs and potential technical challenges, such as system failures or limitations in the sensors' measuring range, must be carefully assessed. Detailed experimental validation of the proposed automated system is necessary to confirm its effectiveness and reliability in real-world scenarios. This validation should encompass a range of operational conditions to comprehensively evaluate the system's performance and identify any potential limitations. Future research should focus on developing a scale model of the pontoon, equipped with the necessary automation components (PLC, valves, pumps), to conduct performance testing under various operational conditions.

Moreover, comparative studies with existing manual systems and other automated solutions could provide valuable insights into the advantages of the proposed approach. Understanding the practical implications of system failures and the limitations of the sensors' measuring range will be vital for developing robust contingency plans. Overall, the establishment of a detailed framework for experimental validation and performance assessment will not only support the proposed automation system's implementation but also foster ongoing improvements in the design and operational protocols of pontoon ballast system. This comprehensive approach will ultimately ensure a smoother transition to automated systems while maintaining high safety and operational standards.

Acknowledgement
This publication has been funded with support from the European Commission. This publication reflects the views only of the author, and the Commission cannot be held responsible for any use which may be made of the information contained therein. Project Greenport, Number: 101139879.

References

[1] Freeman E., Splinter K., Cox R., Flocard F., "Dynamic Motions of Piled Floating Pontoons Due to Boat Wake and Their Impact on Postural Stability and Safety," Journal of Marine Science and Engineering, p. 1, 02 11 2022. https://doi.org/10.3390/jmse10111633

[2] Ministry of Transport, "Order of the Minister of Transport No. 1.472/2018 for the approval of technical requirements for inland navigation vessels," Official Gazette, Bucharest, 2018.

[3] Karczewski A., Piatek L., "The influence of the cuboid floats parameters on the stability of a floating building," Polish Maritime Research, vol. 27, no. 3, pp. 16-21, 2020. https://doi.org/10.2478/pomr-2020-0042

[4] I. R. Kusuma, "Design and Simulation of Automatic Ballast System on Catamaran Ship Based on Programmable Logic Control," International Journal of Marine Engineering Innovation and Research, vol. Vol. 1(3), pp. 161-167, 06 2017. https://doi.org/10.12962/j25481479.v1i3.2076

[5] Halus A., Severin I., Acomi N., "Simulation of ballast transfer for pontoons", IManEE, 2023.

[6] "Codesys," Codesys, [Online]. Available: https://www.codesys.com. [Accessed 2023 12 24].

[7] B.R. Mehta, Y.J. Reddy, "Batch automation systems," in Industrial Process Automation Systems, Oxford, Butterworth-Heinemann, 2015, pp. 135-156. https://doi.org/10.1016/B978-0-12-800939-0.00003-6

[8] Anthony K. Ho, "Structured Text Programming," Center, PDH Online | PDH, Washington, 2020.

[9] "Omniinstruments," Senix, [Online]. Available: https://www.omniinstruments.co.uk/level-distance-sensors/ultrasonic-level-distance-transmitter/senix-toughsonic-tspc-ultrasonic-level-sensors.html. [Accessed 27 01 2024].

[10] Omni Instruments Ltd, "Omniinstruments," 01 12 2015. [Online]. Available: https://www.omniinstruments.co.uk/fileuploader/download/download/?d=0&file=custom%2Fupload%2FFile-1453299006.pdf. [Accessed 27 01 2024].

[11] "NAPA Stability," NAPA, [Online]. Available: https://www.napa.fi/software-and-services/ship-operations/napa-stability/. [Accessed 2024 01 28].

[12] "GE Digital Solution," General Electric, [Online]. Available: https://www.ge.com/digital/documentation/predix-apm/latest/hmi-mimic-mngmt.html. [Accessed 30 03 2024].

Innovative Manufacturing Engineering and Energy - IManEE2024 Materials Research Forum LLC
Materials Research Proceedings 46 (2024) 354-361 https://doi.org/10.21741/9781644903377-46

Digital modeling and AI-driven optimization of ball burnishing process parameters

Diyan M. DIMITROV[1,a] , Stoyan SLAVOV[2,b*] and Lyubomir Si Bao VAN [2,c]

[1] Technical University of Varna, Department of Mechanics and Machine Elements, 9010 Varna, Bulgaria

[2] Technical University of Varna, Department of Manufacturing Technologies and Machine Tools, 9010 Varna, Bulgaria

[a]dm_dimitrov@tu-varna.bg, [b]sdslavov@tu-varna.bg, [c] lubomir.van@tu-varna.bg

Keywords: Ball Burnishing, Surface Hardness, Surface Roughness, Bayesian Approach, Digital Model, Optimization, LLM, Function Calling

Abstract. In this paper, a digital model of the ball burnishing process, developed using data from tests on bronze and aluminum alloys, to create regular micro-reliefs on a cylindrical surface." is presented. The experiments were carried out on a CNC lathe using a cylindrical stock. The input features were burnishing force, feedrate, machine condition and material type, while the results were surface roughness and hardness. A Bayesian approach was used to infer the parameters of the digital model, enabling the creation of distributions that directly account for uncertainty caused to regression coefficients and by random factors incorporated in machine condition feature. The proposed digital model was subsequently used to run simulations and check various optimization algorithms. Additionally, a large language model NexusRaven-V2 that were fine tunned for function calling is combined with the functions from the digital model.

Introduction

Ball burnishing finishing process employs surface plastic deformation of the material, aiming to reduce the machine parts' surface roughness, result in specific regular relief forming, and increase the hardness in the surface layer. Nowadays, using the CNC machines' capabilities, it is possible to achieve programmatically the complex toolpath of a burnishing tool to produce certain patterns onto the shaft's surface called regular reliefs (RR) [1,2]. This type of surface treatment is called "vibratory burnishing" by some researchers [3,4,5,6]. The types of regular reliefs (RR) generated by introducing an additional oscillatory movement of the burnishing tool in the lateral direction of the primary motion are classified in the classical work by Schneider [4]. Particular attention is given to the fourth type of RR, in which the overlapping of sinusoidal traces allows the formation of a cell structure, resembling tetragonal or hexagonal patterns. An overview of this type of finishing operations, including tool path calculation and writing of program code for CNC machines for various types of parts' surfaces, can be found in [2].

Main parameters of ball burnishing operation are:

-the ball tool diameter and burnishing force, from which the applied pressure in the contact zone can be calculated;

- feedrate, giving the plastic strain rate in the surface layer;

- amplitude and the wavelength (frequency) of additionally applied lateral oscillatory movement of the ball tool.

In the context of Industry 4.0 application of digital twin of real industry equipment plays a key role in many industries. Based on the published sources review in the last 10 years, manufacturing industry is among the most popular categories of application of digital twin [7]. The benefits of using digital twin are in area of Production Planning and Control, Predictive Maintenance and

Layout planning, etc. Three types of digital twin approaches can be distinguished today, based on the level of integration between the physical and digital objects [8]. The first type is the digital model (DM), where no automatic data flow occurs between the objects. The second is the digital shadow, which involves a one-way automatic data flow from the physical to the digital object. The final type is the real digital twin, characterized by a fully automatic, bidirectional data flow between both sides.

Building of digital twin of a CNC processing machine include equipping the machine with additional sensors, capable to capture: sound emissions, accelerations, cutting forces, loading electrical currents, etc., and transfer them in real time to the NC-controller. The digital twin model should incorporate artificial intelligence (AI) algorithms to dynamically adjust model parameters and provide real-time feedback to the physical machine. Consequently, the objective of this work is to develop an AI-optimized digital twin model for the burnishing operation, capable of recommending appropriate values for the key burnishing process parameters to achieve specific surface roughness and other desired characteristics in rotary parts.

Methods and Materials

Data collection

Burnishing process is provided on a CNC lathe machine (CT201), see Fig.1. Experiments are carried out by using central composite design with two features (burnishing force F [N], coded as a X_1 and federate f [mm/min], coded as a X_2). Other regime parameters, like ball-tool diameter, the amplitude of the lateral ball tool oscillation; the number of reciprocating movements per minute are fixed to the values, given in the Table 2.

Fig 1. Ball burnishing operation, carried on by CNC lathe.

After burnishing the roughness Rz and hardness HV are measured using Mitutoyo Surf Test 310 roughness tester, and Vickers hardness tester. The response surface in the present case were of quadratic type. The design matrix with four edge points, four axial points and five repetitions in the central point of the experimental plan is shown in Table 1. A diagram of the feature space points coordinates is shown on Fig. 2.

Regime parameters and initial values of material properties are summarized in Table 2. The available results are for the roughness parameter *Rz* and surface hardness HV for two materials (bronze alloy ISO CuAl8Fe3 and Aluminum alloy 2024).

Table 1. Design matrix of the experiment $F=X_1$, $f=X_2$.

№	X_0	X_1	X_2	X_{12}	X_1^2	X_2^2
1	1	-1	-1	+1	+1	+1
2	1	+1	-1	-1	+1	+1
3	1	-1	+1	-1	+1	+1
4	1	+1	+1	+1	+1	+1
5	1	$-\sqrt{2}$	0	0	+2	0
6	1	$\sqrt{2}$	0	0	+2	0
7	1	0	$-\sqrt{2}$	0	0	+2
8	1	0	$\sqrt{2}$	0	0	+2
9	1	0	0	0	0	0
10	1	0	0	0	0	0
11	1	0	0	0	0	0
12	1	0	0	0	0	0
13	1	0	0	0	0	0

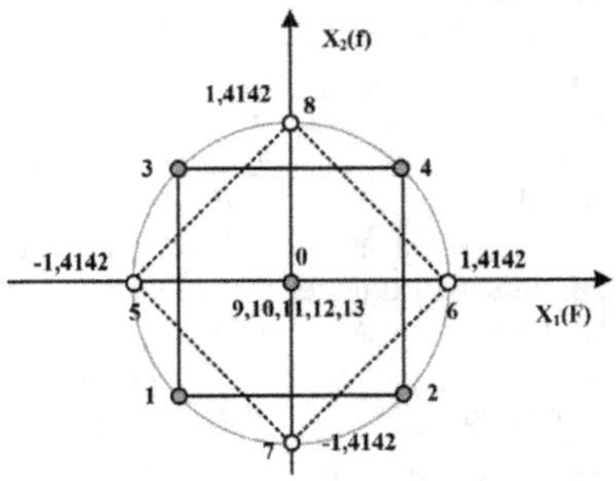

Fig 2. Central composite design

Table 2. Burnishing regime parameters

Materials	Bronze CuAl8Fe3	AA 2024
Burnishing parameters		
Force ,[N]	500 ± 212	
Feedrate, [mm/min]	300 ±109	
Amplitude of oscillation, [mm]	1.5	1
Frequency of oscillation, [min^{-1}]	1100	2200
Ball diameter, [mm]	Ø16	Ø12
Initial values of the material properties		
Roughness Rz [um]	15	8.2
Surface hardness HV	140	-

Building the digital model

To build the digital model of burnishing process the uncertainty should be incorporated. That's why regression is treated in Bayesian way. Resultant y follows a Normal distribution with a mean of $\bar{y} = X\beta$ and a variance of σ^2.

$$y \sim N(X\beta, \sigma^2 I) \tag{1}$$

The regression coefficients and variance are treated as unknown, so conjugate Normal-Inverce Gamma model with noninformative prior distribution should is used (2).

$$p(\beta, \sigma^2) \propto \frac{1}{s^2} , \tag{2}$$

where s-hyperparameter.

The problem for the posterior distributions has a closed form solution:

$$\sigma^2|y \sim Inv\ Gamma\left(\frac{n-k}{2}, s_n{}^2\frac{n-k}{2}\right) \tag{3}$$

$$\beta|\sigma^2, y \sim N(\hat{\beta}, \sigma_n^2 V_\beta) , \tag{4}$$

Materials Research Proceedings 46 (2024) 354-361 https://doi.org/10.21741/9781644903377-46

where $n-k$ – degrees of freedom; $s_n, \sigma_n^2 V_\beta$ – posterior variances

At the end, a digital model is built using posterior predictive distribution, which is *t*-distribution incorporating the uncertainty of the regression coefficients and uncertainty of the model. Block scheme of the digital model is shown on *Fig.3.*

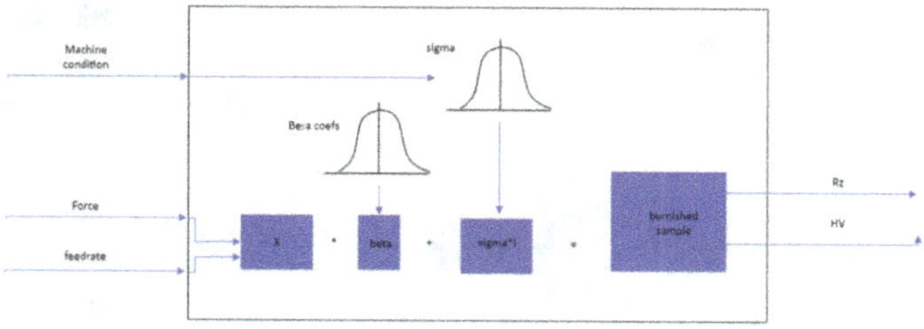

Fig. 3 Digital model of burnishing process

Results and Discussion

Response surfaces

Response surfaces, generated using MLE values of the regression coefficients, are given on Fig.4. Here the roughness Rz and surface hardness HV are converted to relative values $y_{rel} = \frac{y_{burnished}}{y_{initial}}$. In general, the roughness increases with increasing the *feedrate* and for constant *feedrate* reaches a minimum at an intermediate value of the applied *Force*. The hardness increase with increasing the *Force* and decreasing the *feedrate*. Such behavior is typical for the burnishing process in its most varieties [2,3,9,10,11,12].

Optimization

In the most common case burnishing process should ensure target value of one of the properties (Rz , HV). In such case the response surface function should be replaced by objective function and minimization algorithm to be used for calculating optimal burnishing parameters. A possible objective function is squared difference's function:

$$\min_{F,f}\{\bar{y}(F,f) - target\_value\}^2 \quad (5)$$

For solving the optimization problem *minimize* method from the python package *scipy.optimize* is used. Various algorithms are tested (*'Nelder-Mead', 'COBYLA', 'Powell'*). Solved optimization problem is presented on *Fig.5*. Material is bronze alloy, target value is *Rz=4μm* and starting point is varied.

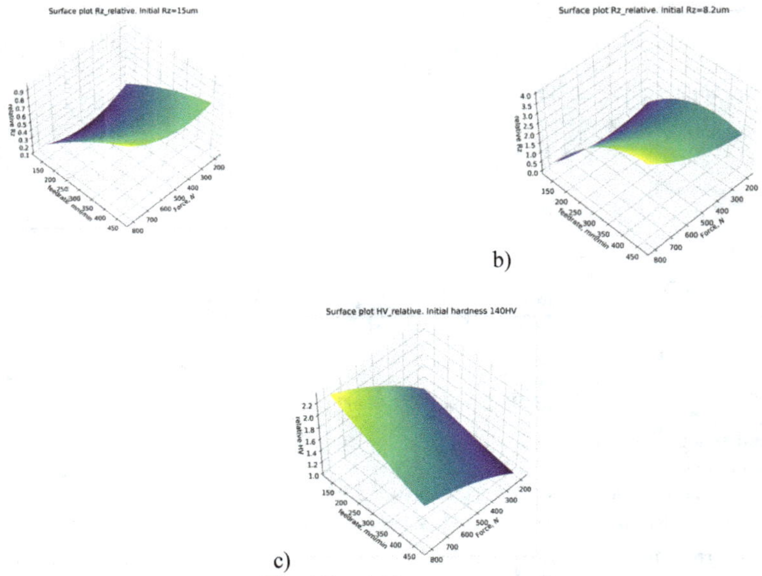

a) b) c)

Fig. 4 Burnishing response surfaces
a) Rz relative (Bronze CuAl8Fe3) b) Rz relative (AA 2024),c) HV relative (Bronze CuAl8Fe3)

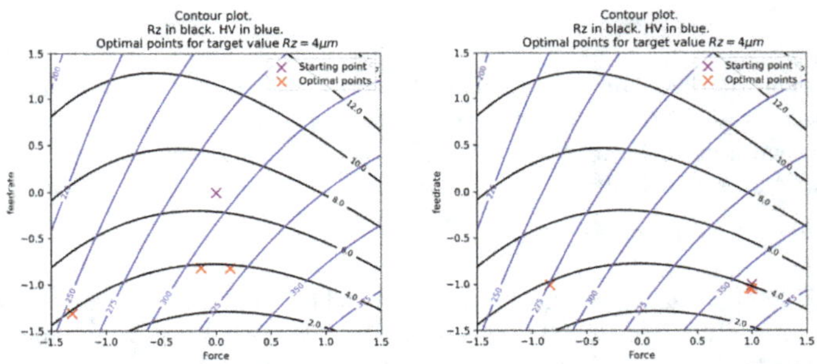

a) x_0=[0,0] b) x_0 =[1,-1]

Fig. 5 Optimization results from various starting points. Target value Rz=4um.

It is seen that all the algorithms capture the desired *Rz* value in various places of the feature space, and global optimum point can be chosen using additional constraints. If the criteria is shortest machining time the point with highest *feedrate* is to be chosen and so on. Created digital model is used to check if the machine condition is enough good to meet the tolerance. For example, 6000 runs are simulated in optimal points with highest f*eedrate* Fig.5a. and the percent of samples out of the tolerance are compared. For ±5.0% tolerance percent of out of tolerance samples is similar for both points, and hardness criteria can be considered for final choice.

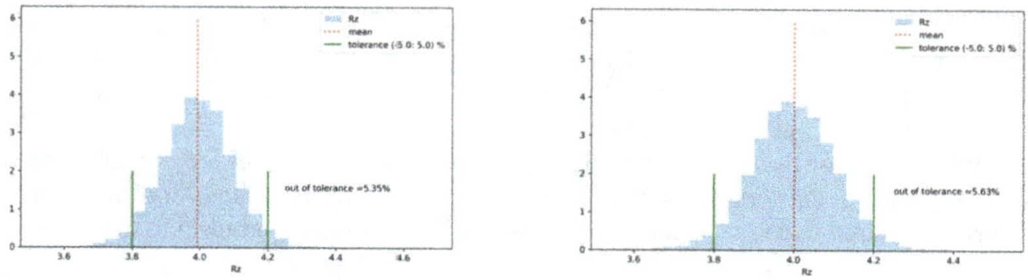

a) F,f = [525.7 N, 210.8 mm/min] b) F,f = [472.7 N, 211.3 mm/min]

Fig. 6. Roughness Rz distribution of 6000 samples burnished with optimal regime parameters

Interaction of digital model with LLM

LLMs are capable to understand and generate human-like text with natural language. The potential for integrating artificial intelligence, in engineering is widely explored now-a-days. For example, in [13] interaction with 2D and 3D open-source software to support design practice is given. The abilities of the AI models for calculation can be customised by learning the external function and extract the inputs of the function in the written text. This ability is called function-calling and represent a significant advancement of Large Language Models (LLMs) [14]. One of the small models fine tunned for function calling is NexusRaven-V2 with 13B parameters. [15].

The database with materials and their burnishing properties is structured in *.json* format. New material can be easily added following the keys in the file. . Function called „burnish" is presenting the digital model. It takes as an input material, *force*, *federate* and machine condition, "burnish" required number of samples and returns the array with Rz, and HV values of each of the samples. Other function called „burnish_results" , takes the results array and returns the mean and standard deviation of the burnished samples. Functions are written with detailed description in a way that makes it capable to be used with Raven model. The sample code is shown on Fig.7. It creates *raven_prompt* from the *function list* and *raven message*. With this prompt *raven_call* is created, and after evaluating it show the results.

```
In [33]: function_list = [burnish, burnish_results]

In [34]: raven_msg = "Hey, can you burnish 10 samples from Bronze_Al8 material." \
         "burnish parameters force and feedrate are x=[0,0]" \
         "show me the mean and standard deviation of the burnished samples from the burnish result dictionary"

In [49]: raven_prompt = build_raven_prompt(function_list, raven_msg)

In [36]: from utils import query_raven
         raven_call = query_raven(raven_prompt)
         print (raven_call)

burnish_results(results=burnish(material_name='Bronze_Al8', x=[0, 0], n_samples=10))

In [37]: eval(raven_call)

Rz 6.68 0.06
HV 231.37 0.74
```

Fig. 7. Sample code representing function-calling abilities of the LLM

Conclusions

Digital model of the ball burnishing process to form regular micro-reliefs on a cylindrical surface is presented. The features are *force*, *feedrate*, machine condition and material type, while the output parameters are roughness parameter Rz and surface hardness HV.

A Bayesian approach was used to infer the parameters of the digital model, enabling the creation of distributions that directly account for uncertainty caused by regression coefficients and by random factors incorporated in machine condition parameter.

The proposed response surfaces were successfully used to solve optimization problem of finding the optimal regime parameters capable to produce samples with desired roughness or hardness, using various optimization algorithms. Digital model is capable to run subsequently burnishing simulations, to form the distribution of the measured roughness of simulated samples and calculate percent of out of tolerance samples.

Additionally, the functions and materials data from the digital model are prepared for usage with large language model NexusRaven-V2, fine tunned for function calling. This incorporates the ability to create a natural language conversation with digital model of burnishing machine.

Data

The data files are available in https://github.com/DMDimitrovJ/Burnishing-Digital-Model

Acknowledgments

This research was funded by the Bulgarian National Science Fund (BNSF), grant number КП-06-H57/6, and the APC was funded by grant contract КП-06-H57/6, entitled "Theoretical and experimental research of models and algorithms for formation and control of specific relief textures on different types of functional surfaces".

References

[1] S.D. Slavov, D.M. Dimitrov, Modelling the dependence between regular reliefs ridges height and the burnishing regime's parameters for 2024 aluminum alloy processed by using CNC-lathe machine, IOP Conference Series: Materials Science and Engineering, 1037 (2021) 012016. https://doi.org/10.1088/1757-899X/1037/1/012016

[2] S.D. Slavov, D.M. Dimitrov, M.I. Konsulova-Bakalova, Advances in burnishing technology, in: Advanced Machining and Finishing, Elsevier, Amsterdam, The Netherlands, 2021, pp. 481-525. https://doi.org/10.1016/B978-0-12-817452-4.00002-6

[3] G. Nagit, O. Dodun, L. Slatineanu, M. Ripanu, A. Mihalache, A. Hrituc, Influence of some process input factors on the main dimensions of the grooves generated during vibroburnishing, IOP Conference Series: Materials Science and Engineering, 968 (2020) 012007. https://doi.org/10.1088/1757-899X/968/1/012007

[4] Yu.G. Schneider, Operational properties of parts with regular microrelief, IVA Publishing, St. Petersburg, 2001. ISBN 5-7577-0166-8.

[5] S.S. Pande, S.M. Patel, Investigations on vibratory burnishing process, Int. J. Mach. Tool Des. Res. 24 (1984) 195-206. https://doi.org/10.1016/0020-7357(84)90004-0

[6] V. Dzyura, P. Maruschak, O. Prentkovskis, Determining optimal parameters of regular microrelief formed on the end surfaces of rotary bodies, Algorithms 14 (2021) 46. https://doi.org/10.3390/a14020046

[7] E.M. Frazzon, A. Albrecht, P.A. Hurtado, Simulation-based optimization for the integrated scheduling of production and logistic systems, IFAC-PapersOnLine 49 (2016) 1050-1055. https://doi.org/10.1016/j.ifacol.2016.07.581

[8] M. Kritzinger, G. Karner, J. Traar, J. Henjes, W. Sihn, Digital Twin in manufacturing: A categorical literature review and classification, IFAC-PapersOnLine 51 (2018) 1016-1022. https://doi.org/10.1016/j.ifacol.2018.08.474

[9] E. Velázquez-Corral, J. Llumà, R. Jerez-Mesa, V. Wagner, G. Dessein, J.A. Travieso-Rodriguez, Fatigue enhancement and hardening effect through ultrasonic vibration-assisted ball-burnishing process on AISI 1045 steel, Fatigue Fract. Eng. Mater. Struct. 47 (2024) 203-219. https://doi.org/10.1111/ffe.14180

[10] C. Felhő, F. Tesfom, G. Varga, ANOVA Analysis and L9 Taguchi Design for Examination of Flat Slide Burnishing of Unalloyed Structural Carbon Steel, J. Manuf. Mater. Process. 7 (2023) 136. https://doi.org/10.3390/jmmp7040136

[11] H. Yamazaki, J. Zhu, T. Tanaka, Study on the Surface Enhancement of Thin-Walled Metallic Materials Using a Novel Double-Side Burnishing Tool, Int. J. Autom. Technol. 17 (2023) 458-468. https://doi.org/10.20965/ijat.2023.p0458

[12] P. Pawlus, W. Koszela, R. Reizer, Surface texturing of cylinder liners: A review, Materials 15 (2022) 8629. https://doi.org/10.3390/ma15238629

[13] T. Dovramadjiev, D. Dobreva,T. Murzova,M. Murzova, V. Markov, I. Iliev, K. Cankova, G. Jecheva, G. Staneva, Interaction Between Artificial Intelligence, 2D and 3D Open Source Software, and Additive Technologies for the Needs of Design Practice, InWorld Conference on Information Systems for Business Management vol.834 pp. 339-350 (2023). Singapore: Springer Nature Singapore. https://doi.org/10.1007/978-981-99-8349-0_26

[14] Z. Liu, T. Hoang, J. Zhang, M. Zhu, T. Lan, S. Kokane, J. Tan, W. Yao, Z. Liu, Y. Feng, R. Murthy, Apigen: Automated pipeline for generating verifiable and diverse function-calling datasets, arXiv preprint arXiv:2406.18518, 2024.

[15] V.K. Srinivasan, Z. Dong, B. Zhu, B. Yu, D. Mosk-Aoyama, K. Keutzer, J. Jiao, J. Zhang, Nexusraven: A commercially-permissive language model for function calling, in: NeurIPS 2023 Foundation Models for Decision Making Workshop, 2023.

Design and optimization of the centrifugal pump impeller for wastewater

Viorel BOSTAN[1,a], Andrei PETCO[2,3,b*]

[1]Faculty of Computers, Informatics and Microelectronics, Technical University of Moldova, Chisinau, Republic of Moldova

[2]Faculty of Mechanical, Industrial and Transport Engineering, Technical University of Moldova, Chisinau, Republic of Moldova

[3]S.R.L. "CRIS", Chisinau, Republic of Moldova

[a]viorel.bostan@adm.utm.md, [b]andrei.petco@tcm.utm.md

Keywords: Wastewater Centrifugal Pumps, Centrifugal Pump Impeller, Geometry Optimization, Computational Fluid Dynamics

Abstract In the given paper the optimization process of centrifugal wastewater pump impeller is presented. The initial geometrical model was designed in CFturbo software. The actual optimization process was carried out in ANSYS environment. The geometric model was realized in DesignModeler, with the application of Blade Editor tools and discretized by applying ANSYS Mesher. The calculations of the flow characteristics were performed in CFX and the application of the optimization algorithms was performed in optiSLang. Evolutionary Algorithm and NLPQL algorithm were compared as optimization algorithms. The best result was achieved by Evolutionary Algorithm, the application of which resulted in an impeller with an efficiency of 78.5, which satisfies the task imposed in the specification.

Introduction

Wastewater pumps are a type of specialized centrifugal pumps used for pumping wastewater, including liquids with various types of inclusions. The principle of operation of a wastewater centrifugal pump is similar to that of conventional centrifugal pumps, but due to the fact that wastewater contains solid residues, which can cause impeller blockage of the pump, the impeller geometry of the centrifugal pump must be modified to increase the flow capacity of the impeller [1]. The selection of an optimal pump impeller geometry is essential to ensure a reliable and efficient operation. In view of accelerating urbanization trends and increasing efforts of the global community environmental protection, optimizing the components of wastewater pumps to enhance its characteristics including energy efficiency is becoming paramount for centrifugal pump manufacturers [2].

Computational Fluid Dynamics (CFD) has a decisive role in the optimization of pump components, which has already become an industry standard for optimization problems of pumps' working parts of various types [3,4], including centrifugal ones [5]. At the same time, it should be mentioned that the application of an iterative optimization process, which is increasingly performed based on optimization algorithms and coupled with Computational Fluid Dynamics (CFD) methods [6-8], often eliminates physical experiment-based or mixed models in industrial applications [9-11], due to low cost and implementation time.

Establishing the initial geometric model

In the given study the following task was set: To create the working parts of a pump for pumping wastewater with the characteristics shown in Table 1. As an optimization criterion was selected the maximization of pump efficiency, with the desired pump efficiency proposed in the specification of $\eta p = 69\%$. As a restriction, the diameter of particles that can pass through the impeller dp = 80 mm and maintaining head at Best Efficiency Point BEP of H=10 mH2O was selected.

Table 1. Pumps parameters

Parameters	Value	Unit
Impeller rotational speed, n	1800	[min^{-1}]
Head at BEP, H	10	[m]
Volume flow rate, Q	110	[m^3 /h]
Maximum particle size, dp	80	[mm]
Specific speed	59	

It should be noted that although there are wastewater pumps with impellers with a similar number of blades to classic impellers [12], often the number of blades is reduced to allow large particles to pass through the flow zone [2,].

In given study, the possibility of application of wastewater impellers with 1÷3 blades, with an increased wrap angle θ, was verified. The application of this type of impeller, with reduced number of blades, as well with an increased wrap angle θ allows pumping of polluted liquids with large particles [14].

For the purpose of obtaining the initial geometric model the CFturbo working parts design software was selected. Obtaining the flow area geometry automatically is difficult to achieve due to the optimization criterion related to the maximum particle diameter. After testing, it was found that the single blade pump impeller is the only one, which allows to operate the pump at the prescribed hydraulic parameters and to pass through the impeller particles with maximum diameter dp = 80 mm (Fig.1). It should also be mentioned that the pump volute was also realized in CFturbo environment. The modeling of the pump body volute was performed according to the Stepanoff model (Stepanoff number ks = 0.327). As the shape of the volute section, the circular section was selected.

From the study we can see that although the flow parameters, namely the pumping head of the pump is approximately equal to the preset H = 9.38 mH2O, the pump efficiency does not exceed ηp = 56.6%. So, the decision was made to apply a process of the wastewater pump impeller optimization.

The optimization scheme is depicted in Fig.2. The steps of the optimization process are shown below.

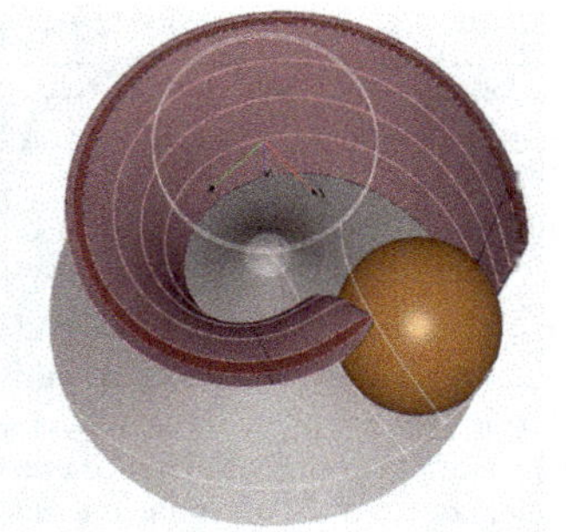

Fig. 1. Maximum particle diameter determination realized in CFturbo

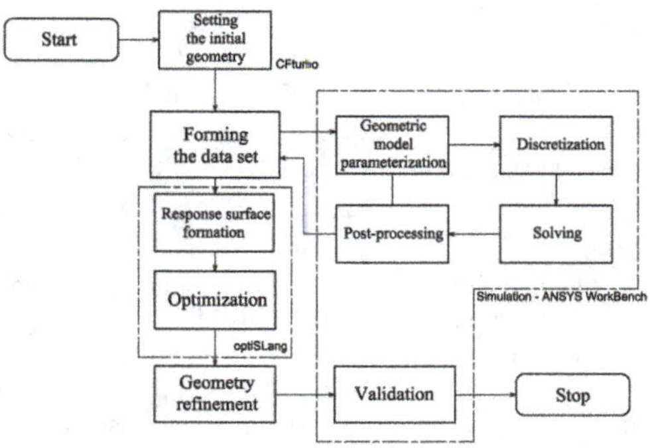

Fig. 2. Single blade impeller optimization scheme

Obtaining the geometric model

ANSYS DesignModeler was used to obtain the geometric model. The blade geometry and the hub and shroud surfaces were optimized. ANSYS BladeEditor was applied to obtain the parameterized geometric model of the impeller. In Fig. 3 (a) the geometry of the optimized impeller is shown.

The blade geometry was parameterized (Fig. 3 (b)) by varying the angle β of the blade at 3 points n1...n3, with the blade thickness distribution from 9 mm at the inlet to 7 mm at the outlet. The impeller of the wastewater pump contains a single blade, has variable curvature contained between their Leading Edge LE and Trailing Edge TE inscribed in the wrap angle $\theta=\sum\Delta\theta_i$ (Fig. 3 (b)), located on the inner diameter D_1 and outer diameter D_2. The shape of the impeller blades is represented by the line of curvature, which passes through the points n1...n3 determined by the angular coordinates $\Delta\theta_2$, $\Delta\theta_3$ and $\Delta\theta_4$ and the blade angles β_1, β_2, β_3 calculated by the software, respectively.

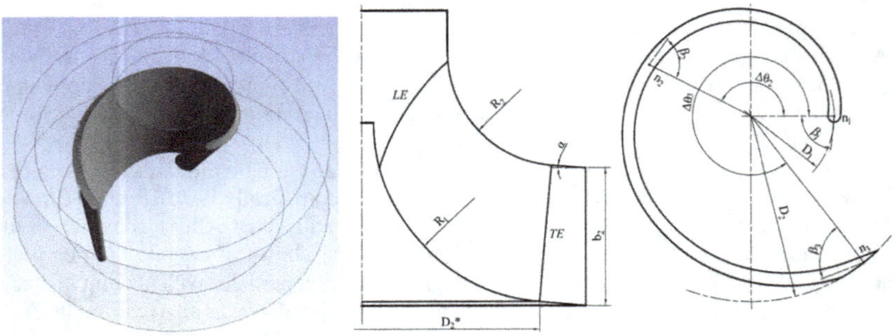

(a) (b)

Fig. 3. The geometric model: a) Model realization in Ansys DM,
b) Single blade wastewater impeller parameterization model.

Geometric model discretization

The ANSYS TurboGrid module does not support grid generation for single-blade impellers, also the application of the ANSYS ICEM module [13,15], which can be applied to obtain a structured grid, was not relevant due to the problems related to the mesh rebuild at each cycle of the iteration process. Due to this, the decision was made to apply the unstructured mesh generated (Table 2) in the ANSYS Mesher module (Fig. 4 (a)).

Table 2. Grid settings used in the study

Calculation type	Max. finite volume size	Thickness of inflation layer / nr. of finite vol.	Nr. of finite volumes
Optimization cycle calculations	$\Delta S = 2$ [mm]	2 [mm] / 28	$1.5 - 2 \cdot 10^6$
Validation calculations	$\Delta S = 1$ [mm]	1 [mm] / 28	$14 \cdot 10^6$

Setting initial and boundary conditions

At the Inlet in the impeller, the flow is all liquid water, with low turbulence (1%) and total pressure $P_{inlet} = 10^6$ Pa. At the Outlet, the flowrate $Q_{out} = 31$ kg/s is applied. As the flow characteristics associated with turbulence and cavitation at the outlet of the system are given Zero gradient, i.e. calculated by the solver, a speed of 1800 min^{-1} is applied to the impeller. The reference pressure is indicated equal to zero (Pref = 0 atm) at a constant fluid temperature of 25°C. In the initial state, the fluid in the domain consists of water in liquid state.

Wall-type boundary conditions apply to walls with no slip specifications at zero velocity in the immediate vicinity of the wall. The Physical Timescale option was chosen as the timing control model, equal to 0.0053s. Frozen rotor, in turn, was chosen as the inter-domain interaction interface.

Materials Research Proceedings 46 (2024) 361-368 https://doi.org/10.21741/9781644903377-47

Processing and post-processing stage

The iterative process is stopped either after reaching 400 computational iterations or after reaching a residual error tolerance of 10^{-5} (Fig. 4(b)). It should also be noted that the value of the parameter y^+ is in the range of units.

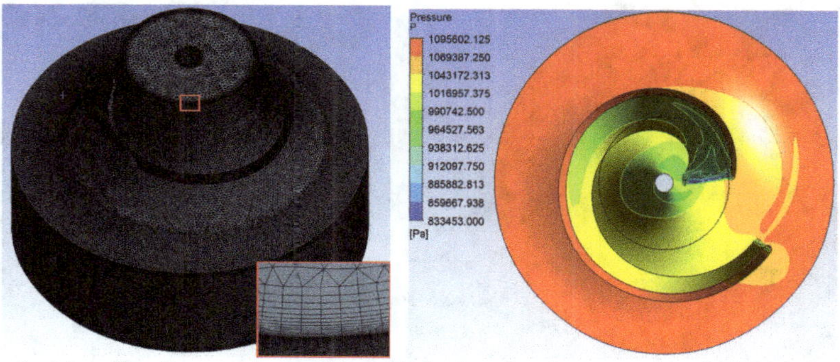

Fig. 4. a) Impeller discretization grid (at optimization phase) generated in ANSYS Mesher, b) Pressure field distribution in the pump impeller (on the blade and hub surface)

Setting the optimization process

The actual optimization process was performed in ANSYS optiSLang. The parameter variation limits are represented in Table 3.

Table 3. Limits of variation of optimization parameters

Impeller diameter D [mm]	Width of the impeller canal output b_2 [mm]	Hub radius R_1 [mm]	Shroud radius R_2 [mm]
140 ÷ 220	80 ÷ 105	90 ÷ 120	50 ÷ 80
Position p.n_2 M, % M	Wrap angle $\Delta\theta_2$ °	Wrap angle $\Delta\theta_3$ °	Shroud angle α °
25 ÷ 75	50 ÷ 250	135 ÷ 330	165÷180

Results analysis

Sampling yielded 600 geometric models of the wastewater pump impeller. Due to the application of large variation limits, an increased number of incorrect geometries were rejected. A total of 168 successful pump impeller geometrical patterns were obtained. The simulation results were stored in a dataset.

Fig. 5 shows the response surface obtained from the simulation results. Linear regression was applied to obtain the response surface.

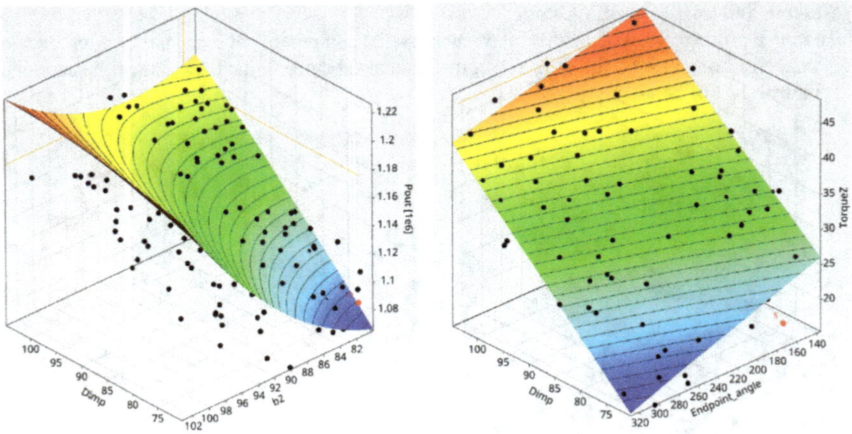

(a) (b)

Fig. 5. Response surface: (a) Influence of impeller diameter and blade width on the pumping head, (b) Influence of impeller diameter and wrap angle θ3 on the impeller torque

In order to obtain the optimal geometry, EA (Evolutionary algorithm) and NLPQL (Non-Linear Programming by Quadratic Lagrangian) based algorithms were applied to the response surface. It can be noted that in the case of single-blade impeller optimization, the algorithms used can provide optimization of the geometry with a given number and types of geometric parameters. The results were compared and the best result was shown by EA.

From the Coefficient of Performance matrix (Fig. 6) it can be seen that the greatest influence on the pumping head and torque has the diameter D of the impeller. The pumping head is also influenced by the width b2 of the impeller channel. We can also observe that influence on the result have 6 out of 8 optimization parameters. We can observe that the influence of the geometrical parameters of the hub radius R1 and the α angle of the shroud can be neglected.

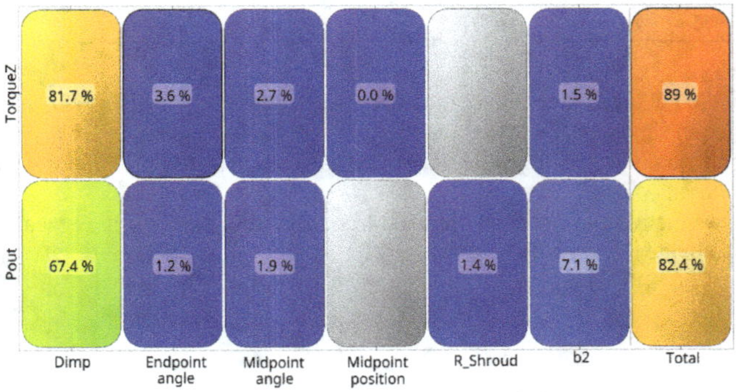

Fig. 6. Coefficient of Performance (COP) matrix

The geometrical parameters of the optimized impeller were slightly modified in terms of technological and design requirements. The final impeller's parameters are presented in Table 4.

Materials Research Proceedings 46 (2024) 361-368 https://doi.org/10.21741/9781644903377-47

Table 4. Geometric parameters of the obtained impeller

Impeller diameter D, [mm]	Width of impeller channel at outlet b_2 , [mm]	Hub radius R_1 , [mm]	Shroud radius R_2 , [mm]
160	81,5	94,5	63,2
Position p.n_2 M% M	Wrap angle $\Delta\theta_2$ °	Wrap angle $\Delta\theta_3$ °	Shroud angle α °
59	59	280	180

Validation of the optimized pump's geometric model is shown in Fig. 7. The pump efficiency increased from η = 56.6% (the efficiency of the initial geometric model obtained in CFturbo) to η = 71.9% (the efficiency of the optimized pump).

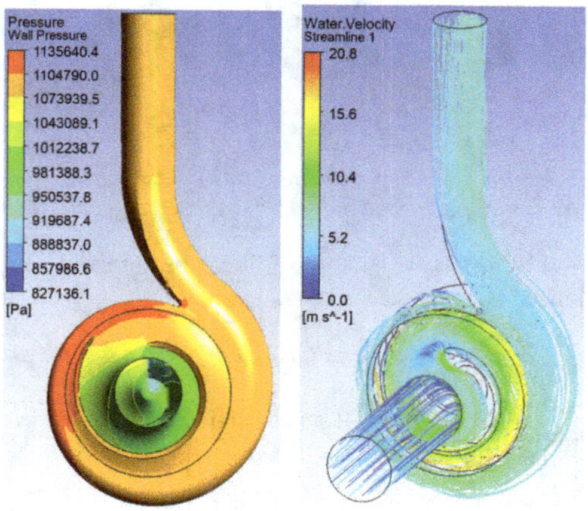

(a) *(b)*

Fig. 7 Modeling results of fluid flow in the optimized pump
(a) Pressure field distribution (b) Current lines, with indication of speed

Fig. 8 shows the part of the flow area where the velocity is less than 1.5 m/s, it represents the flow separation area. For the reference impeller, obtained in CFturbo, a large flow separation area was formed along the inner wall of the impeller blade, as shown in Fig. 8(a). In contrast, the flow separation area was significantly reduced in the flow area of the optimized impeller, as shown in Fig. 8(b). We can also observe a relatively reduced flux intensity in the optimized impeller Fig. 9.

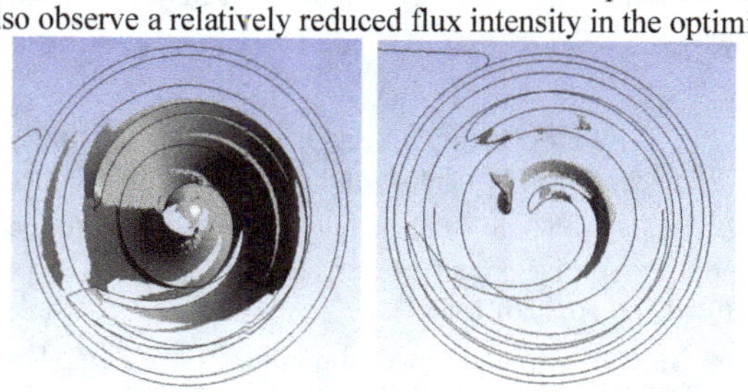

(a) *(b)*

Fig. 8 Impeller area with flow velocity below 1.5m/s
(a) initial geometric model, (b) optimized geometric model

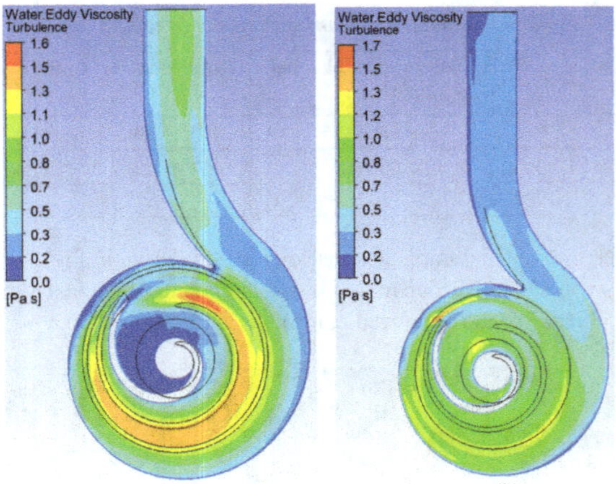

(a) *(b)*

Fig. 9. Water eddy viscosity
(a) initial geometric model, (b) optimized geometric model

The optimization results are shown in Table 5. It can be observed a significant increase in the efficiency of the optimized pump at the set head.

Table 5. Comparison of the initial and optimized pump's parameters

Impeller geometry	Head, [m]	Torque, [Nm]	Impeller efficiency	Pump efficiency
Originally	9,38	24,1	0,629	0,566
Optimized	10,7	22	0,785	0,707

Conclusion

The geometrical model of a single blade centrifugal wastewater pump impeller was obtained. Following the optimization process, we can observe an increase in the hydraulic efficiency of the centrifugal pump's impeller, compared to the initial geometrical model obtained in CFturbo, from 62.9% to 78.5% or in the pump efficiency from 56.6% to 70.7%. This is due to the optimization of the impeller surfaces, which led to the decrease of the hydraulic resistance of the impeller and as a consequence it increased the flow velocity in the flow area of the impeller and decreased the turbulent flow intensity.

References

[1] T. Yonar, Wastewater and Water Quality, IntechOpen, London ,2018. https://doi.org/10.5772/intechopen.71219. https://doi.org/10.5772/intechopen.71219.

[2] J. Chen, W. Shi, D. Zhang, Influence of blade inlet angle on the performance of a single blade centrifugal pump, Eng. Appl. Comput. Fluid Mech. (2021) 462-475. https://doi.org/10.1080/19942060.2020.1868341

[3] I. Malael, F. Costea, M. Draghici, Flow Evaluation of the Lobe Pump Using Numerical Methods, (2019) 301-309. https://doi.org/10.1007/978-3-030-21507-1_42

[4] S.-M. Lee, C.-M. Jang, Development of the microbubble pump having 'S-shape' impeller using experimental and numerical analysis, J. Phys. Conf. Ser. 2217 (2022) 012004. https://doi.org/10.1088/1742-6596/2217/1/012004

[5] V. Bostan, A. Petco, Determining optimal simulation settings for the centrifugal pump parts optimization process, J. Eng. Sci. (2023) 8-22. https://doi.org/10.52326/jes.utm.2023.30(2).01.

[6] Y. Zheng, L. Meng, G. Zhang, P. Xue, X. Wang, C. Zhang, Y. Tian, Study on Impeller Optimization and Operation Method of Variable Speed Centrifugal Pump with Large Flow and Wide Head Variation, Water 16 (2024) 812. https://doi.org/10.3390/w16060812

[7] R. Ramirez, E. Avila, L. Lopez, A. Bula, J. Duarte Forero, CFD characterization and optimization of the cavitation phenomenon in dredging centrifugal pumps, Alex. Eng. J. 59 (2020) 291-309. https://doi.org/10.1016/j.aej.2019.12.041.

[8] I. Malael, G.B. Gherman, I. Porumbel, Numerical investigation of a new LH2 centrifugal pump concept used in space propulsion, INCAS Bull. 10 (2018) 65-74. https://doi.org/10.13111/2066-8201.2018.10.2.7

[9] L. Tan, W. Shi, D. Zhang, C. Wang, L. Zhou, E. Mahmoud, Numerical and experimental investigations on the hydrodynamic radial force of single-channel pumps, J. Mech. Mech. Sci. Technol. 32 (2018) 4571-4581. https://doi.org/10.1007/s12206-018-0903-4

[10] B.D. Souza, A. Niven, J. Daly, Single-blade impeller development using the design of experiments method in combination with numerical simulation, Proc. Inst. Mech. Eng. Part E J. Process Mech. Eng. (2008). https://doi.org/10.1243/09544089JPME218

[11] Pei, J., Yuan, S. & Yuan, J. Numerical analysis of periodic flow unsteadiness in a single-blade centrifugal pump. Sci. China Technol. Sci. 56 (2013) 212–221. https://doi.org/10.1007/s11431-012-5044-x

[12] E. Pagayona, J. Honra, Multi-Criteria Response Surface Optimization of Centrifugal Pump Performance Using CFD for Wastewater Application, Modelling 5 (2024) 673-693. https://doi.org/10.3390/modelling5030036.

[13] F. Benra, H. Dohmen, Investigation on the Time-Variant Flow in a Single-Blade Centrifugal Pump, in:2007. https://www.semanticscholar.org/paper/Investigation-on-the-Time-Variant-Flow-in-a-Pump-Benra-Dohmen/eda20b5979616d709ac4eeb3c0c8645681e8a55a (accessed September 14, 2024).

[14] Yuan S., Huang Q., Zhang J., Zhang X. The effect of blade outlet angle on the performance of chemical centrifugal pumps. J. Drain. Irrig. Mach. Eng. (2019), 37:185–191

[15] J.-H. Kim, B.-M. Cho, Y.-S. Kim, Y.-S. Choi, K.-Y. Kim, J.-H. Kim, Y. Cho, Optimization of a Single-Channel Pump Impeller for Wastewater Treatment, Int. J. Fluid Mach. Syst. 9 (2016) 370-381. https://doi.org/10.5293/IJFMS.2016.9.4.370.

Materials Research Proceedings 46 (2024) 370-384 https://doi.org/10.21741/9781644903377-48

Axiomatic design theory as a design thinking tool for mastering industrial process variables inventorying

Vasile MERTICARU[1,a *], Marius-Andrei MIHALACHE[1,b], Marius-Ionut RÎPANU[1,c], Eugen MERTICARU[2,d], Bogdan RUSU[3,e] and Vasile ERMOLAI[1,f]

[1]"Gheorghe Asachi" Technical University of Iași, Department of Machine Manufacturing Technology, Blvd. Dimitrie Mangeron 59A, Iași, Romania

[2]"Gheorghe Asachi" Technical University of Iași, Department of Mechanical Engineering, Mechatronics and Robotics, Blvd. Dimitrie Mangeron 43, Iași, Romania

[3]"Gheorghe Asachi" Technical University of Iași, Department of Engineering and Management, Blvd. Dimitrie Mangeron 29, Iași, Romania

[a]vasile.merticaru@academic.tuiasi.ro, [b]marius-andrei.mihalache@academic.tuiasi.ro, [c]marius-ionut.ripanu@academic.tuiasi.ro, [d]eugen.merticaru@academic.tuiasi.ro, [e]bogdan.rusu@academic.tuiasi.ro, [f]ermolai.vasile@gmail.com

Keywords: Design Thinking, Axiomatic Design, Process Variables Inventory, Design for Sustainability, Design for Resilience, Green Production, Wood Pellets Micro-Production, Energetic Willow Crop

Abstract. Activities within LCSA approaches targeting Sustainable Product, Process or Project Development, and eventually Societal Resilience usually involve multidisciplinary complexity in inventorying process variables in terms of defining materials, processes, energy consumptions, different categories of costs, or social factors. As long as Design Thinking is gaining field as a general philosophy in many kinds of businesses, for dealing with the more complex problems of design management, the particular idea of the presented research is to consider Axiomatic Design theory as a valuable tool for mastering industrial process variables inventorying. The included case study approaches the problem of analytically inventorying the process variables for sustainably developing a project for a wood pellets micro-production activity supplied from willow crops or alternative biomass resources.

Introduction

After experiencing global pandemics, regional climate disasters, and regional conflictual escalations, communities and scholars deeply turned toward the problem of *Societal Resilience*, as an adaption capacity after disturbances or adversities, both organizational and community levels being targeted. Norris et al. [1] discuss the importance of dealing with complexity within the theory of resilience which comprises reformulated understandings of stress, adaptation, wellness, and, of course, resource dynamics. Inventory, hierarchizing, and quantifying complex sets of adaptive capacities, defined as resources with dynamic attributes, which can be identified within areas like Community Competence, Economic Development, Social Capital, Information and Communication, are complex tasks to tackle.

Under the conditions of continuous features reshaping and boundaries changing for *Sustainability* as an increasingly recognized or even an essential concept, as Ren et al. [2] remarks for multiple uses and manifestations in producing, tracing, and trusting various *green* industrial products, place brands, tourism or retail market products, development policies, and also corporate or educational profiles, responsibly investing public and private funds, for maintaining a balanced triple bottom line in so many and different kinds of businesses, ambiguities are to be avoided, and complexity is to be mastered.

Materials Research Proceedings 46 (2024) 370-384 https://doi.org/10.21741/9781644903377-48

On the other hand, *Design Thinking* (DT) is gaining field as a general philosophy involved in many kinds of businesses, the concept being comprehensively defined by Çeviker-Çınar et al. [3] as referring to various sets of cognitive/applicative methods and eventually patterns employed by the designers, as tools defining their ways to think in an innovative, emphatic and holistic manner, for dealing with the more and more complex problems of design management. They go further on approaching *Design Thinking* more than as a valuable tool for fostering innovation in the business environment but also as a novel and prominent approach within business education. Rusu [4] proposes customized workshops, constructed on the five phases of Stanford's model: empathize, define, ideate, prototype, and test, for integrating DT in business academic education.

Under the wider umbrella of DT as a general philosophy, several design philosophies, theories, methods, and instruments have been inventoried as tools for solving *New Product Design* (NPD) tasks [5], such as Holistic Design, Integrated Engineering, Functional Analysis, Modular Product Design, Reconfigurable Product Design, Design for Excellence (DfX), etc.

More than two decades have passed since Suh [6] introduced the *Axiomatic Design* (AD) theory and principles [6] and has substantiated it as a general scientific basis for the systematic designing of complex systems [7]. AD is widely recognized today as a valuable tool for approaching the design of systems, software, materials and materials processing, manufacturing systems, and new product development, practically for any product or process to be developed, no matter its complexity [8].

Research conceptual framework substantiation

To understand the complexity of sustainability, Palmberg et al. [9] propose an approach based on systems thinking and focus on a system's interrelated parts. Following an approach grounded in the general systems theory, Czelusniak-Serviss [10] explored corporate sustainability strategies and processes, for achieving competitive advantages in the United States.

The identified current research problem was to explore SMEs (Small and Medium Enterprises) business design management problems within the context of complexity, and need for sustainability and societal resilience.

Multidisciplinary complexity usually involved in project management activities and particularly in project development processes can be successfully tackled by considering the project as a product and its development as a process, the last having to be approached from both engineering and business perspectives.

Approaching PPD (Product and Process Development) as a complex system, respectively as being the *Process* and the associated *Structure* for delivering a *New Product* or *New Process* on the market, it can be represented simplified as in Fig. 1, where NP is the vector of New Product/Process attributes, R is the vector defining the input Resources, and T is the transformation (process) matrix. Actually issues involved by such an approach are not so simple, questions arising about R or T matrices highlighting the need to deal with complexity.

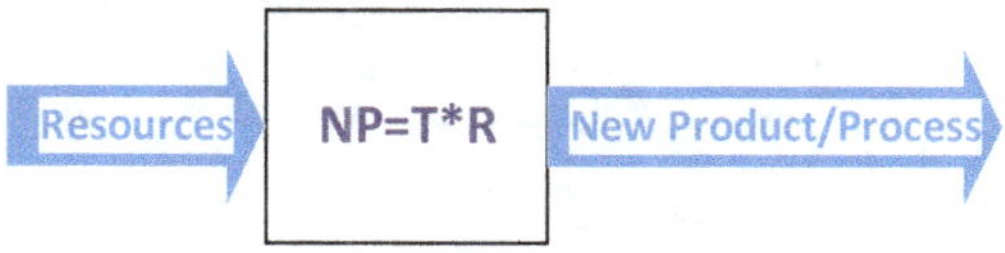

Fig. 1. PPD as system

Within PPD from both engineering and business perspectives, the new product or process has to be seen as merchandise, and the classic approach of the Industrial Design process proposed by

Kressy [11] can be translated and extended to PPD as in the model from Fig. 2, even considering the specific circular economy loops.

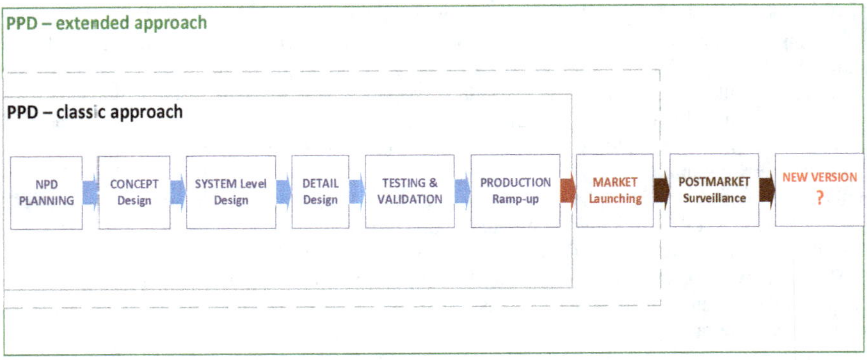

Fig. 2. PPD extended approach model

New Product Development (NPD) involves design levels from general to detail. AD is often considered a tool useful in facilitating innovative and holistic thinking within the Concept Design stage, for an optimized selection of product concept solutions among various sets of others. Pista et al. [12] propose enhancing the effectiveness and efficiency of design activities through the embedment of AD and DfX /DFMA (Design for Manufacturing and Assembly) principles within the optimization of assembled products and assembly processes. They recommend AD application in the Concept Design phase to establish, in a sequential way, the customer's requests, functional requirements, design structural parameters, and process variables. For the System-Level Design stage, they approached several DFMA principles, such as Functional Analysis, Reconfigurable Structure (Multiple Functionality), Reduced Manufacturing, Transport and Manipulation Operations, Maximum Standardization, Part Count minimizing, Joining Efficiency, Top-Down Assembly, Assembly Modularity, etc.

Actually, scholars agree that AD can be used in solving any level of product or process design, starting from the Concept Design, getting through System Level Design, and up to the Detailed Design phase, as all of this depends on the in-depth decomposition approach, the essential aspect for effectively applying AD theory and principles in New Product Development (NPD) is to think an appropriate decomposition process, as it has been shown by Brown [13].

Further on, concerning the Project Development process, Project Planning can be considered to refer to the *Design* and *Execution* of a PDP (Project Development Plan). The particular idea of the presented research approach is to consider Project Management activities in relation to the design management principles and philosophies and to investigate the adequacy of AD theory as a tool for streamlining a project development process, respectively for workflow optimization in terms of effectiveness and efficiency.

On the other hand, business process improvement requires *Process Streamlining* (PS), which involves trimming of waste and excess, least resistance to progress, and performance acquired with minimum effort, as Hall [14] said. JM Assoc. [15] consider that PS is about *Performance Improvement*, which is achieved by improving the individual steps of the process. Streamlining is nominated as one of the six steps within Business Process Improvement (BPI) [14], which are: Setting the Stage, Organising, Flowcharting, Streamlining, Understanding Process Characteristics, and Feeding Back. As goals for PS, the following can be nominated [15]:

- Time to complete reducing (each step and overall process)
- Complexity reducing
- Bureaucracy reducing
- Adaptability increasing
- Client expectations meeting

As a conclusion that arises from the above considerations, PS can be seen as workflow optimization for achieving *Effectiveness* and *Efficiency*, which can be mathematically expressed in Eq. 1 and Eq. 2.

$$Efficiency = \frac{Effect}{Effort} \tag{1}$$

$$Effectiveness = \text{MTR} \tag{2}$$

where MTR = Measure of Target Reaching = DGA (Degree of Goals Accomplishment) and DGA = $Degree\ of\ Goals\ Accomplishment$
The arising problem is how to find a way to quantify MTR. One can propose the solution to consider:

$$\text{MTR} = Benefit\ from\ Objectives\ Accomplishment \tag{3}$$

$$Benefit = Intrinsic\ Effect = \text{VPE} - \text{VNE} \tag{4}$$

where: VPE = Value of Positive Effects (*Gains*); VNE = Value of Negative Effects (*Losses*)
The important idea is to consider *Effectiveness* and *Efficiency* together, not separate, and also how the *Perceived Value* is defined becomes essential.

Engelhardt & Nordlund [16], discuss the use of AD as tool for strategisc planning, in this situation, for instance in developing a technology strategy plan or a business plan, after identifying *Customer Needs* the AD specific domain of *Functional Requirements* (FRs) turns into the domain of *Goals*, the AD specific domain of *Design Parameters* (DPs) turns into the domain of *Strategies* and the AD specific domain of *Process Variables* (PVs) turns into the domain of *Activities.* The zigzaging process for mapping the acceptable solutions will be done between these new domains. They recommend the acceptance of coupling between the technology strategy and market strategy, in the case of a business strategy, if no better uncoupled solution is available.

The problem of algorithmic way versus AD in project management has also been a concerning for many scholars. Steward & Tate [17] explored the integration of AD into the process of project planning and task assignment for software development teams, for dealing with the complexity of involving numerous and different programmers, consultants and support personnel, quality assurance protocols, various supplemental technologies, etc.

It is well-known that a strategy improves the *competitive advantages* for a project, no matter the kind of the involved business. A project specific *strategy* encompasses: a technology strategy plan, a business plan, the design of the strategic content, and the strategy process.

The development of a sustainable product, process or project requires research activities and eventually quantified reports obtaining for Life Cycle Sustainability Assessment (LCSA), covering the triple bottom line, respectively Environmental Life Cycle Assessment (e-LCA), Life Cycle Costing (LCC), and Social Life Cycle Assessment (SLCA). As Kloepffer [18] remarks, LCA is already an internationally standardized tool, based on an algorithm that supposes inventorying materials, processes, transportation, and energy consumptions along the stages of product life-cycle. For LCC, the Society of Environmental Toxicology and Chemistry (SETAC) developed a guideline proposed to represent a basis for future standardization, which takes into account the use-phase and end-of-life phase of the product life-cycle and also considers the hidden cost, surpassing the purely economic cost calculation. On the other hand, different qualitative and semi-quantitative approaches are used nowadays as substitutes for a full, quantitative SLCA. In assessing the social impact, the main problem is to relate the social indicators to the functional unit of the product system and to develop a hierarchized selection of so many existing social indicators to a reasonable number that can be managed within the SLCA model.

Case study – mastering the process variables inventorying by AD theory means

The particular idea of the presented research is to consider Axiomatic Design theory as a valuable tool for mastering industrial process variables inventorying. The included case study involves multidisciplinary aspects, from project management, production engineering, agricultural sciences, and renewable energy resources, coming to solve, on a systematic basis, the problem of analytically inventorying the process variables for sustainably developing a project for a wood pellets micro-production activity integrated with self-providing crops of energetic willow.

AD theory and principles were considered to ensure that goals and visions are related to the action plans. An optimal arrangement of the action plan's process was intended to reduce iterations. On the other hand, one of Béhar's 7 Principles of Holistic Design [19] recommends rethinking or reinventing the business model, and the 360-degree design process requires constant improvement and iteration.

Develop a successful project for a sustainable entrepreneurial activity	(DPs) Strategies	Strategy for optimizing the planning workflow and the project plan content	Optimized alignment of the proje design and management to the goals	Project Development strategy	Financing strategy	Project plan design (specification strategy	Sustainability analysis (LCSA) & strategy	Business & Technology strategy	LEAN development	Optimized resources planning
(FRs) Goals		St0	St1	St2	St1.1	St1.2	St1.3	St2.1	St2.2	St2.3
Getting effectiveness and efficiency in sustainable project planning/development process	G0									
Maximize the effectiveness of PP process (DGA) (project success rate)	G1		X							
Maximize the efficiency of PP process (Effects/Efforts)	G2		X	X						
Optimize financing solution	G1.1				X					
Optimize project specific goals and activities specifications	G1.2					X				
Provide project sustainability	G1.3					X	X			
Maximize (Reach target) gains	G2.1					X		X		
Minimize losses	G2.2					X		X	X	
Minimize (keep in limits) resource consumption	G2.3					X		X		X

Fig. 3. Goals-Strategies Design matrix of second-order decomposition

Several steps were developed within the case study, for deploying AD theory. In the first one, as general customer need (CN) / attribute (CA), the following was identified: To develop a successful project for a sustainable entrepreneurial activity. The result of the second-order decomposition process and the *DPs* mapping is shown in the design matrix from Fig. 3. At the system level (design intent), the following were considered:

FR0 (Goal 0): Getting effectiveness and efficiency in sustainable project planning/development process

DP0 (Strategy0): Optimize the workflow and project plan content

Following the above considerations on *Effectiveness* and *Efficiency*, and defining the *FRs* applicable to all project planning problems from the project details to the abstraction of the PD process, the first-order FRs were stipulated as a *max & max* problem:

FR1 – maximize the effectiveness of PP process (DGA) – project success rate

FR2 – maximize the efficiency of the PP process (Effects/Efforts)

The constraints considered were: resource limitations (financial, time deadlines, space, HR, etc.) and performance indicators. The *Input constraints* were those imposed by design specifications, and overall acting, while the *System constraints* were those imposed by the system where the solution must function. All higher-level decisions act as constraints for lower levels.

The problem can derive into a *max & min* problem under the conditions of defining:

Effects = gains or (gains - losses)

Efforts = resource consumption

Therefore, further on the *FRs* can be:

- Maximize effects & minimize efforts (maximize gains, minimize losses, and minimize resource consumption)
- Maximize effects in conditions of keeping the resources consumption at an affordable (available) level
- Reaching an acceptable level for effects (gains) and minimizing the losses and resource consumption

Project Planning involves dealing with multidisciplinary complexity and collaborative efforts. Collaborative work with supplemental technologies, consultants, and support personnel is needed. More than that, rigorous documentation, task assignments for the project development team, Gantt chart, and resource allocations are to be done.

The result of the third-order decomposition process for *FR1* (*Goal1*) and the corresponding *DPs* (*St1*) mapping can be seen in the extended design matrix from Fig. 4. In Fig. 4, there can be observed that the result is a diagonal matrix. The way approached in defining the *FRs*, the decomposition process and the way the mapping for *DPs* was accomplished, all of these under the rules of *FRs* independence and of information minimizing, have been found as essential in obtaining such a result.

Further on, the mapping for *PVs* (*Activities*), by zigzagging between the *DPs* (*Strategies*) and *PVs* domains, was developed. Some relevant characteristics of project developer (entrepreneurial entity) were considered, for example the business characteristics demanded by the financing solution, as dos Santos et al. [20] recommend (e.g. venture capital, project financing from EU funds, etc.). For optimizing the assignments of project sub-components and workflow management, respectively for tasks identification and for matching the available resource (human, equipment, etc.) to those tasks, the principle of aligning the best options to the needs was considered. On the other hand, some degree of identification of dependencies between tasks identified in the project plan are involved by project planning. Timing and order of tasks were therefore important to be consider. More than that, timing the intermediate milestones can provide a mechanism for monitoring progress and adaptive measures can be enabled.

Develop a successful project for a sustainable entrepreneurial activity	(FRs) Goals	G0 Getting effectiveness and efficiency in sustainable project planning/development process	G1 Maximize the effectiveness of PP process (DGA) (project success rate)	G2 Maximize the efficiency of PP process (Effects/Efforts)	G1.1 Optimize financing solution	G1.2 Optimize project specific goals and workflow process specifications	G1.3 Provide project sustainability	G2.1 Maximize (Reach target) gains	G2.2 Minimize losses	G2.3 Minimize (keep in limits) resource consumption	G1.1.1 Get cheapest financing solution	G1.1.2 Maximize financing funds	G1.2.1 Aware and minimize project risks	G1.2.2 Optimize the project tasks assignment & resources allocation	G1.2.3 Optimize workflow, milestones and deadlines	G1.3.1 Improve project environmental impact	G1.3.2 Improve project economic impact	G1.3.3 Improve project social impact
(DPs) Strategies																		
Strategy for optimizing the planning workflow and the project plan content	St0																	
Optimized alignment of the project design and management to the goals	St1		X	X														
Project Development strategy	St2			X														
Financing strategy	St1.1				X													
Project plan design (specifications) strategy	St1.2					X	X	X	X	X								
Sustainability analysis (LCSA) & strategy	St1.3						X											
Business & Technology strategy	St2.1							X	X	X								
LEAN design (development)	St2.2								X									
Optimized resources planning	St2.3									X								
EU/national development funds	St1.1.1										X							
Competitive advantages for funding selection	St1.1.2											X						
Risk management	St1.2.1												X					
Gantt chart	St1.2.2													X				
Workflow and resources management	St1.2.3														X			
Life Cycle (Environmental Impact) Assessment - LCA	St1.3.1															X		
Life Cycle Cost - LCC	St1.3.2																X	
Social Life Cycle Assessment - SLCA	St1.3.3																	X

Fig. 4. Extended Goals-Strategies Design matrix

The result of the second-order decomposition process, by zigzagging between the *DPs* (*Strategies*) and *PVs* (*Activities*) domains and the *PVs* mapping is shown in the design matrix from Fig. 5.

Develop a successful project for a sustainable entrepreneurial activity	(PVs) Activities	Apply strategy for design & development of the project	Design the project plan	Define project development variables	Apply financing strategy	Define project plan content	Apply sustainability analysis	Define specifications for bussiness & technology processes	Apply LEAN design techniques in project	Implement resources planning tools
(DPs) Strategies		PV0	PV1	PV2	PV1.1	PV1.2	PV1.3	PV2.1	PV2.2	PV2.3
Strategy for optimizing the planning workflow and the project plan content	St0									
Optimized alignment of the project design and management to the goals	St1		X							
Project Development strategy	St2			X						
Financing strategy	St1.1				X					
Project plan design (specifications) strategy	St1.2					X				
Sustainability analysis (LCSA) & strategy	St1.3						X			
Business & Technology strategy	St2.1							X		
LEAN design (development)	St2.2								X	
Optimized process resources planning	St2.3									X

Fig. 5. Strategies-Activities Design matrix of second-order decomposition

The result of the third-order decomposition process for *DP1,2* (*St1,2*) and the corresponding *PVs* (*Activities*) mapping can be seen in the extracts of the extended design matrix from Fig. 6 and Fig. 7.

Develop a successful project for a sustainable entrepreneurial activity	(PVs) Activities	Apply for NPRD sub-measure 6.2	Define competitive advantages for funding selection	Identity & evaluate risks and define contro and prevention	Define activity tasks and resources	Refine project plan	Evaluate and improve the project's environmental impact	Evaluate and improve the project's economic impact	Evaluate and improve the project's social
(DPs) Strategies		PV1.1.1	PV1.1.2	PV1.2.1	PV1.2.2	PV1.2.3	PV1.3.1	PV1.3.2	PV1.3.3
EU/national development funds	St1.1.1	X							
Competitive advantages for funding selection	St1.1.2		X						
Risk management	St1.2.1			X					
Gantt chart	St1.2.2				X				
Workflow and resources management	St1.2.3					X			
Life Cycle (Environmental Impact) Assessment - LCA (E-LCA)	St1.3.1						X		
Life Cycle Cost - LCC	St1.3.2						X	X	
Social Life Cycle Assessment - SLCA	St1.3.3						X	X	X

Fig. 6. Extract of Strategies-Activities Design matrix of third-order decomposition

Materials Research Forum LLC
https://doi.org/10.21741/9781644903377-48

Develop a successful project for a sustainable entrepreneurial activity	(PVs) Activities	Define bussiness process variables	Define technology process variables	Exploit specific knowledge databases	Define quality system for bussines & technological process	Apply SCM techniques	Implement a simple ERP solution
(DPs) Strategies		PV2.1.1	PV2.1.2	PV2.2.1	PV2.2.2	PV2.3.1	PV2.3.2
Allign bussiness process to the project goals	St2.1.1	X					
Align the best technology to the needs of bussiness process	St2.1.2	X	X				
Knowledge management	St2.2.1			X			
Quality system	St2.2.2			X	X		
Supply chain management	St2.3.1					X	
Computerized ERP	St2.3.2						X

Fig. 7. Extract of Strategies-Activities Design matrix of third-order decomposition

Develop a successful project for a sustainable entrepreneurial activity	(PVs) Activities	Identity & evaluate business risks and define controls and prevention	Identity & evaluate technology risks and define controls and prevention	Identity & evaluate context risks and define controls and prevention	Define main project steps, activities, responsibilities & milestones	Activities & Resources scheduling	Elaborate Gantt chart within MsProject application	Analyze operational variables within MsProject application	Establish project progress monitoring plan
(DPs) Strategies		PV1.2.1.1	PV1.2.1.2	PV1.2.1.3	PV1.2.2.1	PV1.2.2.2	PV1.2.2.3	PV1.2.3.1	PV1.2.3.2
Business risk management	St1.2.1.1	X							
Technology risk management	St1.2.1.2		X						
Context risk management	St1.2.1.3			X					
General project planning strategy	St1.2.2.1				X				
Critical Path Method (CPM)	St1.2.2.2					X			
MsProject editor	St1.2.2.3						X		
MsProject application capabilities	St1.2.3.1							X	
Enable adaptive/corrective measures	St1.2.3.2							X	X

Fig. 8. Extract of Strategies-Activities Design matrix of fourth-order decomposition

The result of the fourth-order decomposition process for *DPs* (*Strategies*) and the corresponding *PVs* (*Activities*) mapping can be seen in the extracts of the extended design matrix from Fig. 8 and Fig. 9.

Develop a successful project for a sustainable entrepreneurial activity	(PVs) Activities	Inventory of materials, processes, transportation, energy consumption	Calculate eco-indicators scoring	Modify variables on materials, processes, transportation, energy consumption	Inventory of activities determining direct costs or benefits	Calculate costs & benefits within time horizon scope	Modify business & technology activity variables	Collect & inventory data classified both by stakeholder category and by impact categories	Evaluate aggregate social indicators scoring or casual relationships	Modify activity variables on impact categories
(DPs) Strategies		PV1.3.1.1	PV1.3.1.2	PV1.3.1.3	PV1.3.2.1	PV1.3.2.2	PV1.3.2.3	PV1.3.3.1	PV1.3.3.2	PV1.3.3.3
LCA Inventory strategy	St1.3.1.1	X								
eco-indicators calculation methodology	St1.3.1.2		X							
environmental impact improvement strategy	St1.3.1.3			X						
LCC Inventory strategy	St1.3.2.1				X					
cost effectiveness calculation methodology	St1.3.2.2	X		X		X				
economic impact improvement strategy	St1.3.2.3						X			
SLCA Inventory strategy	St1.3.3.1							X		
social indicators calculation methodology	St1.3.3.2			X			X		X	
social impact improvement strategy	St1.3.3.3									X

Fig. 9. Extract of Strategies-Activities Design matrix of fourth-order decomposition

Further on, the technological processes within the bussines process were defined and analyzed by means of AD theory. The result of the first-order decomposition process and the *DPs* mapping is shown in the design matrix from Fig. 10. The result of the second-order decomposition process and the *DPs* mapping is shown in the design matrix from Fig. 11. Fig.12 shows an extract of superior-orders decomposition process.

				Small factory for micro-production of wood pellets integrated with self-providing crops of energetic willow		
				DP0		
				DP1	DP2	DP3
				Raw material input technology	Wood pellets production technology	Finished product output technology
FR0	Produce wood pellets from self-provided energetic willow biomass and/or from alternative biomass, for own heating energy needs of a micro-farm and for selling on local market (Align the best technology to the needs of bussiness process)	FR1	Input the raw material	X		
		FR2	Process raw biomass into wood pellets		X	
		FR3	Output the finished product			X

Fig. 10. First-order decomposition FRs-DPs Design matrix for technological processes

						Small factory for micro-production of wood pellets integrated with self-providing crops of energetic willow													
						DP0													
						DP1			DP2								DP3		
						Raw material input technology			Wood pellets production technology								Finished product output technology		
						DP1.1	DP1.2	DP1.3	DP2.1	DP2.2	DP2.3	DP2.4	DP2.5	DP2.6	DP2.7	DP2.8	DP3.1	DP3.2	DP3.3
						Raw biomass supply	Biomass storage facilities	Raw biomass feeding equipment	Drying technology	Grinding technology	Intermediary feeding technology	Conditioning technology	Pelletizing technology	Cooling technology	Dust collecting technology	Screening technology	Packaging technology	Finished product storage facilities	Finished product delivery facilities
FR0	Produce wood pellets from self-provided energetic willow biomass and/or from alternative biomass, for own heating energy needs of a micro-farm and for selling on local market (Align the best technology to the needs of bussiness process)	FR1	Input the raw material	FR1.1	Supply the process with willow stems and/or alternative raw biomass	X													
				FR1.2	Store the raw biomass		X												
				FR1.3	Feed (transfer) the raw biomass into the technological system			X											
		FR2	Process raw biomass into wood pellets	FR2.1	Dry the biomass				X										
				FR2.2	Grind the biomass					X									
				FR2.3	Feed (transfer) the biomass						X								
				FR2.4	Condition the wood (biomass)							X							
				FR2.5	Pelletize the wood (biomass)								X						
				FR2.6	Cool the wood pellets									X					
				FR2.7	Collect the dust										X				
				FR2.8	Screen the pellets mass											X			
		FR3	Output the finished product	FR3.1	Package the pellets												X		
				FR3.2	Store the finished product													X	
				FR3.3	Deliver the finished product														X

Fig. 11. FRs-DPs Master Design matrix – second-order decomposition

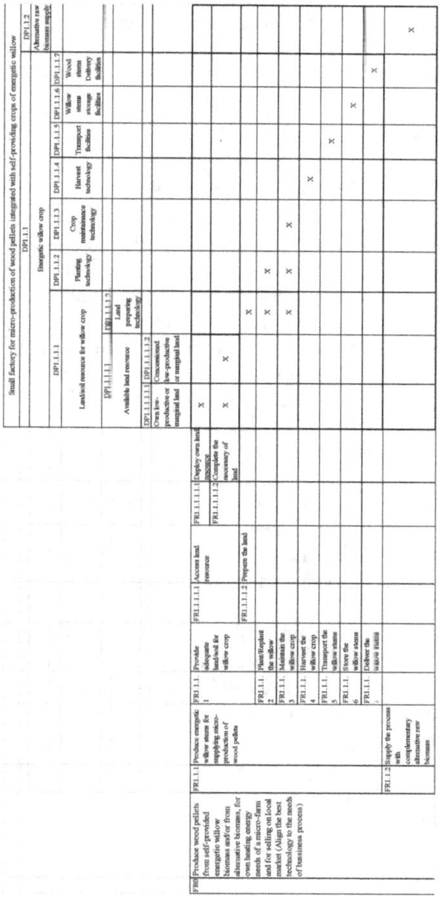

Fig. 12. FRs-DPs Master Design matrix extract– superior-orders decomposition

As general customer needs (CNs) / attributes (CAs), the following were identified for the technological processes analysis: the need for renewable heating energy resources; own needs of heating energy (for a micro-farm); selling opportunities on the local market for ROI/profit; the need for cost-effective operation of some available unproductive or low productive lands (marginal land, river banks, etc.).

As a constraint, (Cs), the following has been considered: *System constraint Cs1*: specific conditions of the implementing location - existing location (specified).

At the system level (design intent), the following were considered:

FR0: Produce wood pellets from self-provided energetic willow biomass or from alternative biomass, for own heating energy needs of a micro-farm and for selling on the local market (Align the best technology to the needs of the business process)

DP0: Small factory for micro-production of wood pellets integrated with self-providing crops of energetic willow.

Further on, for the AD analysis of the energetic willow crop technological process, the technical recommendations of SUNY-ESF [21] were considered, and for the study of the wood pellets micro-production technological process, the technical data published by Piraglia et al. [22] were used as referential.

Discussions. Conclusions
The included case study was chosen from the perspective that biomass pellets, specifically wood pellets gain more and more importance as a renewable energy resource and as a solution for energetic independence within societal resilience aims [23]. If considering and balancing the positive and the negative arguments [24], energetic willow (*Salix viminalis*) outstands as a very performant wood-type resource for biomass pellets, and also from the Societal Resilience point of view, willow crop systems bring other value-added benefits to the related ecosystems and communities [21], along with producing sustainable energy.

The case study approaches the project as a product and its development as a process, the last having to be approached from both engineering and business perspectives. The deployment of AD theory and principles allowed an in-depth study of the functional requirements, and goals, together with mapping of design parameters, activities, and process variables. The assignments of project sub-components and workflow management, task identification, and matching the available resources (human, equipment, etc.) to those tasks, were analyzed as project planning activities, under the principle of aligning the best options to the needs. Also, extended risk management was considered, under the principles of sustainability and societal resilience.

A multitude of multidisciplinary aspects, from project management, production engineering, agricultural sciences, and renewable energy resources, have been identified and hierarchized under the deployment of AD means.

The adopting of Axiomatic Design principles and tools in analytically inventorying, classifying, and selecting the process variables to be considered in translating process requirements into process specifications for providing sustainability of a project or of a small business as in the above-discussed case study, proved to be more effective than other systematic analysis tools, like Ishikawa charts, etc. [26].

Further research approaches intend to develop some simple Excel-based inventory models, to identify the key process input variables.

On the other hand, Social Life Cycle Assessment (SLCA) models are intended to be developed by deployment of AD means, for identifying and selecting reasonable, pertinent, and manageable sets of social indicators.

References

[1] F.H. Norris, S.P, Stevens, B. Pfefferbaum, K.F. Wyche, R.L. Pfefferbaum, Community Resilience as a Metaphor, Theory, Set of Capacities, and Strategy for Disaster Readiness. Am J Community Psychol 41 (2008), 127–150. https://doi.org/10.1007/s10464-007-9156-6

[2] C.B. Ren, T. O´Dell, A. Budeanu, Sustainabilities in the cultural economy, Culture Unbound: J. of Current Cultural Research 6 (2014) 907-912. http://www.cultureunbound.ep.liu.se/v6/cu14v6_Sustainabilities.pdf

[3] G. Çeviker-Çınar, G. Mura, M. Demirbağ-Kaplan, Design thinking: a new road map in business education, The Design J. 20 sup1 (2017) S977-87. https://doi.org/10.1080/14606925.2017.1353042

[4] B. Rusu, Design Thinking Workshops: Uncovering Facilitator and Participants' Experiences, in: G. Prostean, J. Lavios Villahoz, L. Brancu, G. Bakacsi (Eds), Innovation in Sustainable

Management and Entrepreneurship. SIM 2019. Springer Proc. in Business and Econ., Springer, Cham. (2020). https://doi.org/10.1007/978-3-030-44711-3_30

[5] V. Merticaru, E. Merticaru, I. Butnariu, M.M. Cucos, Enhancing Product Structure Design for an automation mechanism, by involving Axiomatic Design theory. MATEC Web of Conferences 178 (2018), 05003. https://doi.org/10.1051/matecconf/201817805003

[6] N.P. Suh, The Principles of Design (Oxford Series on Advanced Manufacturing), 1st ed., Oxford University Press, New York, 1990, ISBN10:0195043456, ISBN13:978-0195043457

[7] N.P. Suh, Complexity: Theory and Applications, (MIT - Pappalardo Series in Mechanical Engineering), 1st ed., Oxford University Press, New York, 2005, ISBN10:0195178769, ISBN13:978-0195178760

[8] N.P. Suh, Axiomatic Design: Advances and Applications, Oxford University Press, New York, 2001, ISBN10:0195134664, ISBN13:978-0195134667

[9] I. Palmberg, M. Hofman-Bergholm, E. Jeronen, E. Yli-Panula, Systems thinking for understanding sustainability? Nordic student teachers' views on the relationship between species identification, biodiversity and sustainable development. Educ. Sci. 7(3) (2017), 72. https://doi.org/10.3390/educsci7030072

[10] C.J. Czelusniak-Serviss, Construction Industry: Profitable Corporate Sustainability Strategies and Processes Achieve Competitive Advantage, Walden Dissertations and Doctoral Studies. 14070 (2023). https://scholarworks.waldenu.edu/dissertations/14070

[11] M. Kressy, Industrial Design, Rhode Island School of Design, Designturn Inc., 2002, https://dspace.mit.edu/bitstream/handle/1721.1/34891/15-783JSpring-2002/NR/rdonlyres/Sloan-School-of-Management/15-783JProduct-Design-and-DevelopmentSpring2002/A3B94580-C4EA-49F3-B89C-736B87D2EAFB/0/intro_to_id_lecture_28feb02.pdf

[12] I.M. Pista, V. Merticaru, G. Nagit, M.I. Ripanu, Advanced engineering design capabilities applied for developing a technological device for automated assembly, MATEC Web of Conferences 137, 04006 (2017), https://doi.org/10.1051/matecconf/201713704006

[13] C.A. Brown, Axiomatic Design for Understanding Manufacturing Engineering as a Science, in: Thompson, M.K. (Ed.) Interdisciplinary Design, Proc. of the 21st CIRP Design Conf., Korea 2011, pp. 173–175

[14] A. Hall, Process streamlining, Businessport (2017). http://www.businessport.net/wp-content/uploads/2017/05/Process_Streamlining.pdf (accessed 2018)

[15] JM Assoc., Overview of Process Streamlining, JM Assoc. (2006), http://www.higheredinfo.com/docs/Process%20Streamlining%20(PS)%20Overview-JM%20Associates.pdf (accessed 2018)

[16] F. Engelhardt, M. Nordlund, Strategic planning based on axiomatic design, Proceed. of ICAD2000, First International Conf. on Axiomatic Design, Cambridge, MA – June 21-23, 2000, ICAD038

[17] D. Steward, D. Tate, Integration of axiomatic design and project planning, Proceed. of ICAD2000, First Int. Conf. on Axiomatic Design, Cambridge, MA – June 21-23, 2000, ICAD062

[18] W. Kloepffer, Life Cycle Sustainability Assessment of Products (with Comments by Helias A. Udo de Haes). Int J LCA 13 (2) (2008) 89–95, http://dx.doi.org/10.1065/lca2008.02.376.

[19] K. Scharwath, Yves Béhar's 7 Principles of Holistic Product Design, TriplePundit's coverage of the 2011 Opportunity Green Conference in Los Angeles, California, November 12th, 2011, https://www.triplepundit.com/story/2011/yves-behars-7-principles-holistic-product-design/71721

[20] J.P dos Santos, A. Gonçalves-Coelho, A. Mourão, Designing the design of new businesses for raising venture capital, Proceedings of ICAD2011, The Sixth Int. Conf. on Axiomatic Design, Daejeon – March 30-31, 2011, ICAD-2011-16

[21] SUNY-ESF, Willow Biomass Producers Handbook, SUNY-ESF, (2017), http://www.esf.edu/willow/documents/CropProducersHandbook2017_000.pdf.

[22] A. Piraglia, R. Gonzalez, D. Saloni, Techno-Economical Analysis of Wood Pellets Production for U.S. Manufacturers, BioResources 5(4) (2010), 2374, https://www.ncsu.edu/bioresources/BioRes_05/BioRes_05_4_2374_Pirraglia_GS_Techno_Econ_Anal_Wood_Pellets_US_Prodn_1108.pdf

[23] C. Verhoest, Y. Ryckmans, Industrial Wood Pellets Report, PellCert (2012), http://www.enplus-pellets.eu/wp-content/uploads/2012/04/Industrial-pellets-report_PellCert_2012_secured.pdf

[24] A. Olejniczak, A. Cyganiuk, A. Kucińska, J.P. Łukaszewicz, Energetic Willow (Salix viminalis) – Unconventional Applications, in Sustain. Growth and Applic. in Renewable Energy Sources, Ed. M. Nayeripour, InTech (2011), http://cdn.intechopen.com/pdfs-wm/24431.pdf

[25] A.M. Onencan, L.E. Liu, B. Van de Walle, Design for Societal Resilience: The Risk Evaluation Diversity-Aiding Approach (RED-A), Sustainability 12 (2020), 5461

[26] V. Merticaru, B. Rusu, R.M. Ciobanu, Analytical inventory of process variables for sustainable development of a small business for integrated production of wood pellets, MATEC Web of Conf. 137 (2017) 07004,

Innovative Manufacturing Engineering and Energy - IManEE2024 Materials Research Forum LLC
Materials Research Proceedings 46 (2024) 385-392 https://doi.org/10.21741/9781644903377-49

Experimental research on the analysis of the stress and strain state of a pressure pipe

Ciocoiu GHEORGHE-COSMIN[1,a*], Flintașu (Oprea) GEORGIANA-DANIELA[2,b], Cherciu IONUȚ[3,c]

[1]National University of Science and Technology POLITEHNICA Bucharest, Romania

[2] Distrigaz Sud Rețele SRL, Romania

[3] National Company for the Control of Boilers, Lifting Installations and Pressure Vessels S.A., Romania

[a]cgcosmin@gmail.com

Keywords: Stress State, Tensionometry, Finite Element Method, Pressure Equipment, Geometric Discontinuity

Abstract. The paper represents an experimental analysis of the state of mechanical stresses and specific linear deformations specific to a cylindrical structure, with defects. The main purpose of the work is to highlight the state of stress and strain at the pre-established measurement points, in order to identify areas with defects or discontinuities where the equipment can fail. The damage to the structural integrity of a pressure equipment can be influenced by the material and geometry of the equipment, working parameters, structural defects or discontinuities, and knowing the stress condition is also a target for safe operation. In this case, electroresistive tensometry and the finite element method (FEM) were used.

Introduction

Mechanical structures for industrial processes are made up of components, which are introduced into manufacturing through various technological operations: casting, pressing, forging, rolling, welding, riveting , mechanical processing by turning, milling, planing, etc. Most of the time, the origin of damage to materials or mechanical structures is the presence of defects that occur during manufacturing processes [1 - 4].

The mechanical performance of equipment intended for use in process industries depends on the materials of which the components are made. Thus, the applications range from aircraft and car structures, to the interconnections of microelectronic circuits.

Defects, under certain conditions, can reduce fatigue resistance and thus cause premature breakage of the mechanical structure. This is the case, for example, of pipelines for the transport of oil or natural gas, which may contain defects both in manufacture and in assembly or operation [5 - 8].

Some damage, e.g. due to excessive deformation, can occur when transporting equipment, especially oversized equipment, on railway platforms or car platforms [9]. The consequences of defects are damage or accidents, and these can be prevented by monitoring machines, installations, etc. and establishing the residual service life.

The geometry of mechanical structures under pressure, as a result of exploitation and mechanical, thermal and chemical loads, as well as various combinations of these processes, can give rise to surface defects that lead to changes in the structural integrity of the surfaces [10 - 13].

Excessive increase in the stress state in a cylindrical structure can lead to performance deterioration and catastrophic failure of the structural system. Therefore, stress estimation in cylindrical structures is essential to ensure the safety and performance of the structural system [14 - 17].

Electroresistive tensometry is used in the analysis of the stress states present in pressure equipment from different technical fields, made of different construction materials [18 - 25].

Materials and methods

In order to eliminate the risks of failure of mechanical structures subjected to internal pressure (pressure pipes), it is necessary to quantify the local effects of stresses (in areas with stress concentrators). To determine the state of deformation, the electrotensometric method was used.

The experimental model consists of a pipe with the following characteristics: outer diameter $D_e = 89\ mm$, length $L = 610\ mm$, wall thickness $\delta = 3\ mm$, manufacturing material - P265GH (according to EN 10216). The material of model is considered homogeneous, isotropic and continuous. The request is considered to be in the elastic field.

The mechanical properties of the material are represented by the longitudinal modulus of elasticity (Young) $E = 2.1 \cdot 10^5$ [N/mm^2] and the coefficient of transverse contraction (Poisson) $\mu = 0.3$.

The existing defects on the surface of the pipe – experimental model, are characterized by the dimensions presented in Table 1.

Table 1. Defect parameters

	Longitudinal opening L_i* [mm]	Depth h_i* [mm]	Width l_i* [mm]
Defect 1 (D1)	50	1.2	5
Defect 2 (D2)	30	1.2	3

*i=1, 2

The experiment was carried out on a pressure pipe (pipe) whose dimensional sketch (with measuring points T_j, j = 1, 2, ..., 22) is shown in figure 1. Two electrotensiometric strain gauges were installed on the equipment for each main stress direction (axial, odd and annular, even).

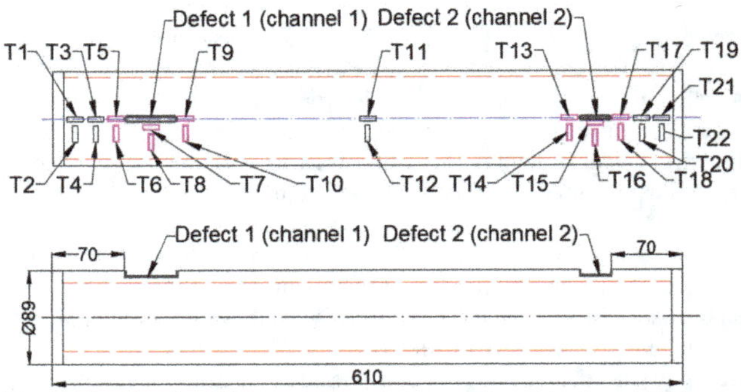

Fig. 1 The sketch of the measuring points and positioning of defects

Materials Research Proceedings 46 (2024) 385-392 https://doi.org/10.21741/9781644903377-49

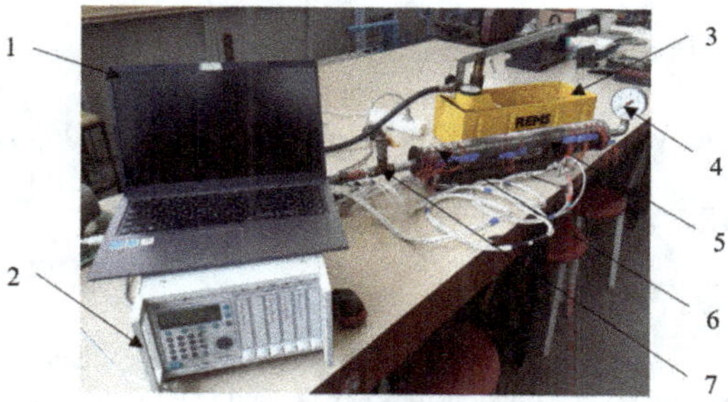

Fig. 2 The experimental stand used for the experiments

The experimental stand consists of the following equipments:
1. Computer for the acquisition of experimental data (with specialized Catman Easy software);
2. MGC plus tensometric bridge, required for acquisition of specific linear deformations, radial and annular directions, manufactured by Hottinger;
3. Pump with pressure gauge;
4. Manometer, $p_{\max} = 4MPa$, manufactured by Afriso;
5. Experimental model, design and manufactured by the main author;
6. Electrotensometer brands with type 6/120 LY 11 foil, manufactured by Hottinger;
7. Tap for equipment supply.

Expressions of axial stress σ_{1j} and the annular stress σ_{2j}, to the forms [26 - 28]:

$$\sigma_{1j} = \frac{p \cdot R_{mj}}{2 \cdot \delta_j},\ \sigma_{2j} = \frac{p \cdot R_{mj}}{\delta_j}, \tag{1}$$

where p is the internal pressure, respectively:

$$R_{mj} = \frac{R_{ej} + R_i}{2}, \tag{2}$$

R_{mj} - radius of the cylindrical part, and R_{ej} - outer radius of the section "j", R_i - inner radius of the section "j"; the thickness of the pipe wall has the expression:

$$\delta_j = R_{ei} - R_i. \tag{3}$$

Using the specific linear deformations – axial and annular $-\varepsilon_{1j}$ [μm/m] and ε_{2j} [μm/m] measured, determined on the length of the cylindrical section "j" [$\mu m/m$], the stresses are calculated [26 - 28]:

$$\sigma_{1j} = \frac{E}{1-\mu^2} \cdot (\varepsilon_{1j} + \mu \cdot \varepsilon_{2j})\ [\text{N/mm}^2], \tag{4}$$

$$\sigma_{2j} = \frac{E}{1-\mu^2} \cdot (\varepsilon_{2j} + \mu \cdot \varepsilon_{1j}) \text{ [N/mm}^2\text{]}, \tag{5}$$

The equivalent von Mises stress, was calculated with the formula [26 - 28]:

$$\sigma_{ech\,j}^{IV} = \sqrt{\sigma_{1j}^2 + \sigma_{2j}^2 - \sigma_{1j} \cdot \sigma_{2j}} \text{ [N/mm}^2\text{]}. \tag{6}$$

Results and discussion

Theoretical and experimental equivalent stresses resulting, $\sigma_{ech\,t}$ and $\sigma_{ech\,exp}$, calculated with the formula (6), respectively the relationships (1), (4) and (5), for the internal pressures 1 MPa, 2 MPa and 3 MPa, are compared (Table 2).

Table 2. Results of equivalent stresses

p [MPa]	$\sigma_{ech\,t}$ [MPa]	$\sigma_{ech\,exp}$ [MPa]	
		Defect 1	Defect 2
$p_1 = 1$	12.41	6.90	6.12
		9.02	6.55
		8.69	5.89
$p_2 = 2$	24.83	16.52	15.91
		21.42	16.18
		17.24	10.99
$p_3 = 3$	37.24	25.14	24.31
		28.55	23.94
		27.31	16.57

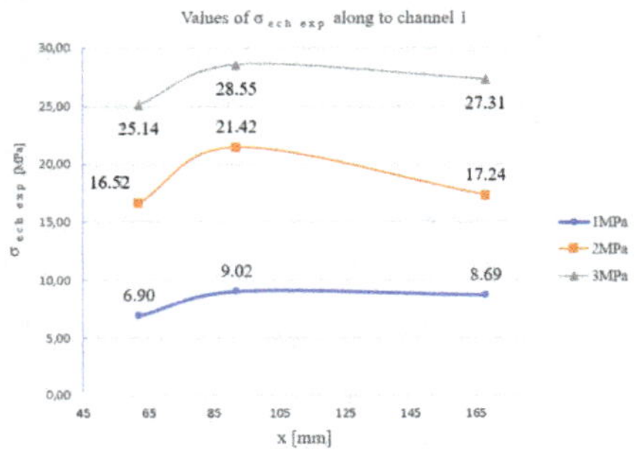

Fig. 3 Experimental equivalent stresses values, along to channel 1

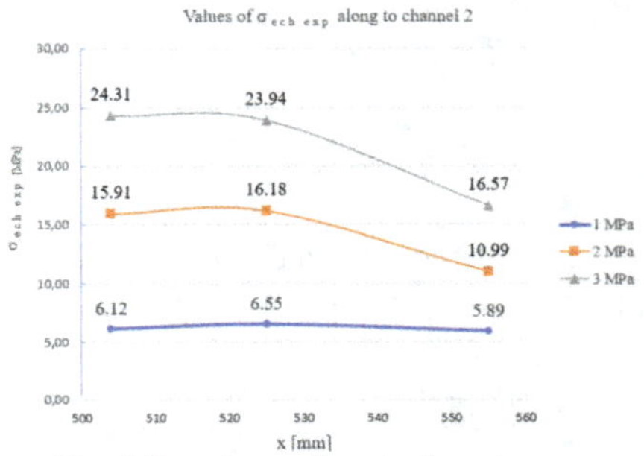

Fig. 4 Experimental equivalent stresses values, along to channel 2

The figures 3 and 4 show the variations of the experimental equivalent stresses, $\sigma_{ech\,exp}$, calculated with the formula (6), along to channel 1 (strain gauges 5, 6, ... 10) and along to channel 2 (strain gauges 13, 14, ... 18).

The specific linear deformations along the two directions, axial and annular, were automatically recorded using the tensometric bridge. The stresses were further calculated axial and annular, respectively the equivalent stresses. It is observed that the values obtained experimentally are lower than those obtained theoretically.

Finite element model and analysis

Finite element analysis (FEA) using Nastran software [29, 33] is a method of evaluating the behavior of materials and structures under different loads.

The geometric model was built in Inventor Professional and then imported into Nastran In-CAD for the numerical calculation of the object. The load is given by the internal pressure, $p = 3MPa$, and the condition of ensuring a convergence of the results has been imposed $< 2\%$.

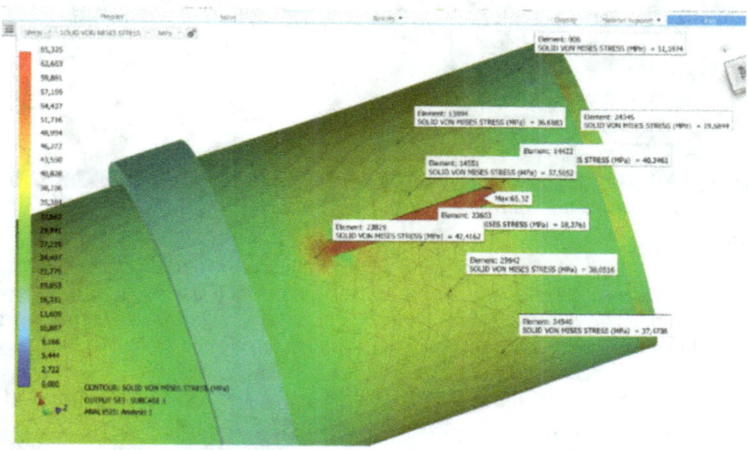

Fig. 5 Experimental equivalent stress, along to channel 1, pressure p=3 MPa

The values obtained by the Nastran numerical method show a maximum of 65.32 MPa for defect 1 and a maximum of 54.99 MPa for defect 2. It can be indicated that in those areas the equivalent stresses are maximum and can influence the further development of defects, with the probability of accentuated equipment failure.

From the analysis of figures 5 and 6, corresponding to the 2 defects, it can be suggested that the stress states do not conform to a linear trajectory. The concentration of stresses is predominant in the 2 analyzed areas. The increase in stress concentration corresponds to a more geometrically pronounced defect or the difference between the wall thicknesses. It is particularly interesting to pass from one wall thickness of the cylindrical shell to another – sudden passage or by connection, respectively the passage by linear variation.

Materials Research Proceedings 46 (2024) 385-392 https://doi.org/10.21741/9781644903377-49

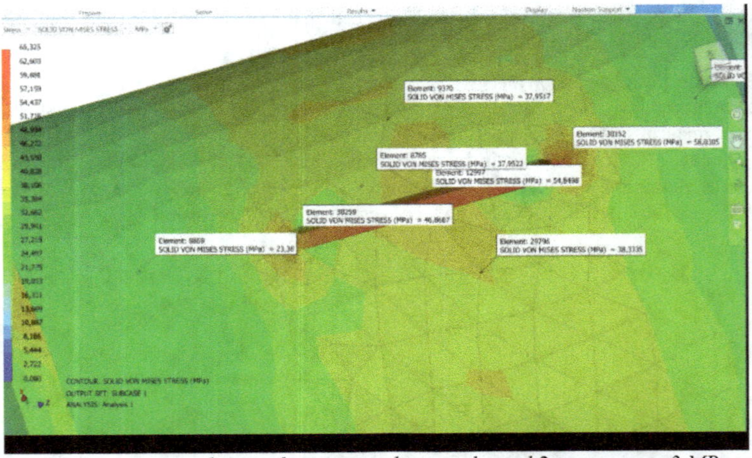

Fig. 6 Experimental equivalent stress, along to channel 2, pressure p=3 MPa

Conclusions

In the practice of designing the components of the equipment of the process industries under pressure, as well as in industrial use, when the existence of defects produced by corrosion or manipulation is found, in order to remedy them, it is necessary to use theoretical or experimental methods. In this regard, the subsequent operation of the respective constructions is certified.

Note: Theoretically, various ways of analyzing stress states are used, such as the finite element method or variants that direct the transition areas between two cylindrical areas of different thicknesses, by connection or linear passage, by lamellar elements or short cylinders.

The present study uses the experimental analysis, by electroresistive tensometry, of the stress states at the existence of defects, under internal pressure. An experimental model with an appropriate geometry is used for the analysis. Theoretical stresses are compared with experimental stresses values.

For the future, it is intended to make models that will highlight the transition through connection or linear, from a greater thickness to a smaller one, experimental testing and analysis with theoretical methods of the type mentioned above. Regardless of the results, they certify the safety of the mechanical structure under the conditions accepted for use under pressure. The existence of a temperature of use or different construction materials is also not excluded.

References

[1] Ghavamian A., Mustapha F., Baharudin B.T H. T., Yidris N., *Detection, Localisation and Assessment of Defects in Pipes Using Guided Wave Techniques: A Review*, Sensors 2018. https://doi.org/10.3390/s18124470

[2] Kharrat M., *Design and development of a torsion guided wave inspection system for detecting and sizing defects in pipes*, Phd Thezis, Ecole Centrale de Lyon, Écully, France, 2012.

[3] Diana Cătălina Cazangiu, *Studies and research on the detection of defects in structures*, Phd Thesis, Braşov, 2015.

[4] Lai A., Jia J., Zhou Z., Xu X., Lim C.W., *Homotopic analysis for post-buckling of cylindrical shells with local thickness defects*, Acta Astronautica 193, 2022. https://doi.org/10.1016/j.actaastro.2022.01.005

[5] Junsang P., Jaturanonda C., Wuttipornpun T., Watcharejyothin M., Liquefied Natural Gas Logistics *Management Through Optimal Road-Rail Intermodal Logistics Planning Considering Community Safety: A Case Study in Thailand.* International Journal of Knowledge and Systems Science, 2023. https://doi.org/10.4018/IJKSS.320486

[6] Peshkov R., Varkentin V., Ismagilov D., Mikhailov E., *Study of Strength Properties of the Designed Tank for Marine Transportation of Liquefied Natural Gas in the Arctic Conditions,* Transportation Research Procedia, 2023. https://doi.org/10.1016/j.trpro.2023.02.115

[7] Paul W., Frittelli J., *Rail Transportation of Liquefied Natural Gas: Safety and Regulation,* Congressional Research Service, 2020.

[8] Zhangabay N., et al., *Experimental research of the stress-strain state of prestressed cylindrical shells taking into account temperature effects,* Case Studies in Construction Materials, 2023. https://doi.org/10.1016/j.cscm.2022.e01776

[9] Iatan I. R., Vasilescu I., *Transportation of oversized technological equipment,* Publishing Matrix Rom, Bucharest, 2002.

[10] Yan Z., Zhang S., Zhou W., *Model Error Assessment of Burst Capacity Models for Energy Pipelines Containing Surface Cracks.* International Journal of Pressure Vessels and Piping, Volumes 120–121, 2014. https://doi.org/10.1016/j.ijpvp.2014.05.007

[11] Shan C., Sun J., Jia X., Wu R., Lu X., *The mechanical behavior of cylindrical shell with composite corrugations under uniform and concentrated pressure,* Ocean Engineering 298, 2024. https://doi.org/10.1016/j.oceaneng.2024.117148

[12] Zelin Z., Xinyang W., Lei W., Xuhui X., Surface defect detection method for discarded mechanical parts under heavy rust coverage, Scientific Reports, volume 14, 2024. https://doi.org/10.1038/s41598-024-58620-8

[13] Xiangqian X., Xing L., *Research on surface defect detection algorithm of pipeline weld based on YOLOv7,* Scientific Reports volume 14, 2024. https://doi.org/10.1038/s41598-024-52451-3

[14] Zhangabay N., Suleimenov U., et. all, *Experimental research of the stress-strain state of prestressed cylindrical shells taking into account temperature effects,* Construction Materials, 18, 2023. https://doi.org/10.1016/j.cscm.2022.e01776

[15] Teixeira M. Jr., Zilio G., Mortean M.V.V., Paiva K.V., Oliveira J.L.G., *Experimental and numerical analysis of transient thermal stresses on thick-walled cylinder,* International Journal of Pressure Vessels and Piping, 202, 2023. https://doi.org/10.1016/j.ijpvp.2023.104884

[16] Moustabchir H., Arbaoui J., et. all, *Experimental / numerical investigation of mechanical behaviour of internally pressurized cylindrical shells with external longitudinal and circumferential semi-elliptical defects,* Alexandria Engineering Journal, 57, 2018. https://doi.org/10.1016/j.aej.2017.05.022

[17] Deloševič M., Petrovič D., Bižic M., *Identification of the stress – strain state of a cylindrical tank with walls of variable thickness,* FME Transaction, 39, 2011.

[18] Pástor M., Trebuňa F., Lengvarský P., Bocko J., *Possibility of using of tensometry in deformation analysis in areas with sudden change of geometry,* American Journal of Mechanical Engineering, Vol. 4, No. 7, 2016.

[19] Ghimbaşeanu I., *Experimental research study on the use of a resistive tensometric sensor,* International Scientific Journal - Machines. Technologies. Materials, year IX, issue 1, 2015.

[20] Ghimbaşeanu I., *Comparative method for determining the mechanical stresses, Review of the Air Force Academy,* No 1 (28), 2015.

[21] Pástor M., Živcák J., Puškár M., Lengvarský P., Klacková I., *Application of Advanced Measuring Methods for Identification of Stresses and Deformations of Automotive Structures*, Applied Science, 2020. https://doi.org/10.3390/app10217510

[22] Matos H., Ngwa A. N., Chaudhary B., Shukla A., *Review of Implosion Design Considerations for Underwater Composite Pressure Vessels*, Journal of Marine Science and Engineering, 2024. https://doi.org/10.3390/jmse12091468

[23] Kumar A., Rajagopalan V., *Influence of geometric variations and stacking sequencing of a composite marine propeller*, Ocean Engineering, Volume 312, Part 2, 2024. https://doi.org/10.1016/j.oceaneng.2024.119106

[24] Ikeda C. M., Wilkerling J., Duncan J. H., *The implosion of cylindrical shell structures in a high-pressure water environment*, 2013. https://doi.org/10.1098/rspa.2013.0443

[25] Ngwa A., Rossell V., Matos H., Shukla A., *Effects of ring stiffeners on the underwater collapse behavior of carbon/epoxy composite tubes,* Journal of Composite Materials, Vol. 57 (10) 2023. https://doi.org/10.1177/00219983231162605

[26] Buzdugan, Gh., *Material resistance*, Technical Publishing House, Bucharest, 1980.

[27] Pavel Al., *Pipes. Piping. Tubular components. Tubular columns*, Ilex Publishing House, Bucharest, 2003.

[28] Tripa P., *Experimental methods for determining deformations and mechanical stresses*, Mirton Publishing House, Timişoara, 2010.

[29] Batalu D., *Finite element analysis in Inventor Nastran*, Politehnica Press Publishing, 2022.

[30] Butnariu S., Mogan G., *Finite element analysis in mechanical engineering. Practical Applications in ANSYS*, Transilvania University of Braşov Publishing House, 2014.

[31] Ji W., Ma H., Liu F., Sun W., Wang D., *Dynamic analysis of cracked pipe elbows: Numerical and experimental studies*, International Journal of Mechanical Sciences, Volume 281, 2024. https://doi.org/10.1016/j.ijmecsci.2024.109580

[32] Zhu Y., Lu D., Liu Y., Li D., *Equivalent mechanical model for liquid sloshing in irregular annular cylindrical containers of liquid metal reactors*, Annals of Nuclear Energy, 2024. https://doi.org/10.1016/j.anucene.2024.110425

[33] Xu S., Wang W., Y Chang., *Using FEM to predict residual stresses in girth welding joint of layered cylindrical vessels*, International Journal of Pressure Vessels and Piping 119, 2014. https://doi.org/10.1016/j.ijpvp.2014.02.002

Materials Research Proceedings 46 (2024) 393-404 https://doi.org/10.21741/9781644903377-50

Design, analysis and topology optimization of a car's suspension lower control arm

Nikolaos MANINAS[1,a], Ioannis PAPANTONIOU[1,2,b*] and Nikolaos VAXEVANIDIS[1,c]

[1]School of Pedagogical and Technological Education, Dept. of Mechanical Engineering Educators, Amarousion, Greece

[2]School of Mechanical Engineering, Section of Manufacturing Technology, National Technical University of Athens, Greece

[a]9nik.mani@mec.aspete.gr, [b]ipapanto@central.ntua.gr [c]vaxev@aspete.gr

Keywords: Control Arm Design, Material Selection, Static Analysis, Topology Optimization, Metal 3D Printing, Fatigue Analysis

Abstract. The modern automotive industry aims to design durable yet lightweight components that meet high performance and material efficiency goals, utilizing technologies like metal 3D printing. This study focuses on the design, analysis, and topology optimization of a MacPherson-type lower control arm, with a particular emphasis on static analysis and mass reduction achieved through topology optimization. The design process began with reverse engineering an existing control arm to acquire basic dimensions, followed by the development of a new model in SolidWorks. SolidWorks Simulation tools were employed for both stress analysis and topology optimization. The analysis considered the four primary stress forces, applied both parallel and perpendicular to the vehicle's longitudinal axis. The forces were calculated using basic mechanics and vehicle dynamics formulas, and the results were examined through finite element analysis (FEA), capturing Von Mises stresses. The selected materials were aluminum 7075-T6 and titanium Ti-6Al-4V, both suitable for 3D printing. The highest stresses (196 MPa) occurred during braking at maximum speed, and based on this scenario, topology optimization was performed. The optimized arm achieved a mass reduction of 38% compared to the non-optimized model, with SolidWorks Simulation used to evaluate mass reduction, strength retention, and fatigue analysis based on material type, underscoring the benefits of this approach.

Introduction

For the modern automotive industry, the continuous effort to produce high-performance vehicles that meet safety and resource-saving requirements is the driving force behind the implementation of new technologies. These technologies pave the way for designing lightweight, reliable, and economical cars [1]. With these goals in mind, manufacturers have determined that the key strategy in car design is the reduction of the weight of components while maintaining their strength. This contributes to the creation of vehicles with lower mass, resulting in better driving performance and reduced fuel consumption [2].

One of the most important systems in a car is the suspension system, which plays a crucial role in ensuring smooth and safe movement on the road. A key component of this system is the control arm, whose role is to support the suspension while also holding the wheels in place on the car's body, bearing road loads during motion. Traditionally, steel has been the primary material used for manufacturing control arms, and in some cases, aluminum [3,4]. However, with the development of material technology and the use of new production and design technologies, such as 3D metal printing [5,6] combined with topology optimization studies, new paths have been opened for creating complex geometries using advanced metals. These metals offer increased strength and reduced weight. Topology optimization is the method by which a component can be lightened without negatively affecting its mechanical strength [2]. When combining this method with the selection of high-performance materials, such as aluminum 7075-T6 and titanium Ti-6Al-4V, materials compatible with 3D printers, it is possible to produce lightweight control arms with enhanced mechanical strength.

Several studies have demonstrated the potential for weight reduction and performance optimization in control arms. Sookchanchai et al. [7] used the homogenization or solid isotropic material with penalization (SIMP) method to address the topology optimization problem, achieving a 68% reduction in mass through finite element analysis (FEA) of the loads acting on the control arm. Gururaj et al. [3] investigated the structural stability and strength of a control arm using Hypermesh and Abaqus. By comparing the weight-to-strength ratio of steel and aluminum, they found that the steel arm weighed 7.6 kg, while the aluminum arm weighed 2.6 kg. Pachapuri et al. [8] used Autodesk Inventor to design a basic control arm model with aluminum A357 as the material. They conducted load calculations for different loading scenarios, followed by FEA static and fatigue analyses, and applied topology optimization, reducing the mass from 2.07 kg to 1.61 kg. Finally, Rahman [2] designed a front lower control arm using CATIA V5 software and analyzed it using HyperWorks software. Using aluminum LM25 as the material, they performed case studies, topology optimization, and fatigue analysis, which resulted in a 25% reduction in mass while meeting the factor of safety requirements.

This study aims to design, analyze, and optimize the topology of a MacPherson-type lower control arm. The design process began by reverse engineering an existing control arm to derive the basic dimensions. Subsequently, a new control arm model was developed and further refined. Static analysis was conducted using SolidWorks Simulation, followed by topology optimization through finite element analysis. Additionally, static and fatigue analyses were performed on the topology-optimized arm to assess its strength under loading conditions, as well as its life expectancy before and after weight reduction. A key achievement of this study is the significant reduction in control arm mass while maintaining structural strength under high-stress conditions, as indicated by Von Mises stress analysis. The optimized arm was evaluated based on its weight reduction, mechanical behavior, and fatigue resistance, with particular focus on material-dependent variations in performance.

Control arm model design

The design of the control arm model was performed using SolidWorks® software, incorporating a combination of reverse engineering and creative adaptation from newer control arm designs. Initially, the basic dimensions were obtained from an existing BMW E30 318i control arm (Fig. 1a), while design inspiration was also drawn from more recent control arm models, resulting in a completely new design (Fig. 1a&b). Key measurements gathered included the length, width, thickness, and the diameters of both the ball joint mounting hole and the solid lower projection. Reverse engineering was chosen to ensure the highest possible accuracy in replicating real-world dimensions, allowing the designed model to withstand realistic load conditions effectively.

Materials Research Proceedings 46 (2024) 393-404 https://doi.org/10.21741/9781644903377-50

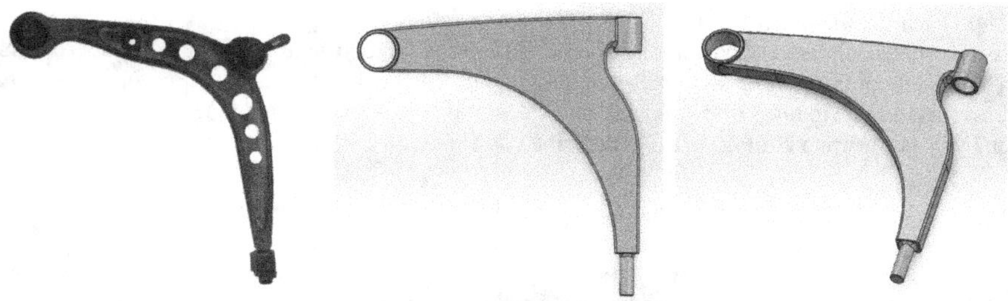

Fig. 1: a) BMW E30 318i control arm, b) designed control arm (front view), c) designed control arm (isometric view)

The control arm model is not a direct replica but includes several modifications to create a visually and geometrically refined version that meets the specific requirements of this study. Initially, sharp angles that could concentrate stress and potentially lead to premature material failure were rounded to minimize stress concentrations. Curves were then incorporated into the main body of the control arm to achieve a smoother stress distribution. A key feature of the design is the cylindrical hole for mounting a silent block, positioned perpendicular to the control arm's design plane. This hole was designed to be concentric with the lower projection, ensuring a geometry that would facilitate accurate static analysis of the control arm.

Load calculations

The calculation of loads exerted on the control arm is a crucial aspect of this study, as the static analysis for each stress scenario will be based on these results. Loads are transmitted to the control arm from the wheels during the vehicle's movement, specifically due to changes in speed and mass displacement. Such situations occur during acceleration, deceleration, and turning, which result in load transfers to the respective wheels. In this study, four stress scenarios are analyzed: braking at maximum speed while moving forward, braking at maximum speed in reverse, and lateral loads exerted on the control arms during steady turns at maximum speed with the wheels positioned on either the outer or inner side of the turn. These scenarios are considered along two axes—parallel and perpendicular to the vehicle's longitudinal axis—with a zero-angle application on the control arm to simplify calculations.

Car Specifications:

- Wheelbase (L): 2,851 mm
- Track Width (t): 1,827 mm
- Curb Weight (Mc): 1,500 kg
- Maximum Weight (Mmax): 2,000 kg
- Maximum Load: 500 kg
- Drive Type: Rear Wheel Drive (RWD) with 50-50 weight distribution
- Center of Gravity Height (h): 720 mm
- Maximum Velocity (Vmax): 225 km/h
- Acceleration (0-100 km/h): 8.9 sec
- Aerodynamic Drag Coefficient (Cd): 0.23
- Aerodynamic Drag Height (haero): 950 mm
- Front Tires: Pirelli P ZERO (PZ4) 225/40 R19 XL 93Y
- Rear Tires: Pirelli P ZERO (PZ4) 255/35 R19 XL 96Y

Load Case 1: Maximum Braking During Forward Motion

In this scenario, the vertical force exerted on the tires due to ground reaction is considered first. This is done by analyzing the forces acting on the car during braking, accounting for longitudinal load transfer due to mass displacement, and calculating the moments about the rear axle. The force analysis is illustrated in Fig. 2 and is described by Equation 1 [9] as follows:

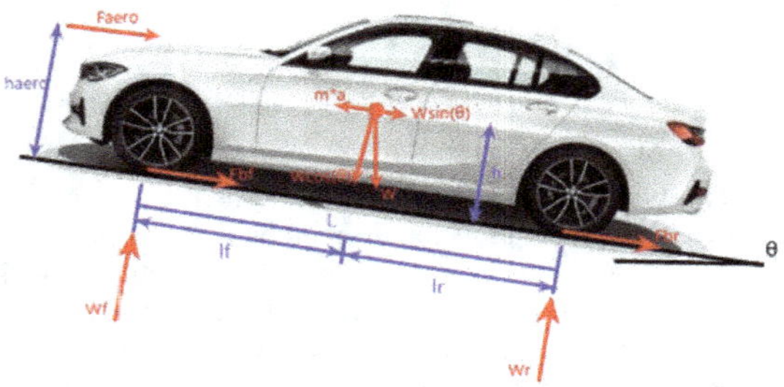

Fig. 2: Force analysis during braking in forward motion

$$W_f * L + F_{aero} * h_{aero} - M_{max} * d_x * h + W\sin(\theta) * h - W\cos(\theta) * \ell_r = 0 \quad (1)$$

To perform these calculations, several assumptions were made. First, the angle of the road surface was assumed to be 1 degree (i.e. $\theta = 1°$) as the road surface is effectively flat. Additionally, the deceleration of the car on a typical road surface while moving at maximum speed (V_{max} = 225 Km/h) was assumed to be d_x = 3.66 m/s^2. The aerodynamic drag (F_{aero}) was determined using Equation 2 [9], in combination with Equation 3 [10]:

$$F_{aero} = \frac{1}{2} \rho\, C_d\, A_F\, (V_X)^2 \quad (2)$$

$$A_F = 1{,}6 + 0{,}00056\,(M_c - 765) \quad (3)$$

By solving Equation 1 for the moments about the rear axle, the vertical load on the front axle (W_f) was found to be 11,200 N. Since this value applies to both control arms, it is divided by two, resulting in a vertical load per control arm of W_{fCA} = 5,600 N.

The force that stresses the control arm during braking is primarily the frictional force between the road surface and the tire. To calculate this frictional force, the formula given in Equation 4 [9] was used, assuming a friction coefficient μ_p = 0,8 for dry asphalt. The resulting braking force applied to each control arm (F_{bf}) is:

$$F_{bf} = \mu_p * W_{fCA} = 4{,}480\text{N} \quad (4)$$

Table 1: Friction coefficient values for different road surface conditions

Surface Condition	Friction Coefficient (μ)
Wet Asphalt	0.5
Dry Asphalt	0.8
Gravel	0.4
Ice	0.1

Load Cases 2-4: Reverse Braking and Lateral Load Analysis During Steady Turns at Maximum Speed

Similar to Load Case 1, the analysis for the other three load cases follows the same principles. The calculations for these scenarios are based on references [9–12]. For shortening the paper, the full analysis of the other three load cases will be provided in the extended version.

Table 2 provides a summary of the four load case scenarios analyzed in this study, highlighting the longitudinal (braking) forces experienced by the control arms and the lateral forces acting on the wheels during different driving conditions. The values indicate the braking forces during forward and reverse motion, as well as the lateral forces on the outer and inner wheels during cornering.

Table 2: Summary of Load Cases

Load Case	Description	Longitudinal (Braking) Forces (N)	Lateral (Cornering) Forces (N)
1	Forward Braking	4,430	0
2	Reverse Braking	-2,537	0
3	Outer Wheel Lateral	0	6,582
4	Inner Wheel Lateral	0	-1,475

Materials

The choice of materials for the design and construction of the control arm was made with the objective of achieving a robust structure capable of withstanding significant stresses and strains while maintaining minimal mass. Aluminum alloy 7075-T6 was selected as the primary material for this study due to its excellent mechanical properties, high resistance to external loads, and good fatigue resistance, making it ideal for high-performance automotive components. As previously mentioned, aluminum is a lightweight material with excellent mechanical and chemical properties, which contributes to its preference in automotive applications.

In addition to aluminum, titanium was also considered, despite being an unconventional choice for a suspension control arm. Titanium is known for its exceptional strength-to-weight ratio and outstanding corrosion resistance, which makes it suitable for applications in aviation, medicine, and high-performance racing car components. Specifically, titanium alloy Ti-6Al-4V was chosen for its superior mechanical properties, including a yield strength of 827 MPa, surpassing that of aluminum 7075-T6 by 322 MPa. Although titanium has a higher density of 4428 kg/m^3 compared to aluminum, its advantages in strength and fatigue performance make it an attractive option. The material properties for both aluminum 7075-T6 and titanium Ti-6Al-4V are summarized in Table 3.

Table 3: Material Properties of Aluminum 7075-T6 and Titanium Ti-6Al-4V

Property	*Aluminum 7075-T6*	*Units*	*Titanium Ti-6Al-4V*
Elastic Modulus	104800	MPa	72000
Poisson's Ratio	0.31	-	0.33
Mass Density	4428	Kg/m^3	26900
Tensile Strength	1050	MPa	2810
Yield Strength	827	MPa	570

The primary rationale for selecting these two materials is their compatibility with 3D metal printing technologies, as both are commercially available in powder form. Selective Laser Melting (SLM) enables the fabrication of intricate geometries derived from topology optimization or generative design, resulting in lightweight components with enhanced mechanical strength due to the unique metal alloys used. The advent of 3D metal printing represents a significant advancement in component production methods for the automotive industry, and its adoption is expected to grow as manufacturers prioritize environmental sustainability and resource conservation.

Static analysis, topology optimization and fatigue studies

All studies were conducted using the SolidWorks® Simulation tool, employing the finite element analysis (FEA) method. Each stress scenario and construction material was analyzed separately when necessary, allowing for a comprehensive examination of the control arm's overall performance and strength.

The static analysis of the control arm is a critical component of this study, as it provides insights into stress distribution within the model and identifies areas that experience significantly higher stresses. The primary objective is to ensure the control arm's construction material does not fracture, thus guaranteeing the reliability of the component. The analysis focuses on four basic load cases, with detailed calculations of the forces acting on the control arm provided in previous sections.

Notably, a solid cylinder made of aluminum 6061-T6 was integrated into the arm model to simulate the ball joint. This cylinder aids in more evenly distributing stresses across the entire length of the arm and prevents excessive stress concentration at the ball joint hole—a detail crucial to understanding the stressed areas of the arm. The regions where the control arm connects to the vehicle's chassis were treated as fixed supports, considering the directions of the applied forces. Additionally, all structural analyses were performed with a preferable factor of safety (FOS) of over 2.5 to account for any sudden shock loads that may increase the stresses and strains experienced by the control arm.

Based on the data derived from the static analyses, which will be discussed in the subsequent section, it becomes possible to conduct a topological optimization of the control arm—one of the main objectives of this paper. The methodology was straightforward: the topology study aimed to identify unnecessary material based on the stress distribution on the control arm, allowing for selective removal using the extrude cut command. Certain areas were designated as retained, as it was not appropriate to reduce mass in these regions, specifically where the arm is hinged to the chassis and the suspension system, thereby focusing solely on the main body of the arm.

The fatigue analysis was centered on the most demanding stress scenario—braking at maximum speed during forward motion. This study aimed to reveal the percentage of damage occurring in the control arm based on the applied load cycles and to estimate the overall lifespan of the component. This scenario represents an extreme and infrequent stress case, and given that the loading is constant, the minimum lifespan values for the control arm were anticipated. The analysis was configured as zero-based with a scale of 2, ensuring that the stress amplitude equaled that of a fully reversed analysis. The fatigue results indicate minor differences between the two materials used for the control arm, as well as variations before and after topological optimization, particularly concerning the component's lifespan. This lifespan is influenced by the applied stresses and the corresponding damage percentages accumulated in the material over the load cycles.

Results and discussion

The initial static studies were conducted on the non-optimized model of the control arm using both selected construction materials to identify areas with the highest Von Mises stresses ass illustrated in Fig. 3.

Materials Research Forum LLC
https://doi.org/10.21741/9781644903377-50

	Aluminum 7075-T6	Titanium Ti-6Al-4V
Case 1: Front braking	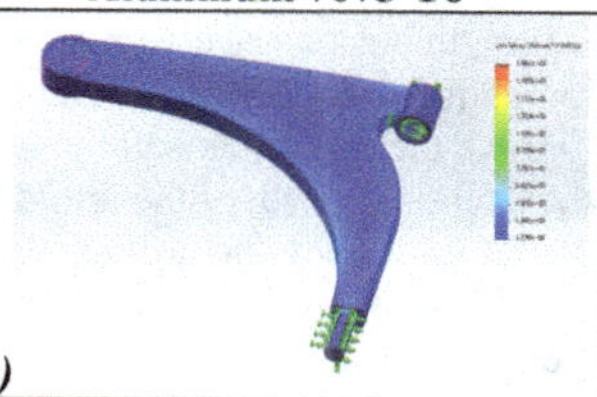 *(a)*	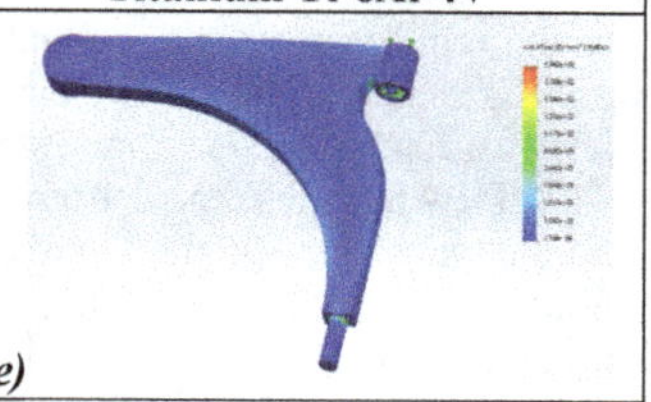 *(e)*
Case 2: Reverse braking	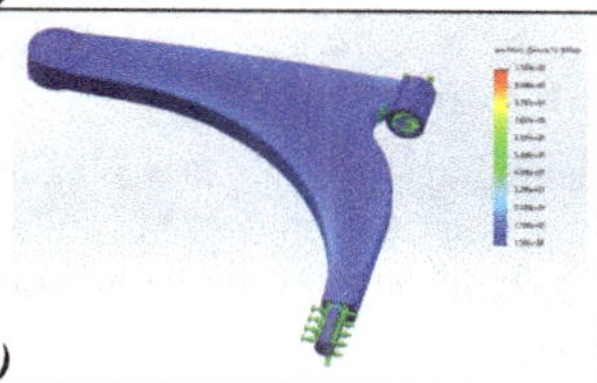 *(b)*	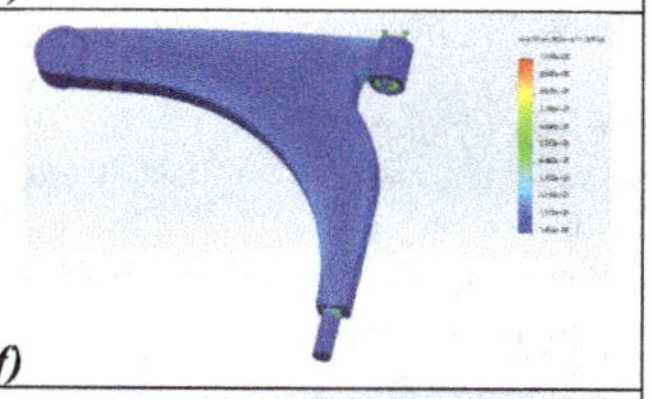 *(f)*
Case 3: Cornering (outer wheel)	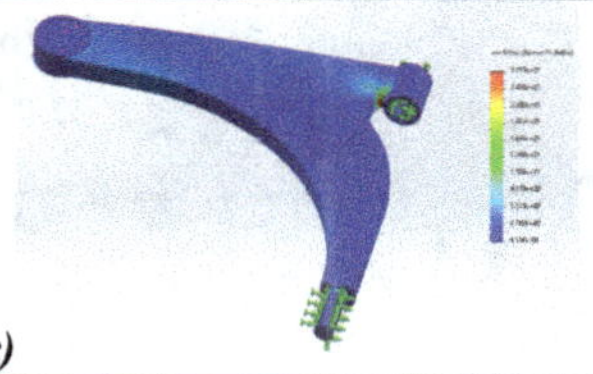 *(c)*	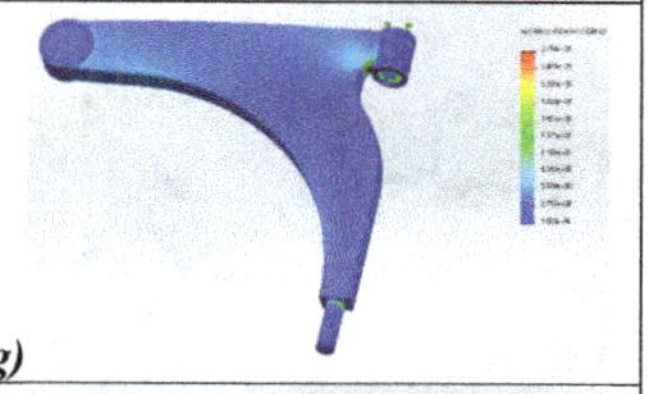 *(g)*
Case 4: Cornering (inner wheel)	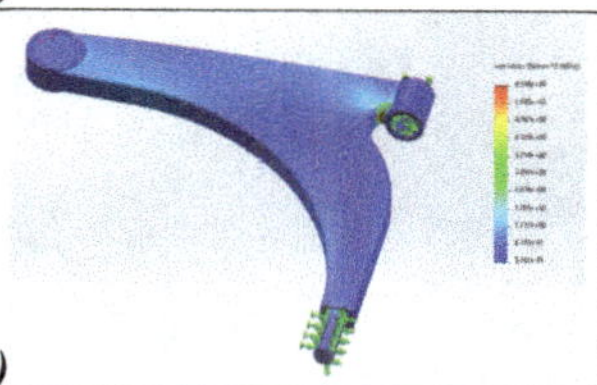 *(d)*	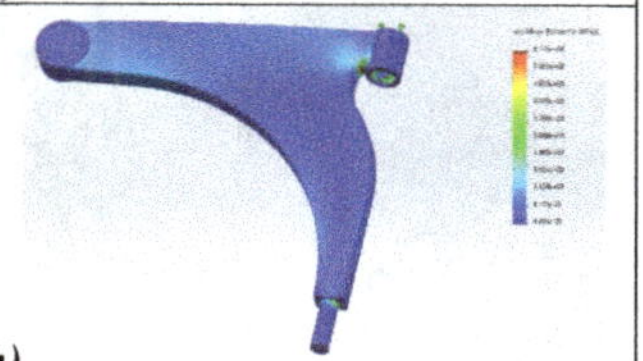 *(h)*

Fig. 3: Non-optimized control arm static analysis, a-d) 7075-T6 arm Cases 1-4, e-h) Ti-6Al-4V arm Cases 1-4

Forward Braking Analysis
During forward braking, the force vector acts backward, creating an inward bending moment on the arm. The maximum stress recorded is 194 MPa for the aluminum control arm and 196 MPa for the titanium model. The highest stress concentration occurs at the lower extension of the arm, which serves as the connection point to the vehicle's chassis. To mitigate the stresses at this critical junction, as well as where the silent block socket cylinder attaches to the arm, the radius of the angle joining the solid metal extension to the arm was increased in the model to create a smoother connection.

Reverse Braking Analysis
In the case of reverse braking, the force vector is aligned with that of forward braking but in the opposite direction, causing an outward bending moment on the arm. Given that reverse braking typically occurs at lower speeds, the maximum stresses generated are also lower, measuring 110 MPa for the aluminum arm and 111 MPa for the titanium arm. While these two loading scenarios present the highest stress concentrations, they remain well below the yield points of the respective materials: aluminum 7075-T6 at 505 MPa and titanium Ti-6Al-4V at 827 MPa.

Lateral Load Analysis
When the vehicle executes a constant right turn, the outer wheel arm experiences a force directed toward the center of the turning circle, coinciding with the centripetal force that maintains the vehicle's trajectory. The maximum stress for both the aluminum and titanium arms in this scenario is 27.6 MPa, which poses no concern for the overall strength of the arm. Conversely, during turns where the wheel is positioned on the inside of the curve, the arm experiences an outward pull, resulting in a different stress distribution due to its geometry. In this case, a minimal stress of 6.2 MPa is observed for both materials, representing the lowest stress among the four scenarios studied.

Topology Optimization
The scenario with the highest stress concentrations—braking at maximum speed during forward motion—serves as the basis for a topology optimization study. This study aims to remove 50% of the material, resulting in a unique geometry for the arm, as illustrated in Fig. 4, which depicts the final optimized model.

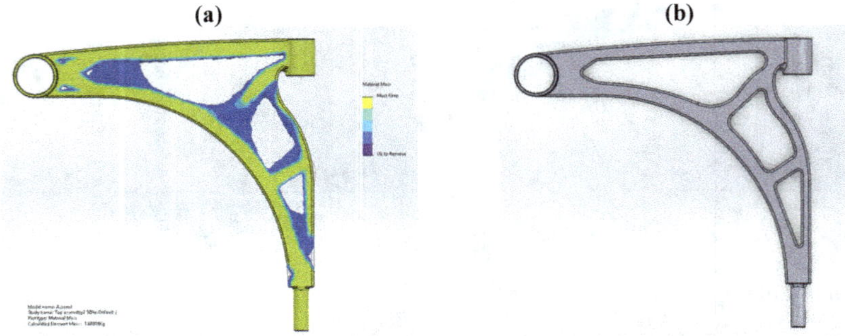

Fig. 4: a) Material Mass Plot (50% topology optimization), b) Final optimized control arm model

To evaluate whether the topology optimization for the highest stress scenario yields appropriate results in the other three cases, the optimized arm was subjected to static analysis across all scenarios. The results indicated that the differences in maximum stress values were minimal and did not adversely affect the performance of the design. This suggests that the weight reduction achieved through optimization did not compromise the structural integrity of the control arm, maintaining its reliability and functionality across various loading conditions, as shown in Fig. 5.

Materials Research Proceedings 46 (2024) 393-404 https://doi.org/10.21741/9781644903377-50

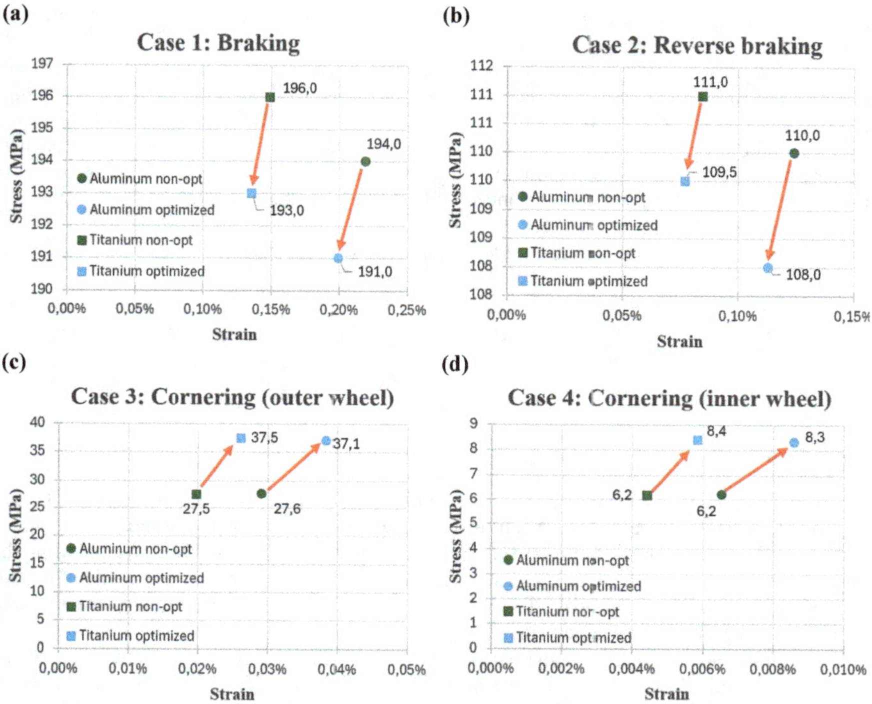

Fig. 5: Stress and strain diagrams for both materials before and after topology optimization a) Case 1: Front braking, b) Case 2: Reverse braking, c & d) Case 3-4: Cornering (outer-inner wheel)

To gain a clearer understanding of the results, a comparison of the mass values before and after the topology optimization is essential. Table 4 presents a comparison between the non-optimized control arm and the topologically optimized arm for each material, along with the calculated mass reduction percentages. As indicated by the values, the mass of the arm was reduced by 37.5% for the aluminum 7075-T6 and by 38.2% for the titanium Ti-6Al-4V. These percentages are particularly satisfactory concerning the initial design goals.

Therefore, if aluminum is selected as the construction material, the total weight of the pair of control arms mounted on the front suspension of the vehicle will be approximately 4.318 kg. In contrast, if titanium is chosen, the total weight will amount to around 6.634 kg. Both cases yield the construction of very strong and lightweight control arms compared to traditional steel arms, which can weigh as much as 8 kg.

Additionally, it is crucial to evaluate the new safety factors following topology optimization. The static studies were initially performed on the non-optimized model of the arm using both selected materials to identify areas of highest Von Mises stresses. The calculated new safety factors demonstrate the reliability of each material: for aluminum 7075-T6, the safety factor is approximately 2.64, while for titanium Ti-6Al-4V, the safety factor is around 4.28.

These results emphasize the effectiveness of the materials in providing robust designs for the control arms, aligning with the overall objectives of this study. Moreover, the titanium structure, due to its high safety factor, presents opportunities for further optimization, potentially enhancing

Materials Research Proceedings 46 (2024) 393-404 https://doi.org/10.21741/9781644903377-50

performance and reducing weight even more, making it a compelling choice for high-performance applications.

While the present study focused on optimizing performance through advanced materials like titanium Ti-6Al-4V, it is important to acknowledge the practical limitations, particularly the high cost of such materials. In many industrial applications, cost-effectiveness is a critical factor. Titanium's high cost generally limits its use to fields like aerospace and motorsports, where performance is prioritized over cost. Future studies could explore the use of hybrid materials or alternative alloys that offer a more balanced trade-off between performance and cost, especially for mass production.

Table 4: Mass reduction and safety factor analysis

Material	*Non-optimized arm (kg)*	*Optimized arm (kg)*	*Mass reduction (%)*	*Non-optimized arm Safety Factor*	*Optimized arm Safety Factor*
7075-T6	3.457	2.159	37.5	2.60	2.64
Ti-6Al-4V	5.363	3.317	38.2	4.22	4.28

Fatigue Study

The fatigue study assessed the lifespan of the control arms made from aluminum 7075-T6 and titanium Ti-6Al-4V, revealing significant insights into their performance. For the non-optimized version of the aluminum control arm, the lifespan is approximately $3{,}465*10^5$ cycles, while the optimized version shows a slight decrease to $3{,}414*10^5$ cycles. In contrast, the titanium Ti-6Al-4V control arm has a non-optimized lifespan of about $3{,}340*10^5$ cycles, which further reduces to $3{,}218*10^5$ cycles in the optimized version.

The results indicate that both materials exhibit high fatigue resistance, contributing to the overall safety and reliability of the control arm's construction. Notably, the differences in lifespan before and after topological optimization are minimal, suggesting that the reduction in weight did not significantly compromise the material's strength.

Fig. 6 illustrates the data regarding damage percentages and the maximum load factor across a range of load cycles from 10^3 to $2.5*10^5$.This data highlights the impact of alternating stresses on the control arm as the number of load cycles increases. Importantly, it is crucial to note that when the load factor falls below 1, material failure is imminent.

Materials Research Proceedings 46 (2024) 393-404 https://doi.org/10.21741/9781644903377-50

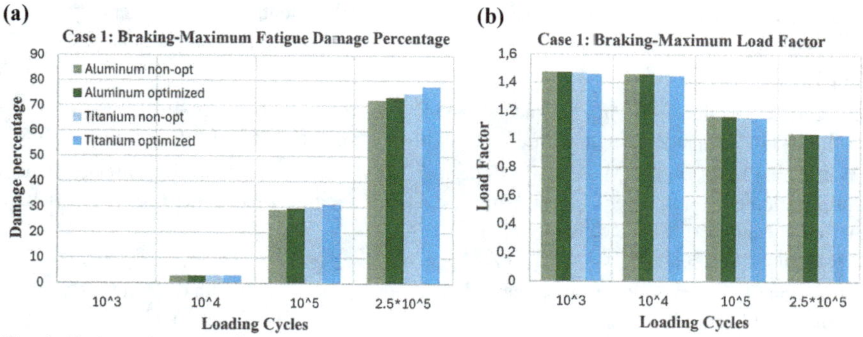

Fig. 6: Fatigue diagrams for Case 1: a) Maximum fatigue damage b) Maximum Load Factor to material failure

Conclusion

The successful analysis and topology optimization of the suspension control arm, combined with the design of the three-dimensional model, have resulted in the production of a lightweight control arm that maintains excellent mechanical strength and durability. The choice of materials, specifically aluminum 7075-T6 and titanium Ti-6Al-4V, has been instrumental in achieving the study's objectives due to their high yield strength, ultimate tensile strength, and significantly low densities. These properties ensure that the control arm can withstand intense loading conditions while reducing the overall weight of the design by 37.5% for aluminum and 38.2% for titanium.

In terms of stress performance, the maximum stresses recorded during the static analysis were 194 MPa for the aluminum control arm (with a safety factor of 2.6) and 196 MPa for the titanium arm (with a safety factor of 4.2) under forward braking conditions. Following topology optimization, the stresses increased by only 2.5% for aluminum and 3.0% for titanium, indicating that the optimization process effectively maintained the strength characteristics of the designs. Furthermore, the fatigue analysis demonstrated that the optimized arm is not significantly affected, showcasing its robustness under cyclic loading conditions.

This research study demonstrates how innovative manufacturing technologies can effectively enhance component performance by integrating fundamental mechanical principles, reverse engineering, CAD design, computer-aided engineering, topology optimization, and 3D printing design approaches. Leveraging these technologies not only streamlines design and production processes but also facilitates the creation of complex geometries that optimize material usage without compromising structural integrity. Furthermore, the findings underscore the potential for innovation in the modern automotive industry. As manufacturers prioritize safety and performance, the adoption of lightweight materials and advanced manufacturing techniques will play a crucial role in achieving these goals. Additionally, minimizing the use of raw materials and reducing emissions will help the industry align with global sustainability initiatives.

Moreover, while this study focused primarily on design and topology optimization based on static analysis, we recognize the importance of dynamic loading conditions, which are critical in automotive applications. We intend to extend this research by conducting dynamic analysis on both the initial and optimized control arms, further validating the robustness of the designs under real-world conditions.

In summary, this research contributes to a growing body of knowledge on the application of advanced materials and manufacturing techniques in automotive engineering. It paves the way for

Materials Research Proceedings 46 (2024) 393-404 https://doi.org/10.21741/9781644903377-50

future studies and developments that can lead to safer, more efficient, and environmentally friendly vehicles.

References

[1] Kale A, Tadamalle A, Patti N. Analysis and Optimization of Lower Control Arm. Evolutions in Mechanical Engineering 2018;1. https://doi.org/10.31031/EME.2018.01.000505

[2] Rahman MHA, Salleh MS, Mohd MS, Kamal MRM, Marjom Z, Anuar L, et al. Design and Optimization of Front Lower Control Arm (FLCA) for C-Segment Passenger Car. International Journal of Engineering & Technology 2018;7:71–5. https://doi.org/10.14419/IJET.V7I3.17.16625

[3] Dhanu G, Kattimani RS. Comparision study of lower control arm with different material. International Research Journal of Engineering and Technology (IRJET) 2016;03

[4] Song T, Xu S, He S. Analysis of material selection trend of the automobile control arm. J Phys Conf Ser 2022;2390. https://doi.org/10.1088/1742-6596/2390/1/012046

[5] Aboulkhair NT, Simonelli M, Parry L, Ashcroft I, Tuck C, Hague R. 3D printing of Aluminium alloys: Additive Manufacturing of Aluminium alloys using selective laser melting. Prog Mater Sci 2019; 106:100578. https://doi.org/10.1016/J.PMATSCI.2019.100578

[6] Jhinkwan A, Kalsi S, Pankaj. An Overview on 3d Metal Printing Technology in Automobile Industry. AIP Conf Proc 2023;2558. https://doi.org/10.1063/5.0120028/2873967

[7] Ling Y, Li A, Liu R, Gauthier S, Abarzhi SI, Sreenivasan KR, et al. Lightweight design of an automotive lower control arm using topology optimization for forming process. IOP Conf Ser Mater Sci Eng 2021;1157:012083. https://doi.org/10.1088/1757-899X/1157/1/012083

[8] Pachapuri MSA, Lingannavar RG, Kelageri NK, Phadate KK. Design and analysis of lower control arm of suspension system. Mater Today Proc 2021;47:2949–56. https://doi.org/10.1016/J.MATPR.2021.05.035

[9] Gillespie T. D. Fundamentals of Vehicle Dynamics, Revised Edition. Pennsylvania, USA: SAE International; 2021.

[10] Wong JY. Theory of ground vehicles. Third Edition. Ottawa, Canada: John Wiley & Sons Inc.; 2001.

[11] Rajamani R. Vehicle Dynamics and Control. Second Edition. Boston, MA: Springer US; 2012. https://doi.org/10.1007/978-1-4614-1433-9

[12] Milliken WF., Milliken DL. Race car vehicle dynamics. Warrendale, Pennsylvania, USA: SAE International; 1995.

Innovative Manufacturing Engineering and Energy - IManEE2024 Materials Research Forum LLC
Materials Research Proceedings 46 (2024) 405-411 https://doi.org/10.21741/9781644903377-51

Role of parametrised design from development process to product manufacturing

Andreea ISTRATE[a] * and Oana DODUN[b]

Gheorghe Asachi Technical University or Iași, Faculty of Machine Manufacturing and Industrial Management, Bdul. Mangeron, 59A, 700050, Iași, România

[a]andreea.cretu@student.tuiasi.ro, [b]oana.dodun-des-perrieres@academic.tuiasi.ro

Keywords: CAD, Parametrized Design, Manufacturing, Assembling

Abstract. The role of this paper is to demonstrate the use of design parameters control in CAD modelling. The case study consists of a 3D model able to be easily customized in multiple design variants to be tested or simulated thermally or mechanically. Such a design can be used from the beginning of the project if the designer can anticipate most of the problems/changes that could occur in the product. The scope of the case study is to see how much the design method helps and until which degree of design complexity/parameters control method complexity it worth that this design method to be used.

Introduction

DFMA (Design for Manufacturing and Assembly) term is used to describe the methods of design improvement and optimization in order to minimize the manufacturing and assembling challenges together with the costs [1]. In the DFMA there is needed to consider right from the start of the design process all the characteristics of your product that could impact manufacturing or production in any way.

When starting a new product design, the first step is to build a concept. We first need to think about the requirements and functions we need to fulfill. Then the concept is presented to the customer and suppliers and the optimization is starting. The customer will care about its requirements and functions and about assembling (when there is the case) in its interface, wherever the product is used. The suppliers care about how they will manufacture the components and if it is the case also about how they are assembling some of the components. We, as suppliers for our customers, will consider our assembly line, our testing, storage, etc. Form simplicity usually reduces assembly time and the costs implied by manufacturing and assembling [2].

Manufacturing refers to all the steps needed to build a component, starting from raw material, processing of raw material (like machining processes, casting, cold forming, injection molding, rheocasting, plasma arc machining, laser cutting/welding, etc), deburring, trimming (when needed), surface treatments/coatings, cleaning, etc.

As found in the literature, the manufacturing activities is taking a 74.7% from the entire energy that the industrializing process is consuming, this will mean that an important part of the product cost [3].

DFP (Design for Production) is taking care of the production equipments and processes and the supply chain. In order to optimize the relation between design, manufacturing and production line and increase time and cost efficiency, platform design is preferred. This approach is known as DFV (Design for Variety) and implies designing a platform product which will be used as start model for different variants of a product, like a family of products [4]. In this way, the geometries used in fixtures or for assembling, the manufacturing characteristics, etc. will be reused as much as possible in order to avoid changes that will need new equipments, new processes or new type of material.

Materials Research Proceedings 46 (2024) 405-411 https://doi.org/10.21741/9781644903377-51

Another similar approach for some companies is also try to use old designs for new ones (not necessarily platform products, but already produced designs that worked good and have good costs). The new product will be more or less similar to the old one, but improved and fulfilling new requirements. The role of DFMA is to lower as much as possible the impact of the changes on the manufacturing and assembly process [2].

Fig. 1 presents a scheme with the factors that could affect the design.

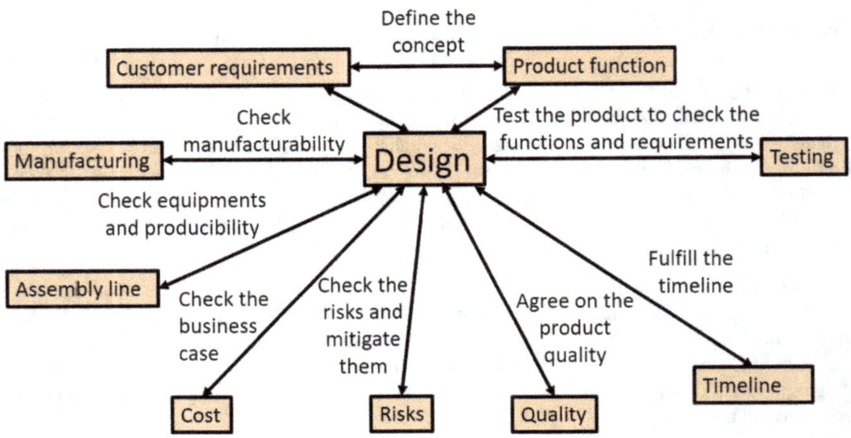

Fig. 1. Factors that influence the product design

Issues in the DFMA process appear when there is no proper communication between those responsible of the processes reminded in Fig. 1. In this case, the designers are not able to build a good design. In some cases, the design could not be manufactured or assembled and in that case big changes are needed with additional loss of time and money [5].

Application

The chosen case study will be the design of a product that would allow easy change depending on the manufacturing supplier feedback referring to the product manufacturability, cost, timeline, cycle time and product quality. The chosen design is a simple ECU (Electronic Control Unit) assemble design which contains two enclosures and a PCB (Printed Circuit Board) assembly. ECU-s are used in automotive industry to control the electronic side of the vehicle or different features in the vehicles like active or passive safety systems. An ECU has two components: the software and hardware of the system. The hardware contains the mechanical components and the PCB. A PCB is an electronic part made out of more layers used to link all the PCB electronic components. If in the old vehicles there was in the past only one ECU which controlled the whole electronic side, nowadays, all the new features included in the vehicle could come with a different ECU, so the ECU-s number in the vehicle increased.

The assembly model for the chosen design was created using a Skeleton part in order to be able to modify all components at once when needed. When choosing the parameters, it is needed to think about any possible changes that could be needed either due to the manufacturer feedback, customer changes, failed testing, production feedback, etc. The Skeleton parameters of this case study model are showed in Fig. 2, including: dimensions, material and fins variant. Also the sketches for components and screws were included in Skeleton in order to be able to use them in all components needed and be able to modify them all at once. The parameters and sketches from skeleton were published and copied with link in the components where these were needed. Whenever a change is needed, the update is made in Skeleton and afterwards the ECU assembly just need an update.

Redesigning a model could take up to two weeks depending on the model's complexity. The parametrization could help in this case in two manners:

- In the case where the designer anticipated all the changes, the update could take between a few seconds up to a few minutes. The updates which include dress-up features linkage will increase the time up to a few hours if the number of these ones is big.
- In the case where the designer did not anticipate the changes, if the parametrised model is correctly built, the update could take up to one to two days depending of the change needed.

Either way, an average of the saved time will be aroud 70-80%.

The most important advantage of the parametrized models is that these ones can be reused in similar projects and decrease the development time through decreased design buiding time (the model will not be built from scratch, an existing model will be used with only some changes). Ussually, when a design is similar, also the simulations from the old model could be carried over if the changes are considering as not affecting the behaviour (thermally/ mechanically).

When improving the design for manufacturing we can think about material filling, pores, welding lines, material cooling, parting lines, etc (for molding), radii/chamfers (cutting tool life and also product strength), thicknesses (important in molding processes and in product weights also important in thermal dissipation or defects occurrence in plastic molding).

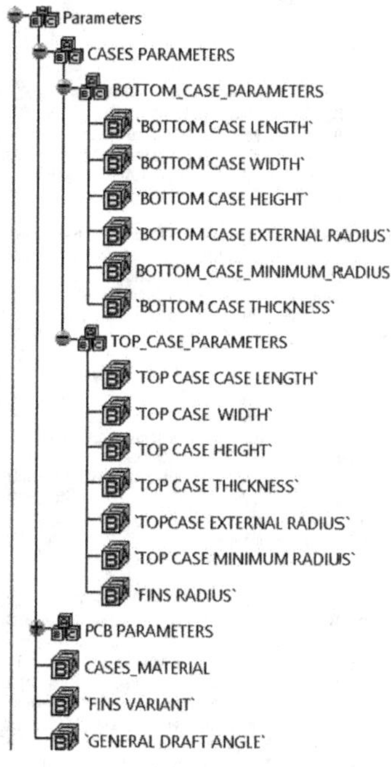

Fig. 2. Skeleton parameters

In this sense, the thickness, draft angle, radii, material parameters and fins variant were included into Skeleton. Some rules were needed to control some of these parameters. The proposed model is built to allow the change of these parameters listed before together with the size of the parts.

Proposed Model Feature for the Material

To enable material modification all at once, some rules were included in the mechanical parts, the rules are similar in both parts. The variants of material to choose can be found in Fig. 3.

Materials Research Proceedings 46 (2024) 405-411 https://doi.org/10.21741/9781644903377-51

Material change rule can be observed in Fig. 4:

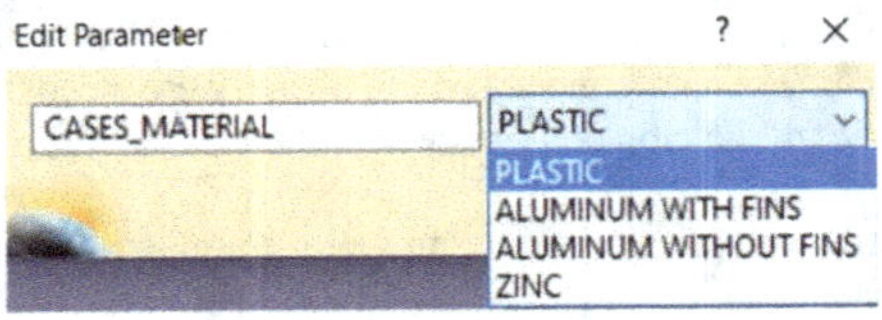

Fig. 3. Material variants

```
"if `External Parameters\CASES_MATERIAL` =="ALUMINUM WITHOUT FINS" or `External Parameters\CASES_MATERIAL` =="ALUMINUM WITH FINS"
{PartBody\Material ="Aluminium"}
else if `External Parameters\CASES_MATERIAL` =="PLASTIC"
{PartBody\Material ="Plastic"}
else if `External Parameters\CASES_MATERIAL` =="ZINC"
{PartBody\Material ="Zinc"}"
```

Fig. 4. Material change rule

```
"if `External Parameters\GENERAL DRAFT ANGLE` ==0deg
{    PartBody\Assemble.1\Body.2\Draft.1\Activity =false
     PartBody\Assemble.2\Body.3\Draft.2\Activity=false
     PartBody\Assemble.3\Body.4\Draft.3\Activity =false
     PartBody\Trim.1\Body.5\Draft.4\Activity=false
     PartBody\Trim.2\SCREWS\Draft.5\Activity =false}
else
{    PartBody\Assemble.1\Body.2\Draft.1\Activity =true
     PartBody\Assemble.2\Body.3\Draft.2\Activity=true
     PartBody\Assemble.3\Body.4\Draft.3\Activity =true
     PartBody\Trim.1\Body.5\Draft.4\Activity=true
     PartBody\Trim.2\SCREWS\Draft.5\Activity =true}"
Top case part 0 degrees draft rule:
"if `External Parameters\GENERAL DRAFT ANGLE` ==0deg
{    PartBody\Assemble.1\BASE\Assemble.3\Body.4\Draft.1\Activity  =false
     PartBody\Assemble.1\BASE\Assemble.2\Body.3\Draft.3\Activity =false
     PartBody\Assemble.1\BASE\Draft.2\Activity =false
     PartBody\Assemble.1\BASE\Draft.12\Activity  =false
     `PartBody\Assemble.4\COOLING-TOWERS\Draft.4\Activity`  =false
     `PartBody\Assemble.4\COOLING-TOWERS\Draft.5\Activity`=false
     `PartBody\Assemble.4\COOLING-TOWERS\Draft.6\Activity`=false
     `PartBody\Assemble.4\COOLING-TOWERS\Draft.7\Activity`=false
     PartBody\Assemble.7\SCREWS\Draft.13\Activity=false
     `PartBody\Remove.1\SCREWS-CUT\Draft.14\Activity`=false}
else
{    PartBody\Assemble.1\BASE\Assemble.3\Body.4\Draft.1\Activity  =true
     PartBody\Assemble.1\BASE\Assemble.2\Body.3\Draft.3\Activity =true
     PartBody\Assemble.1\BASE\Draft.2\Activity =true
     PartBody\Assemble.1\BASE\Draft.12\Activity  =true
     `PartBody\Assemble.4\COOLING-TOWERS\Draft.4\Activity`  =true
     `PartBody\Assemble.4\COOLING-TOWERS\Draft.5\Activity`=true
     `PartBody\Assemble.4\COOLING-TOWERS\Draft.6\Activity`=true
     `PartBody\Assemble.4\COOLING-TOWERS\Draft.7\Activity`=true
     PartBocy\Assemble.7\SCREWS\Draft.13\Activity=true
     `PartBody\Remove.1\SCREWS-CUT\Draft.14\Activity`=true}"
```

Fig. 5 Draft angle rule to avoid 0 degrees draft angle error

```
"if CASES_MATERIAL == "ALUMINUM WITH FINS" or CASES_MATERIAL =="ALUMINUM WITHOUT FINS"
{    Message("Advice:|If you choose for cast aluminum, the draft angle should be at least 1.5 degrees,|If you choose CNC machined aluminum, then set the draft angle to 0.")}
else if CASES_MATERIAL == "PLASTIC"
{    Message("Advice:|When using plastic, the draft angle should be at least 1.5 degrees.")}
else if CASES_MATERIAL == "ZINC"
{    Message("Advice:|If you choose casted zinc, the draft angle should be at least 0.5 degrees.")}"
```

Fig. 6. Draft angle advices rule

Proposed Model Feature for the Draft Angle

Draft angle is the inclination angle of the part walls. In casting and plastic injection molding, the draft angle is needed to allow part removal from the mold.

There is needed to think about manufacturing process: either we machine the parts (maybe more expensive, but possible), either we cast them (when the production volume is big enough, this approach is a better one). Another case is when both processes are needed: even if the casting is preferred because of the production volumes, in order to prove that the design will meet the requirements and functions, some testing will be needed. Then physical parts are to be prepared. CNC prototyping is a good variant. When the design is proven as a good one, then the casting tool can be ordered (changes in the casting tool are to be avoided as they are expensive and could affect tooling lifetime). Therefore, was needed that a rule to be included in order to avoid errors when setting the draft angle to 0 degrees.

Bottom case part 0 degrees draft rule is presented in Fig. 5.

Some draft angle advices for the model user are also added as a rule as it can be seen in Fig.6. The result can be observed in Fig. 7. The fins are designed only for aluminum material, but is possible to be included in to other materials also.

Proposed Model Feature for the Fins Number
In order to be able to choose the fins variant, one of the 3 values of the parameter must be chosen (see Fig. 8). Also a rule was included in order to control the fins geometries:

The proposed rule is made for the designer to be able to choose between two solutions of fins design, or the solution without fins. All 3 designs can be used for simulations/tests in order to decide which design meets the requirements.

Different variant of the fins in the model can be seen in Fig. 9.

A more advanced solution, with a more complex model will consist of a formula which will calculate the number of fins automatically depending of the minimum distance between fins. This minimum distance will be dictated by the manufacturing process. An example of a formula for fins number is stated in Eq. 1.

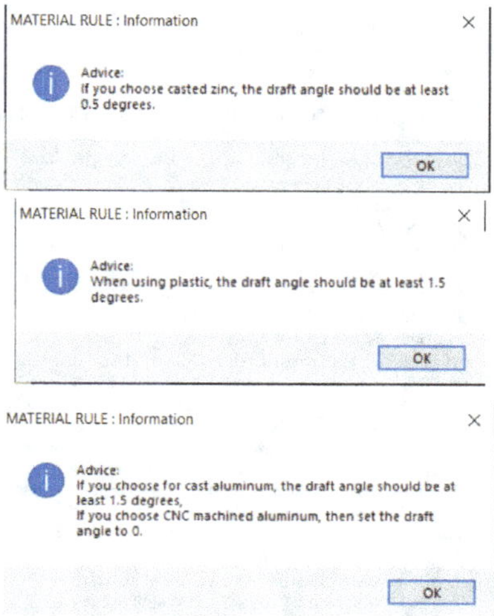

Fig. 7 Draft angle advices

n= round(L:(h+0.5*t)*2)+1 (1)

where:
n= the number of the fins
L= the length of the cover
h= the height of the fin
t= the thickness of the fin
round = operator which rounds the value of the number to integer value.
"*" = multiplying operator
In code the only used brackets are round ones and the order of operations is given by the brackets writing order.
Limitations of the formula: the most of the times, the number, position, height and form of the fins are on the mechanical engineer appreciation and depend on other features position also.

Conclusion
The use of parametrized design can be a strategic advantage. By applying parametric modelling, designers can proactively anticipate potential changes or modifications that may be required later in the development process. This approach allows for the creation of multiple design variants in one model, each easily adjustable to meet the evolving requirements of the project.

By having a range of design alternatives available, the best-fitting solution for the project can be selected with greater precision. Furthermore, these design variants can also be utilised for testing—whether thermal, mechanical, or structural, allowing for comprehensive validation before committing to a final design version. Through this iterative testing, the most robust design can be identified, significantly reducing the risk of issues arising during production or throughout the product's lifecycle.

Materials Research Proceedings 46 (2024) 405-411 https://doi.org/10.21741/9781644903377-51

```
"if `External Parameters\CASES_MATERIAL` =="ALUMINUM
WITH FINS" and `External Parameters\FINS VARIANT` =="5 FINS"
{   PartBody\Assemble.6\Activity = false
    PartBody\Assemble.6\Body.7\Pad.8\Activity =false
    PartBody\Assemble.6\Body.7\Pad.8\Sketch.15\Activity=false
    PartBody\Assemble.6\Body.7\Draft.9\Activity=false
    PartBody\Assemble.6\Body.7\EdgeFillet.8\Activity =false
    PartBody\EdgeFillet.9\Activity=false
    PartBody\EdgeFillet.10\Activity =false
    PartBody\Assemble.5\Activity=true
    PartBody\Assemble.5\Body.6\Pad.7\Activity=true
    PartBody\Assemble.5\Body.6\Draft.8\Activity=true
    PartBody\Assemble.5\Body.6\EdgeFillet.5\Activity=true
    PartBody\EdgeFillet.6\Activity=true
    PartBody\EdgeFillet.7\Activity=true}
else if `External Parameters\CASES_MATERIAL` =="ALUMINUM
WITH FINS" and `External Parameters\FINS VARIANT` =="7 FINS"
{   PartBody\Assemble.6\Activity = true
    PartBody\Assemble.6\Body.7\Pad.8\Activity =true
    PartBody\Assemble.6\Body.7\Pad.8\Sketch.15\Activity=true
    PartBody\Assemble.6\Body.7\Draft.9\Activity=true
    PartBody\Assemble.6\Body.7\EdgeFillet.8\Activity =true
    PartBody\EdgeFillet.9\Activity=true
    PartBody\EdgeFillet.10\Activity =true
    PartBody\Assemble.5\Activity=false
    PartBody\Assemble.5\Body.6\Pad.7\Activity=false
    PartBody\Assemble.5\Body.6\Draft.8\Activity=false
    PartBody\Assemble.5\Body.6\EdgeFillet.5\Activity=false
    PartBody\EdgeFillet.6\Activity=false
    PartBody\EdgeFillet.7\Activity=false}
else
{   PartBody\Assemble.6\Activity = false
    PartBody\Assemble.6\Body.7\Pad.8\Activity =false
    PartBody\Assemble.6\Body.7\Pad.8\Sketch.15\Activity=false
    PartBody\Assemble.6\Body.7\Draft.9\Activity=false
    PartBody\Assemble.6\Body.7\EdgeFillet.8\Activity =false
    PartBody\EdgeFillet.9\Activity=false
    PartBody\EdgeFillet.10\Activity =false
    PartBody\Assemble.5\Activity=false
    PartBody\Assemble.5\Body.6\Pad.7\Activity=false
    PartBody\Assemble.5\Body.6\Draft.8\Activity=false
    PartBody\Assemble.5\Body.6\EdgeFillet.5\Activity=false
    PartBody\EdgeFillet.6\Activity=false
    PartBody\EdgeFillet.7\Activity=false}"
```

Fig. 8. Rule for fins variant

Fig. 9. Fins variants

Ultimately, the ability to predict and accommodate changes early in the process ensures that the final design is optimised for both performance and manufacturability. Preparing the model through parametrization for more scenarios is a step forward in the development process and helps to save time, money and overwork through avoiding the redesign. Having always more solutions on hand avoid keeping the time plan through having back-up solutions and solve the issues before they occur.

Modifications in later stages increase the costs exponentially – avoiding these modifications minimise potential risks during manufacturing (scrap rate, tooling lifetime affected due to changes or need of different new tool if the change in the actual one is not possible) and product use, thus contributing to a smoother DFMA process and a more reliable end product.

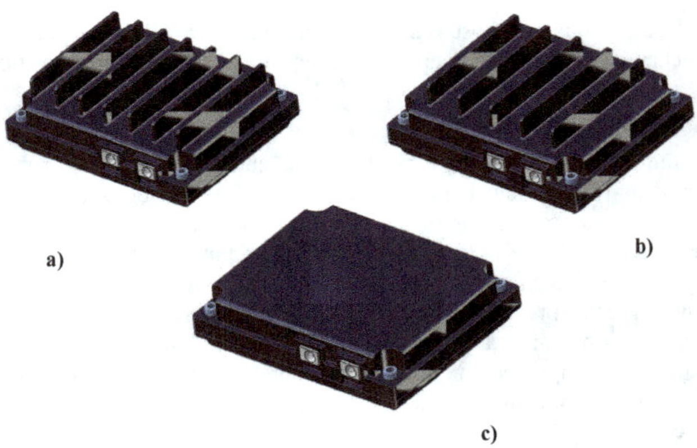

Fig. 10. Model variants for fins, a) five fins, b) seven fins, c) without fins.

Materials Research Proceedings 46 (2024) 405-411 https://doi.org/10.21741/9781644903377-51

References

[1] G. Formentini, N. Boix Rodríguez, C. Favi, Design for manufacturing and assembly methods in the product development process of mechanical products: a systematic literature review. Int J Adv Manuf Technol 120 (2022) 4307-4334 . https://doi.org/10.1007/s00170-022-08837-6

[2] A.I. Juniani, M.L. Singgih, P.D. Karningsih, (2021). Design for Manufacturing, Assembly, and Reliability on Product Redesign: Literature Review and Research Direction, in: Proceedings of the Second Asia Pacific International Conference on Industrial Engineering and Operations Management, Surakata, Indonesia, 2021, pp. 218-231, Information on https://ieomsociety.org/proceedings/2021indonesia/39.pdf

[3] F. Lu, G. Zhou, Z. Chao, Y. Liu, M. Taisch, Integrated optimisation of multi-pass cutting parameters and tool path with hierarchical reinforcement learning towards green manufacturing. Robotics and Computer-Integrated Manufacturing. 91 (2025) 102824 https://doi.org/10.1016/j.rcim.2024.102824.

[4] J. Herrmann, J. Cooper, S. Gupta, Satyandra, C. Hayes, K. Ishii, D. Kazmer, P. Sandborn, W. Wood, (2004). New Directions in Design for Manufacturing, in: Proceedings of the ASME Design Engineering Technical Conferences and Computer and Information in Engineering Conference, Utah, USA, 2004, pp. 853-861. https://doi.org/10.1115/DETC2004-57770.

[5] V. Karandikar, S. Bade, S. Patil, Design for manufacturing and Assembly practices in early stage of product development, in: Proceedings of the International Journal of Current Engineering and Technology, 2021, pp. 509-515. https://doi.org/10.14741/ijcet/v.11.5.3

Integrating digital twins in metal sink factories: A pathway to smart manufacturing

Vassilios KAPPATOS[1,a], Loukas ATHANASAKOS[1,b], Alkiviadis TROMARAS[1,c] and Agathoklis KRIMPENIS [2,d *]

[1]Hellenic Institute of Transport (HIT), Center for Research and Technology Hellas (CERTH), 6th Km Charilaou - Thermi Rd., 57001, Thermi, Thessaloniki, Greece

[2]Mechanical Engineering Dpt., School of Engineering, Hellenic Mediterranean University, Estavromenos, 71410, Heraklion Crete, Greece

[a]vkappatos@certh.gr, [b]lathanasakos@certh.gr, [c]atromaras@certh.gr, [d]akrimpenis@hmu.gr

Keywords: Digital Twin, Monoblock Sink Production, Simulation Modeling, Production Optimization, Manufacturing Process Validation

Abstract. This study presents the development and validation of a digital twin for the production line of monoblock sinks. The digital twin was created using the AnyLogic simulation platform, integrating both 2D and 3D environments to replicate the factory's operations. The model simulates the movement of workers and machinery across key stations in the production process, providing detailed insights into resource availability, machine usage, and potential bottlenecks. Validation of the model was conducted through a comparison of simulated production rates with real-world data for key production stations. The results demonstrated a high degree of accuracy, with discrepancies between simulated and actual production times within a few seconds. The digital twin offers a powerful tool for experimentation and optimization, enabling adjustments in process times, workforce allocation, and equipment utilization. This approach not only enhances production efficiency but also provides a reliable platform for future decision-making and process improvements.

Introduction

With the advent of the Fourth Industrial Revolution (Industry 4.0) [1], businesses are increasingly being driven to enter this new era. Companies are now required to modernize their operations using advanced computing systems and software. Additionally, the globalization of markets has brought significant changes to the way businesses operate, as competition between companies has greatly intensified. As a result, businesses now face competition on a global scale. Therefore, it has become imperative for companies to produce products of the highest quality at the lowest possible cost in order to maintain their market share. One solution aimed at modernizing production units is the concept of Smart Factories [2]. According to the smart factory approach, production units are called upon to integrate modern technologies into their production processes. One category of such technologies is computing systems and software. With these software solutions, companies are able to digitize various activities involved in their operations, such as raw material supply, the various processes during production, sales, and more [3].

A type of software that has been increasingly used in global industry in recent years is process simulation software. These tools allow companies to simulate various activities involved in the production process within virtual environments, which can help them make strategic decisions and design new ideas and products. In reality, implementing such innovations can be quite difficult, and sometimes even impossible.

The objective of this research is to create the Digital Twin of a production line [4], thus forming a digital representation of the entire production process and methods.

Materials and Methods

Modeling is the process of creating a model that represents a real-world system. A model is typically a simplified version of the real system, allowing for analysis of its behavior. Generally, the term 'model' refers to a mathematical model executed with the help of process simulation software. A mathematical model can be deterministic, where inputs and outputs are fixed values; stochastic, where at least one input or output is random; or dynamic, where variables change over time. Most simulation models are either stochastic or dynamic [5]. According to Shannon [6], simulation refers to designing the model of a physical system and experimenting with it so as to understand its behavior and to evaluate alternative strategies, which ideally leads to its optimal operation.

Process simulation software has been used for many years to solve real-world problems. In some cases, conducting real-world experiments to solve problems or finding optimal solutions is not feasible, either due to high costs, danger, or impracticality. For example, when designing a new bridge, engineers cannot afford mistakes. They must model the bridge using specialized software to ensure it can withstand various loads, and without such software, the bridge design process would be much more difficult, time-consuming, and risky.

Modeling software can be used to simulate various systems, such as supply chain problems, traffic networks, queues, banks, airports, ports, industrial production, hospitals, and any system involving processes. According to Banks [7], the main advantages of simulation are summarized in: i) finding the best solution by testing alternative scenarios without using real resources, ii) enabling the acceleration or deceleration of time, helping to better understand the system and its phenomena, iii) helping explain system phenomena through microscopic examination, iv) enabling the exploration of new policies, procedures, or methods without disrupting the real system, v) helping the understanding of interactions between variables in complex systems with many components, vi) identifying bottlenecks, vii) improving the understanding of a system's operation, viii) visualizing processes in 2D and 3D, helping test new ideas or detect potential issues in the current plan, and ix) Simulation models can be used for staff training and improvement.

Model Types. For systems simulation [8], there are three main modeling methods: System Dynamics Modeling (SD) [9], Discrete Event Modeling (DE) [10] and Agent Based Modeling (AB) [11]. As shown in Fig.1, each modeling method has a certain level of abstraction, referring to the amount of information required by each method to simulate a system. Naturally, the more information provided as input, the more reliable the model will be. Thus, each method is suited for specific types of systems. It should be noted that modern software packages (e.g., AnyLogic) allow for the creation of hybrid models, which simultaneously use all three modeling methods.

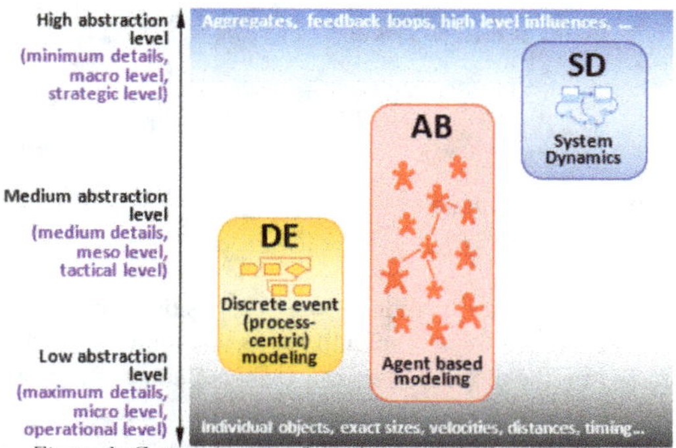

Figure 1: Categorization of simulation methods according to the level of abstraction

Modeling procedure. According to Shannon [6], the following 12 steps are needed to create a simulation model of a system: 1) Clarify the goals or problems to be solved by the model, 2) Identify the resources needed (personnel, equipment, etc.) before model design, 3) Define the boundaries and constraints of the model representing the real system, 4) Create an initial design, such as a flowchart or pseudocode, to define variables and interactions, 5) Conduct a preliminary experimental design, specifying the desired data and outcomes, 6) Collect input data for the model, 7) Develop the model using simulation software, 8) Verify and validate the model by testing with known scenarios, 9) Run a final experiment and make adjustments if necessary, 10) Once validated, the model is ready for simulations, 11) Analyze the data produced by the model, and 12) Document the results for improving the real system.

AnyLogic software. Following the above process, we used the AnyLogic software to produce an accurate Digital Twin of the entire process of a metal sink production facility. AnyLogic software, developed by 'The AnyLogic Company,' can simulate systems using all three methods (SD, DE, AB) [12]. It is widely used in various industries and allows both 2D and 3D modeling. It includes ready-made libraries for users to create logic diagrams of systems, along with numerous built-in tools for simulation [13]. Written in Java, it also allows users to add custom code.

Figure 2: Logic diagram example within AnyLogic

Users select appropriate blocks from the available libraries and arrange them on the design board to represent the system's logic. For example, in Fig.2, agents enter the model through the 'source' block and exit through the 'sink' block, simulating entry and exit of agents like consumers, patients, vehicles, or products. The delay block determines how long agents remain in the model. Fig.3 shows examples of agents. The simulation model in this project simulates a specific sink product entering through the 'source.' In complex models, multiple agent types can be used, such as products, pallets, forklifts, and trucks in a warehouse. AnyLogic also supports creating populations or groups of agents, like the population of a country.

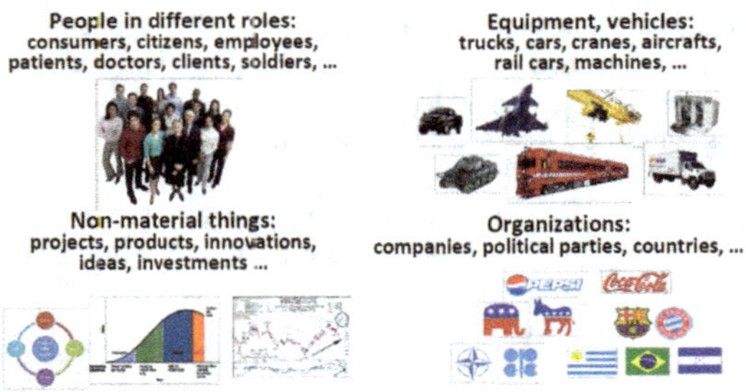

Figure 3: Examples of Agents

In short, the entire Digital Twin is created by selecting and connecting specific predefined blocks, as described below and as visualized in Fig.4:

The '*Source*' block typically starts a simulation model's logic diagram, representing the agents' entry into the model. Users can set agent entry rate, number, type, and location (in 2D or 3D), and they can also add custom code. The '*Sink*' block simulates agents exiting the model, while the '*Delay*' block keeps agents for a defined or random time, after which they proceed to the next block. '*Queue*' simulates waiting lines (e.g., in a supermarket or a production line) where agents wait before continuing and '*ResourcePool*' simulates system resources (e.g., workers or machines), allowing users to define the type and number of available resources. '*Seize*' assigns resources to agents, and '*Release*' frees them once the task is complete. '*Service*' combines the seize-delay-release process to simulate tasks like loading materials. The '*Batch*' block groups agents into a new one (e.g., stacking products on a pallet), and '*Unbatch*' separates them (e.g., unpacking a box). '*SelectOutput*' directs agents based on logic or probability. '*Time Measure Start*' and '*Time Measure End*' record time within the diagram, while '*Hold*' controls whether agents can pass or not,

Materials Research Proceedings 46 (2024) 412-419 https://doi.org/10.21741/9781644903377-52

depending on conditions. Finally, the simulation model in this study, also uses additional blocks for agent movement and intermediate storage between stations.

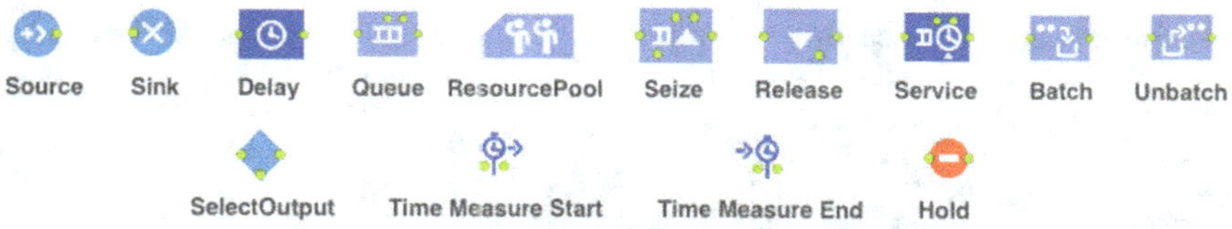

Figure 4: Icons of various blocks within AnyLogic

Simulation modeling. The simulation model of the production line was based off of an analysis of a kitchen product factory. The company's monoblock sink production line was documented through observation. This documentation was the first step in creating the digital twin of the production process. During the documentation phase, each production station's various stages were recorded. All production steps and movements were documented by an observer, and detailed production stages were created for each station. A preliminary version of the simulation model was also developed at this stage. For each station involved in the monoblock sink production, time measurements were made using a stopwatch or were provided by the company's existing data. Separate measurements were taken for each product at stations referenced with codenames E1 and E2, while for other stations, random product data were used as the times are relatively constant regardless of the product. To this end, a comprehensive Digital Twin of the production line was made, consisting of the following stations:

Uncoiling machine station. The uncoiling machine station is the first station for sink production. Here, a steel sheet roll is placed in the uncoiling machine, where it unwinds and a special PVC film is applied to one side. This film is essential for protection during the deep drawing process in stations E1 and E2. Afterward, the roll is cut to desired dimensions, and each steel sheet is placed on a pallet by a robot.

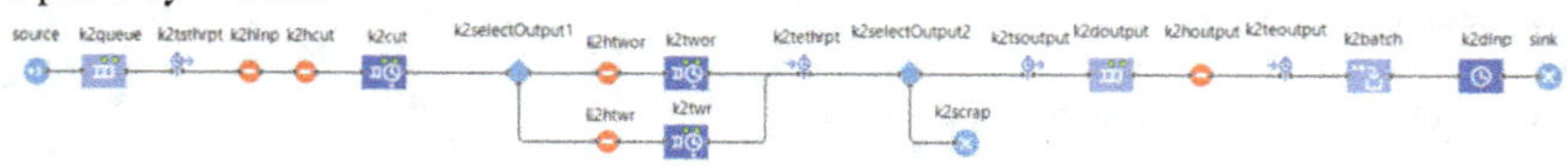
Figure 5: Diagram of the uncoiling machine station within AnyLogic

Tilter station. Once the pallet with the steel sheets is fully loaded, it is transferred to the tilter station, where it is rotated 180 degrees for proper orientation before entering station E1.
Stations E1 & E2. After the tilter, pallets are often stored temporarily before being moved to station E1. In the production line simulation model, intermediate storage is simulated when station E1 is not ready to receive a new pallet. After station E1, the sinks are transferred to intermediate storage and then to station E2 for final shaping. The production model also simulates this intermediate storage.

Washing & quality control station. The next station is washing and quality control. The sinks are transferred from E2 to temporary storage before moving to the washing station. At the wash station exit, a worker performs quality control.

Polyurethane station. After quality control, the sinks are transferred to the polyurethane station, often after a brief period of intermediate storage.

Packaging station. The final station is packaging, where two workers pack the sinks and place them on a pallet. Once full, a worker secures the pallet, and the sinks are ready for shipping.

Complete production line model. The logic diagram of the sink production line at a kitchen product factory combines all the stations discussed earlier, with additional blocks simulating the movement and temporary storage of sinks between stations. A graphical representation of the production line is presented in Fig.6. The process times for each sink model are input into the model via a Microsoft Excel file, along with other variables such as: 1) Total production quantity (units), 2) Number of sheets per pallet from the uncoiling machine, 3) Number of sinks per cart from station E1, E2, quality control, polyurethane, and packaging, 4) Defect rates for the uncoiling machine, E1, E2, quality control, and polyurethane stations. The user also specifies the production method at stations E1 and E2, including whether a robotic arm is used, if holes for faucets are drilled, and if the edge is welded. After entering all the data, pressing a button starts the simulation for monoblock sink production.

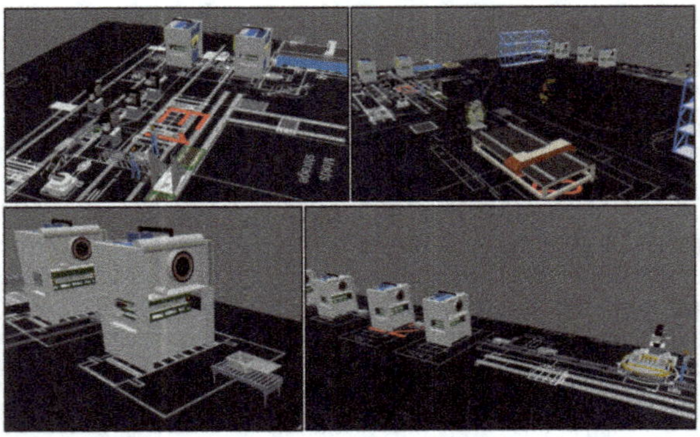

Figure 6: 3D representation of the monoblock sink production line

To enhance visualization and improve the accuracy of sink and machine movements in the factory's production area, both 2D and 3D environments have been designed for the simulation model. Fig. 6 shows snapshots of the 3D simulation environment. During modeling, the user can monitor the availability and utilization of the factory's workers and machines at each station using icons. Fig.7 provides a snapshot of the blocks representing the workers and machines. Specifically, on the left side of the image, the blocks display the availability (at a given moment) of forklifts, the cart transporting semi-finished sinks from the end of station E1 to a pickup point for the forklift (E1_Wagon), and the pallet truck servicing the end of the production line at the washing, polyurethane, and packaging stations (Pallet_Truck). On the right side, the blocks depict the availability of workers at each station (at a given moment). In both cases, these blocks allow the user to view the number of available machines and workers, as well as their usage rates. Additionally, the user can adjust the number of resources for each block, define the resource type (machine, worker, etc.), set their speed, and more. Lastly, the 2D and 3D environments offer the

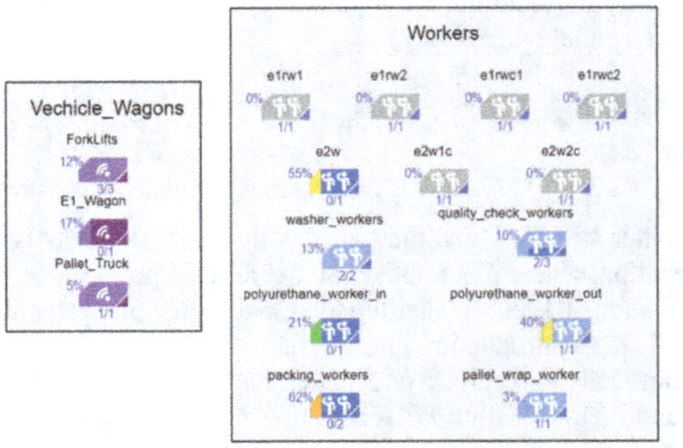

Figure 7: Snapshot of the simulation model with workers and machinery

ability to visualize production bottlenecks. From simulations conducted on the product codes recorded as part of this project, bottlenecks were observed at stations E1 and E2, particularly during the deep drawing processes using presses. This is logical since the processing times of the presses are significantly longer compared to other operations at stations E1 and E2.

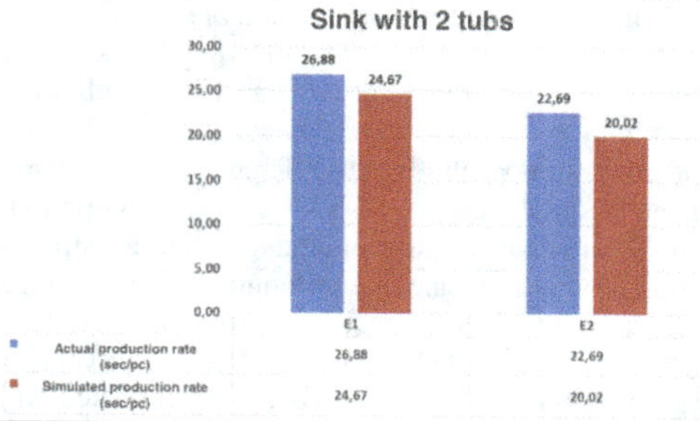

Figure 10: Comparison of actual and simulated production rate for a 2-tub sink in stations E1 and E2

Results

Once the model was created, it was validated by simulating a scenario with a known outcome. The validation of the production line model was performed, specifically for stations E1 and E2. The model is validated separately for each station, as the data provided by the company concerns individual stations rather than the entire production process. Since each station follows a different production plan, comparing the results of the unified model with actual production data would be incorrect. Validation for stations like the washing, quality control, and polyurethane stations is not performed, as there are no corresponding real data from the company. Similarly, although random measurements were taken for the uncoiling and packaging stations, it was decided to use the company's provided data as inputs for the model, so no further validation is required. The company collects data from each batch of products for the uncoiling, E1, E2, and packaging stations. This data includes total production time, net production time (excluding breaks), total sink quantities produced, and a breakdown of 'good', 'defective', and 'returnable' units (those requiring reprocessing). In the simulation model, defective or returnable sinks are considered discarded. Using this data, production rates for each product at stations E1 and E2 were calculated. Figs. 9, 10, and 11 compare the simulated production rates with actual production rates for three product codes at stations E1 and E2. The actual production rates

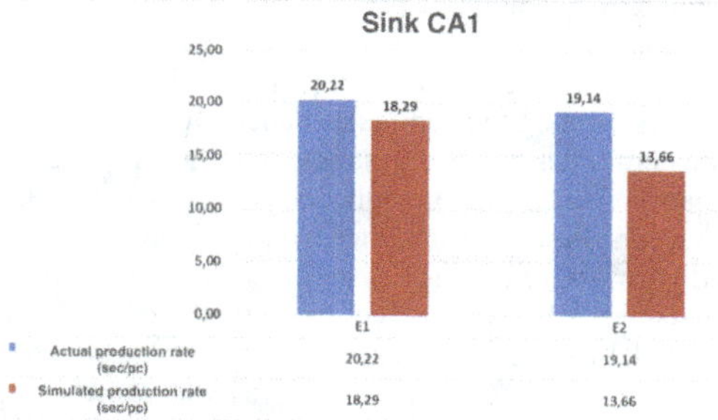

Figure 9: Comparison of actual and simulated production rate for a CA1 sink in stations E1 and E2

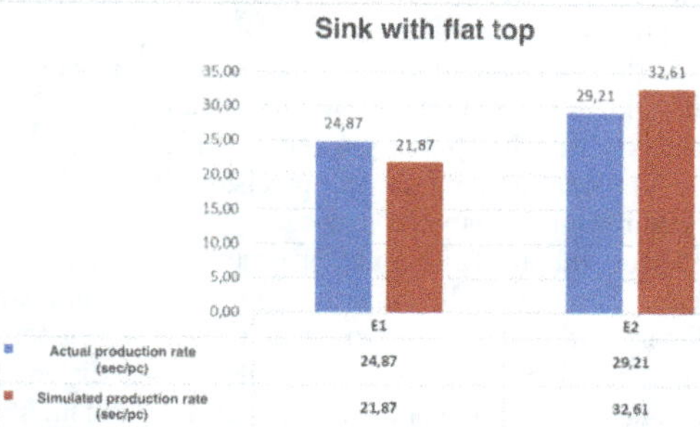

Figure 8: Comparison of actual and simulated production rate for a sink with a flat top in stations E1 and E2

Materials Research Proceedings 46 (2024) 412-419 https://doi.org/10.21741/9781644903377-52

were calculated by dividing the net production time by the number of items produced at each station. In most cases, the model's production rates closely match the actual rates, with differences of 2 to 3 seconds. The only significant discrepancy occurs for the CA1 sinks at station E2 (Fig.9). However, these differences are small and are explained further below.

The company provided production data that includes total production quantities for each station and product. While real production rates were calculated using total production quantities, this approach does not fully reflect reality since it includes defective and returnable items. In the simulation model, defective and returnable sinks were simulated with corresponding blocks at the end of stations E1 and E2. However, a defective item can emerge at any substation within the station, meaning it may not always contribute to total production time. The company's data does not provide detailed information on this. Another factor contributing to differences in model and actual production rates is that the measurements were taken over a 2-3 hour production period. Since certain operations at stations E1 and E2 involve workers, human variability affects production rates. For example, worker performance may fluctuate due to fatigue over the course of a shift. Moreover, a production batch may span multiple shifts, requiring different workers to handle the same task, introducing further variability. Such fluctuations are difficult to capture precisely in observational measurements. In conclusion, the differences between simulated and actual production rates are reasonable, and the model for stations E1 and E2 accurately reflects reality to a high degree.

Conclusion

In this study, a simulation model of the production line for a sink factory was implemented and benchmarked against real data. As a reference, details were carefully documented from a real production line for metal sinks. The entirety of the production line was simulated, included workers, waiting time and redundancies to form a complete Digital Twin that can be exploited in a modern Industry 4.0 infrastructure. Using AnyLogic software, a production line model for each station involved in the production of monoblock sinks was detailed, followed by the integrated production system, represented in both 2D and 3D environments, including all production stations and the movement of agents between them. Once completed, the simulation model was validated for stations E1 and E2 by comparing the production rates for three product codes obtained from the model with the actual production rates provided by the company. The comparison showed that the simulation model accurately reflects reality for these stations. Considering that certain limitations apply to the digital twin output accuracy in terms of realism, such as human variability, machinery accuracy over time and sudden disruption events (e.g. power shortage), further modelling factors could improve the overall accuracy of the system, but with diminishing returns. Other limitations are related to the overall complexity of the modeled factory, where slight changes to the physical production line should be continuously monitored and re-modelled, alongside additional measurements and physical benchmarking to maintain the overall output accuracy.

An accurate digital twin for a monoblock sink production factory offers significant benefits, providing a virtual replica that mirrors the real production process with precision. By validating the digital twin against real-world production data, the model's accuracy is proven, with production rates closely aligning with actual factory output. This validation ensures that the digital twin can reliably simulate various production scenarios, allowing the company to optimize operations, identify bottlenecks, and experiment with changes in worker allocation, machinery usage, and process times. Digital twins offer a flexible architecture that scales across different manufacturing contexts by dynamically adjusting to the level of automation in a factory. In highly automated environments, they can integrate real-time IoT data, robotics, and AI-driven analytics to enhance process efficiency and predictive maintenance. Ultimately, this leads to enhanced efficiency, reduced production costs, and informed decision-making, all while minimizing the risks associated with physical experimentation on the factory floor.

Acknowledgements
The current research has received funding under the Greek Innovation Investment Fund with code ΟΠΣ 4228 from the Operational Plan of the "Region of Central Macedonia 2014-2020".

References

[1] Schwab, Klaus. The fourth industrial revolution. Crown Currency, 2017. https://books.google.gr/books?id=9hgXvgAACAAJ

[2] Grabowska, Sandra. "Smart factories in the age of Industry 4.0." Management systems in production engineering 28.2 (2020): 90-96. https://DOI 10.2478/mspe-2020-0014

[3] Ghobakhloo, Morteza. "Industry 4.0, digitization, and opportunities for sustainability." Journal of cleaner production (2020). https://doi.org/10.1016/j.jclepro.2019.119869

[4] Leng, Jiewu, et al. "Digital twins-based smart manufacturing system design in Industry 4.0: A review." Journal of manufacturing systems 60 (2021): 119-137. https://doi.org/10.1016/j.jmsy.2021.05.011

[5] Bossel, Hartmut. Modeling and simulation. AK Peters/CRC Press, 2018. https://doi.org/10.1201/9781315275574

[6] R. E. Shannon, Systems simulation: the art and science. Englewood Cliffs, N.J.: Prentice-Hall, 1975.

[7] Banks, Jerry. "Introduction to simulation." Proceedings of the 31st conference on Winter simulation: Simulation---a bridge to the future-Volume 1. 1999. https://doi.org/10.1145/324138.324142

[8] Zeigler, Bernard P., et al. Guide to modeling and simulation of systems of systems. Springer London, 2013. https://doi.org/10.1007/978-3-319-64134-8

[9] J. W. Forrester, "Dynamic models of economic systems and industrial organizations," System Dynamics Review: The Journal of the System Dynamics Society, vol. 19, no. 4, pp. 329–345, 2003. https://doi.org/10.1002/sdr.284

[10] Goldsman, D., & Goldsman, P. Discrete-event simulation. In Modeling and Simulation in the Systems Engineering Life Cycle: Core Concepts and Accompanying Lectures (pp. 103-109). London: Springer London. 2015. https://doi.org/10.1007/978-1-4471-5634-5_10

[11] Macal, Charles M. "Everything you need to know about agent-based modelling and simulation." Journal of Simulation 10.2 (2016): 144-156. https://doi.org/10.1057/jos.2016.7

[12] I. Grygoryev, AnyLogic in three days: A quick course in simulation modelling, 5th ed. 2018.

[13] Mahdavi A. The art of process-centric modeling with AnyLogic. 2020. Online: https://www. anylogic. com/resources/books/the-art-of-process-centric-modeling-withanylogic.

An industrial environment background noise reduction technique used in the measurements of the tapered roller bearings

Bogdan Alexandru SABO[1,a *], Ștefan VOLOACĂ[2,b]

[1]Timken Romania S.A., Ploiesti, Romania

[2]National University of Science and Technology POLITEHNICA Bucharest, Romania

[a]bogdan.sabo@timken.com, [b]stefan.voloaca@upb.ro

Keywords: Bearing Noise Measurement, NVH, Tapered Roller Bearing, Industrial Noise

Abstract. This paper investigates the feasibility of noise measurements for automotive tapered roller bearings in an industrial production facility. The European Union regulations regarding the NVH have indirectly led to improvements in the field of the bearings industry, affecting both industrial and automotive markets. To achieve the improvements of the bearings in terms of NVH, a new relevant study methodology must be explored. The goal of this study is to examine the technical possibilities of reducing industrial background noise by defining a relevant noise measurement in the production environment for tapered roller bearings. Each roller bearing production site has different sources of noise coming from various machines. Successfully measuring noise in automotive tapered roller bearings in production requires an enclosure where the production assemblies are not influenced by the background noise of the plant. The enclosure must be covered with isolation materials. Foams and sponges must be selected to achieve the desired noise attenuation in the target range of frequencies.

Introduction

In the prototyping period or after manufacturing, roller bearings are tested by various methods to ensure the quality and performance of the systems that use them. Common tests include durability, loading and noise, and vibration. The most common tests are the loadings, the loadings and the noise and vibration one. These tests help to quantify efficiency, reliability, and the durability of a roller bearing subject to different conditions.

In the industrial environment, background noise can significantly compromise fault detection. The variable operating conditions, in some cases with intense noise, affect the vibrating features of the roller bearings – making it difficult to diagnose faults with accuracy [1-3]. Background noise in some cases can cover the vibration signal of the bearings, causing further inaccuracies in fault diagnosis [4].

Industrial environment noise interference, be it impulsive noise or white Gaussian noise, presents a challenge regarding the fault classification and diagnosis of roller bearings [5], [6]. The interference from the noise interference in real-world industrial settings creates substantial challenges in collecting accurate signals from the rolling bearings [7], [8].

In some cases, fault detection in the industrial field can be masked by those interference. The influence of the noise can be sufficiently strong that it masks imperfections in the product, so finding alternative methods for extracting data becomes even more critical [9]. To reduce the effects of noise interference, engineers can develop new methods of signal processing – for example, creating special test benches or different measuring methods.

Reducing background noise with signal processing methods, including the test bench, can be effective when tests are performed under varying conditions. Some of those methods include denoising and adaptive denoising autoencoders and residual dilated pyramid networks. These techniques can increase the accuracy of fault diagnosis in noisy environments [10], [11].

Materials Research Proceedings 46 (2024) 420-427 https://doi.org/10.21741/9781644903377-53

Another method of reducing background noise is to use various noise-reducing materials in the test bench environment. For example, composite materials are quite effective at dampening noise and vibrations [12]. Other materials that absorb noise include acoustic foam and soundproofing panels, which also reduce background noise and improve the overall testing environment [13].

Insulation materials with high noise reduction coefficients, such as specialized composite materials, can significantly reduce the impact of background noise on the accuracy of the measurements [14].

Failures of electric motors used in various industrial applications often occur due to unwanted excessive bearing excitation. Little research is available on noise analysis evaluation methods addressing industrial bearing excitation, as studies in this area have been performed mainly in the field of vibration analysis and motor current vibration signal analysis. In one study, Nakamura et al. [15] proposed a diagnostic approach that utilizes the rotating sound of the motor during a no-load test. Different motors with different bearing defect faults magnitudes were measured.

P. Henriquez et al. [16] summarized more than 100 papers discussing condition monitoring methodologies. Eighty-two percent of the works studied vibration measurement analysis, while only 18% studied sound/audio measurement. The lack of interest in sound/audio measurement studies is likely due to the difficulties created by the noisy environments (whether from the test machine itself or from other machines) they must be performed in that compromise the signal quality. Anechoic chambers in the industrial environment are not a practical option. However, the popularity of microphone measurements remains – the measurement is performed at a distance, and it is not directly mounted on the application.

Studies (e.g., [17]) have related the effects of grease contamination to the dynamic performance of ball bearings using acoustic emission. [17], specifically, concluded that this measurement technique is an effective method of early detection.

Mamun et al. [18] proposed a real-time multi-sensing condition monitoring method using both accelerometers and microphones. Their tests mounted six microphones inside a noise-isolating enclosure.

However, even if different measurement techniques are used (such as running different sensors in parallel [19]), background noise reduction is necessary. The desired accuracy in detecting defects within rolling element bearings can only be accomplished through proper noise-isolating bench test design and signal filtration.

Measurement equipment and test preparation

The tests were made on a Vertical Sound Test Machine inside the Romanian Timken Company laboratory. To measure the noise, vibration and harshness (NVH) levels of the tapered roller bearings, a configuration of measuring system like in Fig. 1 was done. It consists of a Bruel and Kjaer (B&K) data acquisition system and two microphones model 378B02 from PCB Piezotronics.

The Data acquisition system (DAQ) is a 4/6 channels Input Module LAN-XI 51,2 kHz Type 3050.

The microphone has a nominal diameter of 1/2", a free-field frequency response characteristic with a frequency range of (±2dB) 3.75 to 20000 Hz (±2dB). It has an inherent noise of 15.5 dB(A) re 20 μPa and a dynamic range of 137 dB re 20 μPa.

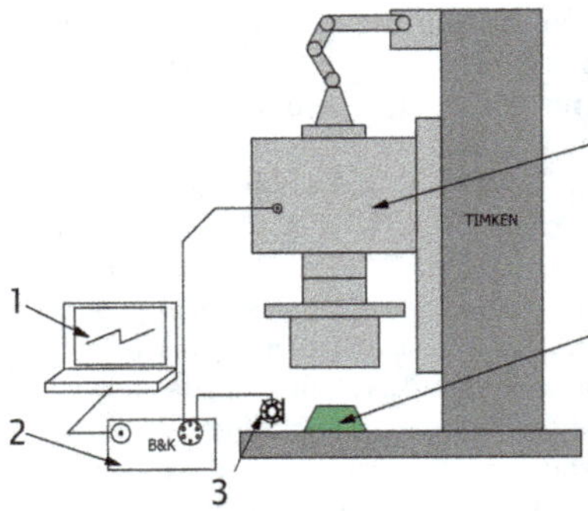

Fig 1. The NVH testing bench of the tapered roller bearings
1. Workstation; 2. DAQ; 3. Microphone; 4. Vertical sound test machine; 5. Tested tapered roller bearing

This study incorporates an enclosure to isolate the bearing from industrial noise. The proposed enclosure encapsulates the Vertical Sound Test machine head and the test bearing itself. The enclosure is constructed from aluminum frames and 8 mm thick plexiglass panels; Fig. 2 shows that all open spaces and edges are sealed.

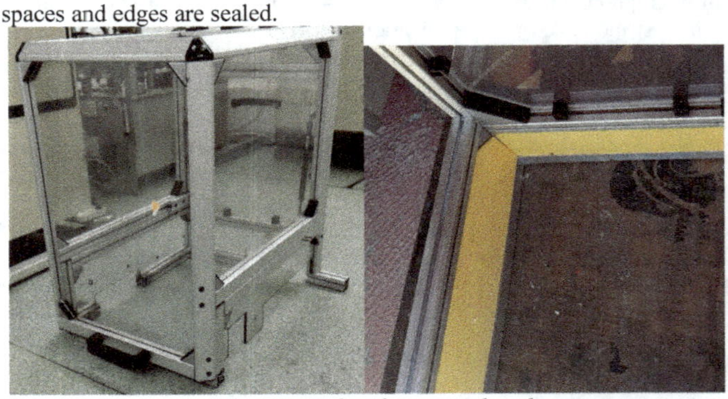

Fig 2. Acoustic enclosure used in the test and enclosing process

The isolating layers stack in the following order and thickness: plexiglass panels of 8 mm, extruded polystyrene of 5 mm, sound isolation membrane of 2.5 mm, and acoustic foam of 50 mm. These inexpensive materials are widely used in building construction and music recording studios.

To achieve a flat surface and accommodate the mounting screws, sheets of extruded polystyrene with a thickness of 5 mm were placed on the plexiglass panels. An acoustic membrane of 2.5 mm thickness, made of a high-density viscoelastic polymer that offers good acoustic insulation, was mounted on the polystyrene. Fig. 3 presents the steps of noise-isolating the cabin. Fig. 4 shows the final state of the cabin.

Materials Research Proceedings 46 (2024) 420-427 https://doi.org/10.21741/9781644903377-53

Fig 3. Extruded polystyrene, acoustic membrane, and acoustic foam applied on the cabin

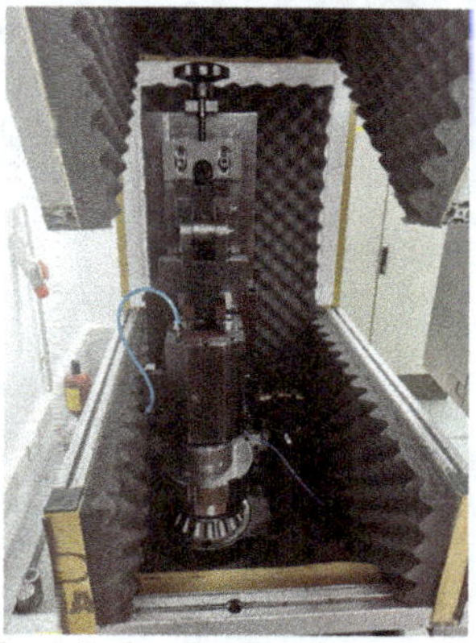

Fig 4. Isolated cabin: final state

Comparative results

The tests consisted of measurements using one reference acoustic point to measure the environment noise, placed near the hydraulic unit, and one microphone placed near the tested bearing. Results from the current state were compared to the results generated after the isolation cabin was placed on the machine enclosure. Fig. 5 presents the two microphones' positions.

Materials Research Proceedings 46 (2024) 420-427 https://doi.org/10.21741/9781644903377-53

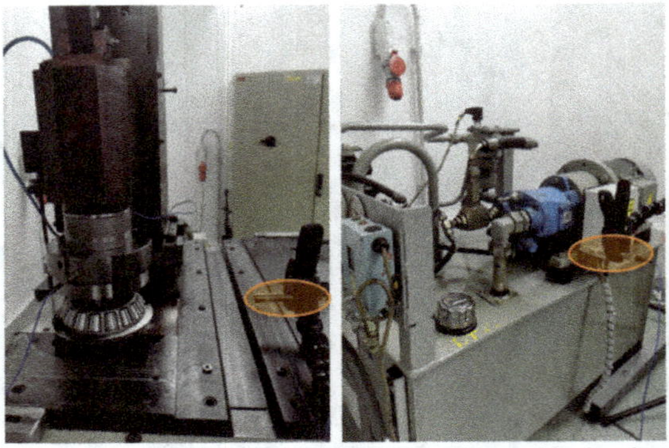

Fig 5. Key measurement points: bearing assembly and hydraulic unit

Fig. 6 plots results obtained from a defect-free bearing. The red line represents the results obtained by placing the microphone within the isolating cabin, and the blue line represents the results obtained by placing the microphone outside the isolating box. The difference is notable: 70.1 dB[A] when the microphone is within the enclosure (red), and 74.2 dB[A] when it is outside and near the hydraulic unit (blue).

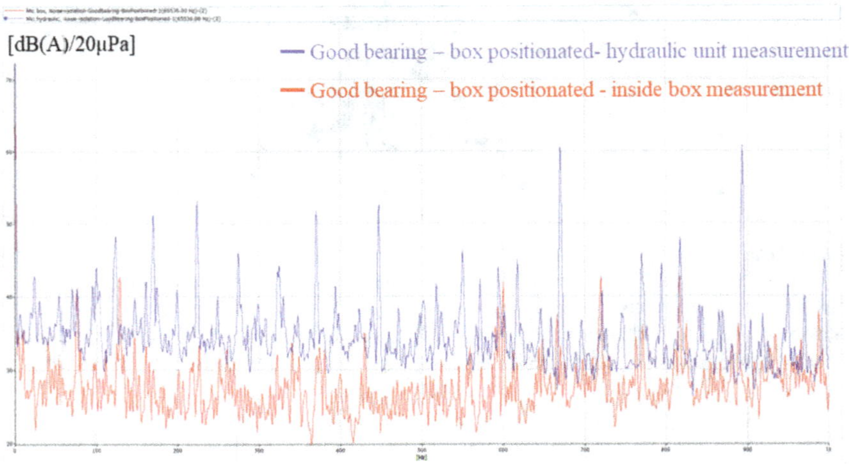

Fig 6. Measurement comparison between hydraulic unit microphone and enclosure microphone with a good bearing

Fig. 7 presents the acoustic response of a bearing. Measurements were taken within (blue) and outside (red) the proposed enclosure, placing the microphone near the tested assembly. The enclosed assembly produces less noise, as presented in Fig. 7.

Materials Research Proceedings 46 (2024) 420-427 https://doi.org/10.21741/9781644903377-53

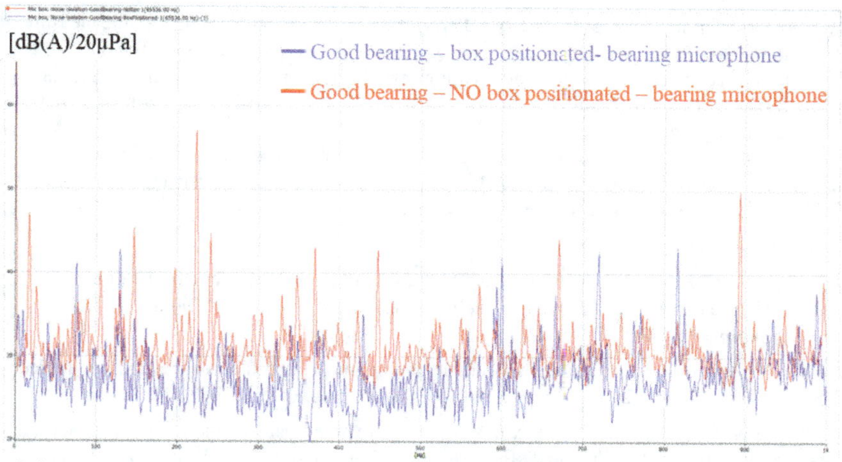

Fig. 7 Acoustic measurement comparison of a bearing assembly within and without the isolation enclosure

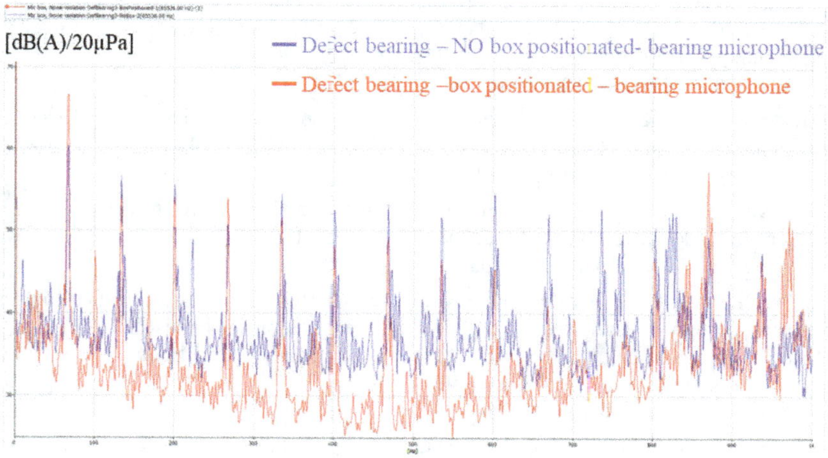

Fig. 8 Acoustic measurement comparison of a defect-bearing assembly within (red) and without (blue) the isolation enclosure

The acoustic signal of the enclosed test exhibits more suitable behavior; it has less background noise and it accurately identifies defect frequencies.

Summary

In the industrial environment, background noise significantly compromises bearing test results and fault diagnosis processes.

The paper introduces a practical method of improving the accuracy of the noise measurements in the industrial environment. Utilizing an acoustic enclosure made from readily available materials such as plexiglass, extruded polystyrene, and acoustic foam, the study achieves a notable reduction in background noise, enhancing the detection of bearing defects. This approach reduces

Materials Research Proceedings 46 (2024) 420-427 https://doi.org/10.21741/9781644903377-53

the noise interference by approximately 4 dB[A], enabling more precise and reliable measurements in a noisy production environment. The technique is simple and cost-effective, making it particularly relevant for industrial applications where conventional anechoic chambers are not present. The paper also highlights the potential of future research over optimized methods of background noise reduction.

This chamber can be used in various environments and noise conditions, increasing both its scientific interest and practical value. Overall, the study makes a significant contribution to the field of noise measurement and industrial testing.

References

[1] X. Zhao and J. Chai, "Adaptive weight-based capsule neural network for bearing fault diagnosis", Measurement Science and Technology, vol. 34, no. 6, p. 065008, 2023. https://doi.org/10.1088/1361-6501/acb9aa

[2] X. Li, "Bearing-fault-feature enhancement and diagnosis based on coarse-grained lattice features", Sensors, vol. 24, no. 11, p. 3540, 2024. https://doi.org/10.3390/s24113540

[3] Y. Wang, S. Zhang, R. Cao, D. Xu, & Y. Fan, "A rolling bearing fault diagnosis method based on the woa-vmd and the gat", Entropy, vol. 25, no. 6, p. 889, 2023. https://doi.org/10.3390/e25060889

[4] X. Wang, G. Sui, J. Xiang, G. Wang, Z. Huo, & Z. Huang, "Multi-domain extreme learning machine for bearing failure detection based on variational modal decomposition and approximate cyclic correntropy", IEEE Access, vol. 8, p. 197711-197729, 2020. https://doi.org/10.1109/access.2020.3034651

[5] Q. Jiang, F. Chang, & B. Sheng, "Bearing fault classification based on convolutional neural network in noise environment", IEEE Access, vol. 7, p. 69795-69807, 2019. https://doi.org/10.1109/access.2019.2919126

[6] H. Lu, K. Zhou, & L. He, "Bearing fault vibration signal denoising based on adaptive denoising autoencoder", Electronics, vol. 13, no. 12, p. 2403, 2024. https://doi.org/10.3390/electronics13122403

[7] V. Patel, N. Tandon, & R. Pandey, "Improving defect detection of rolling element bearings in the presence of external vibrations using adaptive noise cancellation and multiscale morphology", Proceedings of the Institution of Mechanical Engineers, Part J: Journal of Engineering Tribology, vol. 226, no. 2, p. 150-162, 2011. https://doi.org/10.1177/1350650111425750

[8] R. Hu, Y. Zhou, L. Zhang, Z. Guo, & C. Zhou, "Bearing fault diagnosis based on cemdan energy weighting method", DEStech Transactions on Computer Science and Engineering, no. ica, 2019. https://doi.org/10.12783/dtcse/ica2019/30752

[9] H. Shi, J. Chen, J. Si, & C. Zheng, "Fault diagnosis of rolling bearings based on a residual dilated pyramid network and full convolutional denoising autoencoder", Sensors, vol. 20, no. 20, p. 5734, 2020. https://doi.org/10.3390/s20205734

[10] R. Zhu, M. Wang, S. Xu, K. Li, Q. Han, X. Tonget al., "Fault diagnosis of rolling bearing based on singular spectrum analysis and wide convolution kernel neural network", Journal of Low Frequency Noise, Vibration and Active Control, vol. 41, no. 4, p. 1307-1321, 2022. https://doi.org/10.1177/14613484221104639

[11] H. Lu, K. Zhou, & L. He, "Bearing fault vibration signal denoising based on adaptive denoising autoencoder", Electronics, vol. 13, no. 12, p. 2403, 2024. https://doi.org/10.3390/electronics13122403;

[12] P. Goksem, E. Fletcher, & R. Hargreaves, "Measurement of roller speed in a cylindrical roller bearing using a laser anemometer", Journal of Lubrication Technology, vol. 100, no. 4, p. 486-491, 1978. https://doi.org/10.1115/1.3453260

[13] A. Loriemi, G. Jacobs, S. Reisch, D. Bosse, & T. Schröder, "Experimental and simulation-based analysis of asymmetrical spherical roller bearings as main bearings for wind turbines", Forschung Im Ingenieurwesen, vol. 85, no. 2, p. 189-197, 2021. https://doi.org/10.1007/s10010-021-00462-1

[14] W. Chen, R. Mills, & R. Dwyer-Joyce, "Direct load monitoring of rolling bearing contacts using ultrasonic time of flight", Proceedings of the Royal Society A: Mathematical, Physical and Engineering Sciences, vol. 471, no. 2180, p. 20150103, 2015. https://doi.org/10.1098/rspa.2015.0103

[15] H. Nakamura, K. Asano, S. Usuda, Y. Mizuno, A Diagnosis Method of Bearing and Stator Fault in Motor Using Rotating Sound Based on Deep Learning. Energies 2021, 14, 1319. https://doi.org/10.3390/en14051319

[16] R., Issa, G. Clerc, M. Hologne-Carpentier, R. Michaud, E. Lorca, C. Magnette, A. Messadi, Review of Fault Diagnosis Methods for Induction Machines in Railway Traction Applications. Energies 2024, 17, 2728. https://doi.org/10.3390/en17112728

[17] K. A. I. Sheriff, V. Hariharan,, T. Kannan, Analysis of Solid Contamination in Ball Bearing Through Acoustic Emission Signals. Archives of Metallurgy and Materials. 2017, 62, (3), pp. 1871–1874.

[18] A. Al Mamun, M.M. Bappy, A.S. Mudiyanselage, et al. Multi-channel sensor fusion for real-time bearing fault diagnosis by frequency-domain multilinear principal component analysis. Int J Adv Manuf Technol 124, 1321–1334 (2023). https://doi.org/10.1007/s00170-022-10525-4

[19] B. A. Sabo and Ş. Voloacă, Surface Defects Vibration Measurements of Automotive Tapered Roller Bearings, *IOP Conf. Ser.: Mater. Sci. Eng.* **1303** 012022, 2024. https://doi.org/10.1088/1757-899X/1303/1/012022

Innovative Manufacturing Engineering and Energy - IManEE2024 Materials Research Forum LLC
Materials Research Proceedings 46 (2024) 428-435 https://doi.org/10.21741/9781644903377-54

Optimization of the 3D visualization process when adding additional elements in the design of an existing interior

Tsveta TODOROVA[1,a *], Darina DOBREVA[1,b], Tsena MURZOVA[1,c], Mariana MURZOVA[1,d], Iliya ILIEV[1,e] and Ventsislav MARKOV[1,f]

[1]Technical University of Varna, Bulgaria

[a]tsveta.todorova@tu-varna.bg, [b]darina.dobreva@tu-varna.bg, [c]murzova@tu-varna.bg, [d]m.murzova@tu-varna.bg, [e]i.iliev@tu-varna.bg, [f]ventsi_markov@tu-varna.bg

Keywords: 3D Scanning, Photogrammetry, 3D Modeling, Interior

Abstract. This report describes a method for improving the design of an existing interior and its visualization without the need to model all elements anew in a 3D interior project. Through a mobile photogrammetry application, we scan the room in a very short time and then add the desired changes to the interior in a 3D modeling software. In this way, we save a lot of time and value for making the 3D visualization, and the final result is as close as possible to the existing interior.

Introduction

Interior design is a dynamic and constantly developing field, characterized by a rapid pace of development and adaptation to modern trends. If 100 years ago fashion trends lasted for several decades, today we see how interior styles change so much every 3-5 years that something that was modern and relevant until recently is now considered outdated and outdated. That is why a large part of people choose an option in which to rearrange the elements of the interior or to add small but visually effective changes that give a feeling of a new look to the room. The advantages of such solutions are that they save a lot of effort, time, money, and the contribution to the environment is also significant, avoiding excessive consumerism.

In the last few decades, we have observed the following trend: people from big cities live in smaller and smaller dwellings, and since homes have become a permanent place of work for a large part of them, it is now imperative to have a dedicated space for this purpose.

In this report we will look at how we can quickly and easily use "Photogrammetry" and "3D scanning" technology in cases where small corrections, additions or structural solutions need to be made to an existing design, without having to model the entire interior room from scratch in 3D design software. We will give an alternative to doing a major repair when we need to separate a workplace in a small apartment. This approach not only saves a lot of time and unnecessarily high costs, but also provides a much more realistic and detailed visualization of the proposed project.

Problem

Working in the field of industrial and interior design, we increasingly encounter problems of the following types:

- -A bar, café, or restaurant changes ownership, and the new owner wants minor design changes to the room, such as adding decorative grids, without the need for major renovations.
- -In a hotel lobby, the owner wishes to visually separate the relaxation areas so guests can feel more private.
- -In a studio apartment, different dining and relaxation zones need to be separated using a perforated decorative grid.
- -In a beauty salon or spa center, a part of the room needs to be designated as a zone with a new purpose, with the partition not blocking the natural light source.

These and many other similar situations share a common problem: the need for minor corrections, like a decorative partition in an already furnished space, to improve the functionality or aesthetics of the room. [1]

To undertake such an action, every client wants to envision how the space would look before making the change. What they receive as a proposal is the creation of an interior design project with 3D visualizations. However, this entire process has three major drawbacks:

Cost: Creating a complete 3D project is an expensive service, which is not justified for such small interior changes.

Time: Besides being costly, this service is also time-consuming. Building every single element from scratch in an existing interior, considering all the small details that form the final vision, is a process that takes a lot of time even for professionals in the field.

Detail Sensitivity: Small details matter. When a designer is trusted to design a certain room, they provide a professional proposal for a combination of colours, textures, and zoning. However, when the realization of these projects begins, a large part of the clients change some things during the process. Differences in colour shades occur, accent decorations are added, more convenient furniture is chosen that does not match 100% with the prepared 3D project, and so on. Similarly, if an existing interior needs to be recreated in 3D with all the small details like texture specifics, lighting color temperature, reflections at certain times of the day, small decorations, types of paintings, photos, and souvenirs, it would be extremely laborious for the interior designer.

This is why we seek a solution to make the designer's work easier and the service more accessible to clients. The answer lies in the technology of "Photogrammetry".Photogrammetry, traditionally associated with geodetic measurements and mapping, has found new and innovative functions in the field of design in recent years. This scientific discipline, which uses photographic images for the accurate measurement and representation of objects in three-dimensional space, has significantly evolved with technological advancements and has become an integral part of numerous modern design practices. [2,3,4]

Application of Photogrammetry in Interior Design

Today, we have the opportunity to use this technology in an important field of design, as one of the biggest advantages with hundreds of years of research and experimentation in the field, which today does not give much faster and more accurate results in the process of working with technology.

The photogrammetry process begins with taking multiple pictures of a room from different angles. These photos are then processed through specialized software that creates a three-dimensional model of the space. This method allows for quick and accurate collection of geometric data, minimizing the errors that often occur with all measurement methods.

From here we can conclude that the most important are the captured images and a more specific scanning method that we will choose.

There is a professional scanning method, and mobile technology has advanced so much in recent years that even with the help of a few smartphones and a 3D scanning app, we can get good enough results for the whole experiment.[5,6]

Professional 3D scanners

In general, 3D laser or structured light scanners are best suited for scanning indoor spaces and, in particular, the interior.

The reasons for this include high accuracy, speed and the ability to capture complex details. The main advantages of this method are the following:

- **Scan speed:** These scanners can scan large areas quickly and efficiently, which is especially important when scanning large indoor spaces. This greatly reduces the time required to collect a lot of data compared to other methods, such as taking pictures with a DSLR camera and manually assembling a 3D model from the footage in additional software.

- **Easy integration with software:** Most all 3D scanners come with specialized software that automatically processes and joins the scanned data, creating an entire 3D model. This simplifies the scanning process and reduces the need for manual intervention.
- **Multi-sensor systems**: Some 3D scanners use a combination of different technologies (eg lasers and cameras) that allow capturing both the geometry and textures of an object. This is useful for creating realistic 3D models that include both the shape and visual characteristics of the interior.

Of course, there are also disadvantages to consider:

- **Equipment price:** High-quality 3D scanners are often significantly more expensive. This may limit your access to small businesses or individual users. However, prices vary widely depending on the accuracy, resolution and functionality of the scanner. In our experiment, it is not necessary to use the highest class of scanning, which, for example, is used in the construction of virtual museums and landmarks, therefore even betting on the lowest class of hobby (amateur) scanners we will get the desired result. [5,6,7,8]

Scan with mobile devices

Room scanning using mobile devices is gaining popularity thanks to improvements in the cameras and sensors of modern smartphones. This method offers convenience and affordability, making it attractive to a wide range of users, from hobbyists to professionals. After installing a scanning application, we only need to follow the instructions that appear on the screen. This usually involves moving the smartphone slowly and evenly around the room to cover all areas. Some applications require multiple photos to be taken from different angles, while others use a LiDAR sensor for continuous scanning.

Data processing follows. As with 3D scanners and mobile apps, the 3D model assembly software is built-in and automatic, making the final process of creating a three-dimensional 3D model even easier and faster. Depending on the application, it may be possible to edit and improve the model.

Advantages: The advantages of scanning with mobile devices are not few, namely that they are widespread and easily accessible, making 3D scanning accessible to a wider audience, and eliminating the need for expensive equipment. The small volume and lightness of the devices makes them compact and easy to carry and use in different locations. 3D scanning apps usually offer intuitive interfaces and easy-to-follow instructions, and data processing is fast because modern smartphones have enough computing power. [2,3,9]

The disadvantages are as follows: Despite advances in technology, mobile devices cannot match the accuracy of professional 3D scanners. If the volume of the room is too large, there are mirror surfaces and poor lighting, the results may not be satisfactory, but in these cases, the use of 3D scanners is recommended.

We can conclude that 3D room scanning with mobile devices is an affordable and convenient method that offers many advantages, especially for small projects and non-professional applications. Although there are some limitations in terms of accuracy and performance, modern smartphones with powerful cameras and LiDAR sensors can create quite high-quality models. Using appropriate scanning applications and ensuring good lighting conditions can greatly improve results.

Materials Research Proceedings 46 (2024) 428-435 https://doi.org/10.21741/9781644903377-54

Experimental Part

The goal is to compare how much faster and more efficient the method of scanning and correcting an existing design is versus building the interior from scratch. We decided to use a room consisting of a bedroom with bedside tables, as well as a relaxation area composed of a sofa, a coffee table, and also a work area including a desk with a chair (Fig. 1).

Fig. 1. The room to be divided

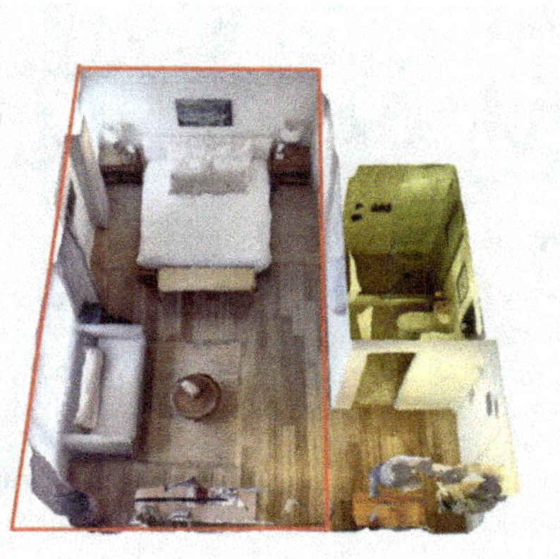

Fig. 2. Decorative divider

The separation of different areas in "studio" type rooms is extremely important in modern interior design, because in order for a single corner to be functional, it has various characteristics that must be respected. Such, for example, are: lighting, arrangement of furniture, colors, etc. In order to create the feeling of a separate bedroom and living room, without stopping the natural light coming from the window, it is planned to place a decorative partition grid, which will be an accent element in the previous design, and will delicately divide the room into two zones with different functions.

One of the biggest advantages of this type of openwork grids, apart from light transmission, is their extreme thinness. In this way, they do not occupy the space with their volume, and their strength can be very high when choosing a suitable material such as metal, aluminium composit panels, pressed first-quality wood and others. (Fig. 2). [10]

The room is scanned with a mobile device equipped with a Lidar sensor, with various smartphone brands offering certain models with this built-in feature. The application used for the scan is PolyCam, and it automatically assembles the captured frames and stitches them into a 3D model that can be exported in different formats, depending on which 3D modeling software we are working with. The scan time is less than 15 minutes and the data processing time is about 10 minutes. The whole process takes less than half an hour and the results are completely satisfactory.

As we can see in Fig. 3 the details of the design of the room are fully preserved and even if we were to build the interior in a 3D modeling program, the textures and shades of elements such as flooring, textiles, and decorations would not be able to be recreated with such accuracy.

Fig. 3. Interior details

Fig. 4. Vector element

This is an extremely important part when visualizing the project in front of the clients, because this way they see completely realistically how the correction made to the interior will affect the rest of the room. Having already a perfect copy of the room, but in a three-dimensional model, the designer can easily visualize countless ideas for the pattern of the decorative grid. In this way, things like the size and density of the holes, the main color of the partition wall, the geometry and style of the pattern can be set most easily. Once the client has approved the design for the decorative wall, It should render correctly with all shadows cast. [11]

The vector element of Fig. 4 is added to the 3D modeling software, where it is given the desired thickness, size and texture as specified by the client.

The divider can be made of different materials, such as wood, metal or plastic, and depending on the material, it is cut from the sheet material, using a CNC router or a laser.

After making the necessary adjustment of the lights that the natural light from the window throws to the room, it is now possible to make a realistic visualization of the interior and see the effect of the decorative grid (Fig. 5).

Various scenes and visualizations can be made from all points of the room as needed. An additional advantage of this method is that it allows the use of virtual reality technology. By using VR and AR, designers and clients can "step into" the 3D models and explore the spaces at a real scale, which contributes to a more realistic visualization.

With a ready-made design for the grille, the time it took to visualize it in a real interior took less than an hour, and the shooting itself was done during the free trial period of the PolyCam app, whose monthly subscription is under $10.

Fig. 5. Decorative grid

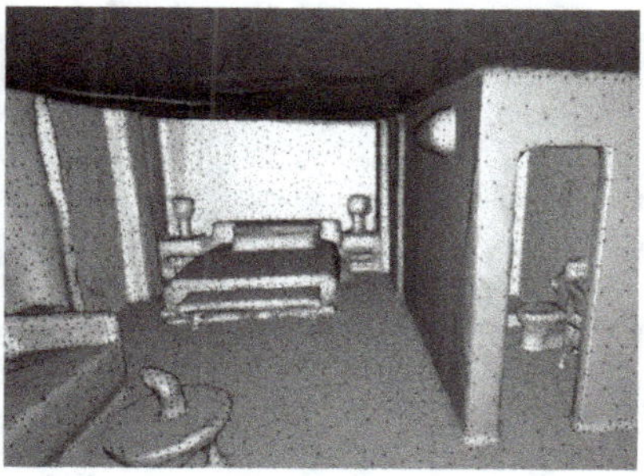

Fig. 6. Polygonal grid of the scanned room

The time to add the decorative grid and render is no more than 15-20 minutes, regardless of the 3D design software we use. All this leads us to the following results:

- Scanning and data processing - less than 30 min;
- Adding a decorative partition and visualization - less than 30 min.

The conclusion is that in less than an hour we manage to create a 1:1 3D model with the existing interior and add the desired elements to the design of the room. Another advantage of the mobile capture method is that it allows even the clients themselves to take the capture, which can eliminate the designer's visit to the site. This is a great opportunity for freelancers who can work with clients from all over the world without being tied down by their location. [12,13,14,15

Regarding the disadvantages - the main one is that this method is applicable in cases where it is not necessary to move furniture and rearrange the room.

If the adjustments require furniture to be moved, this must be done before scanning, which can sometimes be very time-consuming. If the client cannot scan the premises by himself, due to lack of technical literacy or appropriate diligence, the designer must necessarily visit the site and take the survey himself.

Despite this problem, there is a solution for it, and that is if the elements of the interior need to be visualized in a new configuration and need to be moved, then they can be scanned individually so that each element is a separate object in our project. This would of course increase the scanning time, but it opens up a lot of possibilities for working with the individual elements of the interior, such as moving, rearranging, etc. [16,17]

Conclusion

Before drawing a conclusion, we can look at the values in Table 1. The comparison is made between the interior scanned in the experimental part and a 3D-modelled room with the same furniture, made in the Sketchup software.

Table 1: Differences in values when scanning and 3D modelling the same room

	Scanned room	**3D modeled room**
Polygons	57,191	564, 144
Vertices	28,603	212,551
Time spend	< 60 min.	~ 10 hours

It is important to note that when scanning the room, the sanitary room adjacent to the room is also scanned, which includes a shower cabin, a sink and a toilet monoblock, which are not subject to the described redesign, but are included in the total scanning time and the number of polygons and corners from the grid (Fig. 6).

These results show us visually how the whole process is optimized and how much the final 3D object is simplified. Also, we get the textures of the objects automatically, without having to search and select similar ones to map them on the objects. Sometimes it takes hours to make the texture setting.

In Table 1 we can see, that the number of mesh polygons as well as vertices is about 10x fewer when scanning, instead of 3D modelling (Fig. 6). Low poly objects are preferred because they are loading faster. Also, it makes them easier to process and does not require a high computing power when rendering.

The task of the project was to divide a studio-type room into two zones with a separate function by means of a decorative grid. Such walls can visually divide spaces, providing both aesthetic and functional benefits. In order for this modern partition to fit into the existing interior, it must be correctly selected and approved by the client. Visualizing the decorative wall and how it will be present in the existing interior could only be done if the methods of 3D modeling software are used for visualization and rendering.[1]

In order to recreate the interior from the room, in which to then choose the design of the decorative grid, it was necessary to create as quickly and easily as possible a three-dimensional model of the existing interior with the highest possible detail.

In this study, we explored the use of photogrammetry, specifically through a mobile non proffesional application, for room scanning. The results demonstrate that this method is significantly faster and more efficient compared to manually creating a 3D model from scratch using traditional 3D modeling software. The time required to generate a detailed 3D model of a room is drastically reduced, and the polygon and vertex count of the resulting model is up to ten times smaller.

Thanks to new technologies and the advancement of the mobile industry, it is now increasingly easy to use photogrammetry with affordable devices such as smartphones and easy-to-use apps and software. The idea that the client himself can scan the object/objects needed by the designer for the assigned project not only facilitates the designer's work, but also enables him to work remotely. In addition, thanks to this method of scanning the room, the cost of the service for correcting an existing interior or adding a new element should be much lower.

In conclusion, integrating photogrammetry into some of the interior design processes can not only dramatically speed up modeling time, but also improve the overall efficiency of the end result. This makes this technology an invaluable tool in modern architectural and design practices.

Acknowledgement
The present article has been developed with the help of science project 11 *Ergonomic and design research of interior furniture with art design elements.(2024)*

References

[1] A. Behan, (2004). Digital Photogrammetry: Theory and Applications. https://doi.org/10.1111/j.0031-868X.2004.282_1.x

[2] П. Малджански, (2012). Предимства на цифровата фотограметрия пред налоговите фотограметрични технологии, International Jubilee Conference UACEG 2012: Science & Practice,University of architecture. Civil engeneering and geodesy.

[3] И. Кацарски,Технически средства в съвременната фотограметрия.

[4] P. Zlateva, A. Terziev, K. Yordanov, Study of regime parameters of the fermenter in the production of biogas from animal liquid waste materials, (2021) E3S Web of Conferences, 286. https://doi.org/10.1051/e3sconf/202128602010

[5] E.Valero, A. Adán, F.Bosché, (2016). Semantic 3D reconstruction of furnished interiors using laser scanning and RFID technology. Journal of Computing in Civil Engineering, 30(4), 04015053. https://doi.org/10.1061/(ASCE)CP.1943-5487.0000525

[6] M. Koeva, (2015). 3D моделиране в архитектурната фотограметрия.

[7] M.Attene, M. Campen, L. Kobbelt, (2013). Polygon mesh repairing: An application perspective. ACM Computing Surveys (CSUR), 45(2), 1-33. https://doi.org/10.1145/2431211.2431214

[8] G.Hristov, "3D content creation through the use of computer modeling methods19."

[9] W.Linder, (2013). Digital photogrammetry: theory and applications. Springer Science & Business Media.

[10] T. S. Todorova, (2023). Decorative screens as room dividers: their historical use, modern materials, advantages, and applications. ANNUAL JOURNAL OF TECHNICAL UNIVERSITY OF VARNA, BULGARIA, 7(1), 61-67. https://doi.org/10.29114/ajtuv.vol7.iss1.291

[11] K. Markova, T. Dovramadjiev. Method of contrast in designing of interior units

[12] M. Russo, (2006). Polygonal modeling: basic and advanced techniques. Jones & Bartlett Learning.

[13] С.Славов, & Н. Симеонов, Методика за бързо създаване на твърдотелни модели на реални обекти чрез 3D сканиране по метода на структурираната светлина.

[14] Г. Василев: 3D МОДЕЛИРАНЕ. Учебник, Технически университет – Варна, 2012

[15] P.Yakimov, (2020). Технологии и методи за 3D сканиране за нуждите на CGI-нови хоризонти. Годишник Телекомуникации, (7), 41-47.

[16] T. Luhmann, S. Robson, S. Kyle, J. Boehm, (2023). Close-range photogrammetry and 3D imaging. Walter de Gruyter GmbH & Co KG. https://doi.org/10.1515/9783111029672

[17] M. Tachev, T. Dovramadjiev, G. Jecheva, K. Markova (2018). Principle of work in the process of digitization and reconstruction of model of belt buckle by photogrammetry and 3D computer design. https://doi.org/10.29114/ajtuv.vol2.iss2.93

Keyword Index

About the Editor

Angelos P. Markopoulos is a Professor at the School of Mechanical Engineering, National Technical University of Athens (NTUA), Greece. He graduated as Mechanical Engineer from NTUA in 2000 and received his PhD from the same University in 2006. He has served as a teaching and research staff in NTUA since 2014. His research includes topics such as precision and ultraprecision machining processes with special interest in conventional and non-conventional machining, micromachining, additive manufacturing and grinding. Furthermore, he is an expert in manufacturing technology modeling and simulation, including the Finite Element Method, Artificial Intelligence, meshless methods and Molecular Dynamics, among others. He has participated in many national and international scientific projects in the aforementioned areas, as scientific responsible or researcher. He is a member of the editorial board of several scientific journals. He is the author of more than 200 papers in Journals, International Conferences, Books and Book Chapters.

Member

- Technical Chamber of Greece since 2001
- Hellenic Association of Mechanical & Electrical Engineers since 2001
- ASTM member in Committee D30 on Composite Materials and Committee F42 on Additive Manufacturing Technologies

www.ingramcontent.com/pod-product-compliance
Lightning Source LLC
LaVergne TN
LVHW052217200926
848481LV00001B/134